BEITRÄGE ZUR REGIONALEN GEOLOGIE DER ERDE BAND 10

BEITRÄGE ZUR REGIONALEN GEOLOGIE DER ERDE

Herausgegeben von Prof. Dr. H. J. Martini †

Präsident der Bundesanstalt für Bodenforschung

Band 10

Raoul C. Mitchell-Thomé

GEOLOGY OF THE SOUTH ATLANTIC ISLANDS

1970

GEBRÜDER BORNTRAEGER · BERLIN · STUTTGART

GEOLOGY OF THE SOUTH ATLANTIC ISLANDS

by

Dr. Raoul C. Mitchell-Thomé
Consulting Geologist-Geophysicist
Mamer, Gr. D. de Luxembourg

With 106 Figures and 95 Tables
in the text and on 8 Folders

1970

GEBRÜDER BORNTRAEGER · BERLIN · STUTTGART

Alle Rechte, auch die der Übersetzung, des auszugsweisen Nachdrucks, der Herstellung
von Mikrofilmen und der photomechanischen Wiedergabe, vorbehalten
© 1970 by Gebrüder Borntraeger, 1 Berlin 38 · 7 Stuttgart 1
Klischees: Carl Schütte & C. Behling, 1 Berlin 42 (Tempelhof)
Papier: Papierfabrik Scheufelen KG, 7311 Oberlenningen
Druck: Poppe & Neumann, Graph. Betrieb, 775 Konstanz
Schrift: Borgis Garamond — Printed in Germany
ISBN 3 443 11010 X

Preface

The volume herewith treats only of those islands lying in the South Atlantic. There is a valid argument against an artificial dividing of this immense ocean in the manner implied, for fundamental geological entities pay little heed either to human convenience or national boundaries. The Atlantic Ocean is a single major earth feature, initiated something like 200 m. y. ago. Dotted throughout these watery wastes are islands, all characterized, with scarce an exception, by vulcanism, and the vulcanological evolution, rock suites, geomorphological evolution of all show much in common.

It can therefore be justly contended that it were better to treat in one volume the entire collection of islands of the Atlantic Ocean rather than merely those occurring in the southern sector. Against such a view we would, however, note that the total area of the northern islands far exceeds the total area of the southern ones, our geological knowledge, the result of more lengthy studies over a greater period of time, is better for the northern islands. One would also add, as is further stated in the Introduction, that no geologist, living or dead, could claim intimate acquaintance with all the Atlantic islands, and there is every reason to believe this applies even to the southern islands.

In consequence, it has seemed preferrable to consider the southern islands separate, for reasons given above, and no less to furnish a volume of reasonable magnitude.

The islands here discussed, being under the national jurisdiction of four countries – Brazil, Portugal, Spain, Great Britain – have geological bibliographies which also are largely national, with citizens, of the respective countries being usually the principal investigators of said islands. Geologists other than the above nationals, have, in general, shown little interest, and indeed the American J. C. BRANNER and the Belgian, A. RENARD, are the only two outstanding examples of 'outsiders', and of these two, only RENARD showed interest in several islands, but such study was restricted to investigations of rock samples made available to him.

Within the past fifteen years of so, these South Atlantic islands have, as it were, been remembered once again, and there has been a distinct impetus in geological interest shown, e. g. Fernando de Noronha, Trindade, Sao Tomé, Principe, Ascension, St. Helena, Tristan da Cunha, Gough have all been given more recent attention.

The jet plane, the powerful hydroplane, the helicopter, have meant that these out-of-the-way islands of the southern ocean are losing somewhat their remoteness as areas of field study. To hie one's self off to a Tristan or St. Helena is scarce the wildly daring adventure it was in earlier years. Yet the sheer physical characteristics of almost all the islands considered here do indeed call for a robust constitution, hard foot-slogging, strenuous days in the field, frequent weather inclemencies, relative backwardness, lack of many amenities, physical, cultural and mental, and all such has scarce changed in the air age.

Some isotope datings have been carried out on samples from some of the islands, but to date there is almost a total lack of geophysical work to amplify our geological knowledge. Oceanographical geophysical work along the Mid-Atlantic Ridge and in vicinal basins have, of course, been executed, but to date we would keenly await geophysical information gleaned from the islands themselves and their fundaments.

In the following pages we have gathered together the sum total of our present knowledge of the islands in question. The scattered literature, going back some 140 years, often most difficult to obtain, has been studied, discussed, amplified here and there by personal knowledge. In so bringing together in one volume such disseminated findings and opinions, it is the hope that some worth may attach to what follows.

Mamer, Gr. D. de Luxembourg

RAOUL C. MITCHELL-THOMÉ

Table of Contents

Chapter 1: 1
Introduction 1
Acknowledgements 3

The Brazilian Islands

Chapter 2: Saint Paul Rocks 4
General 4
Geology 4

Chapter 3: Fernando de Noronha 12
General 12
Physical Features 14
Climate 15
Geology 16
    Remedios Formation 16
    Quixaba Formation 23
    Sao José Formation 25
Sedimentary Rocks 26
    Old Sediments 26
    Young Sediments 29
Geomorphology 30
Mineralogy 34
Petrography 35
    Remedios Formation 40
    Quixaba Formation 48
    Sao José Formation 50
    Xenoliths and Ejectiles 51
    Pyroclastics 52
    Eruptive Breccias 53
Petrochemistry 53
    Petrochemical Character of the Fernando de Noronha Province 53
    Comparisons 59
Petrogenesis 60
Age of the Volcanics 62
Radiometric Age Determinations 63

Chapter 4: Rocas Atoll 65

Chapter 5: Trindade 67
General 67
Physical Features 68
Climate 69
Geology 70
    Trindade Complex 70

Desejado Sequence	72
Morro Vermelho Formation	73
Valado Formation	74
Paredao Volcano	75
Geomorphology	76
Mineralogy	80
Petrography	81
Trindade Complex	81
Desejado Sequence	87
Morro Vermelho Formation	89
Valado Formation	89
Paredao Volcano	89
Xenoliths and Ejectiles	90
Pyroclastics	92
Petrochemistry	92
Petrogenesis	97
Geological Considerations on the Age of the Trindade Vulcanism	99
Geochronology of the Trindade Volcanics	100
Chapter 6: Martin Vaz Archipelago	102

The Portuguese Islands

Chapter 7: Sao Tomé	105
General	105
Physical Features	105
Climate	108
Geology	108
Igneous Rocks	111
Basaltic Facies	112
Non-Basaltic Feldspathoidal Facies	115
Sedimentary Rocks	121
Palaeontology	123
Chapter 8: Principe	125
General	125
Physical Features	125
Climate	126
Geology	126
Igneous Rocks	128
Sedimentary Rocks	141
Palaeontology	143
Economic Geology	144
Geological Evolution	145

The Spanish Islands

Chapter 9: Fernando Poo	146
General	146
Physical Features	147
Climate	147
Geology	148
Chapter 10: Annobon	156
General	156
Climate	157

Geology	157
Petrography	158

The British Islands

Chapter 11: Ascension	162
General	162
Physical Features	163
Climate	163
Geology	164
Petrography	172
Petrology	189
Isotope Geochemistry	193
Economic Geology	194
Chapter 12: St. Helena	195
General	195
Physical Features	196
Climate	197
Geology	197
Igneous Rocks	198
Sedimentary Rocks	202
Petrography of the Igneous Rocks	203
Basalts	203
Phonolites	206
Form of the SW Magma Chamber	213
Structure	215
Economic Geology	217
Age of St. Helena	219
Geochronology	219
Geological Evolution	221
Chapter 13: Tristan da Cunha Group	224
General	224
Physical Features	226
Tristan da Cunha	226
Inaccessible Island	228
Nightingale Group	229
Climate	229
Geology	229
Tristan da Cunha	229
Inaccessible Island	235
Nightingale Group	238
Petrography	241
Nomenclature	241
Rock Types	242
Norms and Modal Composition	257
Additional Comments	265
Magma Differentiation and Genesis	269
Provincial Relationship	272
Physiography	272
Vulcanology	272
Petrography	273
Chemistry	274

Economic Geology	276
Geochronology of the Tristan Group	277
Geological Evolution	279
The 1961 Eruption	281
Tremor Phase	283
Eruption Phase	283
Morphology	284
Petrography	284

Chapter 14: Gough Island . . . 286

General	286
Physical Features	286
Climate	287
Geology	287
Petrography	289
Isotope Geochemistry	302
Geological Evolution	304
Age Determinations	305

Chapter 15: Falkland Islands . . . 306

General	306
Physical Features	307
Climate	307
Geology	308
Sedimentary Rocks	308
Igneous Rocks	309
Metamorphic Rocks	310
Stratigraphy	310
Precambrian	312
Palaeozoic	313
Lower Devonian	313
Middle Devonian	314
Lower Carboniferous (?)	315
Upper Carboniferous (?)	316
Permo-Carboniferous	316
Permian	318
Palaeozoic-Mesozoic	320
Permo-Triassic	321
Mesozoic	321
Triassic	321
Quaternary	322
Palaeontology	324
Structure	327
Economic Geology	329
Geological Evolution	329
Palaeogeographic Connexions with South America – chiefly Argentina	331
Palaeogeographic Connexions with South Africa	334
Opposing Views of Palaeogeographic Relationship	334
Abstract	336
Zusammenfassung	337
Bibliography	341
Subject Index	351
Index of Fossil Names	356
Locality Index	358
Author Index	365

The oceans of the world have been described as 'the last frontier' of scientific knowledge on this Earth of ours. The remarkable achievements of submarine geophysical surveying techniques and the whole new emphasis and impetus given to Oceanography means that Nature's secrets here are being pried from the depths. Rocks, minerals, fossils are dredged from profound depths, submarine photographs allow us to actually observe the ocean bottom. But only on these small islands can we inspect, study at leisure and with relative ease the rocks in place, their environmental, structural, petrological relations, and thus the islands serve to complement data obtained via geophysical and other submarine techniques. One is still impressed with the fact, in reading the present copious literature, how in the elucidation of the problems of the oceans, these small outposts of **terra firma** play a key role quite out of proportion to their size.

Beginning with the remarkable voyage of the "Beagle", many expeditions have set out to try and unravel the mysteries of the oceans. Today these activities are greater than ever, their findings are more voluminous, the techniques more startling. At the threshold of Man's prodigious leap into Space, it is fitting indeed that the 70% of the Earth's surface represented by the oceans and seas is the scene of great scientific enterprise, that so much of the results of such endeavours must depend upon our knowledge of the islands in question sprinkled throughout these immensities of water.

Acknowledgements

Assistance in the form of correspondence explaining and/or elucidating matters, pointing out references, the sending and/or lending of books, reprints, maps, etc. on the part of many persons has proven of enormous value. The writer wishes herewith to thank personally such many kindnesses to the following, although he must stress that he alone is responsible for any shortcomings: Dr. R. Adie, Exmo. Sr. Jose Luis de Aguilar, First Secretary, Spanish Embassy, London, Prof. F. F. M. de Almeida, Prof. A. Arribas, Prof. C. F. T. Assuncao, Dr. I. Baker, Senhor R. A. Barbosa, Head of Information Section, Brazilian Embassy, London, Dr. L. A. Barros, Prof. T. F. W. Barth, Dr. J. D. Bell, Prof. K. Beurlen, Dr. A. V. Borrello, Prof. A. J. Boucot, Prof. D. S. Coombs, Dr. U. Cordani, Dr. A. E. G. Engel, Dr. L. A. Frakes, Prof. J. M. Fuster, Dr. I. G. Gass, Dr. P. W. Gast, Prof. U. Hafsten, Mr. D. Hart-Davis, Prof. K. Krejci-Graf, Dr. R. W. LeMaitre, Dr. F. Machado, Prof. J. M. Neiva, Senhor Luis Possollo, Dr. J. M. Schopf, Dr. E. P. Scorza, Dr. A. R. Serralheiro, Prof. C. Teixeira, Prof. C. E. Tilley, Miss O. V. Vesentini, Librarian, Brazilian Embassy, London, Exmo. Sr. Don José Diaz de Villegas, Director, Instituto de Estudios Africanos, Madrid, Prof. J. Tuzo Wilson, Dr. J. D. H. Wiseman, Mr. R. Wright.

Frau I. Schneider-Thost and Dr. U. Hellmann, and afterwards Herren J. Nägele and K. Obermiller, of Verlag Gebrüder Borntraeger in their official capacities, have been most friendly and co-operative, and due appreciation is herewith tendered.

THE BRAZILIAN ISLANDS

CHAPTER 2

St. Paul Rocks

General

These tiny islets, variously known as St. Paul Rocks, St. Paul's Rocks and St. Peter and St. Paul, lie ca. 950 km NE of Natal, Brazil, about half-way between the coasts of South America and Africa. The islets are in lat. 00° 56' N, long. 29° 22' W. The group of twelve islets, strung-out in a NNE-SSW direction, have a total length of ca. 350 m and a breadth of some 200 m. The highest point, on the largest islet, is 23 m, rising above a small level platform lying some 5 m above sea level.

Expeditions to St. Paul Rocks were made by H. M. S. BEAGLE, DARWIN landing there in 1832, by the 'Quest' in 1921, by the 'Meteor' in 1925, and specimens were collected when H. M. S. OWEN called there in 1960.

DARWIN (1876) mentioned a total absence of plant life but noted the presence of several insects. Three species of sea fowl are abundant and their guano deposits whiten the rocks and give them an enamel appearance. On approach to the islets one is impressed by the myriads of birds at rest on the rocks and flying around.

The islets rise steeply and deep water lies close to the shores-depths of 185 m occur some 160 m to the N and S. Off-shore rocks occur to the SE and landings at any place are hazardous. DOUGLAS (1923) mentioned several pot holes on the main islet, caused by powerful wave action at high tide.

The islets are uninhabited, but there is a lighthouse on the main islet.

No climatic data are available.

Geology

The rocks of St. Paul are dunites. Megascopically they are very dark in colour, of a slightly greenish-black appearance and, as per RENARD (1882a), have a banded structure. The rocks are dense, compact, almost chert-like, with a somewhat waxy lustre, composed of ca. 74 % very fresh olivine, ca. 24 % pyroxene – chiefly jadeite, enstatite and diopside –

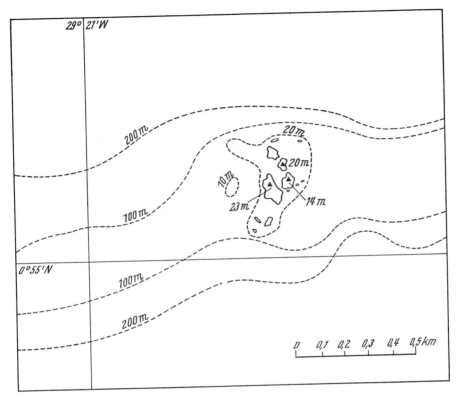

Fig. 2. St. Paul Rocks. (Sheet 51, Serv. Hidra. da Marinha, 1961)

and ca. 2% deep brown picotite (WASHINGTON, 1929, 1930). This author (1930) remarked that were it not for the high specific gravity of the rocks, they might be mistaken for phonolites.

The earlist analyses made of the peridotitic rocks of St. Paul is that given by RENARD (1879, 1882a), as follows:

Table 1   Chemical analysis of peridotitic rock

SiO_2	14.90	Soluble in HCl = 73.53	SiO_2	43.84
Al_2O_3	0.90		Al_2O_3	1.14
Fe_2O_3	5.40		FeO	8.76
MgO	5.26		MgO	44.33
CaO	1.51			
SiO_2	32.25	Insoluble in HCl = 25.97	CaO	1.71
Fe_2O_3	Tr.		Cr_2O_3	0.42
FeO	9.56		MnO	0.12
MgO	31.45		NiO	0.51
$CaSO_4$	0.29		H_2O	1.06
Water loss	0.50			
Total	100.00		Total	101.89

WASHINGTON (1930) gave a chemical analysis and a calculated norm of a specimen from the main islet thus:

Table 2 Chemical analysis and norm of dunite

SiO_2	44.25		
Al_2O_3	2.65		
Fe_2O_3	1.13		
FeO	7.11		
MgO	41.07	Or	0.79
CaO	1.07	Ab	12.58
Na_2O	1.52	An	–
K_2O	0.13	Ac	–
TiO_2	0.25	Di	4.16
P_2O_5	None	Hy	–
S	n.d.	Ol	79.57
Cr_2O_3	0.10	Mt	1.62
MnO	0.09	Il	0.46
NiO	n.d.	Cm	–
H_2O+	0.66	Ap	–
H_2O-	0.17		
Total	100.20		

From this he concluded that the analysis was similar to those wehrlitic, lherzolitic and dunitic peridotites of Timor, Réunion Island and Piedmont, Italy. He therefore affirmed that the specimen, on a modal and chemical basis, was a dunite of wehrlitic type, due to the large amount of pyroxene present.

TILLEY (1947) made a re-study of the rocks collected by the 'Beagle' expedition as well as those of the 'Quest' expedition. Apart from serpentinized specimens and phosphate rocks (q. v. infra) the specimens belonged to two main types: 1. those with olivine and possessing amphiboles, 2. olivine rocks with enstatite. (In RENARD's analysis given in his 1882a publication, a calculation showed the specimen to be ca. 75% olivine, ca. 25% enstatite.) In both types the oustanding feature is the strong development of fine banding and 'fluxion structure' of granulitic base, associated with large grains of olivine, pyroxene, amphibole and spinel. This feature led previous workers, e. g. RENARD, to postulate a volcanic origin for the rocks – 'schistes cristallins' – though RENARD confessed that his examination of the rocks demonstrated 'des faits qui démontrent d'une manière absolue l'origine volcanique de cet îlot' were lacking. At the time of RENARD, GEIKIE (1882) and WADSWORTH (1883) both subscribed to a volcanic origin of the islets, leading to a spirited discussion between RENARD and WADSWORTH. (Vd. RENARD, 1883.) WASHINGTON (1930) and later students have all thought the rocks to be of volcanic origin – 'somewhat metamorphosed peridotites' in the words of WASHINGTON. On the other hand, the first student of the St. Paul rocks, DARWIN (1876) was so greatly puzzled by the nature of the rock exposed on St. Paul that he declined to name it, and categorically stated: 'St. Paul is not of volcanic origin'. This led Sir. JOHN MURRAY, the director of the 'Challenger' studies, to have a special investigation made of the rocks by RENARD (1882a, 1882b, 1885), who devoted considerable space to the question of the origin of the St. Paul rocks, whether they were effusive lavas or belonging to a metamorphic schist series. WADS-

WORTH doubted a primary nature for the small olivine grains in the matrix, and even questioned whether they actually were olivine. (He studied specimens sent him by MURRAY, the same ones as studied by RENARD.) WADSWORTH's criticisms of RENARD were indeed acrimonious but RENARD (1883) replied most tellingly and adroitly. PRATJE (1926) acknowledged the presence of peridotites, but declined to offer opinions as to their origin.

WASHINGTON (1930) discussed in considerable detail the question of the origin of dunite. Peridotites, according to him, can be either: 1. igneous, chiefly plutonic rocks, often with textures like those of schists, with banded structures and even fluidal textures. 2. a type of schist showing evidences of metamorphic deformation. As regards the dunites of St. Paul Rocks, he offered three hypotheses:

1. The dunite is really a metamorphic schist.
2. The dunite is a plutonic, the texture showing modifications due to mechanical stresses.
3. The dunite is an effusive lava from a submarine volcano.

The evidence, as presented by the St. Paul specimens, is more favourable towards postulates 1 and 2, rather than 3.

WASHINGTON summarized his own views by stating that the study of the rock features themselves favour a plutonic-metamorphic origin, whereas the geographical location, coupled with the general characteristics of all other oceanic islands, inclines towards the view that it is an effusive lava. The major difficulty as he saw it of rejecting an effusive origin was in accounting for the uplift of such a tiny speck of land from a position as an integral part of the crust forming the ocean floor in the vicinity. In the neighbourhood of the islets, the floor is unusually irregular, and seismicity is prevalent here. Indeed E. RUDOLPH (1887) had, even eighty years ago, called the narrow zone on either side of the Equator between longs. 16° and 31° W', "die seismische Zone des St. Pauls-Felsen". Instability in this part of the Atlantic could thus conceivably result in the islets formation, which stand on the Mid-Atlantic Ridge, as having been raised to the surface of the ocean consequent upon seismic movement or then orogenic compressional forces. In this scheme of things, the dunite would not be a lava flow but a piece of a massive plutonic representing the oceanic crust, stresses responsible for the upheaval thus accounting for the fluidal textural characteristic.

TILLEY (op. cit.) agreed that the banding and fluidal texture – the 'fluxion structure' of granulitic base – were unquestionably of dynamic origin, and the rocks of St. Paul represented a group of intensely mylonitized dunites and peridotites. He stressed the particular interest of the rocks in that intrinsically they are dynamically metamorphosed ultrabasics and also mylonites, occuring in a deep oceanic environment. In his re-study of the St. Paul rocks, he described them as having porphyroclasts of olivine (forsterite), amphibole and brown spinel (picotite), set in a fine-grained matrix, on occasion striped with bands of coarser grain. There also occur dunites with porphyroclasts of forsterite, enstatite and brown spinel, showing no amphiboles. According to TILLEY, it is in the enstatite that there are the tell-tale evidences of dynamic forces, this mineral varying greatly in size, with well-marked cleavage and showing strong evidence of banding and curling.

Table 3 gives some chemical analyses of dunite specimens collected from the islets, as reported by TILLEY.

Table 3 — Chemical Analyses of St. Paul Dunites, studied by TILLEY (TILLEY, 1947)

	1.	2.	3.	4.	5.
SiO_2	43.97	44.25	–	47.15	43.84
Al_2O_3	2.89	2.65	–	0.90	1.14
Fe_2O_3	1.04	1.13	–	3.40	–
FeO	6.89	7.11	–	36.69	8.76
MgO	41.11	41.07	–	1.61	44.33
CaO	2.35	1.07	–	n.d.	1.71
Na_2O	0.07	1.52	0.09	n.d.	n.d.
K_2O	nil	0.13	nil	0,50	n.d.
TiO_2	0.17	0.25	–	–	–
P_2O_5	n.d.	nil	–	–	–
S	n.d.	n.d.	–	0.19 (SO_3)	–
Cr_2O_3	0.50	0.10	–	–	0.42
MnO	0.13	0.09	–	–	0.12
NiO	0.21	n.d.	–	–	0.51
CoO	tr.	–	–	–	–
CuO	nil	–	–	–	–
NaCl	0.09	–	0.10	–	–
H_2O+	0.35	0.66	–	–	} 1.06
H_2O-	0.20	0.17	–	–	
Total	99.97	100.30		100.00	100.89
S. G.	3.275 at 15° C	3.271 at 28° C	3.30		3.287 at 20° C

1. Dunite with enstatite, from 'Quest' collection, re-studied by TILLEY.
2. Dunite with amphibole, from 'Quest' collection, given by WASHINGTON (vd. Table 1), re-studied by TILLEY.
3. Dunite with amphibole, from 'Quest' collection, re-studied by TILLEY.
4. Dunite with amphibole, from 'Challenger' collection, given by RENARD.
5. Dunite with enstatite, from 'Challenger' collection, given by RENARD.

In his 1966 publication, TILLEY again claimed that the St. Paul ultramafics should be more accurately described as peridotites, as they have significant proportions of pyroxenes, amphiboles and spinel, with a predominant olivine. In this paper he presented two new chemical analysis and norms (Table 3). The Scoon analysis is a re-examination of a 'Quest' sample; the Hess sample is taken from an analysis given by him. No. 1 of this Table is richer in Al_2O_3 and CaO than No. 1, Table 2, but resembles it in its iron oxide content and especially is its low alkalis content.

In 1947 TILLEY had subscribed to the view that the environment of the mylonitic peridotites represented parts of the upper mantle, raised to the surface along a fracture line. In 1966 he drew attention that peridotites of this bulk chemistry had been considered as a potential source of basalt (I. KUSHIRO & H. KUNO 1963). As against such a view he remarked that because of the paucity of alkalis, it scarce seemed likely that such a small percentage of basaltic liquid as would be obtained by the fractional melting of such an assemblage, could be expected to be adequately separated from its solid environment. On the other hand, TILLEY was more kindly disposed to the alternative theory, which postulates that these peridotites may in themselves represent the barren residue of such an extraction process from a less refractory peridotite which had a greater alkali content.

In this 1966 paper, TILLEY was critical of the needless multiplicity of names which have been applied to the peridotite group of rocks. He saw no justification for the coining of the four new rock-names advocated by WISEMAN (vide infra). TILLEY was of the opinion that the varieties referred to by WISEMAN were adequately taken care of by the existing nomenclature of lherzolite, augite-harzburgite and amphibole-enstatite-peridotite. For these St. Paul rocks, with their low alkali content it is obvious how important it is to take into account their continuous saturation by sea water.

To complete available chemical analyses of St. Paul rocks, we give below (Table 4) an analysis of an olivinite quoted by GUIMARAES (1932). ALMEIDA, F. F. M. (1958) compared this specimen with an olivinite xenolith occurring in nepheline-basanites in the islets of Sao José, Cuscuz and de Fora, shown in No. 2.

Table 4   Chemical Analyses, Olivinite (GUIMARAES and ALMEIDA)

1. Olivinite, St. Paul Rocks.
2. Olivinite, xenolith, Fernando de Noronha.

	1.	2.
SiO_2	42.80	41.88
TiO_2	0.04	0.15
Al_2O_3	4.89	4.56
Fe_2O_3	1.27	4.90
FeO	7.40	4.67
MnO	Tr.	0.15
MgO	38.00	39.74
CaO	2.46	2.80
Na_2O	0.89	0.53
K_2O	0.52	0.00
H_2O+	0.72	0.14
H_2O-	0.30	0.52
P_2O_5	0.12	0.06
Total	*99.88	100.10

* Includes: 0.47 Cr_2O_3, 0.30 Cl and 0.17 SnO_2

WISEMAN (1965) yet again subjected previously collected specimens to petrographic analysis, made similar analyses and also isotope studies of specimens collected during the visit of the H. M. S. "Owen" in December, 1960. To him, St. Paul rocks are to be regarded as either modified cumulates within the oceanic crust or then as displaced samples of the upper mantle (original, 'barren' or cumulative in origin but modified by dynamic action). The earlier classification of the rocks into those containing hornblende, with no pyroxene and pyroxene with no hornblende is not satisfactory, for X-ray diffraction examination has shown features not discernable under the microscope, such as the presence of hornblende and pyroxenes. There are also rocks showing forsterite, brown hornblende, enstatite, clinopyroxene and spinel. WISEMAN stated therefore that a purely mineralogical classification of the rocks is not valid, and preferred to adopt a chemical-textural classification into rocks of four types: challengerites, owenites, paulites and questites. (This predilection of coining new rock names and a semewhat chauvenistic over-emphasis in the connotation of the names adopted is open to criticism.) In addition to forsterite, the challengerites contain mylonitized brown hornblende, enstatite, clinopyroxene and spinel.

In bulk chemical analyses made of eight new specimens, WISEMAN found K_2O to be present in all four types. The challengerites have 1447 ppm K, 0.14 ppm U, 0.395 ppm Th, 0.06 % F and 0.09 % S. The average bulk composition of the owenites, paulites and questites approaches that of some rocks which are claimed to indicate a subcrustal origin. The challengerites are richer in Al_2O_3 and K_2O but poorer in MgO than other rock types from the islets, and correspond in composition to a mixture of 1.3–3.0 parts anhydrous dunite with one part average Atlantic floor tholeiitic basalt.

WISEMAN was of the opinion that the islets (especially the challengerites) are derivatives of the upper mantle. He suggested that the mylonitized condition of the rocks occurred at temperatures above 500° C, though he does not rule out the possibility that this took place at a lower temperature. The absence of serpentinization associated with the primary mylonitization, as also the presence of fragmented hornblende is a direct consequence of a water-deficient milieu, but anhydrous conditions in such an environment would seem unlikely.

DARWIN noted that from a distance, the rocks of St. Paul appeared of whitish colour. This is the result of guano deposits, which occur either as greyish-white, glossy encrustations, of pearly lustre and hardness about 5, or than as dull, rusty and earthy material with a very rough feel and which can be easily scratched with a knife (WASHINGTON, 1929). Geologists to whom DARWIN showed these encrustations were of the opinion that such were of volcanic origin, but DARWIN & J. W. BUCHANAN (1874) considered these as being bird excrements of which the insoluble residue had been long exposed to the sun and the waves, the material having becomed concretioned and covered with this 'enamel'. RENARD (1882a) agreed with Darwin, claiming that this explanation applied not only to these encrustations but to all the small veinlets of phosphate formed in the fissures of rocks in the islets. Microscopic examination showed no indications of any natural glass features.

The soft, dull variety was a calcaro-ferruginous stone forming veins, as per DARWIN, a fact also noted by DOUGLAS (1923) and PRATJE (1926) spoke of great cracks in the dunite "mit einem sehr widerstandsfähigen sedimentären Material ausgefüllt, das als Rippen herauswitterte".

In Table 5 an analysis is shown (No. 1) of the glossy, encrustation type, after RENARD (1882a), who considered it to be essentially tricalcium phosphate with some calcium phosphate, perhaps also carbonate of lime, magnesia and iron.

WASHINGTON (1929) made thin sections of several samples and one, filling a vertical crevice in the NE islet was specially studied and the chemical analysis given in No. 2. Comparisons are also taken from WASHINGTON of similar deposits from Norway (No. 3) and France (No. 4). Specimen No. 2 gave a formula $40CaO \cdot 10P_2O_5 \cdot 8H_2O \cdot 3CO_2$, which led WASHINGTON to name the substance as being dahllite.

STUBBS (1965), WRIGHT (1965) and VAN BEMMELEN (1966) all make reference to the significance of isotope datings of St. Paul Rocks samplings carried out by the Carnegie Institute, Washington (1964).

	1.	2.	3.	4.
P_2O_5	33.61	35.75	38.44	38.40
CaO	50.51	56.47	53.00	53.65
MgO	tr.	0.30	n.d.	n.d.
$(AlFe)_2O_3$		n.d.	–	0.57
FeO	tr.	n.d.	0.79	n.d.
$(NaK)_2O$		n.d.	0.11	n.d.
SiO_2		0.01	–	–
H_2O+		3.91	1.37	2.10
H_2O-		0.53		
CO_2		3.36	6.29	5.30
SO_3	tr.	0.36	–	–
Total		100.69	100.89	100.02

Table 5 Chemical Analyses of Guano Deposits (RENARD, 1882a, WASHINGTON, 1929)

1. St. Paul Rocks. (RENARD specimen)
2. St. Paul Rocks. (WASHINGTON specimen)
3. Specimen from Bamle, Norway.
4. Specimen from Mauillac, Quercy, France.

The Institute used the Ru/Sr method of analysis, which depends upon the decay of Ru87 to Sr87, a decay having a half-life of 46 000 million years. The method requires accuracy in the determinations of the ratios of Sr87 to Sr86, the latter a stable isotope not formed as a result of radioactive decay. This method gives age values of about 4500 million years for meteoric fragments. Presuming that the Sun had a cold, accretionary origin, cosmologists agree that the Sun and various parts of the Solar System must surely have all been formed approximately together at about this time.

If the Sr87/Sr86 ratios of the St. Paul samples are similar to those of suitable meteorites, the assumption can be made that the rocks are an unaltered part of the Earth's mantle.

The isotope samplings from St. Paul Rocks would bear out this contention. From the five samples, four of these indicate ages of ca. 3500 million years. The fifth has an Sr ration higher than normal basalt rocks of the Mid-Atlantic Ridge, and thus could not have been formed by normal processes of oceanic vulcanism. Its ratio value can only be intelligible if it is assumed to be ca. 4500 million years. (The Carnegie Institute gave a revised age for the Earth of 4700 million years.)

On these microscopic specks of land in the vast Atlantic, we thus appear to have rocks which represent an unaltered part of the mantle, rocks of great depth origin and rocks which, up to the present, are the oldest known on this globe – a unique distinction for tiny St. Paul Rocks.

CHAPTER 3

Fernando de Noronha

General

Lying in latitude 3° 50′ S, longitude 32° 15′ W and some 345 km NE of Cabo Sao Roque, Brazil and ca. 2600 km from the coast of Liberia in Africa is the archipelago of Fernando de Noronha. The area of the main island is 16.9 km², and with the other twenty odd islets, the total area of the archipelago is 18.4 km². Fernando de Noronha rises from a base 4000 m deep and 60 km in diameter. Between here and the Brazilian coast depths go down to 5000 m (Figs. 3, 4, 5).

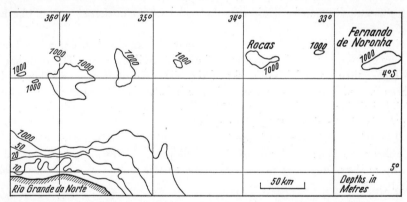

Fig. 3. Bathymetry of the Atlantic Ocean between Fernando de Noronha and the coast of Rio Grande do Norte, Brazil. (From map: Brasil – Costa Norte, do Rio Paraiba a Recife. Serv. Hidro., Min. da Marinha, 1953)

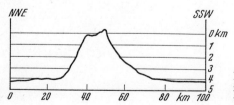

Fig. 4. Bathymetric profile of Fernando de Noronha, showing the large structure erected from the ocean floor. (ALMEIDA, 1958)

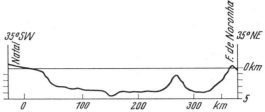

Fig. 5. Bathymetric profile between Natal City and Fernando de Noronha. (ALMEIDA, 1958)

The highest point is Pico, 321 m, near the village of Remedios. This phonolitic dome is quite the most imposing feature of the islands. Morro do Atalaia, on the S coast, is the second ihghest peak, rising to 223 m. Nowhere is the relief as pronounced or as spectacular as in Trindade. In the central part of the main island is an extensive flat area where an airfield has been built.

Almost the entire coastline is rocky and cliffed, with only three localities where seas are calmer and landings possible.

Valleys are few, dry except at times of rain. Fresh water supplies present an ever-present and growing problem.

Soils are fertile but the dryness affects plant-crop growth. In former times the archipelago supported a dense vegetation of both trees and bush, of species similar to those growing on the mainland in Pernambuco. The destruction of the forests and bush dates from the earliest permanent occupation in the earlier part of the 18th. century, when wood was needed for various domestic and constructional uses. The setting aside of pasture areas for goats, sheep and cattle aided in the reduction of the natural vegetation covering. Though this covering today is less dense and less widespread than in former times, the island presents a green and clothed aspect, especially during the wet season, with grasses, brushwood and trees abundant everywhere, even on the summits – except Pico, whose spine is too steep to support plant life.

Fernando de Noronha was the first part of what today we call Brazilian territory to be discovered, it appearing under the name Sao Joao in a map of Juan de la Costa in 1500. However it is generally stated that Gonçalo Coelho was the real discoverer in the year 1503, at which time it was named Quaresma. The name was changed to that of today after a rich conquistador who obtained a monopoly for the exploration of Brazil. In 1612 and again in 1635 it was occupied by the Dutch, who were finally expelled in 1654. In 1736 the French landed, but were driven out a year later, from which date it was effectively occupied and colonized by the Portuguese. In 1770 the archipelago was made part of the Captancy of Pernambuco and later passed under the authority of the Ministry of War, then in 1877 to the Ministry of Justice, and in 1891 once more under the state authority of Pernambuco. From 1738 to 1942, Fernando de Noronha was regarded as a presidium, but after the latter date it became part of the Federal Territory of Brazil, directly under the Ministry of War. At this time, World War Two was in progress, and for defensive reasons, the Brazilians placed the archipelago under military control. Today the population and activities of the archipelago are almost entirely related to the military.

The population of Fernando de Noronha was 1032 in 1957, which included military plus families, fishermen, farmers. There is a penal settlement on the main island but whether the above figure included the convict population is not known. Remedios is the principal village, Quixaba and Tamandaré being merely hamlets.

Local inhabitants find work at the salt-pans, limekilns, raising manioc and millet, work at the flour mills, working on the guano deposits and sundry occupations related to the military and penal establishments.

There are some 20 km of roads on the main island. The airfield has runways almost 2 km in length, and frequent air service for military purposes are maintained with the mainland. There is irregular boat connexions with Recife. The submarine telegraph between Recife and Dakar passes through Fernando de Noronha.

Physical Features

In the main island (throughout references to Fernando de Noronha will relate to the largest island, unless otherwise stated) higher peaks tend to be peripheral, with Pico, Atalaia, Frances, Madeira, Boa Vista, Dois Abraços, Bandeira and S. Antonio, all over 105 m, lying within some 700 m of the shore, the island having a maximum breadth of 3.5 km. In the central area there is quite a large plain, the Quixaba plain, average elevation 45 m, surrounded by peaks on all sides except the W. This plain descends down to a lower coastal plain W of Southeast Bay, continuing westwards along the S coast to the Viracao plain.

Peaks constitute ankaratritic and phonolitic bodies rising prominently above the surrounding terrain. Pico, which BRANNER (1890) termed the most striking landmark to be found anywhere in the entire South Atlantic, is a phonolitic spine or monolith, rising 200 m in perpendicular bare walls above steep lower slopes covered in vegetation.

Apart from individual peaks, the topography is somewhat subdued and rolling, although frequently the lower slopes of hills may be quite steep. The overall incompetency of the drainage network to carve out marked valleys is a notable feature. Nowhere have the ephemeral streams eroded deep valleys, nowhere are streams of any significant length.

Lying N of the main island are a series of islets, forming 'stepping-stones', leading to the largest and farthest removed, Ilha Rata. Ilhas Rasa and Cuzcus are connected to S. Antonio Point, the extreme N salient of the main island, by tombolos, one of which is covered at high tide. Of these northern islets, Sela Gineta is the highest, 128 m, also having steepest slopes. In general the offshore islets present somewhat domal profiles, but on the other hand, some show steep, rugged outlines, such as Frade off the S coast.

The southern coast is much the more indented, being a reflexion of the structure, orientation in relation to the dominant winds and stage of evolution. Headlands are usually composed of phonolite masses, embayments of tuffs. Everywhere marine attack is powerful. The southern coast is exposed to the long 'fetch' of waves from the SE and here erosional effects are strogest, here indentation is greatest, here coastal evolution is most advanced. Wave refractions along the northern coast rob the waves of much of their destructive energy, and here the coastline in general is straighter, beaches commoner. It is along the southern and eastern coasts that Lithothamnium reefs are best developed.

In general the coasts are steep, rugged, precipitous slopes descending right down to the water's edge. Very strong surf makes landings difficult everywhere, the most sheltered region being in Caieira Bay. Currents with appreciable transporting powers are prevalent along the coasts, especially between the southern coast and the offshore islets. Throughout marine erosion is rapid and powerful, denudation slower and less destructive in the interior.

The insular platform surrounding the archipelago is about 3 km broad N off Quixaba beach, water less than 50 m deep, and 3.5 km broad SE of Caracas Pt. water is only 26 m deep. These appear to represent the maximum breadths of the platform. On the other hand, depths of 738 m occur 2.5 km S of the Sapata peninsula and here the platform appears to be narrowest, eastwards as far as Pt. Capim Açu.

Marine terraces occur here and there, suggesting several higher sea levels in previous times and one level lower than that of today.

Climate

There is a meteorological station on Morro S. Antonio, 101 m above mean sea level, where observations have been kept since 1910.

Fernando de Noronha experiences a tropical climate with a well-marked wet season, oceanic influences are paramount, the temperature range is small and there is a uniform distribution of the relative humidity. The climate is healthy, tending to be somewhat semi-arid, with a long dry season.

The average annual temperature is high, 25.4° C, with a range of only 1.5° C. Maximum temperatures experienced are 31° C, minimum, 18° C, with a large daily range during November in the dry season. March is the warmest month, average 29° C, and August the coolest, average 23° C.

The average annual rainfall is 1318 mm, with a distinct wetter period from February to July. April is the wettest month, average 273 mm in 21 days, with the three wettest months providing some 60% of the annual rain. August to January is the dry period, only 8 mm falling on an average in October, and only 31 mm average during the months October to December.

The temperature and rainfall characteristics would class the region as an *Awi* type of climate, as per KÖPPEN.

Relative humidity is uniformly distributed during the year, varying from 81% to 87%, ca. 86% in the wet season, 81% in the dry season. The reduced rainfall and relative humidity, combined with the higher temperatures of the dry season give the islands a distinctly semi-arid appearance at this time of year.

Constant and dominant trade winds blow throughout the year, mostly from the ESE.

In general, the climate of the archipelago is the same as that along the eastern seaboards of Rio Grande do Norte, Brazil, the rainfall being somewhat greater, the dry season not quite so marked on the mainland.

During the long dry period, strong insolation robs the soil of its water content and causes the regolith to have a cracked, dessicated carapace. The climatic conditions provoke grave water problems for the inhabitants, quantities being scarce and the waters being 'hard' and of disagreeable taste, far below accepted hygienic standards. After much thought given to the problem, the authorities believe now that desalinization of sea water offers the only real solution.

Geology

The first geological observations made were those of DARWIN who visited the archipelago in 1832 during the voyage of the 'Beagle' and published his findings in 1839 and 1876. During a stop-over of the H. M. S. 'Bristol' in 1871, RATTRAY (1872) made some observations. The 'Challenger' expedition landed in Fernando de Noronha in 1873 but were prohibited from making worthwhile investigations. However GUMBEL (1880), RENARD (1882, 1886, 1889) and BUCHANAN (1885) wrote papers on the general geology and rocks studied from samples collected. In 1876 the Imperial Geological Survey of Brazil paid a visit and began the scientific gathering of data, e. g. the economic report by DERBY & BARROS (1881). Restrictions against foreign visits were thereafter relaxed, and in 1887 the Royal Society of London and the British Museum explored the islands (LEA 1888), (DAVIES 1890), (RIDLEY 1890, 1891). These collections were later studied by PRIOR (1897) and CAMPBELL SMITH & BURRI (1933). About the same time the American BRANNER (1888, 1889, 1890, 1893, 1919) visited the archipelago and his contributions greatly extended our geological knowledge. Rock samples collected by him were studied by GILL (1888) and WILLIAMS (1889). GUIMARAES (1932 and in MORAES 1928) described rocks in the collection of the Survey of Brazil. In 1947 POUCHAIN visited the islands to assess their economic mineral worth (1948). The visit of ALMEIDA to Fernando de Noronha, along with three colleagues in 1950, was of a rough reconnaissance nature, but in 1952 and 1954 they returned to carry out a detailed survey of the archipelago. ALMEIDA made 250 thin-sections of rocks, 14 chemical analyses, studied 183 aerial photographs of the islands, and his monograph of 1955, revised in 1958, is the most detailed and thorough publication available on the geology of the archipelago.

The writings of DARWIN and BRANNER present the earliest accounts of the general geology. Already DARWIN had noted the prominent phonolitic masses and abundance of pyroclastics, and all further investigations have substantiated the importance of these types of rocks. RATTRAY (1872) was of the opinion that fine-grained, dark, compact basalts were the commonest rocks, that the peaks were composed of light-grey granites, but in both respects he was in error.

The archipelago is of volcanic origin, formed of a substratum of pyroclastics intruded by a variety of alkaline eruptives, which in turn are overlain by two principal types of basaltic flows. The oldest rocks constitute a complex of tuffs and volcanic breccias intruded by phonolitic and trachytic bodies, many dykes and irregular-shaped masses of alkaline ultrabasics. Later ankaratrites flows and dykes and nephelinite dykes are associated with a younger group of pyroclastics. Succeeding are thick nepheline-basanite flows (Fig. 7).

Sediments, of aeolian and marine origin, are restricted to littoral areas, but they cover some 7.5 % of the archipelago area.

The eruptives and pyroclastics can be divided into three formations, here to be described.

Remedios Formation

This, the oldest, forms the highest elevations in the centre of the main island and also the central plain. Coastal outcrops are to be seen along the N coast from the peninula lim-

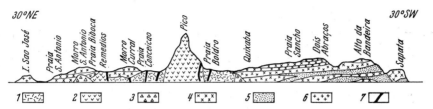

Fig. 7. Schematic profile along N coast of Fernando de Noronha. 1. Remedios tuffs, 2. Phonolites, 3. Trachytes, 4. Ankaratrites flows, 5. Quixaba tuffs, 6. Nepheline-Basanites, 7. Dykes. (ALMEIDA, 1958). (Length of Profile ca. 10 km)

iting Biboca beach to the vicinity of the stream entering Boldro beach; on the S coast, between the E side of the Atalaia embayment to Southeast Bay and further westwards at Morro Branco; on the E coast at Caiera Bay. Phonolites outcrop on the islets of Sela Gineta, Conceicao, Dos Ovos, Cabeluda, Viuva and Leao.

a) Pyroclastics

These are the oldest rocks of the archipelago. They are all deeply weathered and clothed in vegetation. The best exposures are to be seen in the cliffs sections, e. g. at Caieira and Abreu Bays, Cachorro and Boldro beaches. The central plain is formed of pyroclastics which however, seledom outcrop. When not deeply weathered, the rocks can be remarkably hard, have a brownish or buff colour but on weathering acquire reddish, yellowish and greenish tints. Agglomerates, breccias, tuff-breccias, lapilli-tuffs, tuffs, ash and pumice beds are all present, representing a large range of granular sizes. Angular or subrounded blocks, up to a metre in diameter, are quite common, but these are eruptives such as occur elsewhere. Neither epiclastics nor bombs or other ejectiles have been observed. The tuffs comprise only fragments of rocks already consolidated in the course of being hurled skyward, a feature distinguishing these tuffs from those of the next younger formation. At Caieira the pyroclastics are more than 100 m thick, but in general it is difficult to assess thicknesses, due to lack of stratification and extreme weathering. In the Morro S. Antonio-Caieira region, tuffs and lapilli dip 30–40° to the NE, but sotherly outcrops show higher dips, to both NE and NW, amounting to 70°. Extensive fractures, measuring tens of metres in length, are common. It appears that no tectonic disturbances of the pyroclastics has occurred, but this cannot be taken as a dogmatic statement. The chemical and petrographic character of the pyroclastics indicate they are explosive products of phonolite and trachyte emplacements.

b) Eruptives

1. Phonolites. These and trachytes are the essential rock constituents of the archipelago. In most areas of phonolites, a thick regolith and plant covering prohibits the exact demarking of contacts conceals outcrops and frequently yields only deeply-weathered specimens. Good exposures are to be seen at Morro Atalaia, Boa Vista and Meio, where quarrying has laid bare the rocks. Unaltered phonolites are megascopically aphanitic, all of ashen grey-green colour, and individual varieties are distinguished more by structure than mineral content. The commonest type is a tough, compact rock with many small feldspar phenocrysts and small laths giving the rock a planar structure. Such can be seen at Morros Atalaia, Boa Vista, Medeira and Gato. Three occurrences, e. g. at Morro Forte dos Remedios, of aphyric rock shows dendriform structure formed by aegerine crystals. The phonolite of Pico is strongly porphyritic, with phenocrysts of feldspar and nepheline giving a distinct lamination and lineation on occasion. Most of the porphyritic varieties are to be seen in a small hill N of Southeast Bay. The phonolites usually show fluidal and laminar structures, the former giving rise to domal and cylindrical forms with columnar jointings. Lineation is rare. Jointing is common, the most prevalent being parallel to the lamination. Columnar joint systems can be very conspicuous, e. g. the hill N of Sotheast Bay. Vesicles or amygdales have not been noted in the phonolites but are present in phonolite blocks in the tuffs and breccias. Xenoliths are less frequent in phonolites than some other eruptives. Eleven large independent phonolite masses occur in Fernando de Noronha, in the form of domes or plugs. Dykes are not very plentiful. A 10 m high phonolite stack resting on a seemingly insecure pedestal was taken by ALMEIDA to indicate the absence of earthquakes. Pico is the tallest phonolite mass, with a diameter of some 950 m. Close study of this remarkable feature is difficult because of the inaccessible perpendicular walls higher up and the dense vegetal covering lower down the slopes. Atalaia Grande is another spectacular phonolite mass, rising to an elevation of 223 m, the S and E faces of the dome showing clearly the cylindrical form. Boa Vista dome covers the largest area, measuring 1070 m in doameter. Some phonolite dykes are several metres thick, e. g. at Caieira beach, and at Abreu Bay one is ca. 30 m thick. Dyke rocks are similar to those forming domes and plugs, except that they are more porphyritic, with large feldspar and nepheline phenocrysts. Lineation is clearer in the dyke rocks. The larger phonolite masses all are subelliptical in plan, trending ESE along the major axis. In Fernando de Noronha the phonolite masses show no superficial mantle coverings such as have been described from Mont Pelée, Puy de Dôme or the Katmai domes. The angular blocks littering the phonolite slopes in the island are not to be confused with the 'amas rocheux continu, herissé d'asperités, limité par des parois abruptes qui se dressent au milieu de talus d'éboulis' referred to by A. LACROIX (1904). The domal structures of the archipelago phonolites are of 'Zwiebelstruktur' type of E. REYER (1888), but the 'Quellkuppe' type of H. & E. CLOOS have been noted (1927).

Drastic erosion ot the Remedios formation occurred before the Quixaba volcanic phase, which almost completely destroyed the external appearance of the Remedios, and one consequence was that presumably only the intrusive eruptives remained. In this connexion, ALMEIDA believed there was relevance to the question posed by BRANNER (1903) as to whether or not Pico was a spine like that of Mont Pelée. BRANNER thought that if Pico was not a plug then there must have been a vast amount of erosion, which supposition did not seem to be in harmony with present topography. He conceded however that the superficial appearance of Pico to the Pelée spine might be misleading. ALMEIDA took the view that as the Remedios formation was relatively old, it would be almost impossible for a plug like Pico, thrust up in the manner of the ephemeral Pelée spine, to have existed so long, as such volcanic structures have a relatively short life. Pico is merely a small part of a much larger phonolite dome – as ALMEIDA's photo 44, pp. 55 demonstrates – throughout highly fractured, with blocks, etc. scaling off its exterior. ALMEIDA believed that intense fracturing, easing thus drastic erosion, has bared to view this magnificent monolith, that this feature, as well as others in the archipelago of this age, were all subjected to intense erosion for a protracted length of time.

Nowhere can the basal parts of the phonolite masses be seen, hence it is unknown whether such have small depth or are true pluglike intrusions of great vertical extent and terminating in domes, and thus whether the majority of the phonolite bodies are plugs or necks. The tuffs are indicative of explosive expulsion of phonolite and trachyte blocks, and it is most likely that the phonolite bodies attained the surface. It was erosion which completely destroyed the original domal character of these intrusions.

2. **Alkali-Trachytes.** Though less voluminous than phonolites, these rocks form conspicuous features. When fresh the rocks are similar in colour to the phonolites, are phaneritic, some with small prisms of pyroxene occurring as dykes. The abundance of plagioclase phenocrysts is a constant feature. Fluidal structure is distinct, laminar less so. Lineation is difficult to observe. These trachytes are most common as dykes, e. g. at Caieira and Boldro beaches, where at the latter place they are up to 40 m thick, and can be traced for 750 m at low tide at Caieira. Two domal occurrences occur at Leao and Southeast Bay beaches. Petrographically the trachytes show little variation in themselves, such variations as occur referring chiefly to the quantity of feldspar and pyroxene phenocrysts. The best-preserved trachyte feature in the archipelago is the dome of Morro Branco, surface exposures being weathered white, naked of vegetal covering, and attracting attention from far and near. ALMEIDA claimed this was a dome of 'Quellkuppe' type, showing structural affinities with the Drachenfels of the Rhein region.

Like the phonolites, the trachytes are not effusive bodies but rather extrusive domes of increasing endogenism, certainly intrusive into the tuffs.

3. **Kali-Gauteites** occur in about a dozen dykes in the main island. These rocks, usually altered to varying degrees, are porphyritic, with phenocrysts of amphibole and pyroxene. The walls of the dykes have a 1–2 cm thick layering of a black rock like obsidian. Towards the interior, dykes may become amygdaloidal, these trending parallel to the lamination, which latter results from the alignment of microcrystals of feldspar. These rocks are one of the few types present in the archipelago in which lineation is always distinct. Longitudinal jointing and jointing normal to the walls forms columns, but no diagonal joints seem to be present. The dykes may vary in thickness from 0.6 m to 8.4 m, are vertical or nearly so. The trend is usually quite rectilinear, and can be followed for distances up to 460 m at Caieira. These dykes intrude the Remedios pyroclastics, phonolites, porphyritic essexites, alkali-basalts, trachytes and lamprophyres, and because of this wide variation in intruded rocktypes, these form the older rocks of this formation.

4. **Mela-monchiquites** form the most abundant dykes in Fernando de Noronha. The fresh rock is dark in colour, with phenocrysts of amphibole, pyroxene and olivine, on alteration showing a brownish hue. Lamination is usually indistinct, lineation poorly developed. The dykes have a common thickness of some 2.5 m, are frequently ramified, often show abrupt deflexions. In the field, dykes are seen to extend for 250 m but from aerial photographs, is likely they extend to half a kilometre at least.

5. **Furchites** are almost entirely aphanitic, aphyric or with rare phenocrysts of amphibole and pyroxene. The rocks are fine-grained, the vesicles and amygdales suggesting lamination. The rocks are only known in dykes, sometimes protruding strongly above the invaded rocks. At Caieira beach, a furchite dyke is some 190 m in length.

6. **Camptonite** dykes are much more restricted in distribution, e. g. around Morro Meio, where they are up to 1.5 m thick. The dark rocks are non-porphyritic or then a very few phenocrysts of amphibole and plagioclase may be present. Indistinct lineation can be detected, evidenced by amphibole prisms, whilst the vesicles outline the lamination. Dykes outcrop for only a few dozen metres. Like other rocks of the archipelago with plagioclase, they are always deeply weathered.

7. **Sanaites** occur only as two dykes, at Atalaia Point and on the road N of the Bay leading to Remedions. Both outcrops are poorly exposed, the rocks deeply weathered and identification is not too clear.

8. **Tannbuschites** and **Augitites** form the thinnest dykes, some as little as 3 cm in thickness. These rocks are aphanitic, with tiny microphenocrysts of olivine and rarely augite and are easy to confuse with limburgites and the fine-grained varieties of olivne-teschenite. Megascopically they show fluidal structure. At the cliffs at Caieira Bay a thin tannbuschite dyke is notably sinuous in plan, cutting various other dykes of monchiquite, olivine-teschenite, etc. These hypermelanocratic rocks have titanaugite as the dominant mineral.

9. **Limburgites** and **Olivine-Teschenites.** In Fernando de Noronha the former represent a holocrystalline facies of the latter, as is evident from the chemical analyses. The limburgites occur as two thin dykes. The olivine-teschenites have two occurrences, one a thin dyke, sinuous in plan, cutting tuffs in the cliffs at Cachorro beach, the other a somewhat elliptical body, perhaps 6 m thick, highly altered and intruded into phonolites.

10. **Porphyritic Essexites** are of diabasic appearance, finely phaneritic, with phenocrysts of pyroxene and a few olivines. They occupy an area of some 22 hectares to the E of the phonolite dome of Morro Atalaia, where the rock is much altered. Jointing is common here, are usually curved in plan. The outcrop is invaded by numerous and various types of rocks, amongst which can be noted kaligauteite, phonolite, tannbuschite, sanaite, alochetite, furchite and monchiquite, which would indicate that the essexite is perhaps the oldest of the Remedios eruptives.

11. **Glenmuirites** and **Alochetites** are both liable to be confused with the porphyry essexites in the field, having the same diabasic appearance. Alochetite was identified at Atalaia Pt., forming a small dyke intruding porphyry essexite; the glenmuirite occurs both at the village of Remedios and at Atalaia beach, at both places outcropping at the edge of an ankaratritic covering, but detailed field relations cannot be determined.

12. **Olivine-Nephelinites** occur as blocks on the road from the church square in Remedios to Tres Paus, some 650 m from the former. The rock is coarsely granular, of plutonic aspect, rather like a gabbro, with large pyroxene crystals. It is considered as an intrusive rock – the only occurrence of the rock seen in the archipelago. For some 40 m along the road, blocks of this rock lie on the surface.

13. **Alkali-Basalts** occur in three places. The rocks are aphanitic, and weather very easily. The most important occurrence is at Abreu Bay where it forms a small salient on the beach, representing an intrusive body partially covered by the sea. It probably re-

presents a small plug, and is intruded by two kali-gauteite dykes. On the eastern shore of Southeast Bay occur highly weathered alkali-basalts, forming part of the shore and the cliff, intruded by furchite and limburgite dykes. The alkali-basalts are considered older than the furchites, monchiquites and kali-gauteites, but perhaps younger than the porphyritic ess exites.

c) Dyke Associations

ALMEIDA mapped 120 dykes in the archipelago but more occur which are deeply altered and difficult to appraise. Those in the Remedios formation have a dominant NE trend, they may be simple, parallel, rectilinear features, multiple, composite or then arranged concentrically, with radial dykes passing through – annular dykes. The simple variety include principally the lamprophyric rocks. The multiple types consist of successive injections of the same kind of rocks through the same opening, and are rather rare. Fig. 8

Fig. 8. Multiple Dyke, extreme W end of Biboca Beach. 1. Aegerine-Phonolite, 2. Kali-Gauteite, 3. Mela-Monchiquite, 4. Mosean Mela-Monchiquite. (ALMEIDA, 1958)

is a good example of a combined multiple and composite dyke at the western end of Biboca beach where kali-gauteites, nosean-monchiquites and monchiquites, in that order, have been intruded into aegerine-phonolites. The multiple feature measures some 3.5 m in width, extends for some 120 m, is inclined at 80° to the W, with a NE strike. Composite dykes, containing two or more type rocks from the same intrusive body, are quite frequent. The outstanding example of such is seen in a vertical cliff at Caieira Bay, where a large trachyte dyke, up to 12 m wide, in intruded into tuffs (Fig. 9). Cutting through this

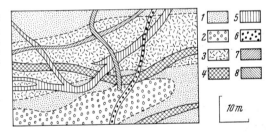

Fig. 9. Dyke relations exposed at the beach at Caieira Bay. 1. Remedios tuffs, 2. Beach pebbles, 3. Alkali-Trachyte, 4. Kali-Gauteite, 5. Hauyne Mela-Monchiquite, 6. Limburgite, 7. Olivine Teschenite, 8. Tannbuschite. (ALMEIDA, 1958)

dyke are hauyne-monchiquite and kali-gauteite dykes, which in turn are diagonally intruded by limburgite and tannbuschite dykes. It is interesting to note here how indifferent the kali-gauteite and monchiquite dykes intrude the trachyte in sinuous patterns. Annular dykes can be seen on the marine abrasion platform at Caieira Bay, but evidently only during the exceptionally low tides associated with the total eclipse of January, 1954. Concentric dykes here are of phonolite, kali-gauteite and monchiquite, with penetrating radial dykes (in the manner of spokes of a wheel) of limburgite, tannbuschite, furchite and monchiquite. All dykes are almost vertical here, the dyke ensemble intruding tuffs striking NW and inclined to the NE, all badly fractured.

d) Chronological Order of Intrusions

Field evidence appears to indicate that the phonolites and trachytes are generally to be regarded as the oldest eruptives of the Remedios, dykes of other eruptives intruding these but not the converse, except for the porphyritic essexite at Atalaia Pt. which is intruded by a phonolite dyke, thus apparently indicating that essexite is older than phonolite. If this indeed is so, then essexite would be the oldest rock of the sequence. Of dyke rocks, kali-gauteites are the oldest, after phonolite, and are seen intruding phonolites, essexites and alkali-basalts at Abreu Bay. The tuffs contain blocks of these kali-gauteites, furchites, alkali-basalts, essexites, phonolites and trachytes. It appears that the lamprophyric rocks – sodic lamprophyres, monchiquites, furchites and camptonites – are intermediate in age between the tuffs and these dyke rocks. The strongly melanocratic rocks – augitites, olivine-teschenites, tannbuschites and limburgites – are younger than the others mentioned above, and from Fig. 8 we see a tannbuschite dyke cutting an olivine-teschenite dyke, hence of younger age. It would thus appear that the intrusions have followed approximately in an order of increasing basicity, first rocks very rich in alkali-feldspar injected into essexites, then medium-rich feldspathic rocks, and finally those in which feldspar is low or wanting.

e) Xenoliths

One of the most striking features of the dykes is the frequency of dark rock inclusions, rich in amphibole, present in all types of gauteites, essexites, phonolites and lamprophyres. Porphyritic phonolite dykes show abundance of these inclusions, but on the other hand they seem to be non-existent in trachyte dykes, though this is not to be taken as a dogmatic statement. The inclusions are angular or then poorly rounded, show a non-orientated internal structure or then distinct lineation of amphibole prisms. The largest xenoliths seem to be about 60 cm in diameter, and in some, amphibole prisms up to 6 cm in length can be observed. Megascopically the only mineral seen is amphibole, occasionally olivine and pyroxene. The inclusions have no mineralogical equivalents in the dyke rocks but are similar to many blocks occurring in tuffs. The xenoliths are autolithic inclusions, resulting from the segregation of Fe-Mg minerals in the magma into aggregates when the magma had become plastic, perhaps solid, and on the point of fragmentation.

f) The Pre-Ankaratritic Erosion Surface

All the Remedios eruptives are of hypabyssal character. Trachyte and phonolite bodies consolidated within tuffaceous masses. As there are numerous intrusives of no great height or volume, such as can be compared to similar intrusives in many parts of the world today, those of Fernando de Noronha, we assume that those of the archipelago have been largely associated with explosive phenomena. In a few places, e. g. Boldro and Biboca, we can observe the contact of the Remedios and Quixaba formations, where ankaratritic flows lie on large trachyte dykes. Similar lavas are also present in depressions overlying the tuffs in which phonolite and trachyte intrusions have taken place. We may conclude therefore that the pre-Quixaba topographic surface underwent severe differential erosion, and is not the result of explosive phenomena. It appears that erosion was effective in all but entirely destroying the external forms of the Remedios intrusives, such erosion having occurred before the onset of the ankaratritic phase. Taking into account the relations of the position of these basal ankaratritic flows and the present altitude of the phonolite peaks and hills, it can be concluded that this erosion stripped off a layer of rock some 300 m thick at least. Less certain however is the actual agency whereby this erosion was effected. No sedimentary rocks occur at the junction of the Remedios and Quixaba formations, but fossil remains indicate subaerial erosive processes, marine erosion being considered of almost no account.

Quixaba Formation

The hamlet of Quixaba lies on a plain formed of black lavas, at an elevation of 70 m, and from here steep, 50 m cliffs drop down to the beaches of Quixaba, Sancho and Carreiro de Pedra. In these cliffs ankaratritic flows, serveral metres thick individually and alternating with pyroclastics also of an ankaratritic nature are to be seen. Such an alternation is typical of the formation which is up to 200 m in thickness, forming a simple, uniform sequence, in contrast to the Remedios formation.

Two isolated occurrences of the Quixaba occur: a western one forming the major peninsula, where heights up to 170 m occur, descending to 80 m in the extreme Sapata peninsula; the second occurrence forms the interior plain between Cachorro and Biboca beaches on the N, Atalaia on the S and all the eastern part of the main island. On Ilha Rata is a partially covered ankaratritic diminutive plateau, and the islets in Caieira Bay also show these lavas. In the interior of the main island, the Quixaba formation forms the scalloped, inclined plain which ends abruptly in marine cliffs. In western Fernando de Noronha, the relief is largely controlled by the angle of repose of flows and pyroclastics.

a) Pyroclastics

These comprise tuffs, lapilli-tuffs, tuff-breccias and agglomerates of ankaratritic composition. The first two types are formed chiefly of lava particles of lapilli size, ash and pumice, in which olivine, and rarely pyroxene, crystals are present. Bombs and driblets of ankaratrites testify to explosive phenomena. Blocks, bombs and driblets represent fragments of earlier, consolidated lava, of various structural form, often of ropy character, always porphyritic and very vesicular. Blocks may be up to 2 m in diameter, bombs up to 50 cm in diameter. The breccias and agglomerates show an amazingly chaotic assemblage. The inaccessible scarps limiting the plain on the N at Viracao show bedding planes a metre thick and several hundred metres in extent. NE of Capim Açu Pt., 30 m of decomposed tuffs are intercalated with flows. The thickness of the pyroclastics between the flows can vary from a few centimetres to a few dozen metres, but in toto the pyroclastics are thicker in the W and are at a minimum at Quixaba beach and Southeast Bay. In the eastern occurrences, pyroclastics are relatively thin. When unaltered the pyroclastics have a dark appearance and almost all are of ankaratritic composition. In the western exposures, the dip nearly always is to the S at angles up to 45°. No epiclastics are present but only ejection products and a few dyke remains.

b) Eruptives

In marked contrast to the Remedios formation, these are almost entirely extrusives and only a few dykes occur. Two kinds of ankaratrites are found, one showing pseudomorphs of melilite, which is subordinate to the other kind which does not have these. Two small nephelinite outcrops are present, and all dykes seem to be ankaratrites.

1. Ankaratrites. The fresh rock shows small phenocrysts of olivine, spheroidal weathering is common resulting in thick, brownish soils. The rock is notably uniform, both in texture and mineral content, with a variable quantity of melilite, vesicles and amygdales abundant. Individual flows range up to 40 m thick, but 5–15 m is the common thickness. Fig. 10 illustrates the structure at Capim Açu peninsula, where a tunnel has been carved by the sea out of alternating flows and pyroclastics. It seems that all the eastern occurrences of the Quixaba formation are formed solely of an ankaratrites flow, extending

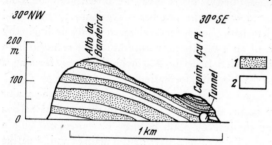

Fig. 10. Generalized geologic section across western area of Fernando de Noronha. 1. Pyroclastics, 2. Ankaratrites flows. (ALMEIDA, 1958)

from Conceicao beach to S. Antonio Pt., S to Pontinha and W to Atalaia embayment. This flow is some 30 m thick at Pedra Alta, perhaps the same at Biboca beach, but certainly decreases to the E. The ankaratritic lavas show original structures which aid in the elucidation of volcanic events. The olivine crystals impose a lineation to the rocks, especially at the top and base of the flow but rarely internally. From this we can deduce the direction of flow, in conjunction with the dips, from which it is seen that these flowed in a southerly direction. Vesicles and amygdales are always small, aligned in the direction of flow. Jointing near the tops and base of flows is characteristic. The flows are free of foreign inclusions but occasionally glenmuirites are found, as e. g. near Pedra Alta, where Remedios glenmuirite outcrops occur. At the extreme point of the S. Antonio peninsula the flows have a pillow-structure appearance, but ALMEIDA interprets this as due to spheroidal weathering. In the vicinity of Tamandaré is a volcanic chimney (Fig. 11) puncturing a flow, of subelliptical shape, with a major diameter of 130 m. The sea has destroyed the ankaratritic wall and penetrates into the chimney at high tide. Ankaratrites dykes tend to be sinuous, vertical, a maximum of 2 m in thickness and essentially similar to the flows petrographically.

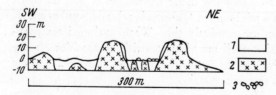

Fig. 11. Geologic Section across the "Chimneys" at S. Antonio Peninsula. 1. Breccias-Agglomerates of the "Chimneys", 2. Ankaratrites, 3. Beach Pebbles. (ALMEIDA, 1958)

2. Nephelinites. The only non-ankaratrites rock noted amongst the Quixaba eruptives is a nephelinite dyke in the cliff between Pontinha and a grotto named 'Captain Kidd'. The rock is coarse-grained, with nepheline and pyroxene visible to the naked eye.

There occur many miarolitic cavities partially filled with analcite and other zeolites. The dyke is ca. one metre in thickness, is vertical, striking N 75 W. Identical material forms a fine filling in joints in an ankaratrites flow at Pt. S. Antonio, opposite the islet Rasa.

c) Volcanic Processes

The western outcrops of the formation suggest conduits near the present coast at Carreiro de Pedra Bay. The dips of the tuffs here imply that at the time of the last emission of the flows now outcropping, the altitude of the ankaratrites structure did not rise much above present sea level, but still some 4300–5000 m above the oceanic base. The lavas must have been fluidal, as flows as thin as 2 cm are to be encountered. Explosive expulsion of lava, due to the high gas content, preceded the extrusion of the flows. The flows were poured forth over eroded depressions lying between the phonolite and trachyte hills and peaks. The imperfection of stratification of the pyroclastics appears to suggest that their deposition did not occur in a submarine environment. During the interval when the Quixaba rocks were being formed, lavas underwent no modification in their composition, with the exception of variation in the quantity of melilite.

Sao José Formation

RIDLEY (1890) had claimed that nepheline-basanites occurred on the islet of Chapeu de Nordeste, but ALMEIDA stated that here there were only rounded fragments whose origin was to be found in the islet of Sao José. WILLIAMS (1889) was the first to draw attention to rocks of this type on this islet, and ALMEIDA not only confirmed this but showed that they also outcrop on the neighbouring islets of Cuscuz and Fora. However he did not support the idea of a geographical continuity between the S. Antonio ankaratrites peninsula and the three basanitic islets. Basanites are absent on the main island, and as the ankaratrites flows of the S. Antonio peninsula dip to the NNE, ALMEIDA considered the basanites of the islets as being younger than the ankaratrites flows of the Quixaba formation, which latter are seen outcropping at low tide on Chapeu de Nordeste. The clear petrographical distinction between the basanites and the ankaratrites led ALMEIDA to include the former as a separate entity which he named the Sao José formation. In this islet and in Cuscuz and Fora a single covering of a nepheline-basanite flow, not less than 25 m thick, outcrops. The dark greyish rocks show small phenocrysts of parallel-arranged olivine, giving a fluidal linear structure oriented towards the SE. Neither the base nor the top of the flow outcrops, the upper part being covered by sea and having clearly been subjected to marine erosion. Occasionally vesicles and anygdales are seen, also olivinite xenoliths, angular or subrounded, some as large as 23 cm in diameter. These inclusions give no hint as to how they were originally formed, and ALMEIDA categorically stated that they are not of nodular origin but rather fragments of a large body which was engulfed by the flow, torn up at the surface into blocks of varying size and abraded and made smaller by flow transport. The islet of Sao José is connected by a tombolo at low tide with Chapeu de Nordeste and on occasion one can cross on dry land to Cuscuz.

Sedimentary Rocks

Old Sediments

Since the cessation of vulcanism and the continued powerful marine erosion to which the archipelago is subjected, it is not to be expected that many opportunities for sedimentation conditions prevailed. However some sediments are worthy of note, the products of wave disintegration and fluvial erosion. Within the archipelago, the following sequence of sediments are likely products of the Pleistocene: 1. Caracas arenites, 2. marine limestones, 3. terrace psephitic deposits, and 4. alluvium of fluvial origin.

1. The old **calcareous** arenites of aeolian origin were discussed by LEA (1888), BRANNER (1889, 1890) and RIDLEY (1890, 1891). LEA called them 'reef rocks' a reef formation laid bare at low tide, comparable to exposures on the shore at Recife (= reef) at Pernambuco, Brazil. BRANNER claimed these arenaceous rocks, on the basis of their gross structure, indicated that they were deposited as sand dunes, microscopic examination showing that the rock was consolidated by means of interstitial cementation of carbonate of lime, the sandstones displaying characteristics of wind deposition. In places the sandstone was composed entirely of organic matter, calcareously cemented, with some sand and ash derived from igneous rocks and thrown up by the sea along with the calcareous matter. BRANNER (1890) gave a chemical analysis (Table 6) of samples from Southeast Bay

Table 6 Chemical Analyses of Aeolian Sandstones (BRANNER, 1890a)

1. Southeast Bay sample.
2. Ilheu Rapta sample.

	1.	2.
SiO_2	2.20	
Al_2O_2	0.79	0.45
Fe_2O_3	0.87	0.13
CaO	0.27	
MgO	0.89	0.64
$Ca_3(PO_4)_2$	0.67	0.82
$CaCO_3$	97.27	98.33
$MgCO_3$	0.49	
K_2O	0.15	0.10
Na_2O	0.22	0.20
H_2O	0.25	
Total	104.07	100.70*

* Includes: 0.09 SiO_2 and insoluble matter.

and Ilheu Rapta, microscopic examination of the former showing it to consist entirely of triturated organic remains – pieces of shells, corals, nullipores, foraminifera. He stressed that these aeolian sandstone occurrences were all on the E or SE sides of the archipelago, claiming they had no connexion with any of the small sand beaches now existing. He thought they must formerly have been much more extensive, that these sands were blown upward from a S and SE direction, originating from beach sands on similar sides of the archipelago. (ALMEIDA quotes exposures at Caracas Pt., along the W side of Southeast Bay where they cover 5 hectares, and in the Atalaia embayment, at Tamandaré, on the islets of Chapeu de Nordeste, Chapeu des Sueste, Rata, Sao José, Meio and on Raza where they occupy some 38 hectares.) BRANNER pointed out the S and SE coasts were on the windward side, that the N and NW coasts were barren of such aeolian sandstones not only because these were the lee coasts but also because of the general NE-SW trend of higher ground and the basic structural trend of the archipelago.

LEA, DAVIES and RIDLEY were not in agreement with BRANNER. To them, the 'aeolian sandstone' of BRANNER is a reef rock, packed with organic remains, and at low tide level, full of living remains. Inland from the coasts, the rock becomes an amorphous, compact mass of carbonate of lime, with mud in varying proportions. They pointed out the very high calcium carbonate content, very low silica content of BRANNER's two analyses and queried how he could claim such rocks were sandstones. They did not deny the presence of aeolian sandstone in the archipelago, e. g. on S. José, at S. Antonio Bay, but these occurrences, according to theme, are very different from the reef rock. This latter is quite unlike coral reef of the open ocean atolls, for it is not largely built of larger corals, but of nullipores, seaweeds, broken shells, echinus stems, worm tubes and crab remains, with nullipore growth dominant. This reef rock only occurs where small streams carry water to the sea, bringing down silt enough to form a firm base for the nullipores. This would account not only for the silica deposited in the reef rock, but also for the occasional larger percentages of SiO_2, occurring in those places where greater quantities of erosive products were washed down to the sea. RIDLEY believed the reef rocks were merely raised reefs of different ages, as they occur in strata lying flat on each other, and he asked how could sand dune form flat strata like this?

The arenites lie on both pyroclastics and eruptives, are finely granular and show unmistakable cross-bedding. Stratification is of dune type, with large cuniform or prismatic bodies inclined at 30°, with laminae of concave form at the lower parts and truncated at the upper parts. ALMEIDA claims they doubtless correspond to frontal portions of migrating dunes, so inclined as a result of dominant SE winds. On the islets of Rasa and Meio the arenites are not less than 20 m thick. Fig. 12 shows an occurrence at Chapeu de Nordeste, where a thickness of some 5 m are exposed, resting on an ankaratrites flow. Table 7 gives six analyses of the Caracas arenites mentioned by POUCHAIN (1948) and by ALMEIDA. Whenever the rock shows more than 90 % $CaCO_3$, it has never undergone any dolomitization, even partial. ALMEIDA termed the rocks arenites and recognized their aeolian origin.

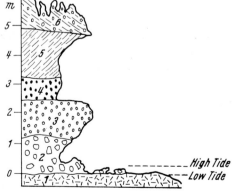

Fig. 12. Geologic Section, Ilheu Chapeu de Nordeste. 1. Ankaratrites, 2. Conglomerate, calcareous Arenite, large boulders, 3. Same, with small boulders, 4. Calcarous arenite boulders, with gastropods and other marine organic remains, 5. Calcareous arenite of uniform particle size, aeolian stratification, 6. Modern beach rubble. (ALMEIDA, 1958)

Table 7 Chemical Analyses of the Caracas Arenite (Nos. 1–5) from Caracas Pt. (ALMEIDA, 1958), and No. 6 from Ataiaia Beach (POUCHAIN, 1948)

	1	2	3	4	5	6
Ignit. Loss	44.23	40.19	42.48	40.85	41.06	43.2
Insol. Residue	0.48	2.86	2.12	4.22	4.42	1.3
R_2O_3	0.55	2.25	1.90	2.35	1.80	0.3
CaO	53.75	52.20	52.15	50.55	51.65	55.2
MgO	0.63	1.86	1.08	1.91	0.61	Tr.
Total	99.64	99.36	99.73	99.88	99.54	100.0

2. Marine Limestones. Near Porto, crowning an ankaratritic cliff, limestones are present which formerly must have had a more extensive occurrence. They lie on ankaratritic rocks or then pyroclastics, with an erosional contact at the boundary. They form thin beds – some 6 cm thick – with a total thickness of perhaps 6 m. The cream or pale yellow rocks are firmly cemented, may contain rare inclusions of poorly-rounded ankaratrites fragments up to 5 cm in diameter, have an arenaceous texture but are not to be confused with the Caracas arenites. Under the microscope the detrital character is clearly evident. The rock is finely granular, grains well rounded, some particles of augite, olivine and hornblende often present, and contain more than 90 % $CaCO_3$. The calcareous grains represent remains of marine organisms – bryozoan colonies, lamellibranch valves, crustacean carapaces, foraminiferal tests and algal remains. *Lithophyllum*, approximating to the genus *Amphiora*, can be recognized, but no evidences of *Lithothamnium* were observed, and hence no Lithothamnium reefs were evidently being formed when this limestone was being deposited.

3. Terrace Psephitic Deposits. In various localities, conglomerates with a calcareous, sandy cement are found in positions indicating that they are marine terrace deposits. These can only be seen in the coastal cliffs and never in the interior of any of the islands. The pebbles, etc. are chiefly composed of ankaratrites, e. g. at Tamandaré and Atalaia beach, or of phonolite, e. g. at the extreme W end of Cachorro beach. Pebbles and cobbles are well rounded, suggestive of the work of marine abrasion. When a sandy matrix is present, this comprises grains of calcareous material originating from the destruction of carapaces and other hard parts of marine organisms.

4. Fluvial Alluvium. The few small valleys show only minor quantities of psephitic material, probably resulting from selective weathering from blocks moved down the slopes by gravitational means. Only in the inner sections of valleys leading down to the Atalaia embayment is there alluvium of any worthwhile extent, covering an area of some 5 hectares. This is ca 5 m thick, comprising coarse sands alternating with fine conglomeratic material which includes fragments of ankaratritis, perhaps other eruptives, and tuffs of the Remedios formation. On the interior sides the alluvium terminates against ankaratritic slopes, rounded pebbles and subrounded fragments of this rock occurring as a raised storm beach at the shore. It seems unlikely that the central plain underwent complete submersion by the sea, but the storm beach and the interior alluvium occurrence were probably formed at a time when sea level stood about a metre higher than now.

Young Sediments

Present sedimentation is largely related to the littoral and areas near the shores. Several types of sediments are accumulating.

1. Beach Deposits. These are principally conditioned by geomorphic processes which will be mentioned later, and here only the lithology will be remarked upon. A hand lens shows that the deposits comprise grains which are fragments of calcareous algae, bryozoans, worm tubes, crustaceans, echinoderms, molluscs and foraminiferal tests. Such grains total ca. 75 % or more of these deposits, with the following chemical analysis: SiO_2 7.63, Fe_2O_3 2.34, Al_2O_3 4.26, CaO 42.02, MgO 6.53, ignition loss 37.45 (IMBIRIBA, 1948). Psammitic beach deposits are of medium-fine grain, and samples from Conceicao beach, quoted by ALMEIDA, showed: insoluble residue, 37.7, of which four-fifths comprised augite, olivine, magnetite and limonite. Psephitic sediments are nearly always intimately related to rocks outcropping at the beaches from which they have been derived, mostly being phonolites, ankararrites, trachytes and monchiquites. These deposits comprise an appreciable part of the tombolos. Pebbles show little evolution, being actually a selection of the types of rocks of which they are products.

2. Talus Deposits. These are principally associated with ankaratritic escarpments showing columnar jointing. Slope evolution is more advanced along the western coastal areas, where wave attack has loosened great blocks, some weighting as much as 150 tons and with a volume of 100 000 m³. At Capim Açu a marine terrace is carved out of Quixaba tuffs, which today is almost entirely destroyed by cliff recession but which can still be observed, thanks to the covering of ankaratrites blocks.

3. Active Dunes. In several places these can be seen, formed and moved by the dominant trade winds. They comprise chiefly calcareous grains of marine origin, with contents of $CaCO_3$ of up to 70 %. Few indications of aeolian abrasion can be noted except high polishing of grains. Movements of up to 400 m of the dunes seem evident. The dune sand characteristics show very close analogy with those of the sand beaches from where they originated. Good examples of such dunes are to be seen at Atalaia, Leao and Southeast beaches and along the S. Antonio peninsula coastal stretch.

4. Lithothamnium Reefs. The Fernando de Noronha reefs are not of coral but of calcareous algae of the family Corallinaceae, and are abundantly developed along the S and E shores of the main island, forming a series of reefs. Algae of the genus *Lithothamnium*, with which are probably associated other Melobesias on a smaller scale, form calcareous structures due to the worms Polychaetae and other marine organisms. Reefs are especially developed along coasts exposed to the SE winds, where they form fringing and barrier structures. The latter are particularly well developed at Caieira Bay, where small lagoons occur which can be crossed on foot at low tide. The calcareous rock, of cream or white colour with a rose tinting, is porous at the surface but much less so towards the interior, where secondary cementation takes place. The rock is very resistant to the hammer. All the reefs of the archipelago seem to be of similar type and origin, all are in active process of growth.

5. Zoogenic Phosphates. In the S part of the islet of Rata, the Caracas arenites have their upper parts partially substituted by calcium phosphate of bird excreta origin. DERBY

& BARROS (1881) were the first to describe these deposits in detail, estimating a reserve of some million tons. They named the mineral a phosphatic guano lacking ammonia, alkalis and organic acids, due to the leaching effects of rains. In 1947 POUCHAIN (1948) visited the occurrence and calculated the tonnage as more than half a million tons, taking an average tenor of P_2O_5 of 19.6 %. POUCHAIN stated that there were two types of phosphates, colophanite and vavelite. Difficulties of access to the islet allowed ALMEIDA to make only a hurried inspection.

Geomorphology

The Interior

As of the present, it is the rock constitution and the structure which controls the geomorphologic development of the archipelago, rather than climate. Coastal areas are dominantly under marine attack, whereas in the interior, fluvial action, combined with chemical weathering, takes control. Rocky slopes tend to be abrupt, and the development of coarse block talus proceeds rapidly. On gentler slopes lower down, detritus is of much finer size. Ravining is strong in tuff areas, but where eruptives outcrop this is much less so, not solely because the rocks are more resistant, but also due to the fact that ephemeral streams can actively erode for only a few months annually. Fracturing is a common feature of the Fernando de Noronha rocks. According to BRANNER this was chiefly due to temperature variations, but ALMEIDA did not agree on this point, claiming that the freshness of the rocks when studied microscopically indicated that neither temperature nor rainfall variations, annual or over greater periods of time, appeared to have had much effect on the original constitution and appearance of th rocks.

Fig. 13 shows the relations of the profiles of the slopes on Morro Atalaia, with the Narrow ravine eroded out of the softer tuffs.

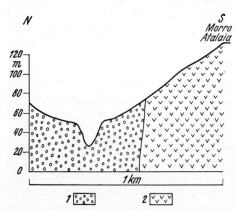

Fig. 13. Transverse profile of Valley leading to Atalaia Beach, showing rupture of concave profile. 1. Tuffs, 2. Phonolites. (ALMEIDA, 1958)

Three smaller morphological regions can be recognized in the archipelago. In the central part of the main island is a plain, comprising Remedios tuffs, extending E-W almost from coast to coast. To the N and S and also within the plain rise up some phonolite hills. The average elevation of the plain is about 45 m, with rather abrupt descents westwards down to Quixaba beach and eastwards down to Atalaia Bay. W of this central plain is a narrower, somewhat higher backbone forming the western end of the main island, dominated by the peaks Dois Abraços, 171 m, and Bandeira, 160 m. Lava flows constitute the geology here. The eastern part of the main island is an irregular plain of perhaps 70 m average elevation, again interrupted by several phonolite masses, reaching a maximum in Morro do Frances, 195 m.

SOARES (1944a) believed that the central plain represented a marine terrace, basing his opinion chiefly on its very level nature. (It is here that the airfield has been constructed.) However ALMEIDA saw no evidences to indicate that this area had been covered by the sea. The profiles of the eruptive masses bordering and within the plain testify to erosion under semi-arid climatic conditions. For marine erosion to construct a plain of these dimensions – some 2.5 km long and up to 2 km in breadth – would surely require relative great stability in sea level, which is refuted by other evidence. The surface is not quite so level as SOARES would have us believe – indeed it is distinctly undulating with a scalloped surface, crossed by valleys leading down to the N and S coasts and ending in abrupt cliffs at the shores. ALMEIDA believed the plain resulted through subaerial processes under climatic conditions perhaps somewhat more arid than now. It is implied that at this time, sea level was some 40 m higher than at present, that the island covered a greater area than now, the relief of the interior regions being not so intimately influenced by sea level oscillations as now. Later lowering of sea level (or emergence of the main island) allowed this interior region to undergo erosion whilst yet maintaining evidences of its earlier erosive phase, i. e. here we have a rejuvenated landscape.

Two points are worth noting regarding the phonolite masses which rise as hills and peaks: they have elliptical plans and are aligned in a WNW direction. Of these elevations, quite the most spectacular is the monolith, Pico, rising some 200 m in almost vertical walls above its base. Weathering along joints has given the feature its remarkable shape, and this fact, rather than any intrinsic rock resistance, is responsible for its form.

The above-noted three morphologic units are characterized by Remedios tuffs in the case of the central plain, and by ankaratrites flows. In the western higher backbone area, Quixaba ankaratrites flows dip gently towards the SSE, the highest elevations, Dois Abraços and Bandeira being formed of upper Quixaba flows. The eastern higher plain shows lower Quixaba ankaratrites flows, reaching to an elevation of 195 m in Morro do Frances.

In Fernando de Noronha, ephemeral streams do not have much competency in transporting material moved into the valley depressions by downhill gravitational movements. The initial dips of the flows have largely determined the longitudinal profiles of the streams, and as dips are not great, as the rainfall is not great, water content during the period of rains is not great, stream removal of detritus is far less effectual. It follows therefore that alluvial deposition is scant, and in fact only at Atalaia beach is this considerable, covering some 5 hectares. River erosion and deposition have had relatively little effect in moulding the present landscape.

The Littoral

The trace of the coastline is dependent upon orientation with respect to the trade winds, the structure and the earlier evolution. The dominant wind direction and direction from which the waves travel is the SE, these waves having a very long 'fetch'. The result is that the southern coastline is more indented, the adaptation of the structure is more advanced. Along the northern coasts the waves have suffered refraction, robbing them of much of their power, enabling tombolo formation to take place here and there. The E and S coasts have long been subjected to rapid recession, with more resistant phonolitic bodies forming off-shore remnants as islets. Structure is a dominant feature of coastal topography, being simply a matter of less resistant tuffs forming embayments, more resistant phonolitic masses and ankaratritic flows constituting more bold headlands. Alternations of tuffs and flows near sea level has favoured marine tunnelling, such as occurs at Portao da Sapata and Capim Açu (Fig. 10, 14). Rock structure determines the

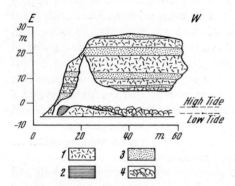

Fig. 14. Geologic section of tunnel at Capim Açu. 1. Ankaratrites flows, 2. Reefs, 3. Pyroclastics, 4. Fallen blocks. (ALMEIDA, 1958)

essential cliffs features and of course the dips of the pyroclastics and flows. On a rock-constitution basis, the islets can be divided into: 1. phonolitic – Sela Gineta, Frade, Ovos, Viuva, Leao, Cabeluda and Conceicao. 2. basaltic flows – Rata, S. José, Dois Irmaos, Lucena, de Fora and Cuscuz. 3. calcareous arenites – Rasa, Meio, Chapeu de Nordeste and Chapeu de Sueste. The littoral of the archipelago shows evidences of recent submergence of a few metres, with aeolian sands covered by the sea, partial drowning of lower valleys on the S coast. The lowered littoral appears to represent a marine abrasion platform resulting from the recession of old cliffs.

The S and E coasts are subjected to powerful wave action and storm beaches occur here, e. g. at Atalaia, where huge boulders, varying between 2 and 3 m diameter, maximum size, litter the lower slopes. Tidal amplitudes are high for an oceanic island varying between 2 m and 3.2 m according to the season. Currents are potent agents of alongshore westward drift of detritus, again most evident along the S and E coasts. Here bouldery and pebbly beaches occur, whilst along the N coasts sandy beaches are more common, a direct result of the relative force of wave attack. Wave refractions, whereby waves from opposite directions may approach each other and deposit their transported loads, are characteristic of the S. Antonio peninsula and the NE islets, where tombolos, all entirely covered at high tide, form connecting links which can be crossed on foot at low tide.

Active dunes bear witness to aeolian processes. As would be expected, the more sandy the beaches, the greater the extent and volume of the dunes originating from them. Perhaps the most favourable region for dune formation is Southeast Bay, where dunes cover some 12 hectares and reach to 12 m above mean sea level. At Leao beach, the dunes have been arrested by natural vegetative growth. Along the N coasts dunes are lacking though here are the most extensive sandy beaches. This is because the dominant winds are blowing off-shore, beach sands being blown into sea and distributed by waves and currents.

As previously remarked, the Caracas arenites are taken to be older aeolian deposits. At the time of formation the dominant winds were as today, viz. from the ESE. The constituent material is of marine origin, possibly raised into dunes and partly included some bottom-water deposits and beach sands. ALMEIDA believed that these dunes probably attained heights above mean sea level no greater than 30 m, and the fact that bottom-water detritus is incorporated presupposes a lower sea level at the time of formation. 60 m is taken as the maximum submersion value for the arenites, more likely 30 m being the actual amount of submersion, which would be adequate to make Southeast Bay, for example, dry land.

Valley erosion-deposition features in the archipelago suggest a lowering of sea level of the order of a dozen metres: A lowering of some 30 m would have allowed of waves and currents to attack sandy detritus of the marine platform, transporting this from the Southeast Bay area northwards towards Caicira and also along the S coast. When subsidence set in, beaches became submerged and their destruction, as well as that of the lowest dunes, became possible. Because the higher dunes had been previously consolidated by cementation, they were able to survive, as the Caracas arenites.

Various marine erosion levels can be recognized in the archipelago. There seem to have been at least two higher positions of sea level in former times. The highest corresponds to the morphologic evolution resulting in the central plain, which today lies at an average elevation of 45 m above present mean sea level. Good evidence of a former level 12 m above present sea level is to be found in the limestones with *Amphiora* which occur in the vicinity of the port on the S. Antonio peninsula, this marine limestone now lying at an average elevation of 8 m above present sea level. Near the hamlet of Tamandaré, calcareous sands, incorporating marine organic remains, lie some 5–7 m above present mean sea level, and near by are rounded boulders resting 9.5 m above this same datum. At Pontinha Pt. rounded boulders in calcareous, arenaceous matrix lie some 10 m above the sea, and between Capim Açu and the Viracao plain, again we find rounded ankaratrites boulders lying in horizontal disposition at ca. 12 m above actual mean sea level. ALMEIDA was led to conclude that in former times, sea level stood some 12 m higher than at present. The Caracas arenites were formed later when sea level stood some 6 m lower than at present, the older sediments of Fernando de Noronha all being formed at this period. In brief, then, former sea levels are postulated at + 40 m, + 12 m, + 1 m and − 6 m, with reference to present datum.

Soundings have roughly determined the extent of a marine platform from which the archipelago rises. At a distance of 3.5 km SE of Caracas Pt., depths of only 26 m occur; for a distance of 2 km off the E coast, depths do not exceed 30 m. To the S of the Sapata peninsula the platform is narrowest, with depths of 740 m occurring 2.5 km from the coast, and 2.1 km to the S of Barro Vermelho Pt., depths attain 530 m. N of the archipelago, the 50 m isobath varies from 1.5 km to 3 km distant from the shore, and it appears that to the N, the marine platform attains its greatest widths.

ALMEIDA has presented a table showing the chrono-geological development of the archipelago, which is here given in Table 8.

Table 8   Chronological Chart of the Fernando de Noronha Archipelago (ALMEIDA, 1958)

Holocene	Present Sea Level	Destruction of beaches, tombolos and part of the dunes of previous cycle. Consolidation of old dunes. Development of littoral border and establishment of reefs at the edge. Trenching of major valleys. Separation of islands.
Pleistocene	Sea Level − 6 m.	Extensive sandy beaches to S and SE of archipelago. Calcareous dunes united to present islands.
	Sea Level ± 1 m.	Accumulation of large boulders and coarse beach sands, with remains of *Lithophyllum*, foraminifera, gastropods, etc.
	Sea Level + 12 m.	Beaches of sand, pebbles, boulders and sea-bottom sands. Partial inundation of the relief.
	Destruction of erosion surface of plateau	Establishment of present major valleys.
	Sea Level ca. + 40 m.	Establishmend of erosion surface of central plateau, under semi-arid climate.
	Erosion	Long erosive cycle destroys external features of the Quixaba and Sao José formations.
Tertiary	S. José form.	Flows of nepheline-basanite.
	Quixaba form.	Explosive vulcanism, accompanied by ankaratrite flows and nephelinite intrusions.
	Erosion	Destruction of volcanic external features of Remedios formation.
Neo-cretaceous?	Remedios formation.	Phonolitic and trachytic vulcanism, with intrusions of ultra-basic rocks.

Mineralogy

Although there is considerable variation in petrographic types in Fernando de Noronha, the mineralogy of these rocks is relatively simple. This is characterized on the one hand by abundance and variety of the clinopyroxenes and feldspars, on the other hand by the simplicity of the amphibole and peridot series.

High tenors in sodium and titanium in the rocks are generally indicated in the composition of the inosilicates. Feldspathoids are relatively constant, even in the ultrabasics. Equally characteristic is the absence of quartz, leucite, pigeonite, orthopyroxenes and peridots more femic than chrysolite. Biotite is rare, occurs usually in ankaratrites, but also some Remedios tuffs blocks it may be conspicuous, as well as orthopyroxenes.

The trend of crystallization of the minerals was briefly as follows. Feldspars obey the general order, with labradorite the most basic present, andesine the most sialic. The latter show mantles of oligoclase, indicating increasing soda. In more advanced stages of differentiation, anorthoclase rather than albite occurs in kali-gauteites, alkali-trachytes and phonolites, often associated with soda-orthoclase. The amphiboles have crystallized along the trend of increasing quantity in alkalis and decrease in titanium oxide. The sequence in the pyroxenes is titanaugite – brown soda-augite – green soda-augite – aegerine-augite – aegerine. The olivines crystallized in the order of progressive increase in $2FeO \cdot SiO_2$.

A mineralogical classification of the archipelago eruptives by ALMEIDA is shown in Table 9.

Petrography

The principal petrographic studies made of the eruptives of Fernando de Noronha are those of GUMBEL (1880), RENARD (1882, 1886, 1889), and BUCHANAN (1885) on the 'Challenger' samples, PRIOR (1897) and CAMPBELL SMITH & BURRI (1933) on the Royal Society of London-British Museum collections, GILL (1888) and WILLIAMS (1889) on the BRANNER collections and ALMEIDA (1958) on samples collected by him during his own investigations.

WILLIAMS was the first to attempt to establish a compilation of the chief rock types occurring, classifying these as belonging to three major groups:

1. Trachytes and Andesites   Hornblende-trachyte
   Hyalo-trachyte
   Hornblende-andesite

2. Phonolites

3. Basalt in Rocks
   Nepheline-basalt
   Nepheline-basanite
   Nephelinite-Dolerite
   Augitite
   Limburgite
   Basalt bombs
   Basalt tuffs

The above classification was made on the basis of microscopic study of 30 specimens.

To date, no one has evidently had an opportunity of studying all the collections made by various expeditions, and opinions expressed by different workers on collections other than their own are generally the views of those who investigated the respective specimens. ALMEIDA, the latest worker, has entered into most detail regarding the petrography, petrochemistry and petrogenesis of the archipelago eruptives, and has naturally been able to consult the greatest number of analyses and descriptions, with the result that his findings are more detailed and cover a wider field. His own study of some 320 thin-sections represents by far the largest investigation up till the present. The petrography discussed here relates the rocks to the three principal volcanoligic units present.

Table 9   Mineralogical Classification of the Eruptives of the Fernando de Noronha Archipelago (ALMEIDA, 1958)

Colour Index	With Feldspathoids				Without Feldspathoids		
	With Feldspars			Without Feldspars		With Feldspars	
	Or > Plag	Or < Plag	Plag			Plag.	Plag. > Or
0 to 30	Alkali-Trachytes Phonolites Kali-Gauteites						
30 to 60		Porphy. Essexite					Glenmuirites
60 to 90			Nepheline-Basanites	Nephelinites Ol.-Nephelinites Ankaratrites Melilite-Ankaratrites Tannbuschites	Augitites	Alkali-Basalts	Olivine-Teschenites (Limburgites)

Lamprophyres
- With Feldspar
  - With Orthoclase: Sanaites
  - With Plagioclase: Camptonites
- Without Feldspar
  - With Olivine: Mela-Monchiquites
  - Without Olivine: Furchites

Petrography

Table 10 — Chemical Analyses (ALMEIDA, 1958)

	1.	2.	3.	4.	5.	6.	7.	8.
SiO_2	54.02	54.70	54.82	52.23	52.86	46.51	48.20	48.66
TiO_2	0.23	0.27	0.50	2.81	0.83	1.82	1.40	2.50
Al_2O_3	22.59	24.48	22.46	16.34	21.59	19.02	19.58	17.43
Fe_2O_3	2.34	1.46	1.84	1.51	3.98	3.72	4.91	5.07
FeO	0.69	0.85	0.72	1.50	0.00	2.90	2.57	1.28
MnO	0.22	0.27	0.12	0.22	Tr.	0.20	0.15	Tr.
MgO	0.56	Tr.	0.07	3.51	0.85	5.26	3.91	2.14
CaO	0.32	0.60	1.42	3.65	2.95	6.82	5.48	5.69
Na_2O	8.32	10.10	10.22	6.75	7.40	4.62	4.84	3.80
K_2O	4.14	4.58	5.93	5.64	5.30	4.25	4.65	4.70
H_2O+			0.82	5.46		5.03	1.60	4.01
H_2O-	1.30	0.96	0.02	n.d.	1.84		2.40	4.65
P_2O_5	0.15	Tr.	0.12	0.43	0.10	0.63	0.42	0.04
CO_2			0.00					
SO_3			0.98					
Cl	0.09		0.28					
ZrO_2								
Ignit. Loss	5.49	2.03			2.68			
Total	100.46	100.30	100.32 a	100.05	100.38	99.98	100.11	99.97
S. G.		2.45	2.58	2.45	2.50	2.72	2.69	2.70

	9.	10.	11.	12.	13.	14.	15.	16.
SiO_2	52.70	60.81	59.13	46.15	43.01	42.23	45.72	46.28
TiO_2	2.13	0.65	1.01	5.29	2.38	3.23	1.00	1.00
Al_2O_3	19.14	18.88	19.62	19.62	16.10	16.84	18.56	18.37
Fe_2O_3	3.17	2.57	1.57	0.00	6.98	3.64	2.78	2.99
FeO	1.28	0.00	0.72	3.49	2.41	7.40	4.31	5.02
MnO	0.03	Tr.	0.02	0.09	0.32	0.21	0.20	0.20
MgO	2.71	0.61	0.41	2.61	2.77	5.97	5.21	5.28
CaO	3.19	1.70	2.71	5.54	5.92	11.37	7.52	7.20
Na_2O	3.54	6.20	5.94	5.20	7.80	3.66	6.62	7.10
K_2O	6.66	5.80	4.65	4.46	3.80	1.93	4.11	2.78
H_2O+	3.30		3.81	6.04	5.60	1.97	1.34	2.00
H_2O-	2.27	2.22	0.27	0.94	2.31		1.26	0.90
P_2O_5	0.06	Tr.	0.02	0.10	0.90	1.34	0.40	0.35
CO_2	0.00		0.00	0.00				
SO_3	0.00		0.00	0.39				
Cl	0.01		0.07	0.20				
ZrO_2								
Ignit. Loss		1.64	0.02					
Total	100.19 a	101.08	99.97	100.12 a	100.30	99.79	100.03	100.01
S. G.	2.43	2.60	2.48	2.81	2.92	2.97	2.74	2.71

Table 10 Chemical Analyses (ALMEIDA, 1958)

	17.	18.	19.	20.	21.	22.	23.	24.
SiO_2	45.05	42.68	44.23	40.80	40.40	40.57	40.48	40.80
TiO_2	2.44	2.00	4.33	1.60	2.20	3.33	2.10	2.20
Al_2O_3	14.85	16.65	10.12	13.92	13.09	12.51	14.68	12.89
Fe_2O_3	3.05	5.08	3.50	2.69	4.55	7.22	3.84	5.00
FeO	5.60	8.11	6.58	11.71	9.24	5.05	7.72	9.05
MnO	0.19	0.20	0.18	0.15	0.15	Tr.	0.30	0.15
MgO	7.82	5.57	11.70	12.38	11.00	8.19	8.94	11.00
CaO	8.92	11.00	11.45	11.00	11.40	11.67	12.80	11.48
Na_2O	5.00	5.04	3.20	3.85	3.94	3.70	3.51	4.04
K_2O	2.19	1.69	1.12	0.78	1.49	2.10	0.67	1.63
H_2O+		0.52	2.04	0.22	1.28		3.18	0.40
H_2O-	0.88	0.78	0.50	0.58	0.72	1.61	1.62	0.60
P_2O_5	0.43	0.67	0.78	0.65	0.80	0.28	0.52	0.80
CO_2			0.31	n.d.				n.d.
SO_3								
Cl								
ZrO_2								
Ignit. Loss	3.86					3.83		
Total	100.28	99.99	100.04 a	100.33	100.26	100.06	100.36	100.04
S. G.	2.78	2.96	3.02	3.08	2.99	2.84	2.81	3.03

	25.	26.	27.	28.	29.	30.	31.	32.
SiO_2	42.18	38.00	37.01	41.58	39.96	38.42	34.60	38.52
TiO_2	1.43	2.50	2.30	3.15	3.03	4.01	2.60	2.60
Al_2O_3	11.91	13.33	11.94	11.86	9.75	13.55	11.90	17.44
Fe_2O_3	7.29	6.67	3.98	7.84	5.98	3.32	5.24	7.63
FeO	5.81	8.19	10.39	5.08	7.61	9.40	9.77	4.81
MnO	0.40	0.15	0.16	0.18	n.d.	0.21	0.18	0.18
MgO	13.23	9.12	14.53	9.70	12.95	12.54	14.41	4.70
CaO	11.12	12.84	13.12	11.14	14.04	11.75	11.20	11.60
Na_2O	2.21	3.45	3.22	3.59	2.86	3.72	4.68	6.36
K_2O	1.32	0.37	0.71	0.50	0.94	0.86	1.87	2.20
H_2O+	3.02	1.40	0.92		1.78	1.15	0.84	1.92
H_2O-	0.53	3.20	1.08	1.70	0.35	n.d.	1.46	0.88
P_2O_5	Tr.	6.82	0.70	1.08	0.79	1.01	1.04	1.34
CO_2	n.d.				n.d.		n.d.	n.d.
SO_3					0.05			
Cl					0.07		n.d.	
ZrO_2								n.d.
Ignit. Loss				2.19				
Total	100.45	100.04	100.06	100.19	100.16 b	99.94	99.79	100.18
S. G.	2.96	2.86	3.13		3.21	3.10	3.09	2.92

Table 10   Chemical Analyses (ALMEIDA, 1958)

	33.	34.	35.	36.	37.	38.	39.
SiO_2	35.72	38.41	36.38	18.72	41.88	43.40	36.44
TiO_2	4.74	3.22	5.01	7.91	0.15	2.00	5.00
Al_2O_3	12.73	14.52	11.41	11.39	4.56	17.04	8.98
Fe_2O_3	9.28	2.64	8.30	28.37	4.90	8.66	15.10
FeO	6.73	9.00	4.92	11.92	4.67	1.72	0.00
MnO	0.24	0.22	0.37	0.25	0.15	0.25	0.16
MgO	7.82	10.60	10.68	9.84	39.74	4.63	8.19
CaO	13.79	13.82	12.20	8.00	2.80	8.00	13.10
Na_2O	4.80	3.21	5.00	1.18	0.53	4.17	0.99
K_2O	1.60	1.32	1.90	0.10	0.00	3.48	0.50
H_2O+	1.10	1.75	2.51	0.93	0.14	5.22	5.31
H_2O-	0.18		0.32	1.37	0.52	0.78	
P_2O_5	1.31	1.43	0.22	0.05	0.06	0.60	2.50
CO_2							
SO_3							
Cl							
ZrO_2							
Ignit. Loss							
Total	100.04	100.14	99.72	100.03	100.10	99.95	100.36*
S. G.	3.09	3.06					

a) CAMPBELL SMITH-BURRI specimens.
b) LACROIX specimen
* Includes: H_2O- and CO_2, 3.99; S, 0.10

1. Trachytoid Phonolite. Fernando de Noronha.
2. Aphyric Phonolite.
   Morro do Forte dos Remedios.
3. Sodalite-Phonolite, trachytic. Ilha Sela Gineta.
4. Porphyritic Phonolite. Morro do Pico.
5. Porphyritic Phonolite.
   Dyke, facing Caieira Bay.
6. Kali-Gauteite. Dyke, beach at Caieira Bay.
7. Kali-Gauteite. Caieira Bay.
8. Kali-Gauteite. Littoral, to N of Caieira Bay.
9. Gauteite. Stream-bed, Central Plateau.
10. Alkali-Trachyte. Littoral, Caieira Bay.
11. Alkali-Trachyte. Caieira Bay.
12. 'Monchiquite' (Furchite). Near Atalaia Pt.
13. Camptonite.
    Road-cutting, S side of Morro do Meio.
14. Mela-Monchiquite. Dyke, littoral, Caieira Bay.
15. Porphyritic Essexite.
    Atalaia Pt. (Fine-grained).
16. Porphyritic Essexite.
    Atalaia Pt. (Coarse-grained).
17. Glenmuirite. Remedios.
18. Alkali-Basalt. Littoral, Abreu Bay.
19. Nepheline-Basanite. Ilha Sao José.
20. Nepheline-Basanite. Ihla Sao José.
21. Olivine-Teschenite.
    Dyke in phonolite, Abreu Bay.
22. Olivine-Teschenite. Littoral, Caieira Bay.
23. Olivine-Teschenite.
    Thin dyke in tuffs, Cachorro beach.
24. Limburgite. Littoral, to E of Southeast Bay.
25. Limburgite. Littoral, Caieira Bay.
26. Augitite. Cutting by the runway,
    S side of Morro do Meio.
27. Tannbuschite. Thin dyke in cliffs, Caieira Bay.
28. Ankaratrite. E of Tamandaré, edge of crater.
29. Ankaratrite. Fernando de Noronha.
30. Ankaratrite. Pontinha.
31. Melilite-Ankaratrite. Coast E of Capim Açu Pt.
32. Nephelinite. Dyke in ankaratrite. Pontinha.
33. Olivine-Nephelinite. S side of Conceicao beach.
34. Olivine-Nephelinite. S side of Conceicao beach.
35. Perkinitic Xenolith in Monchiquite.
    Biboca beach.
36. Perkinitic Xenolith in Porphyritic Essexite.
    Atalaia Pt.
37. Olivinite. Ilha Sao José.
38. Tuff (Remedios Formation).
    S side of Conceicao beach.
39. Ankaratritic Tuff (Quixaba Formation).

Remedios Formation Rocks

Phonolites are the commonest eruptives, and have been recognized since the days of DARWIN. The rocks are aphyric and porphyritic, with numerous phenocrysts always within an aphanitic, holocrystalline matrix, the presence of sodic pyroxene giving a greenish colour. Larger phenocrysts of nepheline, potassic feldspar, perhaps a mineral of the sodalite group, are quite common. There is nearly always a gradation in size between the largest crystals and the matrix constituents in which titanite, magnetite and zircon are found. Biotite and plagioclase are exceptional. No leucite is present and very few sodalite group minerals are seen. The commonest mineral is soda-orthoclase. Feldspar phenocrysts are idiomorphic, usually showing contortions as a result of magmatic corrosion. Nepheline is usually idiomorphic and always abundant, forming the largest crystals. It may show mineral inclusions of the rock other than feldspar and is only rarely altered into analcite or a fibrous zeolite. In quantity, nepheline is always subordinate to feldspar. Pyroxenes are always present, are idiomorphic and almost free of inclusions. Some rocks have crystals of brownish-yellow amphibole, likely titano-hornblende, which invariably is transformed into soda-augite and magnetite. Idiomorphic phenocrysts of titanite, small granules of magnetite and fine needles of apatite are always present. Zircon rarely occurs outside the groundmass. The modal composition of the rocks cannot be precisely determined, due to the aphanitic matrix and the extremely fine granulation, making mineral identification difficult. However microscopic examination indicates that no more than 10 % femic minerals are present and that about 75 % comprises soda-orthoclase. Table 10, Nos. 1–5 shows chemical analyses of some phonolites.

Kali-Gauteites. CAMPBELL SMITH & BURRI recognized gauteite in the collections they studied, occurring on the main island chiefly, and forming dykes which are easy to recognize because of their grey-green colour and abundant phenocrysts of amphibole and pyroxene. ALMEIDA studied 17 thin-sections from dykes and noted their considerable uniformity. They are always porphyritic, with variable number of phenocrysts of soda-augite, syntagmatite, and, more rarely, titanite, anorthoclase and biotite. The matrix comprises soda-orthoclase laths associated with granules of the same minerals as occur as phenocrysts, also aegerine-augite, magnetite and apatite. The feldspar of the groundmass is soda-orthoclase, but it may be that andesine also is present. An isotropic, generally clear and colourless mineral or then with small black inclusions, appears to be sodalite. Pyroxene phenocrysts are idiomorphic, with long prisms sometimes showing hour-glass structure, which optically correspond to titan-augites. Amphiboles are varieties of titani ferous, sodic-magnesian oxy-hornblendes which are identified are syntagmatite. They form long prisms which always show corrosion rims. No true olivine is present, but in two specimens pseudomorphs of the mineral were identified. Biotite also was recognized in two specimens. The accessories include abundant titanite, magnetite and apatite, and as secondary products there are analcite, zeolites, chalcedony and calcite. Nos. 6, 7 and 8 show chemical analyses of these rocks. These Fernando de Noronha rocks have lower tenors in silica and higher values in lime, magnesia and total iron than in the type rocks of Mülhörzen in the Bohemian Mittelgebirge described by J. E. HIBSCH (1898). On the other hand, the rocks of the archipelago show a great chemical affinity with the tahitites of Tahiti (A. LACROIX, 1928). The Fernando de Noronha rocks differ from normal gauteites in the absence, or almost so, of plagioclases, anorthoclase phenocrysts taking their place. This inverse proportion calls for a new name, and ALMEIDA used the term kali-gauteite, adopting a suggestion of A. JOHANSSEN (1939) that the prefix 'kali' indicate the absence or then not more than 5% sodic plagioclase in a rock. The term 'gauteite' is preserved, as the rocks show close similarity, both chemically and mineralogically, to normal gauteites. A chemical analysis of a CAMPBELL SMITH & BURRI sample, which they named gauteite, is shown in No. 9. Two modal compositions are given in Table 11, Nos. A and B, of two kali-gauteites, from which it can be assumed that the injecting magma had already been crystallized in all the hornblende, titanaugite, sodalite and titanite, with no traces of feldspar except rare phenocrysts seen in other specimens.

Alkali-Trachytes. The archipelago rocks have mineralogical features and structures similar to those of other oceanic islands. They are usually slightly porphyritic rocks, with small phenocrysts visible to the naked eye, occurring in an aphanitic, holocrystalline matrix of greyish-green colour, or whitish when alteration has set in. The sub-idimorphic, non-corroded phenocrysts are of soda-orthoclase and andesine, always showing mantlings of anorthoclase. At Morro Branco, the trachytes show anorthoclase phenocrysts co-existing along with andesine. Microphenocrysts of soda-augite, titanite and a colourless mineral of the sodalite group are also present. The matrix comprises a fine aggregate of orthoclase (soda-orthoclase?), which may have a plagioclase nucleus, prisms of soda-augite or aegerine-augite enclosing apatite needles, and grains of magnetite, and, on one occasion zircon was observed, all giving the rock a trachytic texture. Nepheline, quartz, olivine, biotite were not seen. The commonest alteration products are kaolinitic minerals, calcite and zeolites, the last-named being especially notable in the vesicles. The texture is non-hyatal porphyritic, with small phenocrysts in variable percentages contained in a trachytic groundmass, with marked lamination, less distinct lineation. Vesicular or amygdaloidal structures are rare except in tuff blocks. The rocks were named alkali-trachytes by CAMPBELL SMITH & BURRI, No. 11 being an analysis given by them, No. 10 an analysis by ALMEIDA. The former authors called attention to the similarity between their specimen and some trachytes from Mont Doré, Auvergne, France. A modal composition for this specimen is shown in Table 11, No. 11.

Furchites. These rocks are identical to mela-monchiquites, except that they lack olivine. Two varieties can be distinguished in which etther syntagmatite or titanaugite is the dominant mafic mineral. The dykes at Atalaia Pt. and Biboca are of the former type, whilst the second variety occurs as dykes at Forte dos Remedios and Southeast Bay. The phenocrysts are the same minerals as form the groundmass, to which must be added the sodalite group. The pyroxene is titanaugite, the amphibole the same as in the kali-gauteites, and only very rarely are hornblende phenocrysts present. Both varieties of furchites may have small idiomorphic crystals, colourless and very clear, of a mineral of the sodalite group. Accessories are magnetite, apatite, rarely titanite. The structure of the matrix is the same as in the mela-monchiquites, but hornblende appears to be more frequent in the furchites. The texture is holocrystalline, non-hyatal porphyritic. Vesicular or amygdaloidal structure is very common, with pyroxene filling the cavities. Table 10, No. 12 is a CAMPBELL SMITH & BURRI specimen, these authors commenting on the high H_2O+ content which they explained as due to the abundance of analcite and other zeolite inclusions. Worth noting is the lower amount of MgO and FeO, whilst Fe_2O_3 is

Table 11 Modal Compositions (Almeida, 1958)

	11	A	B	13	14	C	16
Aegerine-Augite	7.3						
Amphibole		4.8	4.2	21.3	4.7	39.0	9.8
Analcite	x			54.0c	22.1	46.2	16.3e
Andesine	3.5			12.7			
Apatite	x	0.6		x	x	x	x
Augite							17.7
Calcite					x		
Chlorite							
Chrysolite					5.0		
Hornblende							
Labradorite							31.1
Magnetite	3.7	3.7	1.0	4.0	11.7	5.6	3.9
Nepheline							4.9
Olivine							3.3
Orthoclase	81.8						13.0
Sanidine		74.8a					
Sodalite	2.7	3.5	4.5				
Titanite	1.0	1.7	1.0				
Titanaugite		10.9	6.3	8.0	52.8	6.1	
Titanohornblende							
Zeolites						3.1	
Other			83.0b		d		
Total	100.0	100.0	100.0	100.0	100.0	100.0	100.0

	17	18	20	21	23	24	D
Aegerine-Augite							
Amphibole							
Analcite	5.2	5.0		5.1i	15.9k		
Andesine							
Apatite	1.3			1.3	x		
Augite		41.1		54.1	45.4	46.1	
Calcite	2.4	0.6g		x	4.4		
Chlorite	0.7						
Chrysolite			7.3				
Hornblende				2.6	2.8		
Labradorite	27.4	30.6	19.0h	14.2j	6.5m		
Magnetite	6.6	22.7	7.3	14.3	10.3	13.6	18.0
Nepheline							
Olivine	7.8			8.4	11.1	8.4	3.0
Orthoclase	16.1						
Sanidine							
Sodalite							
Titanite							
Titanaugite	31.1		66.4				78.0
Titanohornblende							
Zeolites					3.6		
Other		f				31.9n	6.0p
Total	100.0	100.0	100.0	100.0	100.0	100.0	

Table 11   Modal Compositions (ALMEIDA, 1958)

	E	30	31	32	33	35
Aegerine-Augite				31.4t		
Amphibole						
Analcite		1.0		15.4	8.4	
Andesine						
Apatite	x	1.3	x	4.2	4.2	0.8
Augite	56.9		61.0r			26.0v
Calcite	x	x		x		18.0
Chlorite				4.8		
Chrysolite		10.8	12.0		8.7	
Hornblende						
Labradorite						
Magnetite	5.4	17.1	21.0	4.2	10.0	4.8
Nepheline	17.2	19.2		38.9	37.4	
Olivine	20.5					
Orthoclase						
Sanidine						
Sodalite						
Titanite				0.4		
Titanaugite		48.2			12.6	
Titanohornblende						43.7
Zeolites	x	1.4		0.7		
Other		q	s		18.7u	6.7w
Total	100.0	100.0	100.0	100.0	100.0	100.0

A   Kali-gauteite, Southeast Bay.
B   Kali-gauteite, dyke, Atalaia Pt.
C   Furchite, dyke, slopes of Atalaia Hill.
D   Augitite, S side of Meio Hill.
E   Tannbuschite, Pontinha.

a   Includes Zeolites
b   Glassy mesostasis with microlites.
c   Includes other fibrous Zeolites.
d   Includes 3.7 Hauyne. Natrolite present.
e   Includes other Zeolites.
f   Includes 1.4 Biotite. Hematite, pyrite, natrolite, aegerine-augite present.
g   Includes Apatite.
h   Includes Nepheline
i   Includes other Zeolites

j   Includes Orthoclase.
k   Includes other Zeolites.
m   Includes Orthoclase.
n   Crystallites and microlites of amphibole, augite, magnetite as a glassy matrix.
p   Includes 6.0 mesostasis.
q   Includes 1.0 Perovskite.
r   Includes Nepheline and perovskite.
s   Includes 6.0 melilite. Biotite present.
t   Includes Titanaugite.
u   Mesostasis with Sanidine, aegerineaugite and zeolites. Biotite, ilmenite, chlorite present.
v   Includes Titanaugite.
w   Isotropic matrix, chlorite and zeolites.
x   Indicates presence.

absent but does occur in all such rocks having magnetite. Table 11, No. C is a modal composition of a furchite from the N slopes of Morro Atalaia.

Camptonites. These are porphyritic rocks, with phenocrysts of amphibole in an aphyric, dark-grey groundmass. Under the microscope phenocrysts of amphibole, pyroxene and apatite, magnetite octahedra and plagioclase laths occur in an isotropic substance, more or less zeolitized. The amphibole is syntagmatite. The pyroxene is titan-augite, possibly sodic, showing zoning and frequently hour-glass structure. The plagioclase phenocrysts are of andesine composition, perfectly idiomorphic and non-corroded. No nepheline or sodalite groups minerals are present. The only accessories recognized are magnetite and apatite. Interstitial material is partially altered into natrolite, but some analcite is always present in varying amounts. The texture is non-hyatal, porphyritic, with clearcut size transitions from phenocrysts to the smallest crystals of the matrix, which latter is panidiomorphic, microlitic and pilotaxitic in some cases. No. 13 is a chemical analysis.

Mela-Monchiquites. These rocks form many dykes in the archipelago, being dark, strongly porphyritic rocks with phenocrysts especially of pyroxene and peridot, the latter nearly always altered. Phenocrysts also include titanaugite, chrysolite and syntagmatite, which can also be present in the groundmass where they are associated with magnetite apatite and sometimes a mineral of the sodalite group. Pyroxenes are varieties of sodic-titaniferous augite, with optical properties identical to those occurring in the furchites. The amphiboles are oxy-hornblendes, sodic-titaniferous and similar to those in the furchites. Chrysolite almost always occurs as phenocrysts and are deeply altered into serpentinous and carbonate material. The matrix has an intersertal texture in which small quantities of a substance taken to be natolite occur between the pyroxene and amphibole. Zeolitization may often be seen. Vesicles and amygdales are common, the latter filled with opal, quartz, chalcedony and natrolite. The chief difference in chemical composition between these rocks (vd. No. 14) and those from the classical area of Caldas de Monchique, Portugal, refers to the greater percentage of mafic minerals in the archipelago rocks.

Porphyritic Essexites. These rocks occur as outcrops on the N and E sides of Morro Atalaia, also as several dykes at Atalaia Pt., and at the edges of the airfield. Microscopically, these greyish rocks have a porphyritic character, non-hyatal, with phenocrysts of pyroxene, olivine and labra-dorite in a groundmass comprising orthoclase, nepheline and some accessories. No ophitic texture occurs as the crystallization of the pyroxenes preceded that of the feldspars. Plagioclases are rare phenocrysts of labradorite and andesine, mantled with orthoclase. Orthoclase and nepheline constitute the major part of the groundmass, both being frequently altered into analcite and a fibrous zeolite. Titan-augite shows many magnetite inclusions. Rare crystals of chrysotile are almost completely altered into bowlingite and a carbonate. There occur abundant microlites and small crystals of brown amphibole – syntagmatite – also needles of apatite and octahedra of magnetite. Some essexite specimens have a much finer texture, are lacking in plagioclase phenocrysts but do have nepheline and analcite phenocrysts and a considerable quantity of carbonate, zeolite and small miarolitic cavities are usually present, giving the rock a certain similarity to sanaites. The chemical composition of the two varieties of porphyritic essexites are given in Nos. 15 and 16. The high soda tenor makes it difficult to compare the analyses with congenital rocks mentioned in the literature but one from Essex County, Massachusetts, U. S. A. shows an almost identical composition, as reported by H. S. WASHINGTON (1899). CAMPBELL SMITH & BURRI classed these Fernando de Noronha rocks as essexites, but ALMEIDA preferred to use the terminology of A. JOHANSSEN (1938), and modified the term to porphyritic essexites, as so used by W. C. BROGGER (1906), who treated the rock as hypabyssal and not plutonic.

Glenmuirites. These rocks have a teschenite character but with abundant orthoclase, and to the naked eye have a diabasic aspect, with dark pyroxene crystals and small laths of plagioclase. The mineral constituents are essentially titanaugite, labradorite, orthoclase and olivine, with analcite and other accessories. The plagioclase has a mantle of potassic feldspar, the nucleus being labradorite and zoned with andesine. Soda-orthoclase forms a large part of the matrix between the plagioclase laths, the groundmass also including a great abundance of crystallites and microlites, apatite needles, aegerine-augite, magnetite, etc., with analcite as the common substitute. The titanaugite occurs as prisms, is automorphic, with strong pleochroism and is very rich in inclusions of magnetite, biotite, apatite and olivine. The olivine generally occurs as aggregates of two or more short, prismatic crystals, with corroded edges. Apatite is abundant as needles and is associated with magnetite. Biotite shows very strong pleochroism, occurs occasionally as plates and is associated with magnetite, olivine and augite as products of deuteritic alteration. Some specimens of rock show microscopic miarolitic cavities lined with chloritic and calcitic fibro-radial material, as well as carbonate pseudomorphs after olivine. The rocks are holocrystalline, but small quantities of green mesostasis may represent original amorphous material. The texture is non-ophitic, as the titanaugite crystals were formed before the plagioclase. Porphyritic types are found with phenocrysts of augite, olivine and magnetite, present in a matrix of plagioclase laths. The order of crystallization is the same as in basalts, but plagioclase developed after the mafic minerals. The Fernando de Noronha rocks are distinguished from normal teschenites by the appreciable presence of potassic feldspar and olivine and by fewer femic minerals. They differ from the porhpyritic essexites in the tenor of analcite and in having olivine, not nepheline, as an essential mineral. A chemical analysis is given in No. 17, Table 10 and the modal composition in No. 17, Table 11.

Alkali-Basalts. Occurrences are known at Abreu Bay, Southeast Bay and on the eastern edge of the airfield, and in all instances the rocks are much altered, with the Abreu outcrops probably being the freshest. They are darkish-grey rocks, entirely aphanitic except at Abreu where rare phenocrysts of pyroxene occur. In thin-section the rocks look like normal basalts, are holocrystalline with a very fine intergranular texture. No olivine is present, but pseudomorphs suggest the possible presence of such. The Abreu occurrence shows a porphyritic rock with phenocrysts of titanaugite, microphenocrysts of plagioclase of labradorite composition and zoned with andesine. A small quantity of oxy-hornblende as microlites is associated with augite at the Southeast Bay exposure. No nepheline or other feldspatoids were noted in any of these basalts. The Abreu outcrop shows some irregular mantling of analcite, a little chlorite and calcite. Small micro-octahedra of magnetite, or more likely, small grains of the mineral and apatite needles are common. No. 18 is a chemical analysis which shows a relatively high quantity of lime, soda and a low tenor in silica, as compared to normal basalts. These archipelago rocks approach quite closely to the feldspathic basalts of the Bohemian Mittelgebirge.

Olivine-Teschenites. These dark rocks have phenocrysts visible to the naked eye of pyroxene and olivine in an entirely aphanitic matrix. In thin-section they are seen to be holocrystalline, with non-hyatal porphyritic texture, showing large titanaugite phenocrysts, also chrysotile, both of which however are absent in the mesostasis. The latter comprises a fine aggregate of prisms of pyroxene and amphibole, magnetite octahedra and apatite needles poikilitically enclosed by plagioclase and a little orthoclase. Analcite mantles are present here and there. No nepheline is seen. Of the feldspars, plagioclase is the most abundant, being labradorite zoned by andesine and clothed by orthoclase, the latter mineral also being present interstitially. The feldspars are partially or totally substituted by analcite and other zeolites. The olivine is generally sub-idiomorphic, with corrosion rims. Chemical analyses are given in Nos. 21, 22 and 23, from which it is seen that these rocks approximate to fasinites described by A. Lacroix (1916) from Madagascar. In a mineral aspect however the Fernando de Noronha rocks differ, for plagioclase is lacking in fasinites but nepheline does occur. These olivine-teschenites are obviously closely related to ankaratrites. In Table 11 the normative composition of two of these rocks are given in Nos. 21 and 23.

Limburgites. Rocks of this type occur as dykes at Caieira Bay, Southeast Bay and at Atalaia Pt. Campbell Smith & Burri mentioned rounded boulders of limburgite as having been collected by Ridley from near Northeast Pt., but it is likely that these were derived from the Caieira occurrences and transported thither by marine currents. These blackish rocks are aphanitic, with small phenocrysts of olivine and titanaugite. In thin-section they show only mafic minerals – augite, olivine and magnetite – distributed throughout a non-hyatal, panidiomorphic, porphyritic texture. Specimens from the first two localities have phenocrysts within a hyalopilitic matrix in which microlites of titanaugite – and a little hornblende in the Southeast Bay occurrences – magnetite granules and apatite needles occur in a translucent, isotropic groundmass, along with globulites having a refringence of 1.54. Some rock types show fluidal linear structure near the margins of the dykes and occasionally amygdales are present containing phillipsite, quartzose material, chloritic and clatitic matter. Chemical analyses of specimens from Southeast Bay and Caieira Bay are given in Nos. 24 and 25 respectively. The latter specimen is practically the same as the olivine-teschenite from Abreu Bay (No. 21). Although the two localities are near, Almeida could not determine whether the limburgite might be a hypocrystalline facies of the Abreu teschenite. The Southeast Bay specimen has a modal composition shown in Table 11, No. 24.

Augitites. At the extreme S end of Morro Meio are some black dyke rocks which are aphanitic, slightly porphyritic under the microscope and are seen to be 80% and more composed of mafic minerals, with pyroxene dominant. Rare phenocrysts of olivine, always much altered, never amount to more than 8% of the rock. The major part of the rock shows granulation less than 0.1 mm, being almost entirely composed of titanaugite prisms, a little magnetite and an isotropic base surrounding these. In all thin-sections examined by Almeida there occurs a colourless isotropic base comprising never more than 10%, with a refractive index lower than that of Canada balsam, having many longulites and globulites. Secondary colouration gives this a dark yellowish tint. This is presumed to be analcite. No nepheline has been identified, but it may occur in the altered state in the isotropic material of the groundmass. Microscopic lineation is present, due to the arrangement of titanaugite prisms and the pattern of the vesicles and amygdales. The last-mentioned show natrolite, analcite, calcite, phillipsite and quartzose material filling the cavities. (Chemical analyses show that ca. 6% of the rock is comprised of these anygdales.) The analysis (No. 26) indicates a close similarity to ankaratrites. A modal quantitative composition of another specimen from the same locality is shown in Table 11, No. D.

Tannbuschites. In the tuff cliffs at Caieira Bay are two dykes of these rocks. They are dark in colour, have phenocrysts of olivine and rare pyroxene in an aphanitic groundmass. In thin-section they show abundant phenocrysts of chrysotile and a few titanaugites in a matrix predominantly composed of small crystals and microlites of titanaugite magnetite grains and some apatite needles, between which is a variable quantity of darkish or colourless mesostasis full of crystallites. The titanaugite phenocrysts are the same type as occur in the limburgites and olivine-teschenites. No nepheline has been identified in these rocks at Caieira Bay but at Pontinha this mineral does occur, poikilitically enclosing pyroxene and magnetite. A mineralogical quantitative determination of a specimen from Pontinha is shown in Table 11, No. E. The chemical analysis (Table 10, No. 27) shows similarities with the nepheline-basalt from Schanzberg, Bohemian Mittelgebirge, described by J. E. HIBSCH (1904). The Fernando de Noronha specimen is a melanocratic nephelinitic basalt belonging to the group 3125 of A. JOHANSSEN (1938), who proposed the name tannbuschite from a locality in the Bohemian Mittelgebirge. These archipelago rocks of the Remedios formation are very similar to ankaratrites, being distinguished from the latter by the absence of biotite.

Olivine-Nephelinites. At Conceicao beach occurs a large dyke which is classified as olivine-nephelinite. The rock is greyish in colour, porphyritic, and has a diabasic appearance, showing coarse granulation, in which prisms of pyroxene, up to a centimetre in length, are conspicuous. In thin-section there occurs nepheline, titanaugite, magnetite and olivine, with a porphyritic, non-hyatal texture, in which nepheline poikilitically includes other minerals. There are abundant xenomorphs of analcite in a mesostasis where also small crystals of sanidine, aegerine-augite, biotite and ilmenite are found. Apatite needles are present throughout. Nepheline occurs as short prisms which enclose femic minerals, the nepheline showing alteration to analcite and natrolite. The mesostasis between the layers of analcite shows microlites of sanidine, biotite, aegerine-augite, magnetite and chrysotile. Chrysotile, apatite and magnetite were the first to crystallize, then nepheline, then pyroxene and finally ilmenite, biotite, aegerine-augite, sanidine and analcite. Nos 33 and 34 are chemical analyses of samples from the above locality, from which it can be considered that the rocks approximate closely to nepheline-basalts of the Bohemian Mittelgebirge on the one hand, on the other hand, to ankaratrites of Madagascar. The archipelago rocks are very similar to the 'Nephelindolerit' of Löbauer Berg, Lobau, Saxony, as described by J. STOCK (1888), a fact already noted by CAMPBELL SMITH & BURRI. The major difference here is the greater olivine continent in the Fernando de Noronha rocks, thus approaching more to the nepheline-basalts of H. ROSENBUSCH (1877). ALMEIDA however preferred to adopt the nomenclature as recommended by the 1921 British Commission on Petrography. The modal composition for one of the above specimens is shown in No. 33, Table 11.

Alochetites. At Atalaia Pt. is a small dyke showing rock of the same general character as glenmuirite, but differing in the presence of nepheline and orthoclase in greater abundance than plagioclase. The rock has a porphyritic texture, with phenocrysts of titanaugite occurring in a groundmass criss-crossed with feldspars, the interstices being of titanaugite and nepheline. These two minerals are sparsely developed but show inclusions of biotite and apatite. The plagioclase is labradorite, which shows slight zoning within orthoclase. A small amount of microperthite can be identified. Titanaugite occurs as sub-idiomorphic phenocrysts and also occurs xenomorphically in the matrix. Olivine is present as small pseudomorphs of serpentine. The modal composition of the rock is a follows: analcite, 4.2%, apatite 1.4, calcite 10.3, labradorite 14.9, magnetite (with biotite) 4.0, olivine (speudomorphs) 2.2, orthoclase (with nepheline) 40.9 and titanaugite 22.1. The high quantity of carbonate, present as an alteration product, does not justify a chemical analysis. The colour index is lower than that for glenmuirite and shows more silica, less magnesia, iron oxide and lime than glenmuirite. On a mineralogical basis, ALMEIDA thought the rock approached closely to the alochetites of A. JOHANSSEN (1938) described from the Tyrol, but differed from these latter in the matrix structure and the microlites present.

Nosean-Sanaites. Intruding the phonolites on the road leading from Remedios to Atalaia beach is a dark, porphyritic dyke rock with small phenocrysts of pyroxene, altered olivine and nosean. In thin-section it has a lamprophyric appearance, showing a predominance of pyroxene rather similar in composition mineralogically to that seen in the alochetites, but distinguished from these latter by having plagioclase and nosean. The texture is porphyritic, non-hyatal, with titanaugite, olivine and hornblende phenocrysts, which also are present in the matrix which also includes orthoclase, nepheline, nosean, magnetite, biotite and apatite, but no nepheline. Orthoclase occurs only in the groundmass, poikilitically including all other minerals except nepheline. Olivine is almost entirely epigenized into serpentine and carbonate products. There are abundant octahedra of magnetite, plates of biotite and needles of apatite. On the beach at Atalaia Pt. is a somewhat similar dyke cutting olivine-teschenites, but here there are more leucocratic minerals present. Due to the degree of alteration, no chemical analysis is possible, but the following gives the quantitative mineralogical composition:

Phenocrysts: Titanaugite 9.8%, nosean 4.0, hornblende (pseudomorphs) 3.9, olivine (pseudomorphs) 2.9, magnetite 0.8, apatite, trace.

Matrix: Orthoclase (with nepheline) 22.2%, augite 30.8, magnetite 8.1, hornblende 0.8, apatite 0.5, analcite, zeolites, calcite 16.1.

The textural and mineralogical features of the rock class it with the alkaline-lamprophyres with feldspar, and on account af the dominant mafic character it would belong to the family 3117 H of A. JOHANSSEN (1938). As per the W. E. TRÖGER classification, the rock belongs to the family of melteigites and shonkinites, where we find sanaites (1935). Because of the presence of nosean, ALMEIDA termed the rocks nosean-sanaites. The degree of alteration prohibited the making of a chemical analysis.

Quixaba Formation Rocks

Ankaratrites. The large flows of the western and eastern regions of the central plain, on Rata islet and some lesser occurrences, are formed solely of hypermelanic, nephelinitic basalts with biotite, which are identified as ankaratrites, as described by A. LACROIX (1916). These rocks of dark hue have small phenocrysts of olivine, but pyroxene is rare, and are set in a completely aphanitic groundmass. In thin-section, they are holocrystalline, non-hyatal, porphyritic in texture, in which phenocrysts are present in a groundmass composed of microlites and small crystals of titanaugite, magnetite, perovskite and apatite, poikilitically immersed in nepheline and biotite. Magnetite is always very abundant, biotite is plentiful also but not as common as the former. Nepheline may be unaltered and clear or then changed into natrolite and analcite. In only one of the 35 thin-sections of these rocks studied by ALMEIDA was nepheline present in grains more or less free of inclusions. Assessories include apatite needles, especially as inclusions in nepheline, and small, xenomorphic grains of perovskite. The rocks are holocrystalline, but what appears to be a vitreous substance can occur as globules. The order of crystallization seems to have been: olivine followed by augite and magnetite approximately contemporaneously, then apatite, perovskite, biotite and nepheline. Especially in the upper part of flows, blocks and smaller fragments of tuffaceous material are full of vesicles and amygdales with natrolite, other fibrous zeolites and carbonates are found. The chemical composition of three samples (Nos. 28, 29, 30) show tenors in silica between 38.42 and 41.58 %, alumina between 9.75 and 13.55 %. Compared to the Madagascar ankaratrites, those of Fernando de Noronha are more alkaline and more aluminous, but contain less magnesia and lime. A. LACROIX (1923) distinguished a type of ankaratrites which he named 'ambohivorona' in which, as in the Fernando de Noronha rocks, the nepheline poikilitically included prisms of titanaugite. The modal composition of the Pontinha specimen is shown in No. 30, Table 11.

Melilite-Ankaratrites. SMITH & BURRI recognized melilite in lavas from Capim Açu and Boldro. In the field the rock does not differ from ordinary ankaratrites, but ALMEIDA believed that melilite was rare in the Fernando de Noronha rocks, as he noted it in only one of 35 thin-sections examined from various parts of the island. Only four thin flows at Capim Açu showed vestiges of melilite. It was doubtless from here that RIDLEY collected samples which were described by CAMPBELL SMITH & BURRI. The rock is holocrystalline, has a porphyritic character, with small phenocrysts of olivine in a dense network of augite, magnetite, nepheline and accessories. The pyroxene is titanaugite, similar in quantity as occurring in the ordinary ankaratrites. Melilite is present in enough quantity as to be considered an essential mineral. It is completely epigenized into zeolite, whilst other crystals show alteration into an isotropic mineral which CAMPBELL SMITH & BURRI thought was deeckeite. Unfortunately no fresh grains of melilite were seen and hence no pronouncement can be made regarding this altered mineral or material. Nepheline is xenomorphic, of very fine grain size. Small grains of magnetite are abundant, also apatite needles and xenomorphic, minute crystals of brownish perovskite with a low birefringence. One or two biotite microlites were detected, especially near amygdales, which latter are filled with natrolite and calcite. Chemically the rock is very like the melilite-ankaratrites of Madagascar, described by LACROIX (1916), but those of the archipelago show an extremely low silica content, compensated by an increase in alkalis and alumina. The archipelago rocks differ from normal ankaratrites in being more basic, contain more melilite, have less nepheline, almost no biotite and display various textures. A chemical analysis is shown in No. 31, and the quantitative mineralogical compisition in No. 31, Table 11.

Nephelinites. A sample from a coarse-grained dyke between Pontinha and Pedra Alta Pt. which intrudes a large ankaratrites flow, showed under the microscope that it was nephelinite (nepheline-dolerite), which CAMPBELL SMITH & BURRI referred to as also occurring at the NE headland of the main island. The same rock is to be found at S. Antonio Pt., where it intrudes ankaratrites which have a structure similar to but not actually pillow lava. The Pontinha specimen has a very coarse texture, with crystals of pyroxene large than a centimetre which are full of miarolitic cavities, projecting pyroxene prisms being layered over with fibrous zeolites. In thin-section the essential minerals appear to be titanaugite and nepheline, between which are sprinkled magnetite, apatite and analcite in a chloritized and zeolitized groundmass. Apatite was evidently the first mineral to crystallize, followed by magnetite, titanaugite, nepheline and analcite in that order, analcite being a late crystallization mineral. A small quantity of titanite can be recognized in the matrix. Miarolitic cavities were successively deposited with natrolite, opal and calcite. The chemical analysis (No. 32) shows a low tenor in magnesia, compensated by a high quantity of alumina and alkalis, features which chemically distinguish the nephelinites from the ankaratrites of the Quixaba Formation. A study of the chemical analyses of these two types of rocks allows us to surmise that the nephelinites correspond to ankaratrites in which olivine is absent, i. e. that nephelinites correspond to the interstitial matrix of the ankaratrites. On a chemical basis these nephelinites show a close relation to the etindites of the Cameroun, Africa, except that the former show less K_2O, which is reflected in the presence of leucite in the rocks from Etinde volcano.

Sao José Formation Rocks

Nepheline-Basanites. The first description of such rocks from the islet of Sao José was given by RENARD (1882) who identified them as 'nephelinitic basalts', describing tham as being black in colour, massive, homogeneous rocks showing distinct conchoidal fracture. A chemical analysis of RENARD is herewith given: SiO_2 42.24, Al_2O_3 20.15, Fe_2O_3 12.17, FeO 4.07, CaO 6.15, MgO 5.22, Na_2O 6.10, K_2O 0.55, H_2O 2.75, Total 99.15. Microscopically he determined that the principal crystals were nepheline with augite interspersed between. Large crystal fragments of peridot were also noted. Of secondary minerals there occurred biotite, much magnetic iron and apatite, but he detected no feldspar. WILLIAMS (1889) recognized a small quantity of nepheline in these rocks and classified them as nepheline-basanites. Later, CAMPBELL SMITH & BURRI confirmed this nomenclature of WILLIAMS. The fresh rock has a porphyritic aspect, with small phenocrysts of pyroxene and olivine in a dark-greyish groundmass which is entirely aphanitic. Many green xenoliths of olivinite are also present. In thin-section the rocks are composed almost 50% of pyroxene, partly as phenocrysts, and a variable quantity of olivine phenocrysts. In the matrix a small quantity of plagioclase, nepheline and accessory magnetite and apatite occurs. The pyroxene is titanaugite, showing strong corrosion where it is changed into bowlingite along the fractures and cleavages. The plagioclase is a variety of labradorite, occurring only as very small laths included in femic minerals. No sanidine has been identified. Nepheline is only present in small quantity and includes the above-mentioned minerals poikilitically, along with cubes of magnetite, also apatite. The texture is holocrystalline, non-hyatal, porphyritic. Some laminae have small vesicles and amygdales. The matrix is very fine, granular and microlitic, with poikilitic development of nepheline. The sequence of crystallization is: apatite and magnetite concurrently, followed by chrysolite, augite, labradorite and nepheline, in that order. Nos. 19 and 20 show chemical analyses, from which it can be seen that, compared to descriptions of these rocks from other regions, the Fernando de Noronha rocks are poorer in alumina, richer in lime and magnesia. The lower tenor in nepheline of archipelago rocks makes them approximate some basanitoids. A modal analysis is given in No. 20, Table 11.

Xenoliths and Ejectiles

The occurrence of these in the Remedios Formation calls for some comment. They include fragments of typical plutonics and material whose origins we know nothing but which are older than the Remedios. We may discuss these xenoliths-ejectiles under two headings:

1. Known eruptives in situ

Trachytes and phonolites are the commonest rocks in the tuffs, occurring usually as blocks rather than matrix material. They are identical to the rocks described here earlier, which are in situ. There does occur, however, an alkali-trachyte with many phenocrysts (xenocrystals?) of chrysolite showing strong corrosion and transformed into magnetite and augite, which was not recognized among the dyke rocks of the island. Porphyritic essexites are present as xenoliths on the borders of the airfield, on the slopes of Morro Atalaia and on the road from Remedios to the airport, which are much larger and more abundant than those occurring in the tuffs. This leads one to surmise that an occurrence of porphyritic essexite extends from the airfield to Morro Atalaia, but this cannot be observed in the field. Glenmuirite blocks in the Remedios tuffs are seen between Conceicao and Morro Forte. Furchites and camptonites form abundant xenoliths in the tuffs at Caieira Bay. Basaltic rock xenoliths also are common: nepheline-basalts with abundant phenocrysts and microphenocrysts of olivine and augite, some approximating the 'basanitoids' of LACROIX, being porphyritic olivine-basalts with a little plagioclase and a matrix of small prisms and microlites of augite. CAMPBELL SMITH & BURRI described from the RIDLEY collection from Caieira Bay and Morro Branco, rocks which they classed as tephritic trachy-basalts, with phenocrysts of hornblende and augite, but ALMEIDA failed to note the occurrence, though he did find between the ejectiles on the strand at Caieira Bay, a mesocratic, strongly porphyritic rock with no olivine, phenocrysts and microphenocrysts of hornblende, augite, andesine, magnetite, titanite and apatite, in a matrix comprising augite, apatite, andesite and a mineral which probably was nepheline.

2. Plutonics

In Fernando de Noronha none of these occur in situ but only as ejectiles. In the phonolite occurring at a small hill N of Southeast Bay, small fragments of nephelinitic syenite are present. The nepheline is mostly analcitized, orthoclase encloses aggregates of aegerine-augite, magnetite and titanite are strongly calcitized. This leucocratic rock has a hypidiomorphic, granular texture, with interstitial nepheline, and probably is the plutonic equivalent of the island phonolites. In the Remedios tuffs at Caieira Bay is a whitish xenolith classed as an aplitic nepheline-syenite, likely a hypabyssal rock associated to the nephelinitic syenites. At Atalaia Pt., Caieira Bay and beside the runway at the airfield are coarse-textured, melanocratic rocks very rich in syntagmatite crystals. The mineralogy is simple, comprising largely big crystals of syntagmatite, which may total some 60% of the rock. There may or may not be small grains of soda-augite, and the remainder of the rock comprises labradorite, not more than 5% apatite, occasionally titano-magnetite encrusted with titanite, perhaps some pyritie. No alkali-feldspars are present, feldspathoids and olivine is lacking. The rock is classed as a hornblende-gabbro. Amongst the most curious facies of lamprophyres, kali-gauteites and phonolites are xenolithic inclusions of macrogranular perkinitic rocks which occur as ejectiles in the Remedios tuffs. The rocks have relatively high density minerals – syntagmatite (and kersutite?), soda-augite, titanaugite, abundant apatite and titanite, scarce in magnetite and only occasional chrysotile. Feldspars and feldspathoids are absent. Some rocks are panidiomorphic and strong idiomorphism of the mafic minerals is clearly evident. The scarce interstitial material can be completely carbonatized, chloritized or zeolitized. A xenolith found in a monchiquite dyke at Biboca beach is analysed in Table 10, No. 35, where the groundmass displays zeolites, chlorite and much calcite. No. 36 is a similar xenolith occurring in a porphyritic essexite at Atalaia Pt. The modal composition of the former is given in Table 11, No. 35. Such xenoliths have not been identified in any other dyke rocks of the archipelago. Hornblende-pyroxenites are present at Caieira beach as xenoliths of coarse, granular texture. In thin-section they are seen to be of jacupiranguite type, with interstitial magnetite, part of the titanaugite altered to hornblende, the amphibole poikilitically enclosing pyroxene. The modal composition is: titanaugite 65.2%, magnetite 18.7, hornblende 14.8, apatite and zeolite 1.3, calcite present. In contrast to the other xenoliths present in the eruptives of Fernando de Noronha, the olivinites and nepheline-basanites of Sao José, Cuscuz and de Fora islets are almost entirely composed of olivine. In thin-section, the former rocks show a dense aggregate of xenomorphic grains of chrysotile, comprising some 90% of the total. Enstatite, picotite and magnetite are also present. The chrysotile is sometimes unaltered, or then in small areas may be changed to philosilicate of low refringence. At times, interstitial material shows labradorite, nepheline, magnetite, augite and fibrous zeolite. The olivinite has undergone complete xenomorphism. A chemical analysis of a Sao José specimen is given in No. 37. With the exception of a lower FeO content and a higher Fe_2O_3 tenor, the analysis is very similar to an olivinite analysed from St. Paul Rocks (q. v.) by D. GUIMARAES (1932).

Pyroclastics

The Remedios and Quixaba tuffs are largely composed of fine particles of microlitic, hypercrystalline, feldspathic lavas, more or less rich in vesicles and always strongly oxidized. The phenocrysts indicate that the tuff matrix is of basaltic, trachytic and phonolitic nature. The largest grains are fragments of trachyte, phonolite, porphyritic essexite and glenmuirite, in descending order of frequency. Also there occur crystal fragments, mostly of plagioclase, orthoclase, augite and hornblende. All these form dense aggregates with a minimum of interstices formed of cryptocrystalline material, presumed to be chalcedony, calcite and chlorite material, in that order of frequency. Most tuffs show varying degrees of alteration. Isolated crystals, on the other hand, show little or no modification, but lava particles are strongly oxidized, with limonitization common. No. 38 is a chemical analysis of a Remedios tuff from the coast at Conceicao beach. In thin-section the specimen shows appreciable amounts of feldspathic basalt particles and fragments of trachyte, phonolite and isolated feldspars.

The Quixaba pyroclastics are relatively uniform in composition, formed of small globular particles of ankaratritic, hypocrystalline lavas, with a few prisms of augite, octahedra of magnetite, and much more rare are large, corroded chrysotile crystals. These lava fragments are always strongly fritted and vesiculated. Between the grains, zeolites and ferric oxides are plentiful. The tuffs are always extremely oxidized, showing hematite pigmentation as well as limonite, and are to be considered as vitric tuffs. A chemical analysis is given in No. 39, Table 10.

Eruptive Breccias

Some comments may be made regarding the matrix of these rocks. A large breccia dyke crossing the runway at the foot of Morro Atalaia has a kali-gauteite groundmass, similar in all respects to that seen in these eruptives, except that has somewhat fewer femic minerals. Another large breccia dyke at Cachorro beach is of distinct petrologic interest, and ALMEIDA studied 9 thin-sections of the rock. The matrix is of two types, one green and of distinct leucocratic type, the other melanocratic, and transitions between these two are not unknown. The first has a composition like a normal kali-gauteite, with microphenocrysts of two varieties of soda-augite, syntagmatite, a sodalite group mineral, titanite, magnetite and apatite. Mafic minerals comprise less than 50 % of the rock. Large phenocrysts include soda-orthoclase and partly zeolitized nepheline, with phonolite fragments also present. The dark type of matrix occurs most commonly in rocks occupying the marginal parts of dykes and shows some features similar to the leucocratic type, with the same microphenocrysts. It is distinguished from the latter however by the absence of feldspars in the matrix, which here has very fine microlites of analcite (?) and pyroxene, with magnetite much more abundant, and hence the melanocratic appearance. Between these two extreme matrix types are transitional varieties with few feldspars occurring. One interesting specimen showed xenocrystals of soda-orthoclase, labradorite, nepheline, 'sanidine', sodalite, soda-augite, titanaugite with aegerine-augite rims, syntagmatite, titano-biotite, chrysotile, titanite, apatite and magnetite. Here we have representatives of all the silicate groups essential in eruptive rocks, a decidedly strange association which would invalidate Gibbs Law.

Petrochemistry

ALMEIDA recognized 20 different types of eruptives in situ in the archipelago. There are available in the literature 28 chemical analyses given by ALMEIDA, 6 by CAMPBELL SMITH & BURRI and perhaps 4 or 5 by other writers — adequate to allow us to draw some conclusions as to the petrochemistry of the rocks.

Petrochemical Character of the Fernando de Noronha Province

Fig. 15 shows the variations in the oxides as a function of silica, from which it can be seen that the archipelago rocks are strongly alkaline, showing large amplitudes of differences as the silica content ranges from 34.6 % to 60.8 %. The alkali-calcic index — the value of silica which corresponds to the point where the CaO and $Na_2O + K_2O$ curves flatten-out, is extremely low, 45.2, thus typifying the province as the most alkaline yet investigated in the world. The alumina curve, as also the alkali curves, attain maximums at a silica value of ca. 54 %, which corresponds to the phonolites of the archipelago.

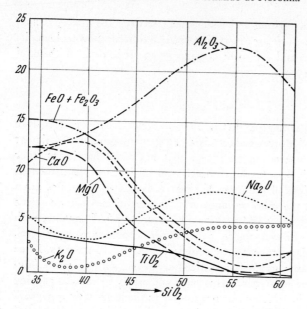

Fig. 15. Variation diagram of chief Oxides as a function of SiO_2. (ALMEIDA, 1958)

Thereafter the curves fall with increase in silica, at the same time that the alkali curves approach each other in the trachyte zone. Mineralogically this is shown by the disapperance of nepheline, with plagioclase taking its place. The alkali curves are approximately parallel for silica values less than ca. 46 %, with a minimum locus at ca. 40 % silica, which value approximately coincides with the maxima for the CaO and MgO curves, and near the minimum for the alumina curve. Between 35–40 % silica, the highest values of the iron oxide and magnesia curves are attained, where are located the ankaratrites, nepheline-basanites, olivine-teschenites and augitites, all rocks having chrysotile and titanaugite in abundance. The zone with silica values less than 39 % is important, not generally noted in similar provinces, given the rarity of petrographic types which are existent. It is notable the antinomical variation in the CaO and $N_2O + K_2O$ curves, the small variation here in the magnesia and alumina curves, whereas the iron oxides curve continues to rise with decreasing silica. It is this extreme alkaline ultra-femic nature which is found in the melilite-ankaratrites, nephelinites, tannbuschites and more basic ankaratritic rocks with akermanite present that the modal composition displays an extreme degree of differentiation in the melilite-ankaratrites. It is also worth pointing out that these are amongst the youngest eruptives in the archipelago. A study of the variation in the principal Niggli values (Fig. 16 and Table 12) leads us to conclude that all the rocks belong to the subsaturated silica series, involving those of intermediate and ultra-basic composition. The qz values are all negative, between –4.4 and –94.9, and two-thirds of the analyses have an index below –50. This is shown mineralogically by the presence in all the rocks of feldspathoids and/or analcite, which, when occurring as primary minerals, corresponds to the furchites and monchiquites or their substitutes when deuteritic. Si varies between 59 and 238, indicating molecular proportions of silica for the sum of the basic oxides between three-fifths and less than seven-eighths. Fm varies between 6.1 and 58.3, being above in three-quarters of the analyses. Al oscillates between 10.5 and 48.7,

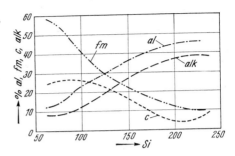

Fig. 16. Variation of Niggli values. (ALMEIDA, 1958)

whilst the c index varies between 1.1 and 28.2, being 17 in about two-thirds of the analyses. K is low, below 0.35 in five-sixths of the rocks studied, only one sample exceeding 0.45, in a gauteite quoted by CAMPBELL SMITH & BURRI, which is represented by 6.66 % K_2O and 3.54 % Na_2O in the chemical analysis – vd. Table 10, No. 9. This high tenor in K_2O is not confirmed in the chemical analyses of the kali-gauteites – vd. Table 10, Nos. 6, 7 and 8 – presented by ALMEIDA, and three further analyses (not recorded in the above table) of these rocks occurring as dykes at Southeast Bay gave values not exceeding 3.77 % K_2O. Mg varies between 0.08 and 0.68, with two-thirds of the analyses having values in excess of 0.50, indicating the predominance of magnesian over ferric types, well seen in the composition of the olivinites. Alk values vary between 5.6 and 45, two-thirds of the analyses being above 9.

The molecular norms (Table 13) show that Ne, Fo or Fa occur in all the analyses. The appearance of akermanite in the norms of the tannbuschites, nephelinites, melilite-ankaratrites and other basic ankaratrites, indicates that we are here dealing with ultra-basic rocks of an extreme chemical composition in which the molecular proportion of silica does not exceed by twenty-nine-fortieths the sum of the basic oxides present.

Referring again to Fig. 15 we see that the eruptives of Fernando de Noronha belong to two group series. One, basic to intermediate, shows increasing values of silica, alumina and alkalis, decreasing values in lime, magnesia and the iron oxides; the other group is ultra-basic, has decreasing values of silica and lime but increasing values in the iron oxides and alkalis. These two series converge where silica values are about 40 %, coinciding with the nepheline-basanites.

In both Figs. 15 and 16 there is no interruption such as has been noted for other Atlantic provinces, as for example by P. ESENWEIN (1929) nor indeed is this the case for the Cabo Verde archipelago. However, this interval in Fernando de Noronha is taken by rocks of quite exceptional character, namely the kali-gauteites and one specimen of furchite.

From a petrochemical standpoint, the eruptives of Fernando de Noronha belong to the following NIGGLI magma types: vesicular polzenitic, normal theralitic, ankaratritic, normal gabbro-theralitic, mela-theralitic, normal ijolitic, normal soda-syenitic, nosicombitic, leuco-syenitic, foyaitic essexitic, lardalitic and bostonitic. ALMEIDA concluded that the Fernando de Noronha province is strongly sodic-alkaline, belonging to the 'Natronreihe' of P. NIGGLI (1936). In broader terms it can be said that the rocks of this archipelago show typical characteristics of the Atlantic Province.

Table 12

NIGGLI Values (ALMEIDA, 1958)

	1.	2.	3.	4.	5.	6.	7.	8.	9.
Si	196.0	185.0	180.0	165.6	172.2	118.0	134.0	153.3	177.6
Qz	−59.2	−87.0	−94.9	−58.4	−64.2	−56.0	−48.8	−32.3	−25.2
Ti	0.6	0.6	1.2	6.7	1.6	3.5	2.9	6.0	5.4
Al	48.1	48.7	43.2	30.6	41.6	28.8	31.2	32.8	37.9
Fm	12.0	6.1	6.7	24.8	13.8	34.0	32.2	26.1	24.7
C	1.1	2.2	4.9	12.4	10.5	18.7	15.9	19.6	11.6
Alk	38.8	43.0	45.0	32.2	34.1	18.5	20.7	21.4	25.7
K	0.25	0.23	0.28	0.35	0.32	0.38	0.38	0.45	0.55
Mg	0.25	0.00	0.06	0.68	0.30	0.59	0.56	0.39	0.53

	10.	11.	12.	13.	14.	15.	16.	17.	18.
Si	238.0	232.0	143.5	115.0	94.1	107.8	108.7	104.0	91.4
Qz	−14.8	−4.4	−55.0	−87.6	−48.3	−77.4	−73.3	−53.6	−59.8
Ti	1.9	3.0	12.3	4.8	5.3	1.8	1.8	4.2	4.0
Al	43.6	45.2	35.8	25.3	22.0	25.6	25.3	20.1	21.0
Fm	11.1	9.4	21.3	30.9	40.2	34.2	36.0	43.5	41.0
C	7.1	11.3	18.4	17.0	27.2	18.9	18.2	22.0	25.2
Alk	38.2	34.1	24.5	26.7	10.6	21.3	20.5	14.4	12.8
K	0.38	0.35	0.36	0.24	0.27	0.29	0.21	0.22	0.18
Mg	0.32	0.25	0.57	0.36	0.50	0.54	0.57	0.62	0.44

	19.	20.	21.	22.	23.	24.	25.	26.	27.
Si	93.2	74.4	77.2	86.9	82.7	77.7	80.4	75.0	63.5
Qz	−39.2	−56.4	−59.7	−55.1	−47.7	−59.9	−42.0	−53.0	−60.9
Ti	6.8	2.2	3.2	5.4	3.2	3.2	2.0	3.7	3.0
Al	12.2	14.9	14.6	15.8	17.5	14.4	13.4	15.3	12.0
Fm	54.0	55.8	52.4	46.9	46.9	52.7	58.3	50.6	57.7
C	25.7	21.5	23.4	26.8	28.0	23.4	22.7	27.1	24.2
Alk	8.1	7.7	9.2	10.5	7.6	9.4	5.6	7.0	6.1
K	0.19	0.12	0.20	0.27	0.10	0.21	0.28	0.06	0.12
Mg	0.68	0.61	0.59	0.56	0.58	0.59	0.65	0.53	0.65

	28.	29.	30.	31.	32.	33.	34.
Si	86.3	73.3	72.7	59.0	81.3	68.4	72.4
Qz	−44.9	−51.5	−58.9	−81.8	−82.7	−75.2	−57.6
Ti	4.9	4.2	5.7	3.4	4.2	6.8	4.5
Al	14.5	10.5	15.1	12.0	21.7	14.3	16.4
Fm	53.0	55.7	53.2	57.9	36.0	46.6	48.5
C	24.7	27.6	23.8	20.5	26.3	28.2	27.7
Alk	7.8	6.2	7.9	9.7	16.0	10.9	7.5
K	0.08	0.19	0.14	0.20	0.18	0.18	0.21
Mg	0.57	0.64	0.65	0.64	0.41	0.48	0.62

Table 13   Molecular Norms (Almeida, 1958)

	1.	2.	3.	4.	5.	6.	7.	8.	9.
Or	25.0	26.5	33.8	34.0	31.3	25.7	28.5	30.4	41.0
Ab	50.5	36.0	21.7	24.1	30.5	12.2	19.0	30.2	32.0
An	0.4	3.0			10.3	19.0	18.0	18.3	13.5
Ac			3.4	4.1					
Ne	15.3	30.9	37.1	19.6	21.5	18.3	15.0	4.1	0.6
C	4.8	1.8							1.7
Le									
Kp									
Di			0.3	9.2		9.0	5.2	7.1	
Wo			2.2		0.4				
En									
Fo	1.2			4.1	1.8	7.9	6.6	2.2	5.7
Fs									
Ol									
Ak									
Mt	1.5	1.5	0.6			3.5	3.6		
Il	0.4	0.4	0.6	2.4		2.6	2.0	2.2	2.0
Pf				1.6	1.1			1.6	1.2
Ap	0.3		0.3	1.9	0.3	1.4	1.1		
Hm	0.6				2.8	0.4	1.1	3.9	2.3
Q									

	10.	11.	12.	13.	14.	15.	16.	17.	18.
Or	34.6	28.1	28.0	23.4	11.7	24.0	16.5	13.0	10.1
Ab	53.8	54.4	18.0	8.0	9.6	0.5	9.5	15.3	5.4
An	6.4	12.8	17.8		24.4	8.5	9.9	11.8	17.9
Ac				3.8					
Ne	1.2		19.0	36.7	14.4	34.8	32.1	18.5	24.2
C		0.2							
Le									
Kp									
Di	0.2		5.1	16.3	19.4	20.9	18.8	22.7	26.3
Wo				1.9					
En		1.2							
Fo	1.2		4.3						
Fs								0.2	
Ol					9.1	5.1	8.5	10.3	6.6
Ak									
Mt				1.2	3.8	3.9	2.7	3.2	5.4
Il		1.2	5.8	3.5	4.6	1.4	1.4	3.4	2.8
Pf	0.9	0.3	2.0						
Ap				1.9	3.0	0.8	0.8	1.6	1.3
Hm	1.8	1.1		3.3					
Q			0.7						

Table 13 Molecular Norms (ALMEIDA, 1958)

	19.	20.	21.	22.	23.	24.	25.	26.	27.
Or	6.8		3.8	12.2	3.5	1.0	8.0	2.0	3.4
Ab	14.3	17.9			3.0		7.0	2.4	
An	9.6	20.4	13.3	12.0	23.5	11.9	19.3	20.9	16.1
Ac									
Ne	9.2		21.4	21.2	17.7	21.8	7.7	18.0	17.5
C									
Le		3.7	4.2	0.5		6.9			0.4
Kp									
Di	36.4	25.3	29.3	37.6	30.8	32.0	29.3	32.1	
Wo									
En									
Fo				4.2					
Fs									2.2
Ol	12.8	26.4	18.1		13.0	16.6	18.9	11.7	29.4
Ak									22.1
Mt	3.8	2.8	4.8	4.9	4.2	5.2	7.8	7.4	4.2
Il	6.1	2.2	3.1	4.9	3.0	3.1	2.0	3.6	3.2
Pf									
Ap	0.9	1.3	1.9	0.6	1.3	1.6		1.9	1.5
Hm				2.0					
Q									

	28.	29.	30.	31.	32.	33.	34.
Or	2.8				1.6		
Ab	20.5						
An	15.5	11.0	17.5	6.0	12.7	8.6	21.2
Ac							
Ne	7.8	15.6	20.2	25.0	34.8	26.3	17.4
C							
Le		0.6	1.2	1.3	9.1	3.4	4.2
Kp		3.1	2.3	5.5		3.2	1.5
Di	27.4	43.4	27.5		26.3	28.4	22.2
Wo					1.6		
En							
Fo						3.4	
Fs				1.7			
Ol	10.7	13.3	20.3	28.2			18.5
Ak		1.4		21.0		8.6	4.6
Mt	8.0	6.4	3.4	5.5	6.2	6.6	2.8
Il	4.5	4.3	5.6	3.6	3.7	6.7	4.5
Pf							
Ap	2.4	0.9	2.1	2.2	2.8	2.7	3.1
Hm	0.4				1.2	2.2	
Q							

Comparisons

CAMPBELL SMITH & BURRI drew attention to the petrographic similarities between Fernando de Noronha and the Bohemian Mittelgebirge area of Czechoslovakia. The latter region has become classic through the studies of J. E. HIBSCH published between 1898 and 1915, some of which have been referred to above, and of H. KNORR (1932). CAMPBELL SMITH & BURRI noted that Fernando de Noronha, like Cabo Verde and the Canary Islands, represented typical sodic- alkaline provinces, whilst in the Bohemian region, many rocks are more potassic than the Fernando de Noronha equivalents, which is shown by the appearance of leucite, a mineral lacking in the archipelago. However, the dominant petrographic types present in Bohemia can be identified with those occurring in Fernando de Noronha, and the order of the Bohemian extrusions shows no essential differences to those postulated for the archipelago. We would note though that the Bohemian sequence is much broader, with rocks ranging from 30.05 % silica (polzenites) to trachytes with 64 % silica, thus making the Bohemian Mittelgebirge an alkaline province showing the most extreme variation anywhere in the world.

From studies in the literature regarding the igneous rocks occurring in the islands of the Atlantic, both North and South, it would appear that those of Cabo Verde, some 2000 km to the NE, bear the closest resemblance to the eruptives of Fernando de Noronha. ALMEIDA, on the basis of 50 chemical analyses available to him for Cabo Verde and 34 for Fernando de Noronha, made calculations of NIGGLI values for each archipelago and constructed diagrams which illustrated the similarities. Table 14, taken from ALMEIDA, shows variation limits for the NIGGLI values for both archipelagos, and Fig. 17 is a

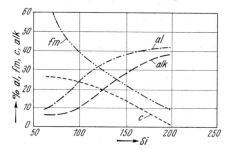

Fig. 17. Variation diagram for principal Niggli values for S. Antao & S. Vicente, Cabo Verde. (ALMEIDA, 1958)

Table 14   Variation of NIGGLI values for Cabo Verde and Fernando de Noronha. (ALMEIDA, 1958)

	Cabo Verde	Fernando de Noronha
Al	10.0 – 43.0	10.5 – 48.7
Fm	9.9 – 65.3	6.1 – 58.3
C	2.6 – 30.7	1.1 – 28.2
Alk	3.8 – 44.9	5.6 – 45.0
Si	61.8 – 195.6	59.0 – 238.0
Qz	−16.6 – −97.6	−4.4 – −94.9
K	0.04 – 0.35	0.07 – 0.45*
Mg	0.02 – 0.70	0.08 – 0.68
Isophalia	117	124

* 0.55 in an analysis of gauteite by CAMPBELL SMITH and BURRI (1933).

variation diagram for the principal NIGGLI parameters for Cabo Verde (islands of S. Antao and S. Vicente), which should be compared with the similar diagram No. 14 for Fernando de Noronha. The alkalinity of the two areas is identical, the alkali-calcic index for the two Cabo Verde islands being 46 and for Fernando de Noronha 45.2. ALMEIDA further substantiated his contention of petrochemical analogies of the areas by constructing BECKE diagrams of the NIGGLI values, and one is indeed impressed by the large degree of similar pattern of the spacings. The diagrams for the two archipelagos both show wide variation for c + fm, whilst k shows considerably less variation. In the two tetrahedral projections there is a notable similarity of corresponding points, equivalating the identical distribution of respective magmatic types. The figures corresponding to the relation of si with respect to c + fm are neatly superimposed on one another, and the equation:

$$230 - si = 205.5 (c + fm)$$

satisfies both diagrams.

Petrogenesis

We must note that petrographic, petrochemical and petrogenetic opinions are based upon some 34 samples occurring within an area of some 9 km^2, which includes all the igneous rock types described. This led ALMEIDA to conclude, in conjunction with other evidence, that here in Fernando de Noronha we have a consanguinous group of rocks yielding an advanced petrogenetic series which resulted from magmatic differentiation.

Fig. 15 shows that the lime and iron oxides curves on the one hand, the alkali curves on the other, have traces close to each other, i.e. they are not dispersed, and for each an average curve would closely approximate their trends. It is also to be noted, on the other hand, that the alumina and magnesia curves are spread apart. In this diagram the traces of the curves to the right of a silica value of 40% agree remarkably well with the variation curves deduced by N. L. BOWEN (1928) to illustrate fractional crystallization in the light of theoretical considerations and the diagrams experimentally derived therefrom. The notable dispersion of the alumina and magnesia curves results chiefly from this process, with the separation of magnesian olivines (chrysotile) frequent in the basic and ultra-basic group of rocks. The marked concavity of the iron oxides, lime and magnesia curves likewise resulted from the same process, with the separation of chrysotile, ferriferous augite (possibly diopsidic in the primary state) and probably, in a modest manner, labradorite. The progressive rising of the alumina and alkalis curves with increasing silica towards a maximum before declining, was explained by BOWEN as a result of fractional crystallization. The pronounced rise of the alumina curve between 40% and 50% SiO$_2$, when compared with similar diagrams for sub-alkaline provinces, demonstrates the reduced subtraction of sesquioxide of this stage, a fact noted in other strongly sodic-alkaline series. Petrographic observation confirms the relatively small importance of calcic plagioclases in the fractionation of alkaline magmas. This stage of evolution, in which the liquid has between 42% and 50% silica, is found in other sodic-alkaline provinces, hornblende being important in the evolution of the series. In Fernando de Noronha, where this amphibole is likely calcic with as much lime as in labradorite, there is a much smaller quantity of alumina, equal or less than the liquid, its separation contributing to the enrichment of this in the sesquioxide.

The femic and ultra-femic states in the archipelago rocks are represented by the hornblende-pyroxenites, hornblende-gabbros, tephritic trachybasalts and porphyritic essexites, with the first two evidently forming later.

The general character of the variation diagrams for Fernando de Noronha and of other sodic-alkaline series is strongly convex for the alumina and soda curves in the vicinity of $SiO_2 = 50–55\%$, which, as per BOWEN, indicates reduced fractionation. Here in the archipelago, this appears to result from the rapid cooling of liquids, corresponding at this point in the diagrams to phonolites and trachytes, the highest zones attained by the eruptives. However, doubtless also the degree of viscosity of the liquids, as well as the relatively reduced volumetric expression in the province of the series olivine-basalt – trachyte – phonolite did not allow of easy movement of crystals. The reduced fractionation of the sialic residues in Fernando de Noronha is clearly reflected in the association porphyritic essexites – glenmuirites, the orthoclase and plagioclase containing calcium, with labradorite ca. 65 % An.

The above paragraphs may be said to be based upon factual data which, in the light of our present knowledge of petrology, allow us to recognize a petrogenetic unity for the Fernando de Noronha eruptives, from which it is possible to arrive at a concept of fractional crystallization for the genesis of the series. There is a strong temptation to draw certain inferences from the data available from the archipelago, allowing us to postulate certain hypotheses. The olivinite and perkinitic xenoliths, the phenocrysts of the ultra-mafics, the ultra-femic alkaline rocks, all these invite speculation, but as factual information is scant here, it is thought better to leave unsaid what might be inferred from hypothetical speculation.

In Fernando de Noronha no olivine-basalts occur, a rock corresponding to a magma that frequently is considered the generator of the alkaline series – always remembering, however, the extremely small area from which samples have been obtained within the archipelago. Unless we admit the concept of an exterior contributing source of alkalis, the originally alkaline basaltic matrix would represent one of high tenor in alkalis, which is a characteristic of all rocks of the series. It appears therefore that the generating magma was of alkali-basalt type. The high tenor in alkalis and low silica content would satisfy the observed facts. This rock is not normally porphyritic, and doubtless in all its occurrences it was intruded in a completely fluid state. The alkali-basalts are considered to be the oldest eruptives in the archipelago. The Abreu Bay occurrence is intruded by a kaligauteite dyke, possibly by phonolite and trachyte at Southeast Bay, and can be recognized amongst the ejectiles in the the Remedios tuffs which at Caieira Bay are intruded by phonolites and trachytes. A comparison of the specimen given in Table 10, No. 18 with a feldspathic basalt from Scharfenstein in the Bohemian Mittelgebirge, described by J. E. HIBSCH (1898) and which was considered by KNORR (1932) to represent the 'Stammagma' of the alkali series in the area in question, such a comparison shows a striking similarity, the sole difference being in the ratio K_2O/Na_2O, which is higher in the Bohemian sample, and perfectly in accord with the more potassic nature of the Bohemian province.

In Oahu, Hawaii, the Honolulu series shows identical features in the linosaite rocks (H. WINCHELL, 1947). In Cabo Verde (J. B. BEBIANO, 1932), and in the Canary Islands, as per data presented by ESENWEIN (vd. reference above), rocks similar in character to the alkali-basalts of Fernando de Noronha are present. All these rocks, from the Bohemian

Mittelgebirge, Hawaii, Cabo Verde, Canaries and Fernando de Noronha, have low tenors in silica and magnesia, high values in the alkalis, alumina and titanium oxide.

The origin of these magmas with a primary olivine-basaltic matrix should be accepted in principle, given the field associations especially as seen in Hawaii and Cabo Verde, but there are difficulties in determining the processes of formation of said rocks. In fact, the first minerals to crystallize in an olivine-basalt magma are not the type to separate from residual liquids so poor in silica as these alkali-basalts. One may suggest initial precipitation of enstatite at the expense of olivine, possibly in consequence of the high pressures, and of strongly diopsidic pyroxene, but facts are lacking to substabtiate this.

The nepheline-basanites of Fernando de Noronha, which CAMPBELL SMITH & BARRI considered as basanitoid types of rocks, are such that the classification of Rosenbusch would incorporate them in the alkali-basalts, and this would also satisfy the conditions of a generating magma for the series. It is possible that this a rock emanating from a magma contaminated with chrysotile crystals, as a simple calculation would suggest. The only significant difference between the nepheline-basanites and the alkali-basalts in Fernando de Noronha is the excess of magnesia in the former to which corresponds a high tenor in the sialic components in the latter. Mineralogically there is a difference in there being a little more than 10 % by weight chrysotile in the basanites. The addition of such an excess of fosterite to an alkali-basaltic magma would not entail modification in the silica and iron oxides tenors, causing lower tenors in alumina, the alkalis and lime. As regards CaO, examination of specimens indicates contamination with pyroxene, the nepheline-basanites having up to 11 % phenocrysts of augite.

ALMEIDA was of the opinion that the nepheline-basanites of the Sao José formation were formed from a magma of alkali-basaltic composition, which was contaminated with chrysotile and possibly augite. It is likely that part of the olivine was reabsorbed by the magma in traversing the dunitic threshold of the magma chamber, contributing to the crystallization of augite in the nepheline-basanites.

Age of the Volcanics

The absence of sediments associated with the volcanics militates against exact age determinations for the volcanic episodes. Only by inferences from comparisons with more or less well-dated congenital phenomena in other Atlantic islands can the problem be tackled.

In Cabo Verde, in the island of Maio, limestones with Barremian fossils were metamorphosed by basaltic flows (BEBIANO – vd. reference above; R. C. MITCHELL-THOMÉ, 1964). Here the phonolite and trachyte intrusions occurred between the Eocene and the Miocene, basaltic activity having continued up to the present time in Ilha Fogo.

At Dakar, Sénégal, the ankaratritic flows of Cap Manuel appear to be post-Aquitanian in age and the Mamelles vulcanism is thought to be of Pleistocene times (R. FURON, 1960).

In the Canary Islands the basalts are of presumed Cretaceous age.

In the E and NE coasts of Brazil no vulcanism is thought to be later than the Barreiras series, which is considered as Pliocene. At Abrolhos Rocks, on Ilha Sta. Barbara, HARTT (1870) described a basaltic flow which had already been recognized by DARWIN (1876), covering beds thought to be Cretaceous, but HARTT's descriptions make it difficult to decide if we should regard the flows as of the same age.

F. VON WOLFF (1931) believed that the vulcanism of Fernando de Noronha began in the Upper Cretaceous, due consideration being taken of its petrographic similarities with Cabo Verde, and the relative similar location of the two archipelagos within the Atlantic.

ALMEIDA thought that in Noronha already in the Pleistocene vulcanism had long since ceased. To him, all we could say was that the vulcanism occurred between the Upper Cretaceous and the Neogene.

Radiometric Age Determinations

CORDANI (in Press) visited the island with the purpose of collecting specimens for radiometric analysis, also using some ALMEIDA specimens. As the potassium-argon method of analysis is especially suitable for young volcanic material, fourteen such determinations were made of twelve Fernando de Noronha rocks (Table 15).

Table 15   Radiometric Age Determinations (Modified after CORDANI (Press))

Sequence	Sample No.	Rock Type	Material	Age (m. y.)
Quixaba	UC-FN-18	Ankaratritic lava	Whole rock	1.73 ± 0.13
	UC-FN-16	Ankaratritic lava	Whole rock	2.38 ± 0.09
	UC-FN-16	Ankaratritic lava	Whole rock	2.86 ± 0.26
	UC-FN-24	Ankaratritic lava	Whole rock	3.19 ± 0.10
Sao José	UC-FN-17	Nepheline-basanite	Whole rock	9.49 ± 0.33
Remedios	UC-FN-1	Phonolitic neck	K-feldspar	8.91 ± 0.27
	FA-FN-380	Phonolitic neck	Whole rock	9.03 ± 0.27 / 9.10 ± 0.27
			K-feldspar	8.96 ± 0.27
	UC-FN-22	Phonolitic neck	Whole rock	10.67 ± 0.32
	FA-FN-9	Volcanic breccia	Matrix	4.30 ± 0.30
	UC-FN-6	Essexite porphyry	Whole rock	8.91 ± 0.27
	FA-FN-265	Essexite porphyry	Whole rock	9.19 ± 0.28
	UC-FN-10	Alkali-basalt	Whole rock	11.79 ± 0.35

The ankaratritic lavas of the Quixaba Formation give ages varying between 1.7 and 3.2 m. y. The youngest specimen is from Rata Island; the oldest specimen is from the eastern plateau. The two other samples are from the western plateau, and show an age difference greater than the interval allowed for analytical error. These samples are from different lava flows, the lower one having the greater age.

The one sample from the Sao José Formation gives an age value well within the range indicated by the Remedios intrusives.

The other seven samples were taken from the Remedios Formation. Results give consistent ages around 9 m. y. or slightly older, with the one exception of the volcanic breccia. The oldest age recorded, near 12 m. y., is from an alkali-basalt.

The 12 m. y. age for the alkali-basalt would place the intrusion in the Lower Pliocene, giving a lower limit for the age of the Remedios pyroclastics. It must be stressed, however, that samplings relate to the subaerial part of the volcanic island, and that igneous activity began earlier than the Lower Pliocene is a logical conclusion.

ALMEIDA, on the basis of field evidence, thought that the phonolites and trachytes of the Remedios Formation – or perhaps porphyritic essexite at Atalaia Pt., represented the oldest eruptives. However, as per the above radiometric data it would appear that the alkali-basalts are older than the essexite and phonolite.

The relationships of the Sao José Formation to the other two formations is not known, but ALMEIDA thought that the Sao José rocks were amongst the youngest in the archipelago. As per the meagre radiometric analyses here presented, this would not be so, the Sao José giving age characteristics typical of the Remedios Formation.

CHAPTER 4

Rocas Atoll

This small atoll lies in lat. 3° 52′ S, long. 33° 49′ W, some 130 km of Fernando de Noronha.

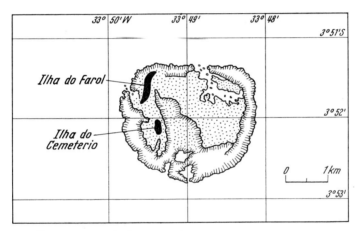

Fig. 18. Rocas Atoll. (Conc. Nac. de Geogr. 1950)

The circular-shaped islet measures 2.4 km E-W and 1.6 km N-S.

The atoll comprises two coral reefs, named Ilha do Farol and Ilha do Cemeterio. On the former there is a light-tower for ship navigation and also the ruins of a house, but today there are no inhabitants here. Farol reaches an altitude of 13 m and has a slight vegetation. The best landing place lies NW of the light-tower, where there is a small coral beach. In times of calm seas, another landing can be made at the entrance across the atoll along the northern coast. Rocas and vicinity are considered extremely dangerous reefs to shipping.

Rocas Atoll represents a great seamont, a coral reef resting on a slope which rises from depths of some 3800 m. The atoll shelves quite rapidly down to depths of 20 m, thereafter more abruptly to 50 m, then very rapidly to depths of 500 m.

It would appear that no geological investigations have been made of Rocas Atoll.

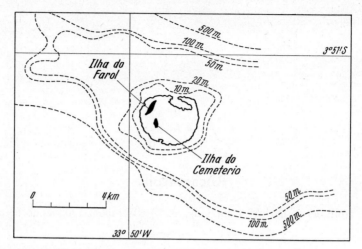

Fig. 19. Rocas Atoll. (Sheet 51, Serv. Hido. da Marinha, 1961)

CHAPTER 5

Trindade

General

The small island of Trindade lies in lat. 20° 30′ S, long. 29° 19′ W, some 1140 km from the Brazil mainland, 48 km W of the islet Martin Vaz and ca. 2400 km SW of the island of Ascension. The length is 4.8 km, breadth 2.6 km, area ca. 8 km².

There appears to be uncertainty as to which is the highest summit. ALMEIDA (1961) believed that neither Pico Trindade nor Pico Desejado were quite as high as 600 m, and was informed that Pico Sao Bonifacio rivalled them in elevation. Which actually is the topmost peak is thus unknown, with perhaps Pico Trindade having best qualifications in this respect.

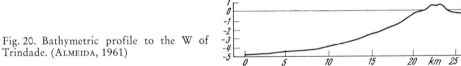

Fig. 20. Bathymetric profile to the W of Trindade. (ALMEIDA, 1961)

The island represents a great volcanic cone rising to a height of some 5500 m above the ocean floor, with a basal diameter of some 50 km. From the island a marine platform extends outwards no more than 3 km from the coasts, covering an area of some 32 km², sloping steeply down to oceanic depths. The coasts are precipitous and coral reefs are particularly developed along the NE coasts, against which the waves break with great force.

Trindade, along with Martin Vaz, already appeared on a map of KUNTSMAN III in 1507, but the date of discovery is not known. It is believed that the first landing on Trindade were made by BELCHIOR DE CARVALHO in 1559, and since that time, permanent occupation has been rare and of short duration. The Portuguese settled in 1756, English occupied the island in 1781, the Portuguese returned during 1783–1795, and during the reign of Napolean I treasure-hunting pirates vere in possession. RIBEIRO (1951) mentioned more than twelve expeditions to Trindade by treasure seekers, beginning in 1885, and in that year, the British again occupied the island. During both World Wars Trindade was under Brazilian military occupation, and between 1924–26, it was a penal settlement.

Beginning with Capt. James Cook in 1775, Trindade has been visited by various explorers and naturalists. Important Brazilian espeditions were those of the National Museum, Manguinhos Institute, Botanical Gardens and National Observatory during World War I, the Joao Alberto Lins de Barrros expedition of 1950, the Hydrography and Navigation Department of the Navy in 1957 in connexion with the setting-up of observations for the International Geophysical Year, and the visit of the geologist F. F. M. de Almeida in 1958 to compile a monograph on the island.

Soils are sparse and vegetation is limited to lowly carpets of grasses, the important exception being the forest of giant samambaias, limited to the highest summits. J. D. Hooker visited the island in 1839 as a member of the Ross Antarctic Expedition, and was the first to give details of the plant life. He listed 71 species of flora, of which many were endemic. According to F. S. Viana (1950), who visited Trindade in 1950, the lower parts are covered in grasses (*Paspalum* sp.) and a Stergulinacea, *Waltheria americana*, half a metre in height. In the valleys there are Ciperaceae, notably an endemic species *Cyperus atlanticus*, occurring on lower slopes. On the axial plateau and the higher peaks, the vegetation consists of various types of brushwood, the genera *Pisonia*, Rapanea and *Bannara* being common. The most notable of the brushwoods, however, and which grows on shaded, humid slopes, is a giant samambaia, native to the island, *Cyathea copelandi*, which attains a height of 6 m. This representative of the Filicineae is unique in the world, suggesting a Palaeozoic forest lost in the ocean. According to Viana, the plant indicates a previous more humid climate than the present.

Since 1927 there have been no permanent settlements on the island, and during Almeida's survey his only companions were scientists staying temporarily there for geophysical investigations, and two assistants.

Physical Features

Structure and climate are the principal factors in the moulding of the relief. The volcanics show considerable chemical resistance but mechanically break down into blocks and smaller fragments: the abundant pyroclastics easily weather chemically and experience powerful disintegration. Strong and persistent winds, heavy pounding by waves, frequent storms causing fearful marine damage, deep, narrow ravines ravaged by torrents, seismic instability, a more or less long and continued phase of vulcanism – all such factors have resulted in enormous destruction of the island and everywhere the dominance of erosion over deposition is manifest.

Trending parallel to the NW-SE alignment of the island is an axis of high land forming the water-divide for the major valleys descending to the NE and SW coasts. Much of this area lies above 350 m, with deep ravines separating scattered peaks rising to heights of some 600 m. As this axis lies somewhat nearer the SW coast, slopes down to the NE shorelines are less steep than those towards the SW side of the island.

At the NW and SE extremities of Trindade are two imposing highland masses. At the former lies Crista de Galo, a serrated, narrow, steep arête trending NW-SE, with the inaccessible summit of Obelisco rising to some 430 m. On either side of this arête are a series of precipitous scarp slopes leading down to wild, rocky shores. At the SE end of

Trindade is the extinct volcano, Paredao, the ruins left by intense marine erosion of the latest vulcanism. The crest of the crater rises 217 m above sea level, with flows outcropping at the coast in the Tartarugas area, and elsewhere mantled in pyroclastics. Here occurs a tunnel carved by the sea, piercing right through the peninsula, measuring 130 m in length, 15 m high and 12 m broad.

Very prominent and spectacular domes and necks of phonolites dot the island, including Monumento, claimed by ALMEIDA to be one of the finest examples in the world of a neck, Picos Preto, Pontudo and Tartarugas and Pao de Açucar, a complete replica of the more famous one at Guanabara, Rio de Janeiro. The highest summits, Picos Trindade, Desejado and Sao Bonifacio also comprise phonolite domes but of somewhat younger age than the above.

Ravining and mass movements of materials are responsible for the modelling of the island slopes. Talus slopes are especially prominent on the NE side of the island and around Principe Bay in the S. Fluvial action is of a torrential character, with very high stream competency. The valleys carry water only during the wet season. The abundant outcrops of highly porous pyroclastics allow of copious infiltration of meteoric waters, and some springs are flowing all the year. Fresh water is never a problem on Trindade. Though streams are ephemeral, rapid recession of the cliffs maintains the drainage network and the steep longitudinal profiles, so that stream capacities are able to move most material right down to sea level and out on to the marine platform, with the result that there is scarce any fluvial deposition on land. The tendency is for the streams to seek the shortest way down from the sources to sea level. An excellent example of stream-capture is to be seen in the Vermelho valley, which originally was directed northeastwards down the ankaratritic flows towards Tartarugas beach. A shorter, steeper, more active stream flowing into Principe Bay worked backwards until it captured the Vermelho stream, which now turn abruptly southwards to reach the sea at Principe beach.

Sand and pebble beaches are commoner along the NE coasts, and indeed all along the W, SW and S coasts the only developed sand beach is at Principe. Lithothamnium reefs likewise are principally round off the NE shores.

Marine terraces and arenite deposits occur ca. 3.5 m above present sea level can be seen at many places.

The island is surrounded by a narrow insular platform, in area about four times that of the island itself. Depths here are generally less than 90 m. Submarine terraces have been carved out of this platform at depths of ca. 44 m and 77 m.

Climate

Principal data relating to the climate of the island come from the observations of S. SERECBRENIK and reported by BARROS (1950). These cover an interrupted period of three and half years during 1941–45.

The average annual temperature of Trindade, 23.2° C, is the same as that at Vitoria, on the coast of Brazil in almost the same latitude. March is the warmest month, July the coolest, with a 5.5° C annual variation. Total annual rainfall amounts to 806 mm, a low value related to the island location with respect to the semi-stationary S. Atlantic anticyclone. The annual variation would seem to be rather irregular. The wettest months are April and May, driest in summer and winter. The relative humidity is comparatively low, with a mean average of 76 % at the station located 12 m above sea level. It is presumed to be higher at greater altitudes, for there is frequent cloud condensation provoked by the relief, resulting in a significant vegetation. Dominant winds throughout the year are from the SE and NE, the latter especially in summer. Throughout most of the year winds are strong, with an average annual velocity of 7.3 m/sec. for those from the SE.

Geology

The first geological observations in Trindade were those of the English botanist HOOKER who accompanied the Ross Antarctic Expedition, and paid a visit to the island in 1839. Since then a few geological visits have been made or then studies of samples collected by others. By far the most important geological contribution is that by ALMEIDA (1961) and his large-scale geological map (1 : 10 000) is most detailed. The report of the island geology presented here is primarily based upon the work of Professor ALMEIDA.

ALMEIDA recognized five distinct volcanic phases, typified by characteristic lava outpourings. These will be described below.

Trindade Complex

Some 80–85 % of the exposed rocks of the island are assigned to this unit, which comprises a heterogeneous assemblage of pyroclastics and eruptives. This, the oldest vulcanism, has been drastically eroded, volcanic structures greatly worn down from their initial prominences, a thick regolith and talus deposits mask contacts, and throughout the rocks have a deeply rotted appearance.

Pyroclastics are predominant in the Complex. The oldest rocks of Trindade are tannbuschitic lapillitic tuffs, bombs and blocks of tannbuschite lava, exposed along the high, inaccessible coasts of Cachoeira Bay. These poorly stratified deposits measure some 120 m in thickness. Other later tannbuschitic pyroclastics are to be found at Portugueses beach, Pico Desejado, Vale Verde and elsewhere, associated with tannbuschite dykes. Phonolitic pyroclastics include lapillitic tuffs, tuff-breccias and breccias, but in lesser amount than tuffs with some lapilli or then ash. These pyroclastics occur as lentils, 'stringers' or beds of variable thickness, are coarsely stratified and have dips up to 30°. In the region of Obeslisco, phonolitic pyroclastics measure some 400 m thick. Indications point to 'nuées ardentes' phenomena during formation. The general character of these rocks points to subaerial deposition of explosive products, but in some localities, e. g. the section from Desejado to Portugueses beach, subaquatic deposition in a lacustrine environment is likely.

Some 16 phonolite domes and necks can be recognized, of which the largest, Morro Branco, is 450 m in diameter. Monumento, an extraordinary structure, rises 400 m sheer from the sea, and was claimed by ALMEIDA to be probably the finest example in the world of a neck. The phonolites may be porphyritic or aphyric. Fluidal structure, especially of laminar type, is common in necks, domes and dykes, but lineation is rarely seen. Xenoliths are abundant, being of a cognate plutonic character, more often of angu-

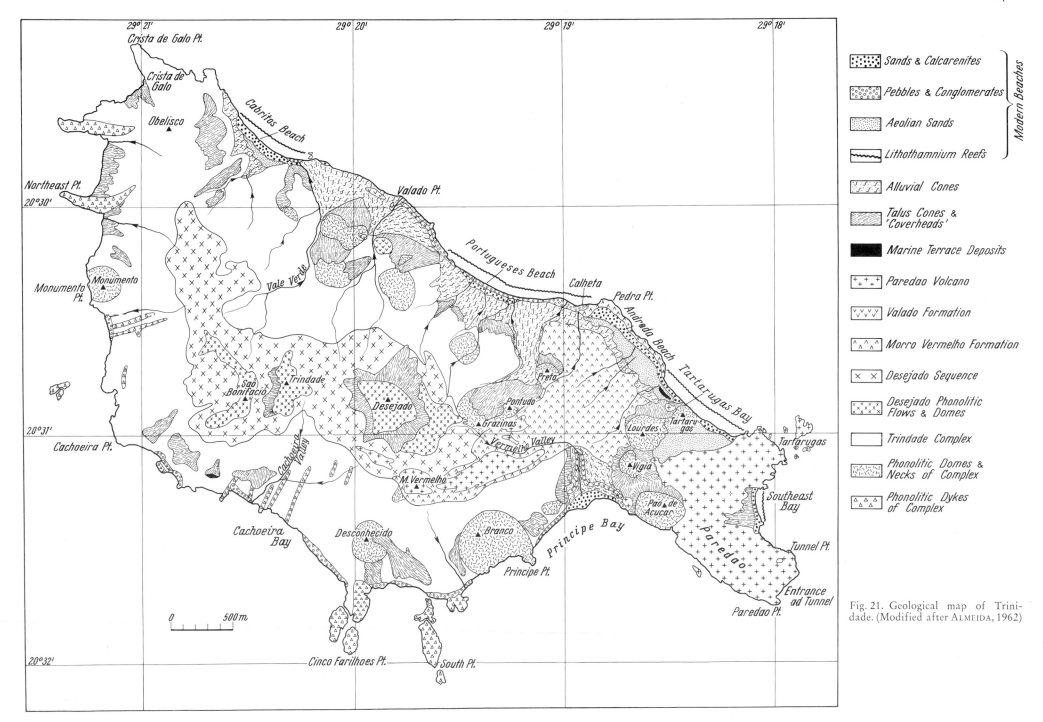

Fig. 21. Geological map of Trindade. (Modified after ALMEIDA, 1962)

lar shape. Columnar jointing is a notable feature of many necks, e. g. Picos Vigia, Preto, Desconhecido, etc., resembling small replicas of the famous Devil's Tower, Wyoming, U. S. A. Vesicularity is absent in the phonolite of necks and domes but present in dykes. Of the various kinds of dykes the phonolitic ones of the Complex are the most distinctive. They can be as much as 50 m thick, and form great coastal salients N of Monumento Point.

Ultrabasics occur in numerous dykes, and are particularly well seen on the intensely eroded slopes leading down to the coast between Cinco Farilhoes Point and Monumento Point. Although these dykes do not invade the succeeding Desejado volcanic phase, ALMEIDA presumed that Desejado ultrabasic flows were fed by some of these dykes. In any case, most of the island dykes were intruded before the flow phase of the Desejado. The ultrabasic dykes are seldom thicker than 5 m, and thus contrast greatly with the phonolite dykes. Some intersting associations are seen in multiple dykes near Crista de Galo Point (Fig. 22).

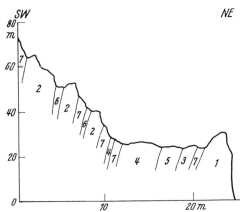

Fig. 22. Multiple dykes exposed on N coast between Cabritos Beach and Crista de Galo Pt. 1. Tinguaite, 2. Phonolite, 3. Analcite-Phonolite, 4. Amphibolitic-Nephelinite, 5. Analcite-Basanite, 6. Tannbuschite, 7. Phonolitic tuff of the complex. (ALMEIDA, 1961)

Rocks of the Complex are intensely fractured. Mostly this is due to local causes associated with the emplacement of eruptives. There are two principal joint systems: 1. radial, especially developed in the western half of the island, centring approximately on Pico Sao Bonifacio and Pico Desejado. Some joints are occupied by dykes and may extend up to 500 m. Radial joints apparently do not affect the Desejado sequence. 2. NW-trending fractures, well displayed in the NE part of the island. (Fig. 21 is a multiple dyke association of this trend.) No important faults occur in the island, but the NE coast, which trends parallel to the NW-orientated joints, suggest it was outlined by older faults.

The oldest volcanic episode, that of the tannbuschites of the SW coast, is represented by fluidal lavas forming fine-grained dykes, but no flows have been seen. The natural glass of much of the tannbuschitic pyroclastics and the abundance of bombs would presuppose a process of fire fountaining. The phonolitic pyroclastics represent cone remains such as mantle domes during their growth. The phonolite peaks seem to have more than one origin. Some appear to be monolithic viscous lava intrusions of cylindrical form, of which Monumento is the best example, but two unnamed peaks looking down on Portugueses beach appear to have a similar origin. The maximum diameter of these necks is 250 m, and they seem to have much in common with the Quellkuppen of the

Drachenfels trachyte described by H. & E. CLOOS (1927). Other pinnacles of phonolite correspond more to the Staukuppen of A. BERGEAT (1927), growing independently where the funnel orifice of the cones has a much greater diameter than the feeding conduits. These bodies have diameters much greater than the monolithic intrusions – up to 450 m. Picos Grazinas-Pontudo, Vigia and Desconhecido may possibly represent Staukuppen. The emission of these very viscous phonolitic lavas to form necks and domes was preceded and accompanied by violent explosions, forming the pyroclastics which constitute the greater volume of the island.

Desejado Sequence

This comprises a succession of phonolite, grazinite and nephelinite flows, intercalated with pyroclastics of the same composition. Between Pico Desejado and Portugueses beach the thickness is ca. 180 m, but it probably thickens to something like 400 m towards the W. Topographically the formation is represented by a plateau of beds only slightly inclined. Rising above this central plateau are the highest peaks of the island, Sao Bonifacio, Trindade and Desejado, formed of phonolitic flows totalliting about 100 m in thickness.

The phonolitic and nephelinitic pyroclastics interbed with similar type flows. Phonolitic pyroclastics are particularly evident in the lower part of the sequence, and at Pico Grazinas, well-stratified tuffs, lapilli and ash beds, with rare blocks but no bombs are to be seen, also minor occurrences of cross-bedding. Nephelinitic pyroclastics are well exposed N of Pico Desejado, where they exceed 50 m in thickness, comprising tuff-breccias, blocks and bombs.

The phonolite and analcite-phonolite flows are resistant rocks and maintain the altitude of the plateau. W and S of Pico Sao Bonifacio nephelinite flows occur. The phonolites, much commoner than the nephelinites, are typically massive, homogeneous rocks, but the top and base of flows become vesiculated and filled with amygdales. Vertical contraction joints are common in the lower part of the sequence. Phonolite flows may be up to 16 m thick, and in the upper Cachoeira valley, flow scarps extend for hundreds of metres with scarce any variation in thickness. Pico Desejado comprises phonolite flows some 160 m thick on the NE face, but much less on the western slope.

Volcanic activity involved explosive extrusion of viscous phonolitic lavas, interspersed with more fluidal nephelinite and grazinite lavas. The basic character of the rocks and quantity of gases suggests an analcitic magma, these phonolites being less viscous than those of the Complex. Pico Desejado appears to be an erosional remnant of a Staukuppe. There are intercalated pyroclastics here, and the peak certainly does not have the form of a neck. Probably Picos Sao Bonifacio and Trindade are similar, but the large volumes of modern talus deposits prohibits a clearer assessment. The nephelinite and grazinite eruptions were emitted during the phonolitic phase, and thus the vulcanism was mixed, with phonolite emissions alternating with effusions of very liquid lavas in the manner of fire fountaining, recognizable in the accumulations of vitric tuffs, lapilli-tuffs and tuff-breccias rich in bombs, driblets and other vesicular and scoriaceous components resulting from the fragmentation of the lavas.

Morro Vermelho Formation

The volcanics and pyroclastics of this phase are the products of a single and continued series of explosions and outpourings from a centre located in the upper Vermelho valley, near Morro Vermelho. After the vulcanism resulting in the Complex, fluvial erosion carved out this valley, bordered by Complex necks and domes rising above marginal interfluvial areas of pyroclastics and eruptives of the Desejado phase. The Vermelho vulcanism filled the valley with more than 200 m of ankaratritic flows and intercalated pyroclastics of the same composition. The flows poured down northeastwards towards Andrade beach area, where they are well exposed in cliff sections. Whilst the Vermelho volcanics are limited to the eastern part of the island, it should be noted that ankaratritic flows also occur at Paredao volcano, these being of younger age and showing petrographical differences.

The pyroclastics are mostly lapilli-tuffs, with blocks and bombs, which change locally into tuff-breccias and agglomerates. At Point Pedra, ashy tuffs are exposed in the cliffs. The granular components of these pyroclastics include ankaratrite fragments of angular shape, resulting from explosive fragmentation, to form driblets, bombs, typical 'Pelée tears' and blocks. Blocks may exceed a metre in diameter, and bombs vary from 3–8 cms. in diameter, the latter having vesicular or even scoriaceous interiors with vitrified crusts more than a centimetre thick. 'Ribbons' or sections of ropy lava also occur sporadically. Near Morro Vermelho the pyroclastics dip at 30°, and those at the coast are slightly inclined seawards. The pyroclastics are highly porous and subject to rapid weathering. In thickness they vary up to 50 m. Fig. 23 shows pyroclastics 25 m thick alternating with ankaratritic flows in a ravine leading down to Calheta.

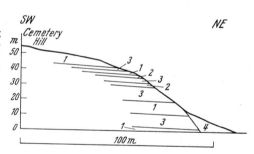

Fig. 23. Section of Cemetery Hill showing alternation of flows and pyroclastics. 1. Analcite-Ankaratrites flows, 2. Tuffs, 3. Lapilli-Tuffs & Tuff-Breccias, 4. Talus deposits. (ALMEIDA, 1961)

Flows are mostly of analcite-ankaratrites. The rocks are dark, aphanitic and contain many xenoliths. The flows largely consist of lava blocks whose surface were covered with a chaotic mantle of agglutinated angular blocks of scoriaceous lava. There is a tendency towards spheroidal weathering, but no pillow lavas occur here or elsewhere in the island. High vesicularity and meagre thickness – seldom more than 4 m – does not favour columnar jointing.

The total thickness of the formation exceeds 230 m.

Vulcanism was centred in the interior of the island in a large depression carved out of the Complex and Desejado rocks, the eruptivity being focussed on the upper slopes of said depression. The ankaratritic magma was very fluid and had a large gas content, emission being largely in the form of pyroclastics whose components were expelled into the air in a pasty condition in the manner of fire fountaining. These products mantle only the extreme eastern part of the central plateau. One might speculate therefore if the dominant winds of the time may not have been from a westerly direction, instead of an easterly direction as at present. Morro Vermelho is an erosion remnant of such an ankaratritic pyroclastic accumulation, but being higher than the level attained by ankaratritic flows, was not covered by these. Small faults, dykes and unconformities in the pyroclastics have accentuated their original inclinations nearer to the focal vent. The flows surging up into the depression spilled downwards as tens of 'rivers' of lava direct to the coast. In consequence of the Vermelho vulcanism, all the area between Pao de Açucar and Pico Desonhecido constitutes high relief, dropping down steeply to the southern coasts.

Valado Formation

In the vicinity of Valado Point, on the NE coast, are dark flows interbedded with deposits of a large cone of dejection. Here there is an extensive stretch of talus deposits leading down to the coast, indicating the proximity of a vent where tannbuschitic filaments, bombs and smaller clotted masses were expelled.

The pyroclastics comprise lava fragments of various forms, varying in length up to 40 cms., of very vesicular or even scoriaceous material. Driblets, on the other hand, show little or no vesiculation. 'Ribbons' of ropy lava can be quite common. The present thickness of the agglutinated pyroclastics is ca. 20 m, but originally must have much more. These agglutinated deposits (Schlackenagglomerat) also occur in the Vermelho formation and at Paredao volcano, of older and younger date respectively.

The relations between the flows and pyroclastics are not clear. The mountain slopes are clothed in talus deposits from a large dejection cone which evidences vulcanism. But it cannot be expressly determined whether both materials are related to the same centre of vulcanism. Fig. 24 is a schematic cross-section from the peak inland from Valado Point

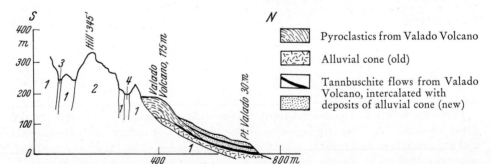

Fig. 24. Section from Hill "345" to Pt. Valado. 1. Tuffs & phonolitic breccias, 2. Phonolitic neck, 3. Dyke of analcitite, 4. Phonolite dyke. (ALMEIDA, 1961)

down to the coast, which shows the relationships of the alluvial cone deposits, pyroclastics and flows of this formation. The accumulation of the Valado pyroclastics occurred during the growth of these dejection cones, and continued with the accumulation of breccias which originated from these cone deposits. The pyroclastics are clearly later than the Vermelho vulcanism, and ALMEIDA attributed them to the Valado phase of vulcanism, which he considered contemporaneous with the dejection cones of the island.

The tannbuschite lava flows outcrop only near the coast, but there is no hint of consolidation under water. The flows vary in thickness from a fraction of a metre up to 4 m, have a massive appearance and include spheroidally weathered phonolite blocks. The flows were emitted in rapid succession, there being almost no time for the interbedded deposition of either pyroclastics of dejection cone deposits.

The rocks of the formation and their spatial relations indicate that the coastline rose sometime before the onsset of this volcanic phase, and a vent appeared through which very fluid lavas in moderate volume were extruded. The emission process corresponded to fire fountaining, with the projection of liquid masses brought about by the ascent of gas bubbles accompanying explosion. The lavas ascended through a fracture, probably parallel to the coast – one of the NW-trending systems. The duration of vulcanism was short, there being no time for the growth of talus deposits from the dejection cone over which the products of vulcanism were spread.

Paredao Volcano

This volcano is the eroded remnant of a structure which was rapidly and profoundly attacked by the sea. The crest of the crater rises some 217 m above sea level. Except in its NE part, where ankaratrite flows are exposed, the structure is formed of pyroclastics.

The pyroclastics chiefly comprise lapillitic tuffs, as blocks, bombs and driblets, associated with tuff-breccias and agglomerates. Ashy tuffs, ash and lapilli are also present. Blocks and bombs are in places highly vesiculated, and become scoriaceous where attacked by the waves. Dips of 40° are visible. The most spectacular exposures are seen in the raised cliffs in the vicinity of the tunnel entrance. Here is a prominent marine terrace, where resistant, quite well stratified beds of lapillitic tuffs, with enclosed bombs and blocks, pass locally into tuff-breccias and breccias, the transition being either gradual or abrupt. The stratification is plano-parallel, but locally may be cross-bedded. Tuff-ash beds my be almost 3 m thick and 1.5 km WNW of the crater edge beds 5 m thick are exposed.

The ankaratrite flows poured down towards the N and are exposed in the Tartarugas Bay area, both along the cliffs and up to 300 m inland. In general these lavas are highly vesiculated, become scoriaceous in the upper part of individual flows, favouring rapid weathering into brownish-yellow earthy material. W. VELTHEIM (1950) maintained that pillow-lava structure occurred in these lavas, but ALMEIDA disagreed, claiming this was merely a form of spheroidal weathering, which indeed did give an appearance superficially of pillow structure. The flows at Tartarugas beach dip at angles of 6–12° towards the sea.

The Paredao rocks display many large vertical joints which can extend the full height of the great walls. In plan, the joints show a radial pattern, presumably related to compressions within the vent. None of these joints were occupied by dykes. It is the close vertical spacing of the joints rather than ease of disintegration of the pyroclastics which enabled the sea to excavate the large tunnel opening here.

The volcano arose on a marine platform to the E of Picos Açucar, Vigia, Lourdes and Tartarugas. The base of the volcanic accumulations occurred near sea level on a terrace raised 3 m above the sea. This small initial elevation was rapidly built upon to form a large 'apron' to the E of these peaks. Paredao was built up as a scoria cone, with which the ankaratrites flows were associated. Indeed, most of the magma volume was expelled as ejectamenta in the manner of fire fountaining, raising a cone some 200 m in height, with a crater radius of some 300 m. The abundance of bombs, driblets, agglomerates and agglutinoids show that the magma, highly charged with gases, burst forth in clots which accumulated as a cone. As blocks are few, one would presume that the explosive action was not violent.

The dynamism of Paredao is presumed to have been of Strombolian type, which likely also applies to the Paracutin, Mexico eruption of 1943 and the Ilha Nova, Azores, eruption of 1957. Such volcanos may be active for weeks, months or years, but Paredao had a short life of activity, probably to be reckoned in terms of months, even weeks. This, the last volcanic phase of Trindade, probably occurred in post-Glacial times.

Geomorphology

Six morphologic units can be recignized in Trindade. 1. Central Plateau, in general above 300–350 m, formed of phonolite, grazinite and nephelinite flows with intercalated pyroclastics. Above the plateau rise phonolitic domes, separated by deep valleys. On the plateau is a median rise forming a water divide. 2. Coastal slopes, reaching from the edge of the Central Plateau down to the shorelines, with phonolitic necks and domes rising conspicuously above the steep declivities. 3. Crista de Galo, at the northern extremity, is really a narrow finger of the Central Plateau and also includes the slopes. From the high axis of the plateau, the land drops steeply down to the E and W in a series of concave-shaped scarps. A distinct NW grain is evident here, fractures, dykes and the topographic crest all trending in this direction. The one outstanding prominence is the inaccessible Obelisco. 4. Ankaratritic Plateau, in the E, is a concordant feature inclined towards the NE, descending from an elevation of some 500 m at Morro Vermelho, to 40 m, the topographic slope reflecting the dip of the lava flows. Several ravines drain directly down this surface to the coast, where ankaratrite cliffs front the sea. As previously mentioned, the upper part of the Vermelho valley is filled with pyroclastics of the Vermelho and Paredao formations. 5. Phonolitic necks and domes arc separate the above plateau from Paredao volcano to the E. Before the building of this volcanic pile, these necks and domes formed the eastern extremity of Trindade. 6. Paredao volcano, a notable morpho-structural volcanic feature, represents the ruins of the last volcanic episode in the island. Marine erosion has been chiefly responsible for the wearing away of this feature.

On the Central and Ankaratritic Plateaux, chemical decomposition has progressed much farther than mechanical disintegration (vd. Table 17, No. 28), resulting in a regolith of moderate thickness mantling the gentler slopes and undulations.

The scarped slopes surrounding the Central Plateau, the phonolitic peaks and Paredao have all evolved through the process of sudden, shock movements downhill of detritus. Whilst the majority of such movements occurred in the dry state, i. e. falls under the action of gravity, running water also played its role in the moving of material. Around nearly all steep necks and domes, aprons of talus deposits are spread out lower down, where irregular shaped, angular blocks of all sizes have moved inexorably downhill. During the 18th. century we read of the island inhabitants having to build strong walls around their dwellings as a protection against damage from falling blocks. Fig. 25 shows

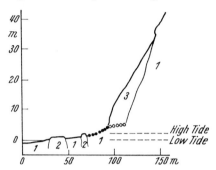

Fig. 25. Coverhead at Eme Beach, by Cachoeira Bay. 1. Tannbuschitic Pyroclastics, 2. Tannbuschite Dykes, 3. Talus Cone. ∘∘∘ Pebble deposits of old marine terrace-coverhead, ••• Pebbles of modern beach-coverhead. (ALMEIDA, 1961)

a composite scarp, the basal area formed of tannbuschitic pyroclastics and dykes of the abrasion platform. The 20 m thick talus deposits rest on marine boulders lying 3 m above the platform, the feature representing a typical 'coverhead', as defined by W. M. DAVIS (1933).

Active marine cutting back of the cliffs maintains steep, longitudinal profiles for the water courses. The valleys carry water only during the wet period, when they become veritable torrents, capable of moving rubble which has fallen down the lateral slopes under the action of gravity, and thenceforward tumbled down the valley by the full, fast-flowing streams. Distinction is not always easy between mass gravitational movements and torrential movements, for the relatively short distances of fluvial transport mean little rounding of detritus, which confusion is well exemplified by talus and alluvial cones. The constituent materials of these cones differ little from the pyroclastics and eruptives from which they originate, being merely the disintegration products thereof (vd. Table 17, No. 29). All alluvial deposits show two characteristics in common: growth is not taking place at present but rather throughout, marine and fluvial erosion is operative, and all show in their higher reaches, but not quite at the apices, interbedding of ultrabasic volcanics and pyroclastics. This would seem to indicate simultaneous formation of these adventitious talus and alluvial cones under morphological conditions lacking at the present time.

The relatively large extents and thicknesses of detritals accumulated towards the shores and today experiencing rapid erosion suggest that previously the island relief was such that mass movements down slopes and torrential erosion were together not sufficiently efficient to evacuate all materials right down to the coasts but deposited them on lower slopes adjacent to the shorelines. Progradation resulted in slopes undergoing abrupt decrease in inclination, with consequent deposition. As at present mass gravitational movements and torrent transport seem competent to move any size of detritus from the highest interiors right down to the coasts, it is assumed that in former times the climate was considerably drier than at present, thus reducing stream volumes, hence stream competency, and also denying a lubricating medium for gravitational movement.

Though aeolian phenomena are minor in importance compared to fluvial and marine, their evidences are present in Trindade. The NE slopes of Morro Vermelho, formed of

Paredao tuffs, are subjected to intense deflation, giving rise to small yardangs. Polishing and characteristic striation of ankaratritic flows can be seen in the Tartarugas region. At Tartarugas, Andrada and Cabritos beaches are sand accumulations of aeolian origin. At the first-named, the volume of such sand is impressive, rising to 45 m above sea level, and here a dune 120 m in length and ca. 10 m high lies at the base of Picos Tartarugas and Lourdes.

The young, irregular coastline is pounded by great waves having a very long 'fetch', with very stormy seas during winter months. The coast was carved principally by subaerial processes when sea level was lower than at present, but the narrow, discontinuous abrasion platform was sculptured at present sea level.

At no time was there vast alluvial deposition by streams, and the littoral is the result rather of current marine processes. Strong wave action on the shelving marine platform, with vigorous cliff recession, facilitated the formation of detritus which was carried seaward on to the platform. Thus beaches are relatively few, narrow, and generally of pebbles accumulated in privileged localities. The scale at which these processes operated naturally varied, and thus along the NE coastline scarps have receded farther, beaches are more frequent and voluminous, here the littoral is more evolved with sandy and pebbly beaches, dunes and reefs also are present. Elsewhere only at Principe Bay is the littoral well developed, although marine erosion has operated differently here. The more resistant phonolitic masses of Cinco Farilhoes and South Points, Morro Branco and Pao de Açucar, all rising steeply from deep waters, have caused significant wave deflexions. Between these masses, the tuffaceous coast is high, scarped, furnishing much detritus which is moved back and forth by the sea to form pebbles. The sand beach at Principe is an anachroism (Fig. 26), lying in front of a dead 40 m cliff, in fact, a sectioned large

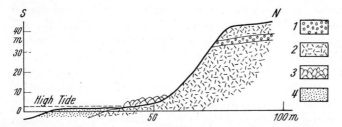

Fig. 26. Geologic profile at Principe Beach. 1. Valado Tuffs, 2. Alluvial Cone, 3. Talus, 4. Beach Sands. (ALMEIDA, 1961)

alluvial cone which once occupied a much larger area of the Bay. The constituent materials were particularly susceptible to marine attack, causing rapid recession of the cliffs and forming an abrasion platform which is exposed in front of the beach. The beach itself results from wave refraction, where beach drifting has occurred, causing sand accumulation, partially composed of calcareous coral fragments, although no reefs are in close proximity.

Deep waters inshore allow the waves to operate with maximum efficiency. However, in spite of this, the abrasion platforms are relatively poorly developed, which must surely testify to the recency of geological events in Trindade.

Between Calheta and Pico Tartarugas is an abrasion platform, exposed above low tide, of ankaratritic lavas upon which lie beach sands at an elevation of 3.2 m above present mean sea level. Lowering of sea level exposed this platform over which were deposited erosion products more than 4 m thick. The beach was formed by agitated seas during high tide bringing detritals from deeper water. This concept is confirmed at Andrada beach where calcareous-cemented arenites, including marine organic remains, lie between 2.8 and 3.5 m above present sea level. Beachrocks or arenites are generally considered to be exclusive formations of the strand, and thus those at Andrada and the more recent replica at Tartarugas beach point to a former sea level ca. 3.5 m above the present one. At Paredao, especially where the tunnel opening occurs, a marine terrace 3.5 m above present sea level also can be seen. ALMEIDA suggested that this terrace was probably formed some 3500–5000 years ago, but a figure of some 5000–7000 years ago seems more appropriate.

Some 290 soundings on the insular platform show this to lie at depths less than 91 m, and only seven soundings gave values over 100 m. BESNARD (1051) obtained 230 soundings over the platform, taken in a radial pattern. Most of these were in water deeper than 30 m, but in general the values were lower than those got by the Brazilian Navy. The lines of soundings extend outward from the coasts to a maximum of 2370 m. ALMEIDA shows 16 profiles taken by BESNARD, all in radial directions, from which the following comments can be made: The platform varies in width from 800–3000 m, with an area of approcimately 32 km². The average slopes vary between 3.7 % and 13 %, being broadest SW of the island, narrowest off the Pt. Valado-Pedra coastal area (Fig. 27). The edge of

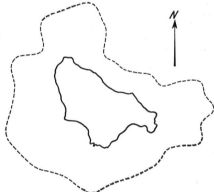

Fig. 27. Approximate position of edge of insular platform, at depth of 110 m. (ALMEIDA, 1961)

the platform occurs at ca. –110 m, from where slopes abruptly increase to 22 %, then gradually flatten out at depth (ALMEIDA, 1960). Platform profiles show breaks in slope, with concave slopes nearer shore, smoothing out into more gently inclined slopes to the platform edge. These abrupt changes most often occur at depths of 59 m and 77 m, corresponding to the lower border of two levels of submarine terraces (Fig. 28). Several profiles give a suggestion of a higher terrace at –47 m. Off Pt. Valado is an interesting profile (Fig. 29) where the Valado alluvial cone extends to the edge of the submarine platform at the –59 m terrace level. It can be deduced therefore that for such a cone to form, sea level must have been some 50 m lower than at present.

The submarine terraces at –47 m, – 59 m and –77 m off Trindade have close correlates in similar terraces on the continental platforms of the world which most frequently occur at depths of 29 m, 47.5 m, 62.5, 79.2, 91.5 m. This, would suggest that these Trindade terraces are of glacio-eustatic, epeirogenic or mixed origin. ALMEIDA believed these terraces were sculptured during oceanic volume fluctuations related to the glacial-interglacial episodes of the Pleistocene. The area of the insular platform is in keeping with the 100 000 odd years separating us from the onset of the last Würm glaciation.

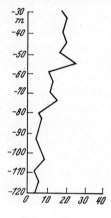

Fig. 28. Frequency curve of depths of the Trindade platform, after BESNARD (1951). (ALMEIDA, 1961)

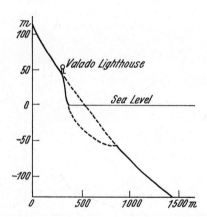

Fig. 29. Bathymetric profile in front of Valado Cone, after BESNARD (1951). (ALMEIDA, 1961)

Mineralogy

The mineral assemblage of the eruptives is characterized by deficiencies in silica and excess of sodium and aluminium. Of the feldspars the most notable fact is the very insignificant role of the plagioclases, actually only occurring in analcite-basanites. Potassic feldspars are usually cryptoperthitic sanidine-anorthoclase. Orthoclase is rare. Nepheline is common, often associated with a mineral of the sodalite group, nosean on occasion can be quite abundant, but usually no leucite is present. Magmatic analcite, playing the role of a feldspathoid, is the distinctive mineralogical feature of the Trindade eruptives. Amphiboles and pyroxenes occur in rocks rich in sodium, aluminium and titanium. Of the clinopyroxenes, diopside and pigeonite are lacking, as also the orthopyrocenes. Alkali-amphiboles likewise are not found. Biotite is present but always in limited quantities. Of the peridots, golden-yellow chrysolite is the common variety, but is absent in rocks rich in iron. Magmatic quartz is nowhere found.

A mineralogical classification of the Trindade eruptives is shown in Table 16.

Table 16   Mineralogical Classification of the Trindade Eruptives (ALMEIDA 1961)

Colour Index	With Feldspars		Without Feldspars	
	Alkaline Feldspars	Plagioclases	With Olivine	Without Olivine
0–30	Kali-Gauteites Tinguaites Phonolites Analcite-Phonolites Nosean-Phonolites			
30–60	Grazinites		Olivine-Nephelinites	Nephelinites Analcitites
60–90		Analcite-Basanites	Monchiquites Ankaratrites Analcite-Ankaratrites Tannbuschites Olivine-Analcitites	

Petrography

The first detailed petrographic studies of Trindade rocks were made by PRIOR (1900) who investigated the collection made by HOOKER during the short visit of the Ross Antarctic Expedition to the island in 1839. PRIOR's general descriptions of the topography and structure were based upon the notes of HOOKER and the writings of KNIGHT (1887, 1890), but according to ALMEIDA there are important errors here.

PRIOR recognized phonolites, phonolites-nephelinites, limburgites, agglomerates and what he presumed were sedimentaries. As we have seen, beach deposits, talus and aeolian sands occur, but no true sedimentary rocks are present in Trindade.

ALMEIDA made the first comprehensive study of the eruptives, and what follows is principally based upon his findings.

Phonolites are the commonest eruptives. Grazinites are intermidiate chemically and mineralogically between the most basic phonolites and the nephelinites. Ankaratrites show a special character.

In the descriptions which follow, the rock classifications of JOHANNSEN and TRÖGGER are followed.

Trindade Complex

1. Nepheline-Phonolites. The phonolites of Trindade are distinguished on structural grounds as well as the relative percentage of essential minerals. Three chief types can be recognized in the Complex: nepheline-phonolites (s. str.), nosean-phonolites and tinguaites.

Table 17 Chemical Analyses (ALMEIDA, 1961)

	1.	2.	3.	4.	5.	6.	7.	8.
SiO_2	50.6	51.0	51.0	45.0	46.8	39.6	38.8	39.0
TiO_2	0.88	0.30	0.50	1.82	2.06	3.00	3.60	2.50
Al_2O_3	21.3	20.25	19.41	19.9	19.0	15.25	14.53	16.05
Fe_2O_3	2.94	2.26	2.97	4.06	4.02	6.51	5.88	5.95
FeO	1.39	2.58	1.58	2.61	2.43	6.75	7.97	5.38
MnO	0.14	0.15	0.15	0.20	0.14	0.10	0.20	0.15
MgO	1.56	0.80	1.44	2.87	2.83	6.72	6.44	5.71
CaO	3.08	2.21	3.40	6.54	6.17	10.72	11.60	9.92
Na_2O	9.17	13.00	8.00	3.58	2.84	6.64	5.42	4.19
K_2O	7.95	5.00	5.26	6.02	6.26	0.54	0.61	1.32
H_2O+	1.31	0.89	1.82	4.56	3.25	1.50	2.35	2.75
H_2O-	0.04	0.55	2.30	3.20	3.08	1.80	1.35	3.85
P_2O_5	0.21	0.18	0.22	0.33	0.28	1.05	1.23	0.78
CO_2	0.20			0.16	1.40			
Ignit. Loss		1.00	1.98					1.40
Total	100.8	100.08	100.08	100.8	100.6	100.18	99.98	99.25*
S. G.	2.62	2.56	2.53	2.61	2.54	2.86	2.90	2.56

	9.	10.	11.	12.	13.	14.	15.	16.
SiO_2	40.60	37.60	40.08	41.00	40.00	37.60	38.60	37.40
TiO_2	3.30	2.50	2.30	2.00	3.20	2.60	2.50	2.50
Al_2O_3	17.03	14.41	15.67	15.55	15.67	10.75	15.09	18.58
Fe_2O_3	4.99	7.47	6.75	4.35	5.63	7.23	5.39	3.96
FeO	6.03	6.75	5.17	7.90	6.46	6.61	7.83	8.26
MnO	0.20	0.20	0.15	0.18	0.20	0.19	0.20	0.15
MgO	5.28	8.39	4.49	5.64	5.43	12.68	9.41	7.93
CaO	9.80	12.40	10.60	10.40	10.00	13.00	11.72	10.00
Na_2O	5.23	3.40	6.49	6.67	6.42	5.24	4.52	3.60
K_2O	0.71	1.19	1.35	2.78	3.39	1.46	1.56	1.90
H_2O+	3.15	2.30	3.40	0.50	1.65	0.60	1.20	1.35
H_2O-	2.85	1.20	2.20	0.50	1.15	1.10	1.00	0.65
P_2O_5	0.67	1.33	1.40	1.15	0.93	0.90	0.93	0.65
CO_2								
Ignit. Loss								2.30
Total	99.84	99.14	100.05	100.02	100.13	99.96	99.95	99.23
S. G.	2.75	3.08	2.84	2.96	2.87	3.01	3.01	2.81

1. Phonolite. Pico Pontudo.
2. Nosean-Phonolite. Dyke in cliff, Cabritos Beach.
3. Tinguaite. Dyke, between Cabritos Beach and Pt. Crista de Galo.
4. Kali-Gauteite. Escarpment to W of old settlement.
5. Kali-Gauteite. Dyke, between Cabritos Beach and Pt. Crista de Galo.
6. Analcite-Basanite. Dyke, NE of Monumento, on Central Plateau.
7. Olivine-Analcitite. Dyke, N coast of Portugueses Bay.
8. Analcitite. Dyke, scarp to SW of Hill 345.
9. Analcitite. Dyke, Central Plateau, before Monumento.
10. Olivine-Nephelinite. Dyke, Vale Verde.
11. Nephelinite. Dyke, scarp to W of old settlement.
12. Nephelinite. Dyke, Cabritos Beach.
13. Nephelinite. Dyke, between Cabritos Beach and Pt. Crista de Galo.
14. Tannbuschite. Valado formation.
15. Tannbuschite. Trindade Complex.
16. Monchiquite. Dyke, near Pt. Crista de Galo.

Table 17 — Chemical Analyses (Almeida, 1961)

	17.	18.	19.	20.	21.	22.	23.	24.
SiO_2	49.40	49.40	44.80	37.40	37.12	37.52	38.00	39.00
TiO_2	0.80	0.80	1.60	3.40	3.20	3.60	2.50	3.60
Al_2O_3	20.32	21.16	17.76	14.29	12.22	12.46	12.99	11.86
Fe_2O_3	3.06	3.29	5.55	7.71	5.47	6.15	5.48	6.20
FeO	2.51	1.58	3.81	6.25	7.18	8.76	8.33	9.55
MnO	0.15	0.15	0.20	0.25	0.20	0.20	0.20	0.19
MgO	1.40	1.52	3.47	8.50	11.36	11.52	12.16	12.31
CaO	3.60	3.52	7.80	11.60	12.40	11.68	11.60	10.40
Na_2O	11.35	8.16	6.87	4.06	6.75	3.37	4.47	3.68
K_2O	5.30	2.75	3.87	0.70	0.61	0.05	2.48	1.80
H_2O+	0.85	2.30	1.10	2.47	1.07	1.70	0.27	0.50
H_2O-	1.15	3.10	2.30	1.83	1.33	1.60	0.11	0.30
P_2O_5	0.29	0.25	0.88	1.42	1.34	0.86	0.79	0.55
CO_2								
Ignit. Loss		1.40						
Total	100.18	99.38	100.01	99.88	100.25	99.92	99.38	99.94
S. G.	2.62	2.47	2.73	2.81	2.97	2.94	2.96	3.12

	25.	26.	27.	28.	29.
SiO_2	49.80	38.30	29.96	21.5	33.0
TiO_2	0.40	4.19	3.00	1.26	2.83
Al_2O_3	22.84	14.80	11.63	39.0	26.9
Fe_2O_3	2.40	5.37	12.52	6.43	9.2
FeO	0.50	6.95	8.47		0.68
MnO	0.10	0.23	0.30		0.13
MgO	1.01	6.85	7.38		0.6
CaO	3.00	12.06	17.12	Tr.	1.5
Na_2O	8.38	2.80	2.67	0.01	1.04
K_2O	5.30	3.81	0.10	6.02	1.85
H_2O+	3.93	3.19	1.20	15.18	10.37
H_2O-	2.37	0.52	1.60	11.98	9.17
P_2O_5	0.10	0.62	2.15	7.8	3.01
CO_2					
Ignit. Loss			1.54		
Total	100.13	99.70	99.97**	99.2	100.2
S. G.	2.42	3.03			

* Includes: Cl 0.30, SO_3 0.04.   ** Includes: Cl 0.30, SO_3 0.03.

17. Phonolite. Morro Desejado.
18. Analcite. Phonolite. W slope of Morro Desejado.
19. Grazinite. Upper valley of Vermelho stream.
20. Nephelinite. Central Plateau, in front of Monumento.
21. Analcite-Ankaratrite. Calheta Beach.
22. Analcite-Ankaratrite. Dyke, Vermelho valley.
23. Ankaratrite. Paredao volcano.
24. Ankaratrite. Paredao volcano.
25. Nepheline-Syenite. Xenolith in phonolite, Pico Vigia.
26. Biotite-Melteigite. Ejectile in phonolitic tuff, Principe Beach.
27. Pyroxenite. Ejectile in tuff, Paredao volcano.
28. Decomposed Phonolite. Summit of Pico Trindade.
29. Breccia Cement. Covering of alluvial cone, Portugueses Beach.

Nepheline-phonolites are characterized by the quantity and size of the phenocrysts, the rocks in nearly all necks being strongly porphyritic. The phenocrysts are alkali-feldspars, along with nepheline and nosean. All specimens show abundant pyroxene, with which is associated a small quantity of hornblende and biotite. In thin-section they are holocrystalline, non-hyatal porphyritic texture with many microphenocrysts. When much feldspar is present, the rocks acquire a trachytoid or trachytic texture or then become tinguatic when the nepheline, like the feldspar, shows saccharoidal structure with prisms of soda-augite arranged without any orientation or pattern. The alkali-feldspars are cryptoperthitic varieties of sanidine-anorthoclase, present as phenocrysts in a microlitic matrix. The phenocrysts may contain inclusions of all minerals which preceded them in crystallization. Zeolitization is a common method of alteration, analcite and natrolite forming. No anorthoclase or plagioclases have been identified. The largest phenocrysts are of nepheline and also is always present in the groundmass, being often altered into fibrous zeolites. Nosean is universally present as microphenocrysts but seldom is present in the matrix. The clino-pyroxenes form essential minerals and are the most common and abundant of the Fe-Mg silicates. Within the matrix, augite varieties are commonest – titanaugite and soda-augite – but aegerine and acmite are lacking. Hornblende occurs as long prisms and is always more abundant than biotite. Secondary products include ferric oxides, calcite, natrolite and thomsonite.

The fine degree of granulation of the matrix content prevents a determination of the mineral content, but a partial analysis of a rock from Pico Tartarugas showed: microlitic matrix 79%, nepheline 9%, zeolitized nosean 7%, sandine 5%, titanaugite and augite less than 1%. A chemical analysis is shown in Table 17, No. 1.

2. Nosean-Phonolite. The principal occurrence is a small plug on the scarped coast near Principe beach. The rock possesses many xenoliths giving it the appearance of a volcanic breccia, but in the interior of the rock there are few xenoliths. The rock has a macroporphyritic appearance, with large sanidine crystals. Under the microscope many microphenocrysts of nosean are seen, with lesser quantities of soda-augite, nepheline, titanite, and exceptionally, hornblende and biotite. The groundmass is a fine mesh comprising sanidine, nepheline, aegerine-augite, nosean, titanaugite and apatite, in order of abundance. The non-oriented arrangement of the pyroxene prisms and the xenomorphic-granular character results in a tinguaitic texture. At Cabritos beach is a 15–20 m thick dyke of the rock which shows neither nepheline nor pyroxene microphenocrysts but on the other hand has many microphenocrysts of a sodalite group mineral completely zeolotized. The uniformly fine, granular matrix has microlites of nepheline, sanidine, aegerine-augite, a sodalite group mineral, titanite and apatite. A chemical analysis is given in No. 2.

The chemical analyses given in Nos. 1 and 2 show that the Trindade phonolites are particularly poor in silica and rich in alkalis.

3. Tinguaites. A specimen from Crista de Galo Pt. shows many phenocrysts of feldspar and nepheline to the naked eye. In thin-section there occur microphenocrysts of soda-augite, titanite and a zeolitized mineral of the sodalite group. Magnetite grains clearly show reabsorption of hornblende. The mesostasis in which these minerals occur is a fine, xenomorphic-granular mosaic with abundant laths of sanidine, the whole showing a trachytoid character. A partial modal composition showed: phenocrysts of nepheline 9%, soda-augite 8%, sanidine 5%, altered 'sodalite' 1.5%, magnetite, titanite, apatite 0.5%, tinguaitic matrix 73%, amygdales with zeolites 3%. No. 3 is a chemical analysis.

4. **Kali-Gauteites.** On the trail from Cabritos beach to Crista de Galo Pt. is a thick dyke which is highly vesiculated and filled with small crystals of analcite. The rock, showing clear fluidal structure, is porphyritic, with detachable phenocrysts of feldspar and hornblende Microscopically the specimen is seen to be a kali-gauteite, such as occur in Fernando de Noronha (q. v.). The rock is mostly holocrystalline, with a small amorphous area partly devitrified. Soda-augite microphenocrysts are abundant, and in lesser quantity occur a sodalite mineral, biotite, titanite, magnetite and apatite. In the mesostasis are many small crystals of perovskite. The feldspar phenocrysts are cryptoperthitic sanidine-anorthoclase, with hornblende much less prominent. The soda-augite varieties are identical with those of the nephelinites and grazinites. Biotite is present only in small amounts as minute, corroded microphenocrysts. Magnetite is the prominent accessory. In the vesicles are perfect crystals of analcite. The chemical analyses shown in Nos. 4 and 5 indicate similarities to the rocks of Fernando de Noronha, but the Trindade specimens have more K_2O and less Na_2O.

5. **Analcite-Basanites.** Four such dykes intruding Complex phonolites and pyroclastics are known, and the special interest here is that they are the only eruptives in the island having plagioclase. The rocks are aphanitic or then have rare, small pyroxene phenocrysts. In thin-section they are holocrystalline, of microphyric texture with many microphenocrysts of augite, fewer chrysolite crystals. The matrix is largely microlitic, showing augite, analcite, plagioclase, magnetite, biotite, apatite and titanite, in decreasing abundance. In the dykes higher up in the interior of Trindade, NE of Monumento and on the slopes of Crista de Galo, occur many pseudomorphs of hornblende phenocrysts almost entirely comprising augite, magnetite, biotite, ilmenite and labradorite. Olivine is present as automorphic crystals or then partially or totally corroded. Except in the dyke NE of Monumento, the olivine is usually more or less altered into serpentinous and ferric oxide material. The plagioclase is a variety of labradorite near andesine, is abundant as twins, is polysyntectically xenomorphic and is only present in the groundmass. The order of crystallization seems to be: olivine, augite, biotite, plagioclase and analcite. Apatite and magnetite are partly of secondary origin. A chemical analysis given in No. 6 from the dyke NE of Monumento shows chemical identity with other ultrabasic analcitites of the island but the tenor in Na_2O is higher than in the olivine-analcities, see below.

6. **Olivine-Analcitites.** A dyke rock some 400 m inland from Portugueses beach was studied which showed visible crystals of augite hornblende and biotite in an aphanitic matrix. Under the microscope it is holocrystalline with phenocrysts of the above minerals plus olivine in a groundmass of analcite crystals and abundant microcrystals of augite, magnetite, biotite and apatite. Pyroxene shows typical hour-glass structure and is a basaltic augite passing into titanaugite around the borders. Titano-biotite is partially reabsorbed at the rims and changes into augite, magnetite and ilmenite. The olivine is a colourless variety of chrysolite, partly serpentinized. The outstanding mineral here is analcite occurring as abundant large globules, full of inclusions. No traces were seen of

either feldspars or feldspathoids. The rocks show a lamprophyric character but is distinguished from the monchiquites on a chemical and mineralogical basis. ALMEIDA preferred to name the rock olivine-analcitite, analagous to the olivine-nephelinites but is distinguished from the analcite-basalts and nepheline-basalts in that feldspars are lacking. A chemical analysis is shown in No. 7.

7. Analcitites. With the disappearance of olivine and the greater abundance of analcite the rocks become analcitites. These are massive, quasi-aphanitic rocks, with rare, small phenocrysts of pyroxene. Microscopically they are seen to be holocrystalline, microphyric, in which microlites and small crystals of augite, analcite, biotite, magnetite, apatite and one or two amphiboles and an odd sodalitic mineral are scattered within a clear, isotropic matrix identified as analcite. Analcite comprises ca. 15 % of the rock. The groundmass has a spongy appearance, similar to the texture seen in leucitic rocks of the Roman magmatic province (H. S. WASHINGTON, 1906). The trapezoidal subautomorphic crystals of analcite, which give the rock its distinctive character, contrary to what is generally the case in Trindade rocks, are all but completely lacking in inclusions, but magnetite may occasionally occur. The pyroxene is basaltic augite passing into soda-augite. Biotite forms small microphenocrysts of automorphic crystals. Small amygdales of natrolite with analcite crystals projecting into the cavities are common. There is little doubt that the analcite crystallized as a primary component. A chemical analyses from a thin dyke high up in the central plateau is given in No. 8. In this vicinity there also occur dyke rocks which at first may be taken for analcitites but on closer inspection are really augitites. No. 9 is from a dyke NE of Monumento. In these analcitites it is difficult to prove the presence of interstitial nepheline. The term analcitie adopted here refers to a rock having augite and analcite as essentials but lacking in feldspars and feldspathoids. These rocks have a panidiomorphic tendency like the lamprophyres. ALMEIDA thought that the magma was almost entirely in a liquid state during intrusion, and for this reason he excluded the rocks from the pyrocenitic furchites.

8. Olivine-Nephelinites. A thick dyke intruding Complex pyroclastics in the lower Vale Verde consists of a porphyritic rock with large phenocrysts of biotite and less augite. Microscopically it corresponds to an olivine-nephelinite. The grain size is medium-fine with microphenocrysts of augite, biotite, hornblende and olivine. In thin-section microphenocrysts of nosean (?) and nepheline occur. The texture inclines towards panidiomorphism and the rock has a lamprophyric appearance. The groundmass comprises augite, magnetite, biotite, primary analcite, a sodalite group mineral, sanidine, titanite and apatite, all of which are included poikilitically in nepheline, analcite and zeolitic products of deuteritic alteration, especially natrolite. Olivine occurs only as microphenocrysts, fragmented or corroded in appearance. The sequence of crystallization appears to be: olivine, magnetite, apatite and titanite. A chemical analysis is given in No. 10. This is the most basic of the Trindade nephelinites, but has more magnesia, less alkalis than the others. The texture is unfavourable for modal composition determination, but from one sample, the following was ascertained: zeolitic matrix 42.3 % 'sodalite' 1.2 %, sanidine 3.2 %, augite and soda-augite 35.5 %, biotite 3.5 %, hornblende 2.3 %, olivine 2.2 %, magnetite 8.5 %, apatite 1.3 %. Hematite and iron pyrites are also traceable.

9. **Nephelinites** are aphanitic or then show small phenocrysts of pyroxene. After the phonolites, these are the most abundant of the Trindade eruptives. A study of 23 thinsections by ALMEIDA showed almost identical microscopic characters. The rocks are holocrystalline porphyritic, with phenocrysts and microphenocrysts of augite and hornblende. There may or may not be nosean, and nepheline and biotite are rare. Occasionally magnetite, apatite or titanite may also occur as phenocrysts. The presence of olivine establishes the transition to olivine-nephelinites. Natrolite is the commonest mineral of the amygdales. Nepheline and a sodalite group mineral when present as phenocrysts are the most conspicuous components of the rocks. The mineral presumed to be nosean is entirely zeolitized and has inclusions of presumed magnetite. The granular matrix is partly microlitic, composed chiefly of augite, magnetite, nepheline mosaic. Chemically the commonest types are shown in Nos. 11, 12 and 13, with No. 12 the most typical. These basic rocks are relatively poor in silica and magnesia, rich in the alkalis, and thus similar to the etindites of Cameroun, West Africa.

10. **Tannbuschites.** These melanocratic nepheline-basalts have principal occurrences in the Complex but also are present in the Valado sequence. The rocks are generally aphanitic or then have small crystals of olivine and pyroxene, and show a more uniform petrographic and structural character amongst themselves than most other eruptives of the island. The simple mineral content comprises augite, olivine, magnetite and nepheline, along with accessories and deuteritic products. The rocks are holocrystalline, porphyritic, with small phenocrysts of olivine and augite in a microlitic mesostasis which poikilitically comprises nepheline or its deuteritic product. Olivine usually present as phenocrysts, but when less abundant, may then occur as small grains in the matrix. The modal composition can not be determined, but the microphenocrysts comprise 22 % chrysolite and 13 % augite in one sample. No. 14 is a chemical analysis of a Valado specimen, and No. 15 one from the Complex.

11. **Monchiquites.** At the N and of Cabritos beach is a kali-gauteite dyke in turn intruded by a dark dyke rock carrying xenoliths and small phenocrysts of pyroxene. Microscopically it shows a panidiomorphic, porphyritic texture, with many phenocrysts of basaltic augite and olivine. There also occur tiny prisms and microlites of amphibole, octahedra of magnetite and a little apatite. The minerals are embedded in a darkish, isotropic groundmass. Olivine is a variety of magnesian chrysolite, partially altered into an indeterminate greenish substance. The amphibole is brown hornblende. The isotropic mesostasis appears to be a glass complete with globulites and crystallites, amongst which natrolite does not seem to be present. Some deuteritic carbonate material is also present. No. 16 is a chemical analysis of this dyke occurrence, showing a high tenor of alumina and a low silica content compared to the monchiquites of the classical Caldas de Monchique area of Portugal. Chemically there is little to distinguish this rock from the analcitites, but in a mineral basis and its porphyritic character, this is a difference.

Desejado Sequence

1. **Phonolites.** These differ from the Complex phonolites even megascopically, for they have a massive, rather glassy appearance, are entirely aphanitic or than show rare, small

feldspar and amphibole phenocrysts. Conchoidal fracturing is common. In thin-section are microphyric, holocrystalline, with a trachytic arrangement of microlitic material of the mesostasis. The microphenocrysts are soda-augite, hornblende, nosean, titanite, apatite and biotite, in order of abundance and frequency. On occasion there may be nepheline crystals also some sanidine, but these are rare as larger microphenocrysts. Titanohornblende and biotite are rare. Apatite, magnetite and titanite occur in the matrix. The mesostasis is packed with microphenocrysts, largely of xenomorphic-granular particles of sanidine and nepheline. The chemical analysis (No. 17) of one of these phonolites shows it to be similar to those of the Complex.

2. **Analcitic-Phonolites** are massive, aphanitic rocks usually having amygdales near the tops and bottoms of flows. Microscopically they are microphyric, with microphenocrysts of soda-augite, nosean, hornblende, sanidine, nepheline and titanite. The matrix shows fine granulation, comprising trapezoids of analcite which give a granulose structure to the rocks. Between the analcite of the matrix occur laths of potassic feldspar, nepheline granules and clinopyroxene prisms. Soda-augite is transitional to aegerine-augite. The chemical analysis (No. 18) shows the same general characters as other phonolites of Trindade, but with a relatively high combined water content, reflecting the presence of analcite.

3. **Grazinites** from aphanitic flows up to 70 m thick, with vesicular and amygdaloidal exteriors. In thin-section they show some features not seen in other soda-alkalis in other similar provinces. The rocks are holocrystalline, microphyric, with small phenocrysts of augite, hornblende, biotite, nepheline and a sodalite group mineral, in a microlitic mesostasis where augite and small magnetite crystals are poikilitically included in sanidine. Apatite prisms are abundant and within the matrix are many trapezoids of zeolitic aggregates representing magmatic analcite. Olivine is lacking. Soda-augite is the most abundant mineral, passing into basaltic augite. Biotite is strongly pleochroic and may be deeply altered. Grazinite is a type of mesocratic phonolite, transitional between phonolites and nephelinites of the island. A chemical analysis is shown in No. 19. The mineral oddities of the rock have a special chemistry in which the relatively low tenor in silica and magnesia is associated with large quantities of alumina, lime and the alkalis. These rocks approcimate the murites described by A. Lacroix from Raratonga, Cook Islands, but differ in having both primary and essential analcite and no olivine (1927).

4. **Nephelinites** are dark and massive in appearance or then may be amygdaloidal. Microscopically are seen to be holocrystalline, of microphyric texture, with microphenocrysts of augite, hornblende and an unidentifiable, completely zeolitized mineral of the sodalite group. Nepheline is rare. The matrix comprises microlites of augite, apatite, magnetite, all poikilitically included. The presence of alkali-feldspars establishes the transition to grazinites. Basaltic augite is the dominant mineral, passing into soda-augite. The chemical analysis (No. 20) is typical of these Desejado rocks which are more basic than those of the Complex, but on the other hand, show close analogies with the Complex olivine-nephelinites.

Morro Vermelho Formation

Analcite-Ankaratrites. The rocks are generally aphanitic, with only very small crystals of olivine. In thin-section they are idintified as soda-alkali basalts with biotite, nepheline and analcite, akin to ankaratrites. They are usually holocrystalline with a porphyritic texture, containing many microphenocrysts of olivine in a microlitic mesostasis containing analcite, clinopyroxene, apatite and magnetite. Olivine is the variety chrysolite, occurring as automorphic crystals. Basaltic augite, passing into titano-augite, occurs as small prisms. Very small crystals of magnetite are abundant, and some apatite usually is present. Analcite is the most characteristic mineral, occurring as small and abundant clear granules and as microlitic inclusions. The order of crystallization appears to be: olivine, accompanied by a little magnetite when the lava reached the surface, followed by magnetite, then pyroxene, accompanied by a little apatite. Biotite, analcite and nepheline, in this order, formed the late crystallization phase. The chemical analysis (No. 21) of a specimen from Calheta beach, shows a relatively high amount of soda but otherwise is very similar to the ankaratrites elsewhere in the island, being essentially a variety which crystallized in the presence of much water. No. 22 is a specimen from a dyke rock in the Vermelho valley.

Valado Formation

Tannbuschites are entirely aphanitic, massive or then with small vesicles and amygdales. Microscopically the rocks are very similar to those of the Complex and only the differences will be referred to here. The Valado tannbuschites have less chrysolite and they are not present as microphenocrysts but as granular filament material of the matrix. Another feature is that small granules of perovskite are present in the matrix. A partial modal composition of a specimen, referring only to the phenocrysts showed: olivine 14 %, augite, about 4 %. A chemical analysis is given in No. 14.

Paredao Volcano

Ankaratrites cannot be distinguished macroscopically from the analcite-ankaratrites of the Vermelho Formation. In thin-section they are holocrystalline, even in the thinnest flows, and vitreous material occurs in the pyroclastic components derived from these rocks. The texture is porphyritic and microphyric, with phenocrysts especially of olivine. The mesostasis is microlitic, containing titanaugite as the dominant mineral, much magnetite and very little apatite. These minerals are present also in the matrix along with biotite and nepheline, with analcitization common. Olivine is the same magnesian chrysolite variety as in the Vermelho flows. In the matrix the characteristic pyroxene is titanaugite. Biotite is the lepidomelane variety, possibly titaniferous. Nepheline may be unaltered or then analcitized. The order of crystallization would appear to be the same as in the analcite-ankaratrites, but here analcite is the only deuteritic mineral. The rocks probably contain between 14–16 % olivine. From the chemical analyses (Nos. 23 and 24) it is seen that they differ from the Vermelho lavas in having more K_2O, reflected in the greater quantity of biotite, but are poorer in soda.

Xenoliths and Ejectiles

No plutonics outcrop in Trindade, such as occur in islands where more profound erosion has laid bear deeper levels in the volcanic structures. The xenoliths are not of equal frequency in all the eruptives; ankaratritic flows usually have them, phonolites and other lavas usually to varying degrees, but the Paredao lavas show much less. Nearly all the xenolithic types also occur as ejectiles in the pyroclastics of the Complex and the Desejado.

Xenoliths and ejectiles can be petrographically classed into four groups: 1. of known in situ eruptives, 2. nepheline-syenites, 3. melteigites, and 4. perkinites.

1. Nepheline-Syenites occur in the Vigia and Pontudo necks, in tannbuschite dykes outcropping at low-tide at Cachoeira beach, in phonolite blocks of unknown origin on the slopes behind Cabritos beach, and in breccias of the scarps behind Principe beach. Here mention is made of the Vigia occurrence. The rock is coarse-grained with tabular crystals of feldspar. The xenolith fragment is of angular shape, with no magmatic reaction rims at the perimeters. In thin-section shows more than three-quarters content of alkali-feldspars, nepheline and a zeolitized mineral of the sodalite group. The feldspar is probably sanidine-anorthoclase, likely representing the sanidization of primary soda-orthoclase of the fragment during its inclusion in the lava which brought it to the surface. This feldspar comprises more than half the rock. Nepheline is lagely altered to analcite and fibrous zeolites, especially natrolite. As accessories are small amounts of aegerine, biotite and hornblende, in order of abundance. The texture is hyautomorphic-granular, with automorphic evolution of the accessories and the feldspars. No. 25 is a chemical analysis. In comparison with nepheline-syenites in general, this specimen seems poorer in silica and richer in lime and magnesia, which is reflected in the high quantity of pyroxenes and nepheline. On the other hand, the specimen shows close similarity with the phonolites of Nos. 1, 2 and 3, and almost certainly is to be regarded as a granular facies of phonolite.

2. Biotite-Melteigites. Some xenoliths and ejectiles are like the melteigites described from Norway by W. C. Brogger (1920). All Trindade specimens have biotite as an essential mineral, and are thus termed biotite-melteigites. There are no indications of feldspars and nepheline is always subordinate to the mafic minerals. All rocks have green pyroxene, biotite and nepheline, with usually large quantities of magnetite, titanite and apatite as accessories. A small quantity of hornblende is usually present. Most frequently the nepheline is unaltered or then may be transformed partially into analcite and cancrinite. The abundance of titanite and apatite is a notable feature of the rocks, being in large part epigenetic minerals. No. 26 is a chemical analysis of specimen. The modal composition of the rock shows: biotite 40.2%, soda-augite 29.0%, nepheline 23.3%, titanite 5.1%, apatite 2.1%, magnetite 0.3%.

3. Perkinites. Many dark inclusions in the Complex and Desejado units are seen microscopically to be perkinites, granular rocks with only mafic minerals as essentials. Large aggregates of crystals of green pyroxene, titano-biotite, titano-hornblende, titanite, magnetite and apatite characterize the rock. They are unusually rich in magnetite, apatite and titanite, but pyroxene and biotite are the dominant minerals. The rocks approximate the biotite-melteigites, but have neither nepheline nor feldspathoids, and are essentially biotite-pyrocenites, similar to the bebedourites of W. E. Tröger (1935). Biotite is always much reabsorbed – titanite less so – with peripheral alteration into ilmenite. In the titanites are small inclusions of perovskite. Small quantities of zeolitic material perhaps represents original nepheline. The modal composition is: aegerine-augite 61.7%, titano-magnetite 12.1%, biotite 8.1%, titanite 9.5%, apatite 3.8%, zeolitic filling 4.8%. There also occur perkinites in which biotite or hornblende predominate over pyroxene, both showing much reabsorption and alteration into augite and magnetite. These are the coarsest rocks in Trindade, with hornblende crystals up to 8 cm in length.

There also occur in Trindade hornblendite xenoliths and ejectiles in which the amphibole may be up to 7 cm in length, the rocks having a pegmatitic appearance.

With the disappearance of biotite, these rocks become true jacupiranguites, formed only of pyroxene, magnetite, titanite and apatite. The modal composition of one of these rocks rich in titanite and apatite is: aegerine-augite 58%, titanite 19%, titano-magnetite 11%, apatite 12%.

The extreme simplicity mineralogically of the Trindade xenoliths is well represented by masses of titano-magnetite, up to 5 cm in length, seen in the island tuffs and phonolites. A chemical analysis of one of these masses gave the following: TiO_2 16.80%, Fe_2O_3 33.19%, FeO 35.79%, which would indicate a xenolith with ca. 80% titano-magnetite (approx. 48% magnetite, 32% ilmenite). There also occur similar petrographic aggregates of magnetite originating from the segregation of extreme differentiation of gabbroic magmas or alkali magmas crystallized under plutonic conditions, which were termed kiirunavaarites by F. Rinne (1921).

The xenoliths mentioned above are in disequilibrium with the original liquid of the rock in which they are enclosed. Such disequilibrium causes important processes of magmatic reabsorption, which modifies the xenolithic minerals and also the composition of the host rock. Crystals of hornblende and biotite are especially subjected to modification, to the point that they are completely changed into pyroxene and magnetite. Almeida believed that probably many 'phenocrysts' of the basic eruptives of Trindade, particularly those of amphibole and biotite, were true xenocrystals, liberated by desegregation of melanocratic inclusions.

4. **Olivinites.** Xenolithic material shows wide variation in the relative percentages of soda-augite, chrysolite and titano-biotite. Some of this material comprises olivinites and even dunites when olivine alone is present. Olivinites have a holocrystalline, xenomorphic-granular texture. When there is a high quantity of pyroxene, this is dispersed between chrysolite grains, substituting for them peripherally along the cleavages and fractures. Associated with the pyroxene are small grains of magnetite.

5. **Peridotites and Pyroxenites.** When the percentage of pyroxene rises to ca. 50%, olivinites pass into peridotites. This increase is accompanied by the association of xenomorphic crystals of biotite which substitute for pyroxene. Only one sample studied showed substitution of soda-augite by a variety of titano-hornblende, but precise determination was not possible. A chemical analysis is shown in No. 27 of a biotite-pyrocenite in which ca. three-quarters of the rock comprises soda-augite, the rest being mostly titanite, titano-magnetite and apatite.

6. Glimerites. At the western entrance to the Paredao tunnel occur curious fragments in the tuffs which have biotite as the sole essential mineral. The rock has a xenomorphic texture, the structure so orientated so as to give a biotite-crystal appearance. In thin-section the rock is formed of a uniaxial type of titano-biotite, with strong pleochroism, and is thought to be some variety of biotite which substitutes for augite in the pyroxenites and peridotites. Chrysolite constitutes no more than 5 % of the rock. It thus appears that in the xenoliths and ejectiles of the Paredao volcano are to be found the same dark minerals as in the ankaratritic lavas, but devoid of pyroxene. It is certain that these rocks crystallized from the same magma as gave origin to the lavas, being largely modifications of the metasomatic processes which caused the olivine to be substituted by soda-augite, and this in turn by titano-biotite to give glimerites.

Pyroclastics

The most basic rocks of Trindade, the ankaratrites and tannbuschites, give rise to vitric tuffs, formed principally of fragments of vitreous or hypocrystalline lava, containing phenocrysts of olivine and, more rarely in the tannbuschites, also pyroxene and microites of augite, biotite, magnetite and large quantities of chrysolite. The degree of devitrification of these fragments, at least those investigated, is generally not great. These vitric tuffs always show microscopic or macroscopic vesicles and amygdales, filled with natrolite or other unidentifiable zeolites. No palagonite or congenital transformation products have been noted. Amongst the glass fragments of the basic tuffs, other constituents of holocrystalline rocks of respective lavas can be identified. Field study shows many isolated crystals in all the ultrabasic pyroclastics, comprising unaltered crystals of augite, biotite, olivine or titano-magnetite, which separated from the magma when liquid during explosive action.

Microscopic examination of nephelinitic or grazinitic tuffs of the Desejado sequence show high proportions of phonolite fragments, as well as pyroxene crystals. Vitreous fragments show the same features as the above, and again no palagonite can be detected. Most of the Complex tuffs are of phonolitic type, being lithic tuffs, but often full of isolated crystal fragments. Viterous tuffs are subordinate, except in the more basic varieties of rocks.

The matrix and cement of the phonolitic tuffs is almost always of zeolitic material, and natrolite is certainly present, very possibly analcite and chalcedony.

Petrochemistry

The petrochemical characteristics of the Trindade rocks can be appreciated from Tables 17, 18 and 19. Fig. 30 is a variation diagram showing the positive correlation of the numerical values. Study of the above shows that the rocks of Trindade represent a continuous sequence of consanguinous terms which have evolved as a result of magmatic differentiation.

The fm values in the diagram fall progressively and rapidly with increase in silica. Al and alk increase simultaneously. The left-hand side of the diagram represents the ultrabasics, where it is seen that the c values increase first to a maximum in the nephelinite

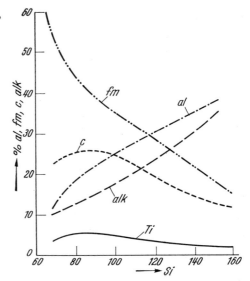

Fig. 30. Variation of Niggli values. (ALMEIDA, 1961)

group before decreasing with increasing silica. The k parameter (Table 19) rises above 0.4 in two cases – in the kali-gauteites which have no nepheline but much potassic feldspar, and in the xenolithic melteigite which has 40% biotite. In the true phonolites, k is relatively low. Si varies between 66 and 160, this restricted interval denoting poverty in silica. The rise of the al and alk parameters, in conjunction with this poverty in silica, is a consequence of extreme subsaturation, with negative values of qz not greater than –28. The equivalent curves for fm and al intercept the si abscissa at 119, thus defining a low isophalia. With the exception of the most basic ankaratrites and tannbuschite rocks, all other rocks show quite high tenors of iron relative to magnesium, with mg parameters not greater than 0.55. The ti curve parallels the the c curve, naturally with less variation.

It may be concluded therefore that the Trindade eruptives are characterized by silica poverty, richness in alumina and soda, relatively poor in potassium and the calc-alkalis.

ALMEIDA compiled a table showing the magmatic parameters of the Atlantic alkaline series, including the islands of Trindade, Fernando de Noronha, S. Tomé, Principe, Cabo Verde, Canaries, Madeira, Azores, Ascension, St. Helena, Tristan da Cunha and Gough, using published petrographic data available to him then. He adopted the parameters al, fm, c, alk, qz, al-alk and c-(al-alk) with increasing values of si. Such a table allows a classification of the Atlantic series into two groups: 1. the hyperalkaline series, including the following islands, with the isophalia rating in brackets: Trindade (119), S. Tomé-Principe (127), Cabo Verde (128), Fernando de Noronha (129), Canaries (133). 2. the mioalkaline series, linking the islands of the Mid-Atlantic Ridge, thus: Madeira (143), Tristan (144), Gough (158), Azores (158), St. Helena (158) and Ascension (192). These groupings of the Atlantic islands show not only petrochemical but also mineralogical and petrographic differences, as well as tectonic differences. The Canaries are actually somewhat transitional between the two groups, with a tendency towards hyperalkalinity.

Fig. 31 is based upon data supplied by ALMEIDA, from which it can be determined that an isophalia of 127 represents an average for the hyperalkaline group and 157 for the

Table 18 Molecular Norms (ALMEIDA, 1961)

	1.	2.	3.	4.	5.	6.	7.	8.	9.
Or	19.5	28.0	32.0	38.0	39.5	3.0	3.5	8.5	4.0
Ab		7.6	22.5	2.0	9.5	8.0	6.7	10.0	17.7
An			1.7	21.3	22.5	10.0	14.0	23.3	22.3
Ne	39.6	41.0	30.0	19.5	10.8	32.0	26.4	18.6	19.0
Le	20.0								
Kp									
Ac	7.8	6.0							
Ns	0.8	8.1							
Ak									
Fo				3.1	3.4				
Wo	0.7		1.5						
En									
Di	10.2	7.2	8.4	8.6	7.2	29.2	29.6	21.6	19.4
Ol		1.4				4.3	5.7	5.4	5.5
He				0.9	1.8				
Mt			3.0	3.0	1.8	6.9	6.5	6.8	5.5
Il	1.2	0.4	0.6	2.8	3.0	4.2	5.0	3.8	5.0
Ap	0.2	0.3	0.3	0.8	0.5	2.4	2.7	1.9	1.6

	10.	11.	12.	13.	14.	15.	16.	17.	18.
Or	7.5	9.0	1.0					28.0	16.5
Ab	0.5	9.0							37.2
An	21.3	9.5	4.5	4.3	1.8	16.0	30.0		13.5
Ne	18.9	30.9	36.7	34.7	27.6	24.6	20.1	45.8	23.3
Le			12.4	15.4	6.8	2.8	8.8	1.2	
Kp				0.5		3.6	0.3		
Ac								8.0	
Ns								3.1	
Ak					17.5				
Fo								0.1	2.4
Wo		1.4							
En									
Di	27.4	26.0	31.6	32.2	18.2	29.2	14.8	12.4	2.4
Ol	9.8		5.0	0.7	15.4	12.8	16.9		
He									0.9
Mt	8.3	8.1	4.7	5.9	7.5	5.7	4.4		2.1
Il	3.6	3.2	2.8	4.5	3.6	3.4	3.6	1.0	1.2
Ap	2.9	2.9	2.4	1.8	1.6	1.9	1.3	0.5	0.5

	19.	20.	21.	22.	23.	24.	25.	26.
Or	20.0	5.0	3.3	1.5		10.5	31.5	
Ab	10.9	6.5	0.5				16.5	
An	7.5	19.0	1.3	18.0	7.8	10.7	8.8	17.3
Ne	31.3	18.9	35.6	18.6	23.4	19.5	35.7	15.9
Le				1.2	4.4			10.8
Kp					5.7			6.0
Ac								
Ns								
Ak			15.0		11.0			
Fo		9.1						
Wo	0.3							
En							0.8	
Di	20.2	24.8	17.0	28.8	16.6	23.4	4.0	32.9
Ol			14.7	19.5	21.0	21.0		3.8
He							1.6	
Mt	5.8	8.5	5.7	6.6	5.7	6.5	0.2	5.9
Il	2.2	5.0	4.4	3.9	2.6	5.0	0.6	6.2
Ap	1.8	3.2	2.7	1.9	1.9	3.4	0.3	1.3

Table 19   Niggli Parameters (Almeida, 1961)

	1.	2.	3.	4.	5.	6.	7.	8.	9.
Si	142.0	145.4	160.2	129.0	142.0	82.1	81.7	90.4	94.7
Al	35.2	33.8	35.9	33.4	33.8	18.7	17.9	21.9	23.4
Fm	16.2	14.5	17.9	27.7	28.5	43.0	44.0	41.0	39.3
C	9.2	6.4	11.4	20.0	20.0	24.0	26.2	25.8	24.4
Alk	39.3	45.3	34.8	18.9	17.7	14.3	11.8	11.3	12.8
K	0.36	0.20	0.30	0.42	0.53	0.04	0.06	0.17	0.07
Mg	0.41	0.23	0.37	0.44	0.55	0.49	0.46	0.48	0.47
C/Fm	0.57	0.44	0.64	0.73	0.70	0.56	0.60	0.64	0.63
Qz	−102.9	−101.3	−79.0	−46.6	−28.8	−75.1	−65.5	−54,8	−56.5
Ti	1.8	0.7	1.1	4.0	4.6	4.7	5.6	4.3	5.9

	10.	11.	12.	13.	14.	15.	16.	17.	18.
Si	75.4	91.2	87.0	87.0	66.9	74.5	77.5	135.5	154.8
Al	17.0	20.9	19.5	20.0	11.2	17.1	22.6	32.6	39.0
Fm	48.2	36.9	39.3	38.7	53.3	48.2	45.2	17.7	19.1
C	26.7	25.9	23.8	23.2	24.9	24.3	22.6	10.5	11.7
Alk	8.1	16.3	17.4	18.1	10.6	10.4	9.7	39.2	30.2
K	0.19	0.12	0.22	0.26	0.16	0.19	0.25	0.23	0.18
Mg	0.52	0.41	0.46	0.45	0.63	0.57	0.55	0.32	0.37
C/Fm	0.55	0.70	0.61	0.60	0.47	0.50	0.50	0.59	0.61
Qz	−57.0	−74.0	−82.6	−85.4	−75.7	−67.1	−61.3	−101.7	−66.0
Ti	3.8	3.9	3.2	5.2	3.5	3.6	3.8	1.6	1.8

	19.	20.	21.	22.	23.	24.	25.	26.
Si	111.5	76.0	67.9	70.9	68.6	72.0	156.3	80.7
Al	25.8	17.0	13.1	13.8	13.7	12.9	42.0	18.6
Fm	31.6	48.6	49.9	55.7	53.2	57.9	12.0	43.3
C	20.7	26.3	24.4	23.8	22.4	20.6	10.1	27.5
Alk	21.9	9.1	12.6	6.7	10.7	8.6	35.9	10.6
K	0.24	0.12	0.05	0.08	0.27	0.24	0.29	0.47
Mg	0.41	0.53	0.62	0.59	0.62	0.59	0.41	0.51
C/Fm	0.65	0.52	0.49	0.43	0.42	0.38	0.84	0.63
Qz	−76.1	−60.4	−82.5	−55.9	−74.2	−62.4	−87.3	−61.7
Ti	3.4	5.3	4.3	5.1	3.4	5.0	0.9	6.7

mioalkaline group. Trindade is an extreme product of hyperalkaline differentiation, with no rocks having an si parameter greater than 200. Values for this parameter go as low as 70 for Fernando de Noronha and Cabo Verde, but is always above 70 in the mioalkaline group – except in some xenoliths of unknown origin.

It is possible to attempt a correlation of the hyperalkaline and mioalkaline series on a tectonic basis, the former being related to islands on fracture zones bordering continental structures and trending towards the Atlantic basins, whereas the mioalkaline group refers

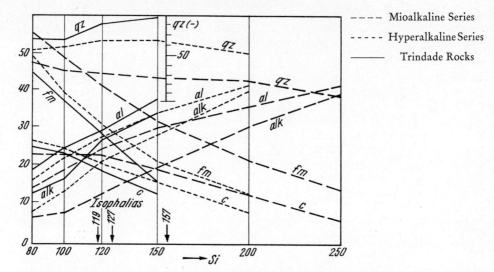

Fig. 31. Comparison of General Diagrams of Variation, drawn as average values of the Principal Parameters of the Atlantik Mioalkaline and Hyperalkaline Series, and those of Trindade Island. (ALMEIDA, 1961)

to islands associated more or less with the Mid-Atlantic Ridge, although St. Helena, Madeira, Tristan and Gough are not actually located on this feature.

The petrographic differences between the two groups seems to reflect a considerable difference in the nature of the oceanic substratum, as has been determined by geophysical work, but this aspect of the problem need not be entered into here.

As regards Trindade, we can state that petrographically it corresponds to an extremely subsaturated alkaline province of the hyperalkaline group. Its isophalia of 119 is the lowest of the islands referred to above, its fm values are lower, its al and alk values higher than in any other Atlantic series. The si interval as also lower than in any of the islands mentioned above.

M. A. PEACOCK (1931) proposed the term alkali-calcic index which is the value of silica to which, in the variation diagram, correspond equal weight percentages of CaO and $Na_2O + K_2O$. In accordance with the internationally applied alkalinity indicated by such an index, PEACOCK distinguished four classes of the eruptives, but ALMEIDA has

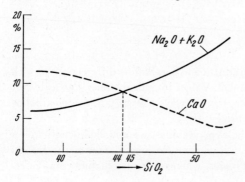

Fig. 32. Alkali-Calcic Index. (ALMEIDA, 1961)

proposed a fifth class, whose alkali-calcic index does not exceed 46. Thus the eruptives can be classed thus: hyperalkaline < 46; alkaline < 51; alkali-calcic < 56; calc-alkaline < 61 < calcic. Trindade would thus have an alkali-calcic index of 44. (Note: the PEACOCK alkali-calcic index is not to be confused with the alkali-calcic index of H. KUNO (1959) which is related to KUNO's index of solidification [SI], and further, KUNO does not use silica as an abscissa in his diagram.)

Petrogenesis

Whatever genetic interpretation is placed upon the eruptives of Trindade, they do show certain peculiarities which are worthy of mention.

We must note that Trindade is indeed a voluminous volcanic structure, rising from depths of some 5000 m on the ocean floor, and isolated within the largest basin in the Atlantic. We thus view with scepticism any interpretation which would have the materials of sialic nature, either crustal or magmatic. Further, it seems reasonable to postulate the presence within this oceanic milieu, of calcareous sediments which were accessible to the magma, provoking a notable desilicification of the series. From a petrochemical point of view, Trindade should be thought of as representing an oceanic volcanic series which is the most subsaturated and sodic-alkaline so far recognized anywhere in the world. Olivine-basalts are lacking and plagioclases are almost non-existent. It therefore appears unlikely that simple fractional crystallization of a universal olivine-basalt magma could explain the origin of the island rocks.

To us, the most probable explanation is a process of fractional crystallization of a magma of basanitic character. Fig. 33 taken from ALMEIDA, shows variations of the principal oxides as a function of silica, and this diagram can be used to demonstrate such a process for Trindade.

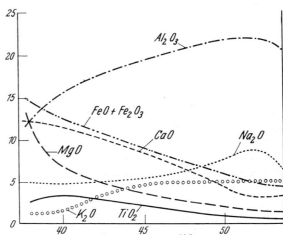

Fig. 33. Variation Diagram of Chief Oxides as a Function of SiO_2. (ALMEIDA, 1961)

The trace of the curves is what one would expect in such crystallization. The alumina, soda and potassic curves rise with increase in silica; the iron, magnesium, titanium and lime curves likewise fall in the same direction. The alumina curve rises rapidly from the most basic stages, the intermediate stages being very basic, with tenors in alumina above 20 %, contrary to what one would expect if there had been important separation of plagioclases in the initial evolutionary stages. (This reduced importance of the plagioclases in the Trindade rocks must always be reckoned with.) The lime curve falls more gradually from the more basic terms. Beyond the nephelinites, its decline is accentuated by increasing silica values, reflecting the abundant crystallization of basaltic augite and soda-augite which have more calcium than the liquid. The curve for the iron oxides is relatively smooth, whilst the magnesia curve falls sharply at the basic end, evidently as a result of separation of chrysolite, then more slowly as the silica content rises.

It was the view of ALMEIDA that fractional crystallization, with the expulsion of liquid residuals, either by pressure, decantation or other means, led to important differentiation within the Trindade series.

As already remarked, in Trindade we have no olivine-basalts, practically no plagioclases which could be the products of fractional crystallization. The rocks display high tenors in alumina and the alkalis, general poverty in silica, suggestive of a near parental magma. The rapid decrease in the MgO/FeO ratio in the more basic terms of the series is significant as regards the separation of magnesian olivine crystals. Such a magma would be rich in titanium, phosphorous and water. Crystallization of plagioclase would not take place or then only on a moderate scale, having very little influence on the differentiation of the series which was characterized tather by the separation of ferromagnesian minerals. A magma which would satisfy these requirements would be one of basanitic type.

In Trindade, no rocks, either outcropping or then as xenoliths or ejectiles, correspond to these deduced characters of a parental magma. The analcite-basanites occurring as dykes in the Complex have plagioclase, and as such, may be said to illustrate a type of advanced differentiation.

The eruptives of both Trindade and Fernando de Noronha show very close petrochemical and mineralogical analogies. In the latter also there are no olivine-basalts, but for this island ALMEIDA surmised that the parental magma was alkali-basalt (q. v.). However, his opinion was that the nepheline-basanites of the Sao José formation of Fernando de Noronha, the youngest sequence in that archipelago, could equally satisfy the conditions and requirements of a parental magma for that island group. The chemical differences between these two types of rocks results from a larger quantity of olivine in the nepheline-basanites, rocks derived from alkali-basalts contaminated with olivine crystals originating from olivinite fragments present in notable quantities. It was the view of ALMEIDA that the nepheline-basanites of the Sao José formation represented the parental magma for the Fernando de Noronha eruptives, having been derived from an alkali-basalt as a consequence of gravitational separation (which can readily be detected in the island lavas) of chrysolite crystals and thus forming nepheline-basanites. He further argued that a nepheline-basanite magma, identical to that as represented by the Sao José rocks, could very possibly be derived and constitute the origin for the majority of the Trindade series, although he admitted not all the Trindade rocks could be explained in this fashion.

Of the various hypotheses to account for the origin of basanitic parental magma, that of P. G. HARRIS seems most appropriate (1957). This writer suggested the formation of subsaturated sodic basalts through fractional crystallization of a primary tholeiitic magma, modified by reactions with olivine and other mantle minerals enriching the liquid in magnesium and sodium. Equally possible is the existence of alkaline femic magmas of primary character, originating from the direct fusion of mantle components and under conditions which can lead to the acquisition of geoglasses. It is not presumed that these magmas derived from olivine-basaltic or tholeiitic origins, could all on a major scale, result from processes directly related to the mantle via the routes of great fracture zones. The composition and the thickness of the crust would influence the nature and the differentiation of such magmas, which would tend to be enriched in potassium in passing through continental masses whilst conserving their original subsaturated, sodic character in the oceanic hyperalkaline series. If such primary magmas occur and if they can be so modified, then Trindade and Fernando de Noronha, because of their particular tectonic setting, would, more than other hyper-alkaline series, be in a condition to furnish indications as regards their nature and origin. We would note, in this connexion, that olivinite xenoliths are relatively abundant and frequent in the nepheline-basanites of the Sao José, Cuzcus and Fora islets of Fernando de Noronha, being generally angular in shape, and leave little doubt that they originated from the fragmentation of consolidated masses, the superficial roof-pendants of magma currents. Neither in the main island of Fernando de Noronha nor in Trindade do we find such xenoliths or ejectiles as inclusions in nepheline-basanite, but on the contrary, xenolithic peridotite bodies occur in both islands. In the ankaratritic lavas of Paredao volcano it is true that we find olivinite xenoliths, but these are associated with fragments of diverse lithology, under conditions which appear to indicate fractional crystallization. It would thus seem that the olivinite fragments in the nepheline-basanites of Fernando de Noronha suggest that they constitute material of the original peridotitic mantle of the earth.

Geological Considerations on the Age of the Trindade Vulcanism

As true sedimentary rocks and fossils are lacking in Trindade, and as ALMEIDA had no radiometric samplings available to him then, the dating of volcanic events was only feasible via indirect means.

Basaltic vulcanism, dating from the Lower Cretaceous, has been recognized in Cabo Verde (BEBIANO, 1932), continuing during the Tertiary, various fossiliferous beds are associated with volcanics in the Canaries, Azores, Madeira and in Cabo Verde, in some instances, e. g. Azores, to within a few years ago. The islands in the Gulf of Guinea indicate vulcanism in Upper Cretaceous times. But as regards islands further south in the Atlantic, geological evidence though more obscure is counterbalanced by a meagre collection of isotopic datings. For Tristan da Cunha (q. v.) GASS (1967) suggested a constructional volcanic phase enduring for less than one million years, succeeded by an erosional phase lasting for some twenty million years, although of course Tristan erupted as late as 1961. On the other hand, for St. Helena (q. v.) BAKER ET AL (1967) believed that all the subaerial volcanic activity occurred within a minimum period of some 7.5 million years, that the building of the volcanic pile began more than 14 million years ago.

For Trindade, ALMEIDA believed that the Desejado sequence was representative of pre-Glacial times. The Morro Vermelho phase is likely of Pleistocene age, whereas the

Valado phase is to be correlated with the Würm glaciation in age. There are no indications whatsoever of the Paredao volcano having been affected by eustatic changes during the Pleistocene glacial-interglacial episodes, and from this it is concluded that the volcano indicates a post-Glacial age.

ALMEIDA concluded that the Trindade vulcanism probably began in the Tertiary, perhaps even earlier. The island certainly existed in the Pleistocene, and before the period of Optimum Climate, some 5000–7000 years ago, the last volcanic episode took place.

Geochronology of the Trindade Volcanics

CORDANI (In Press) has carried out some age determinations for the island, using the potassium-argon method. Twenty one determinations were made on 15 samples, and results are shown in Table 20.

Table 20   Radiometric Age Determinations (Modified after CORDANI (Press)

Sequence	Sample No.	Rock Type	Material	Age (m. y.)
Morro Vermelho	FA-T-9	Ankaratritic lava	Whole rock	< 0.27
	UC-TD-5	Ankaratritic lava	Whole rock	< 0.17
Desejado	UC-TD-27	Phonolitic lava	Whole rock	2.23 ± 0.08 2.30 ± 0.07
Complex	UC-TD-29	Phonolitic dome	Whole rock	2.60 ± 0.08 2.53 ± 0.08
	FA-T-94	Phonolitic neck	K-feldspar Matrix	2.32 ± 0.08 2.21 ± 0.33
	UC-TD-31	Phonolitic neck	K-feldspar	2.38 ± 0.09
	UC-TD-10	Phonolitic neck	K-feldspar	2.57 ± 0.13
	UC-TD-16	Phonolitic dyke	K-feldspar	2.77 ± 0.08
	UC-TD-22	Phonolitic dyke	K-feldspar Biotite	2.96 ± 0.10 2.63 ± 0.37
	UC-TD-23	Nephelinite dyke	Whole rock	1.06 ± 0.08
	UC-TD-20	Nephelinite dyke	Whole rock	1.89 ± 0.07
	UC-TD-33	Olivine-Analcitite	Whole rock	2.38 ± 0.64
	UC-TD-26	Nephelinite dyke	Biotite	2.55 ± 0.38
	UC-TD-14	Tinguaite	K-feldspar	2.55 ± 0.08
	UC-TD-13	Nephelinite dyke	Whole rock	3.63 ± 0.54 3.26 ± 0.42 3.39 ± 0.27

The two Morro Vermelho age values only represent maximum values – less than 17–27 000 years.

The Desejado phonolitic lava analyzed gave an average age determination of 2.27 + 0.75 m. y.

The other thirteen samplings were all of Complex rocks, yielding age values ranging from about 1.1 m. y. to 3.4 m. y. It apparently is worthy of note that much of the vulcanism in Trindade occurred either simultaneously or then just before the onset of the Desejado lava flows. The oldest measured age, 3.6 m. y. for a nephelinite dyke, indicates a lower limit for the age of the pyroclastics of the Complex, thus being likely of Middle-Upper Pliocene age.

It must be stressed that these age values refer only to the subaerial part of the volcanic pile, the vulcanism however having started a considerable time before the pile attained its present elevation above sea level.

CHAPTER 6

Martin Vaz Archipelago

The Martin Vaz archipelago lies in lat. 20° 30′ S, long. 28° 51′ W. The main island and there small ones, of which only two are named, viz. Ilha do Norte and Ilha do Sul, are all considered to be of volcanic origin. To all intents and purposes, they are inaccessible, perhaps never trodden by the foot of man except Ilha do Norte. The islands have a general N-S alignment of some 3.2 km. Martin Vaz attains a maximum elevation of 175 m, Ilha do Norte, 75 m, the main island lying some 48 km E of Trindade. In between, the ocean goes down to depths of more than 1100 m. Martin Vaz itself shows quite a profuse tree coverage.

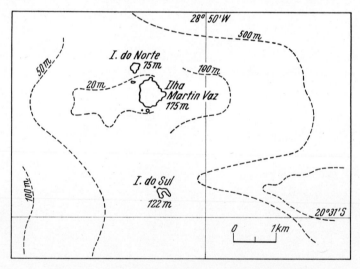

Fig. 34. Martin Vaz Archipelago. (Sheet 21, Serv. Hidra. da Marinha, 1963)

Until a mere matter of a few years ago, it is very doubtful if ever a human being landed in the archipelago, and such geological impressions as were known were obtained from aerial photographs taken by low-flying aircraft. Studies of such photographs suggested that the archipelago was formed of phonolitic necks and tuffs, as in Trindade.

Coastal waters are considered extremely treacherous, not only because of swift currents, rocks and shoals, but also because they are not well chartered.

In 1962 a captain of a naval hydrographic vessel belonging to the Brazilian Navy managed to land on Ilha do Norte and secured some rock specimens which were later given to Dr. E. P. Scorza to study (1964). To date, no samples are available from wither Martin Vaz or Ilha do Sul.

The samples from Ilha do Norte proved to be alkaline volcanics, of two types, ankaratrites and haüynite. The former is a typical member in Fernando de Noronha, Trindade, etc. Haüynite is indeed a rare rock, so far only recognized in the Wiesenthal region of Saxony, Germany, in the vicinity of Morgenberg, Neudorf, Annaberg and Erzgebirge. The rock was first studied by Reinisch, and was considered by W. E. Tröger as belonging to the tawite type.

Petrographic information on the two rock types present in Ilha do Norte follow.

Haüynite

In hand specimen the rock has a greyish colour, vacular structure and contains blue crystals. Microscopically the rock displays a porphyritic texture, involving the minerals titanaugite, haüyne, biotite, titanite, apatite and ferric oxide. Probably some nepheline is also present. The microcrystalline groundmass shows microphenocrysts of haüyne and titanaugite and biotite flakes. This matrix comprises microcrystals of titanaugite, haüyne, ferric oxide and, on occasion, titanite and apatite crystals assume important dimensions. Haüyne often shows inclusions, either irregularly scattered or then along certain trends, crystals of the mineral attaining maximum lengths of 2 mm. Some haüyne is altered into zeolites – natrolite, stilbite, probably analcite. In some parts of the rock specimen, a feldspathoid has also undergone alteration into what appears to be gibbsite. Some haüyne crystals have a clear blue colour, such as can be observed macroscopically, whereas others are colourless or of yellowish tint when altered. Titanaugite has generally a greenish-yellow colour, but usually irregular spread throughout the crystal. From a study of the pleochroism it appears that the pyroxene is a sodic titanaugite, intermediate between titanaugite and aegerine-augite. In one fragment of haüynite studied, olivine is present, thus giving rise to an olivine-haüynite variety.

W. E. Tröger (1935) has given the following information on a German specimen of haüynite (p. 104).

Ankaratrite

Megascopically the rock is dark coloured, formed of an aphanitic matrix in which are seen phenocrysts of olivine and pyroxene. In thin-section it shows a holocrystalline porphyritic texture in which phenocrysts are present in a groundmass comprising microlites and crystallites of titanaugite, nepheline and magnetite. Within this matrix occur phenocrysts of olivine (crysolite) and titanaugite. The olivine occurs as clear crystals, with borders altered into goethite (?), well displayed along cleavages and fractures also.

It is likely that the age of the Martin Vaz archipelago is the same as that of Fernando de Noronha and Trindade, and Scorza agrees with Almeida (q. v.) that this comprises the period between the Upper Cretaceous and the Neogene.

Mineralogical Composition

Titanaugite	54 %
Haüyne ± Nepheline	38 %
Metallic minerals, apatite, biotite	8 %
Glass, with potential plagioclase	±

Chemical Composition

SiO_2	40.43
TiO_2	2.08
Al_2O_3	18.33
Fe_2O_3	5.88
FeO	6.01
MnO	0.21
MgO	4.62
CaO	12.43
Na_2O	5.06
K_2O	2.85
H_2O+ / H_2O-	1.25
P_2O_5	0.82
Cl	0.16
S	0.28
Total	100.41

THE PORTUGUESE ISLANDS

CHAPTER 7

Sao Tomé

General

S. Tomé, the principal Portuguese island in the Gulf of Guinea, lies between longs. 6° 28′ and 6° 45′ E and from the Equator to 0° 24′ N lat. It has an area of 854 km², some 47 km in length and 28 km broad. The highest summit, Pico S. Tomé, 2023 m, and indeed all the higher terrain, lie nearer the W coast.

Much of the island is still covered in dense tropical vegatation. There is a well-developed plantation economy. The chief products are cacao, coffee, rubber, bananas, palm oil, kola nuts, copra and coconuts. Fish canning, palm oil processing and soap manufacturing are the only local industries. Cacao accounts for some 75 % by value of the island exports.

The island population is about 60 000, with a population density of some 70 inhabitants per km². However in the more subdued NE region, the density increases to about four times this figure. S. Tomé, the capital and really only port, has a population approaching 6000.

Physical Features

The NE part of the island has the most subdued topography, being essentially a bevelled platform sloping gently seawards from elevations of 200–300 m. The platform extends from Ribeira Castelo to Agua S. Joao, being larger between Agua de Guadelupe and Rio Manuel Jorge. Rising above this planated surface are several small volcanic cones. In this the driest part of S. Tomé, rainless from July to September, the rivers are almost dried up, the slightly incised valleys appearing as bouldery and sandy trenches. Where valleys debouch into marsh areas, these latter are separated from the sea by banks of white calcareous sands.

In the S part of the island there is also a planated region between the Rio Martim Mendes and the Ribeira Peixe, and a smaller one at the extreme southern tip of S. Thomé.

Elsewhere the island is distinctly mountainous, with a pronounced relief. The highest land runs N-S and NW-SE. In the W there is an older volcanic landscape, contrasting with the younger volcanic forms present in the NE.

Much of southern S. Tomé shows two distinct types of terrain: broad relatively level expanses alternating with narrow, deep, steep valleys. The mountains rising above 600 m result from intensive erosion. Here in the wettest part, stream erosion and 'creep' are most powerful and drastic agents of removal and modelling are operative. The larger development here of phonolites, readily susceptible to weathering, the very high average rainfalls – over 4000 mm even at sea level – and the dominating influence of the SE Trade Winds which are ever warm and moist, all combine to cause profound dissection of the terrain. TENREIRO (1961) indeed claimed that of all the Atlantic volcanic islands, S. Tomé had a more vigorous relief because of significant climatic variations which resulted in a variety of landscape forms.

Impressive cliffs front the ocean along many coastal stretches. On the W coast such are to be seen at Ponte Figo and from Ponte Furada to S. Miguel; on the S coast between Ponta Baleia and Baia do Iogologo. In both localities the cliffs comprise basalts. Less common are cliffs of phonolite, but these do occur, e. g. S of the bay at Praia Grande and between the beaches Io and S of Angra de S. Joao dos Angolares, in the SE part of the island.

Basalt escarpments are present in the interior, forming ledges over which the streams leap in cascades of considerable height, such as, e. g. in the Rios Manuel Jorge, Ouro, Agua Grande, the Ribeira Funda and Agua Castelo.

Steep slopes characterize most of the island, not merely valley sections. Talus deposits and 'creep' phenomena are common, occasionally covered by luxuriant vegetal coverings. At Binda, for example, scree deposits ca. 10 m thick are well consolidated, comprising basalt fragments of many sizes.

Several of the streams in the N and W regions have natural barriers composed of sands, pebbles, cobbles, thus backing-up the rivers and forming swampy areas, e. g. at the confluence of the Rios Guadelupe Sela and S. Joao. Other areas where the drainage is somewhat indeterminate and muddy, such as between Morro Peixe and Fernao Dias, have been caused by invasions of the sea at high tides whereby the waves have scooped out hollows forming small swampy areas with barriers of calcareous sands. Near Melao beach is a depression filled with brackish water and supporting mangroves. At the extreme S of the island, at the Malanza small lake, there is a much larger depression, the contained waters managing to break through a natural barrier of beach sands and thus achieve open contact with Baia do Iogologo.

On a geomorphologic basis, four types of relief occur in S. Tomé (TENREIRO, 1961): 1. the Serranias, highest land, where slopes are precipitous, valleys deep and narrow with many waterfalls along stream courses. In this mountainous volcanic terrain where erosion is most intense, typical volcanic erosional landforms are found. 2. the Morros, lying in the central, southern and north-eastern areas, where many volcanic cones and craters occur. The Morro region is peripheral to the Serranias, intermediate in elevation between the latter and lower areas nearer the coast. Volcanic cones vary in height from 60–1500 m, and all are second generation forms. 3. 'Pain du sucre' landscape, with parabolic profiles of inselberg character, found especially in the central part of the island but also in the S.

Volcanic plugs and necks, of which Cao Grande, 663 m is a truly magnificent and fantastic example of the latter, are also placed in this category. 4. raised beaches and fluvial terraces, common along many coastal stretches and in several valleys.

Evidences of recent epeirogenic uplift are seen in the fluvial terraces and raised beaches. In the lower course of the Rio Maria Luiza, above the road crossing, the steep, narrow valley intersects a 15 m thick layer of basalt, with more than 30 m of detritals lying above. In the valleys of the Rios Anambo and Cantador and at Io Grande, rounded cobbles and pebbles forming fluvial terraces now lie at elevations from 4–7 m above the present river beds.

Raised beach deposits are found in many localities. On the road between Sta. Jenny, near its junction with the Diogo Vaz-Sta. Catarina road, is a cutting showing the following section (NEIVA, 1958):

	Thickness (m)
Basalt mantle	–
Pyroclastics, basalt	0.35
Tuff	1.50
Beach pebbles, sands	0.50–1.0
Basalt	10.00
	Sea level

To the N of Baia de Ana Chaves, near to the road, resting upon spheroidal basalts, is a section comprising the following (NEIVA op. cit.)

	Thickness (m)
Sands	0.5
Rounded basalt pebbles	0.2
Sands	1.3
Coral remains	0.05–0.10
Sands	1.5

Base of lower sands 3 m above sea level.

Old, well-stratified, fine-grained beach deposits, including consolidated shell fragments, sands and granules of basaltic material, can be seen at the extreme N end of Praia Pantufo, where these total some 20–30 cms thick, and also to the S of Ponta de S. Jeronimo where they attain a thickness of 2–3 m, dipping gently seawards.

NEIVA assumed that the raised beaches could be correlated with the topographic platforms of Fernao Dias, Praia Lagarto, Praia Cobo and from Praia de Sta. Catarina to Praia Lemba.

The platform of northeastern S. Tomé incorporates argillaceous laterites, of basaltic derivation, which can be clearly observed at Hospital, S of Praia Lagarto and other places.

Climate

At S. Tomé, capital of the island, on the NE coast, average maximum daily temperatures are 28.8° C; average minimum daily temperatures, 20.7° C.

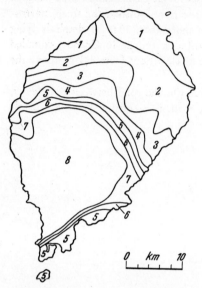

Fig. 35. Rainfall Distribution, São Tomé. (After Teneiro, 1961)

Legend:

1 = < 1000 mm
2 =   1000 – 1500 mm
3 =   1500 – 2000 mm
4 =   2000 – 2500 mm
5 =   2500 – 3000 mm
6 =   3000 – 3500 mm
7 =   3500 – 4000 mm
8 = > 4000 mm

The average annual rainfall is 965 mm. Here the wettest months are March and April, 250 mm rainfall, whilst during the driest months, July and August, only 2.6 mm rain falls on an average. Climatic data for the capital are typical for the NE part of the island, where land below 200 m has a distinctly arid aspect. From here, the driest part of S. Tomé, rainfall increases towards the SW. At Monte Café, 690 m and only 15 km SW of the capital, the average annual rainfall has increased to 2240 mm! Land over 1000 m receives 4000 mm, annual average. It is even reported that snow has fallen on the highest peaks.

The southern part of the island is warm and wet, cooling as one ascends into the mountains.

The climate of S. Tomé is humid and tropical but with a marked dry season from June to October; the rest of the year constitutes the wet season.

Climatically S. Tomé can be divided into three regions: 1. NE, warm, dry at lower elevations, average annual rainfall ca. 1000 mm, average annual temperatures of 25.7° C, but cooler at higher elevations. 2. Central-Western mountainous areas, strong condensation characteristic. Above 1000 m, atmosphere saturated, with very cool nights. Average annual rainfalls from 2500–4000 mm. 3. S, average annual rainfall between 3000 and 4000 mm, warm, SE Trade Winds all year round.

Geology

Rocks of basaltic type predominate throughout S. Tomé and on Ilheu das Rolas. These show both prismatic and spheroidal jointing. Columnar jointing similar to that at Giant's

Causeway. Northern Ireland, occurs at Baia do Iogologo, between Diogo Vaz and Sta. Catarina, near Binda and at Praia Baixa and many other localities. Spheroidal jointing is present S of Almas, near Morro Vigia, between Diogo Vaz and Sta. Catrina ond other places. On occasion it is possible to see the transition from prismatic into spheroidal jointing, a result of the alteration of the basalts. This can be noted, for example, between Sta. Catarina and D. Amelia and in the environs of Ubabudo. Laminar jointing also occurs, and between Sta. Catarina and Diogo Vaz the co-existence of prismatic, spheroidal and laminar jointing can be seen.

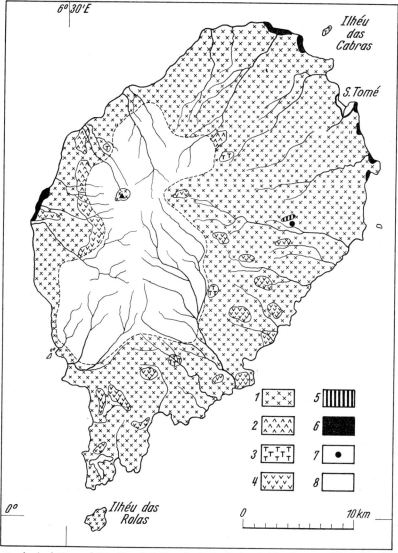

Fig. 36. Geological Map of São Tomé. 1. Basalt, 2. Andesite, 3. Trachyte, 4. Phonolite & Tephrite, 5. Sandstone, 6. Raised Beaches, 7. Oil Seepage, 8. Unmapped. (After NEIVA, 1956)

Volcanoes are present in the northern, central and southern parts of the island. Some of these are of large dimension, and Lagoa Amelia is a crater. Evidences of vulcanism are seen in the volcanic cones, bombs, lapilli, volcanic sand and ropy lavas. In Ilheu das Rolas are two well-preserved cones, rising to heights of 96 m and 95 m respectively, each of which are encased in craters measuring some 90 m in diameter. Here ropy lavas and

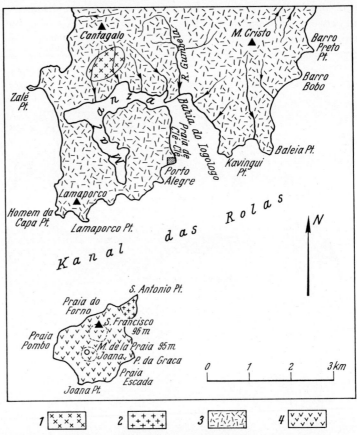

Fig. 37. Geologic Map of Southern São Tomé & Ilheu das Rolas. 1. Phonolites, 2. Basalts, 3. Basalts, Ankaramites, Limburgites, 4. Ankaramites, Limburgites. (NEIVA & ALBUQUERQUE 1962)

scoria line the sides of these craters, today under intensive cacao cultivation. Tuffs, usually well stratified, are relatively abundant throughout the island, especially in the W, from Esprainha to Binda. These are usually basaltic in composition, occurring in small, scattered outcrops. At Barro Bobo a 20 m high section shows basalts and basaltic tuffs alternating, well bedded and dipping at 45°. At Lemba volcanic ash has been liberally sprinkled over plant leaves.

Phonolite necks and plugs, partially eroded, from prominent peaks, e. g. Quinas, Mizambu, Maria Isabel, Monte Sinai, Novo Brasil, Antonio Dias and others. Several of these, e. g. Cao Grande, Cao Pequeno, Maria Fernandes, Cagungué, etc., assume remarkable shapes, standing up as veritable needles from the surrounding countryside.

Volcanic breccias are present here and there. Near Mateus Sampaio occur basalt cobbles, pebbles, with phonolitic material as cement. At the beach at Inhame, basaltic boulders, cobbles and pebbles are bound together by vesicular, spongy basanitoids.

Igneous Rocks

One of the earliest more thorough compilations on the igneous rocks of S. Tomé is that of Carvalho (1921a, 1921b). He classified and described the rocks thus:

A. Phonolites
 1. Trachytoid phonolite or bostonite, with little sphene
 2. Nepheline-phonolite
 3. Hauyne-phonolite
 4. Phonolites with plagioclase

B. Phonolitic Andesites (= Trachyandesites?)
 1. Phonolitoid type:
   a) With hornblende
   b) With hornblende and pyroxene
 2. Basaltoid type:
   a) With phenocrysts of plagioclase – hornblende-trachyandesites (Hornblende-cantalite)
   b) Devoid of plagioclase phenocrysts – camptonite

C. Alkali-Basalts
 1. Trachybasalts:
   a) Basaltoids
   b) Trachydolerite
   c) Trachybasalt of phonolitic type
   d) Passage rocks to limburgites
 2. Nepheline-basanite, phonolitic
 3. Nepheline-tephrite, basaltic
 4. Limburgite
 5. Augite

All later work has proven the predominance of basalts as essential constituents of the island. According to Teixeira (1949), the commonest rock in S. Tomé is a compact, dense, smooth rock with phenocrysts of augite and olivine.

During his petroleum investigations, Mendelsohn (1942) made a large collection of rocks from S. Tomé which were later studied in detail by Assuncao (1957). The latter classed these specimens as belonging to two major types: 1. basaltic facies, of alkaline (sodic) character but often showing a tendency towards calc-alkaline types; 2. non-basaltic feldspathoidal facies, typically alkaline, with a large variety of types, resulting from heteromorphism.

Basaltic Facies

Typical of this facies is labradorite-basalt with olivine, such as occurs at Roça Agua Izé. This dense, compact rock with olivine, augite and magnetite phenocrysts in a microlitic groundmass of labradorite or labradorite-bytownite, with augite, ferric oxides and accessory sphene, is of calc-alkaline character. The high percentage of H_2O is attributale to analcite, making the specimen approximate a basanite or basanitoid (Table 21, No. 1). Rocks studied by NEIVA (1954) from Monte Maculo are somewhat more calc-alkaline, and can be distinguished from the above mineralogically by the presence of hornblende phenocrysts and the absence of olivine and analcite (No. 12). Vesicular basanitoids occur on Monte Muquinqui and on the Rio do Ouro plantation (No. 2) and on Ilheu das Rolas (Nos. 3, 4) (NEIVA, op. cit.). These are black lavas, of vesicular appearance, spongy, on occasion ropy in structure, of porphyritic texture with phenocrysts of olivine and titaniferous augite in a microlitic, in parts glassy matrix. Slight serpentinization can sometimes be noted within fractures and around the edges of some olivines. The parameters of Nos. 3 and 4 show them to be basanites, with a tendency towards ankaratrites. No. 2 is also a basanite but the alkaline nature of the glass sets it somewhat apart. As the actual mineralogical composition of these three specimens is basaltic, and as the norms presume the virtual existence of a large quantity of nepheline, NEIVA classed tham as basanitoids.

No. 5 is a nephelinitic basanite rich in phenocrysts of zoned augite and olivine, some of the latter smaller ones having undergone almost total ferruginous alteration, giving rise to yellowish-golden iddingsite.

Nos. 7 and 8 are also basanites but poorer in feldspathoids. No. 7 shows a tendency towards ankaramites, verified under the microscope by the predominance of mafic elements, at the early stage being represented solely by small crystals of partially serpentinized olivine. The groundmass, which constitutes the major part of the rock, is rich in augite, magnetite and oxides of titanium. Basic plagioclase generally only forms well-defined microlites. Some interstitial nepheline is present, associated probably with analcite. No. 8 is rich in augite and olivine, at the first stage of crystallization these two being associated with phenocrysts of brown augite, often with green nuclei. The matrix comprises feldspars, more abundant and more euhedral than in No. 7. The microlites are of labradorite and interstitial nepheline; accessories include titanium oxide and apatite.

JÉRÉMINE (1943) and NEIVA (1954) described basanitoids similar to the above two specimens. (A sample classified by BOESE [1912] as being a 'trachydolerite' was redefined by JÉRÉMINE as a basanitoid – vd. No. 9, and as alkaline olivine-basalt by FUSTER [1954] more alkaline than any of those analyses studied by him for the Gulf of Guinea islands.)

No. 6 is a limburgite, or then a rock of 'limburgitic type' as per ASSUNCAO, rich in augite and olivine, an almost total absence of feldspar and abundance of glass. There are corroded enclaves more feldspathic which show long, needle-like microlites of calcic plagioclase. The partly glassy matrix includes feldspathoids, formed essentially of nepheline. The rock is thus classed as a limburgite with a small of nepheline – a basanite heteromorph.

No. 11 is a S. Tomé specimen studied by BOESE who classified it as a 'compact basalt', but JÉRÉMINE re-classed the specimen as approximating more to an andesite.

Table 21 — Chemical Composition and Norms of Basaltic Lavas (ASSUNCAO, 1957 Neiva and Albuquerque, 1962)

	1.	2.	3.	4.	5.	6.	7.	8.	9.
SiO_2	41.69	45.56	41.22	41.69	45.05	40.77	41.59	43.03	44.96
Al_2O_3	14.57	13.59	10.19	10.73	16.52	14.08	11.44	13.00	18.06
Fe_2O_3	4.13	6.31	5.67	1.34	3.79	4.05	6.51	3.89	4.04
FeO	8.99	3.35	6.84	8.43	7.12	8.76	6.54	8.20	7.82
MnO	0.14	0.12	0.17	0.07	0.25	0.24	0.13	0.18	0.26
MgO	7.74	6.07	10.79	11.76	7.30	9.37	10.29	10.02	5.58
CaO	11.44	9.06	11.21	10.95	9.96	10.48	11.32	10.08	8.99
Na_2O	2.89	5.88	3.25	3.37	4.34	3.21	3.55	3.80	4.56
K_2O	0.79	2.27	1.15	0.98	1.82	1.90	1.19	1.73	2.54
TiO_2	3.47	4.40	5.16	6.00	2.21	4.06	3.44	2.93	1.94
P_2O_5	0.69	1.38	1.02	1.63	0.67	0.91	0.85	1.31	0.62
SO_3	–	–	0.01	0.03	–	–	–	–	–
Cl	–	0.18	0.06	0.07	–	–	–	–	–
H_2O+	2.15	1.59	1.72	1.51	0.48	0.89	1.33	0.91	0.13
H_2O-	1.39	0.30	1.68	1.15	0.97	1.07	1.64	0.97	0.60
Total	100.08	100.06	100.14	100.31	100.48	99.79	100.02	100.07	100.10
Q	–	–	–	–	–	–	–	–	–
Or	5.00	13.34	7.32	6.12	10.56	11.12	7.23	10.01	15.0
Ab	12.18	20.65	11.00	10.27	13.49	3.88	9.11	11.13	11.4
An	24.19	4.73	9.73	8.62	20.29	18.35	11.68	13.34	21.3
C	–	–	–	–	–	–	–	–	–
Ne	6.75	14.94	8.80	12.33	12.57	12.67	11.25	11.29	17.8
Na_2SO_4	–	–	–	–	–	–	–	–	–
NaCl	–	–	–	–	–	–	–	–	–
Wo	11.13	Di 23.54	Di 31.10	Di 28.83	10.21	11.63	16.70	11.83	Di 15.5
En	7.50				6.90	8.16	14.00	8.30	
Fesil	3.56				2.51	2.69	0.53	2.51	
Fai	3.06	Ol 3.01	Ol 8.82	Ol 14.59	3.16	2.82	0.41	3.78	Ol 11.0
For	8.29				7.91	10.67	8.19	11.69	
Mt	6.03	–	7.66	1.86	5.57	5.80	9.51	5.57	4.2
Hm	–	6.24	0.48	–	–	–	–	–	–
Il	6.69	7.30	9.88	11.40	4.26	7.90	6.54	5.62	2.6
Ap	1.68	3.36	2.35	3.70	1.68	2.02	2.02	3.02	1.2
H_2O total	3.54	1.89	3.40	2.66	1.45	1.96	2.97	1.88	–
Total	100.30	100.30 a	100.57 b	100.64 c	100.56	99.67	100.14	99.97	100.0

a) Total includes also Hl 0.35, Pf 0.95.
b) Total includes also Hl 0.12.
c) Total includes also Hl 0.12, Th 0.14.
d) Total includes also Pf 5.58.

1. Labradoritic Basalt, Roça Agua Izé, S. Tomé.
2. Basanitoid, plantation, R. do Ouro, S. Tomé.
3. Basanitoid, near lighthouse, Ilheu das Rolas.
4. Basanitoid, northern part, Ilheu das Rolas.
5. Nephelinitic Basanite, Roça Diogo Nunes, S. Tomé.
6. Limburgite, Porto Alegre, S. Tomé.
7. Nepheline-Basanite, Praia do Ubabudo, S. Tomé.
8. Nepheline-Basanite, bridge over Rio Manuel Jorge, S. Tomé.

Table 21 Chemical Composition and Norms of Basaltic Lavas
(Assuncao, 1957, Neiva and Albuquerque, 1962)

	10.	11.	12.	13.	14.	15.	16.	17.
SiO_2	41.56	46.76	47.52	40.65	41.29	41.70	42.74	41.47
Al_2O_3	12.36	14.80	13.96	10.86	10.99	10.37	11.37	10.92
Fe_2O_3	1.02	8.71	5.32	5.55	9.64	7.54	3.98	2.51
FeO	9.03	8.32	6.21	7.56	3.15	5.29	7.50	9.92
MnO	0.15	0.34	0.09	0.12	0.17	0.18	0.16	0.17
MgO	10.41	1.70	3.74	10.96	9.45	10.88	10.82	9.50
CaO	10.01	9.46	8.15	11.36	10.75	11.61	10.93	11.57
Na_2O	4.58	2.75	3.89	2.54	2.90	3.88	3.11	3.85
K_2O	2.36	2.48	2.87	0.82	1.62	1.19	0.98	1.10
TiO_2	4.29	2.05	5.20	5.10	7.00	5.10	4.75	5.70
P_2O_5	0.85	0.63	1.28	0.87	1.00	0.93	0.73	0.86
SO_3	–	0.35	0.11	n.d.	n.d.	n.d.	n.d.	n.d.
Cl	–	–	0.20	n.d.	n.d.	n.d.	n.d.	n.d.
H_2O+	2.10	1.45	0.71	1.92	1.08	0.96	1.93	1.33
H_2O-	1.21	0.19	0.66	1.86	1.04	0.54	1.23	1.16
Total	99.88	99.99	99.91	100.17	100.08	100.17	100.22	100.06
Q	–	3.60	–	–	–	–	–	–
Or	8.74	15.01	17.24	5.00	9.45	7.23	6.12	6.67
Ab	–	20.96	30.92	12.32	20.44	7.60	13.62	6.29
An	6.39	21.68	13.07	15.85	12.23	7.23	14.18	9.17
C	4.03	–	–	–	–	–	–	–
Ne	21.02	–	–	4.97	2.27	13.77	6.82	14.20
Na_2SO_4	–	–	0.14	–	–	–	–	–
NaCl	–	–	0.35	–	–	–	–	–
Wo	15.78	9.05	8.35	Di 27.78	Di 18.38	Di 35.21	Di 28.18	Di 34.84
En	10.80	4.20	7.80					
Fesil	3.70	5.41	–	Ol 11.01	Ol 10.50	Ol 7.63	Ol 11.86	Ol 10.25
Fai	3.98	–	–					
For	10.64	–	1.12					
Mt	1.39	12.53	5.10	8.12	–	3.02	5.80	3.71
Hm	–	–	1.76	–	9.60	5.44	–	–
Il	8.21	3.80	9.88	9.73	7.14	9.73	9.12	10.79
Ap	2.02	1.34	3.02	2.02	2.35	2.02	1.68	2.02
H_2O total	3.31	1.64	1.37	3.78	2.12	1.50	3.16	2.49
Total	100.01	99.22	100.12	100.58	100.06	100.38	100.54	100.43

9. Trachydolerite (Basanitoid), Monte Café, S. Tomé.
10. Limburgitic Basanitoid passing into Etindite, Roça Porto Alegra, S. Tomé.
11. Compact Basalt (Andesite), rapid below Monte Café, S. Tomé.
12. Amphibolitic Basalt, Boa Entrada plantation, Monte Maculo, S. Tomé.
13. Basanitic Ankaramites, W slope of S. Francisco, Ilheu das Rolas.
14. Basanitic Ankaramites, Cone of Porto Alegre, southern S. Tomé.
15. Basanitic Ankaramites, Monte da Praia Joana, Ilheu das Rolas.
16. Basanitic Ankaramites, Cone of Lama-Porco, southern S. Tomé.
17. Limburgite, NE slope of S. Francisco Ilheu das Rolas.

ASSUNCAO concluded that the rocks of basaltic facies, though chemically of alkaline (sodic) character, show a predominant tendency towards calc-alkaline types. Commonest are basanites and their heteromorphs – basanitoids and limburgites. The only clearly hypersaturated specimen studied was a partially vitreous type classed by JÉRÉMINE as a limburgitic basanitoid passing into etindite (No. 10). Etindites, rocks relatively rich in potassium, are one of the most characteristic types present in the fracture zone extending from the Gulf of Guinea via Cameroun to the Lake Chad area – the 'Cameroun Line'.

Thus in S. Tomé we have alkaline 'basaltic' lavas and those whose chemistry indicates a calc-alkaline composition, namely basalts properly speaking, and andesites. We do not know the conditions of occurrence of these latter types nor of their relations to the more alkaline basaltic facies, but it is assumed that it is a question of differentiation of the same series and not a case of lavas representing distinct chronological phases. What does appear to be established is that the non-basaltic feldspathoidal lavas (vd. infra) are more recent than the basaltic facies.

Non-Basaltic Feldspathoidal Facies

Rocks of this type, both in S. Tomé and other islands of the Gulf of Guinea, are well enough known from the writings of NEGREIROS (1901), BERG (1903), CHEVALIER (1906), BOESE (1912), CARVALHO (1921), JÉRÉMINE (1943), FUSTER (1954) and NEIVA (1946, 1954, 1955, 1956). Therein the importance in S. Tomé of phonolites is made manifest, although these rocks are less abundant than those of basaltic nature.

The MENDELSOHN collection studied by ASSUNCAO showed not only the presence of phonolites with hauyne but also tahitites and analcitic phonolites. He distinguished the following petrographic types:

1. Lavas with hauyne, equivalent to feldspathoidal monzonites (tahitites).
2. Potassic lavas without hauyne, equivalent to feldspathoidal monzonites.
3. Lavas with analcite and virtual nepheline.
4. Hauyne-phonolites and nepheline-phonolites without hauyne.
5. Alkaline trachytes passing into phonolites.

To this can be added the 'sodalitic trachyte' of BOESE, considered by JÉRÉMINE as a nephelinitic trachyte.

There are rocks whose affinities are close to phonolites, usually containing alkaline and calc-soda feldspars, with hauyne as the typical feldspathoid. They are the equivalents of feldspathoidal monzonites and chemically are very close to phonolites but also ordanchites. An example of such in S. Tomé are the tahitites, fine-grained lavas rich in feldspars, with phenocrysts of feldspar, andesine, some albite or albite-oligoclase, aegerine-augite, hornblende, hauyne, sphene and metallic oxides largely derived from hornblende but also pyroxenes to a lesser extent. The groundmass comprises relatively large microlites, principally feldspathic – sanidine, albite-oligoclase. The S. Tomé specimens are very similar to those of the Canary Islands described by JÉRÉMINE. Amongst the classic tahitites of Tahiti, there are varieties much less crystalline than those of the Atlantic province, in some of which hauyne is the only crystallized mineral, of blue colour, which colour feature is also present in the rocks of Fogo (Cabo Verde). In the table of chemical analyses (Table 22), No. 1 is a tahitite from Cao Pequeno, S. Tomé.

Table 22 Chemical Composition and Norms of S. Tomé Non-Basaltic Feldspathoidal Lavas. (ASSUNCAO, 1957)

	1.	2.	3.	4.	5.	6.	7.
SiO_2	54.24	50.40	55.82	56.16	55.61	58.75	60.34
Al_2O_3	21.96	18.78	20.74	20.36	19.58	20.63	20.69
Fe_2O_3	1.64	2.83	1.91	2.04	1.48	2.29	1.53
FeO	2.15	3.06	1.95	1.79	1.80	0.72	1.15
MnO	0.14	0.20	0.14	0.12	0.11	0.19	n.d.
MgO	0.45	1.67	0.63	1.01	0.07	0.08	0.38
CaO	3.38	5.70	2.62	2.30	2.05	0.98	1.97
Na_2O	9.16	5.07	8.81	7.80	9.15	6.69	8.15
K_2O	4.53	6.15	4.99	5.79	5.02	5.97	3.94
TiO_2	0.64	2.64	0.90	0.88	0.33	0.48	0.62
P_2O_5	0.10	0.36	0.14	0.10	0.13	–	n.d.
SO_3	0.18	–	–	0.13	0.04	–	n.d.
Cl	0.02	–	–	–	0.21	–	–
S	–	–	–	–	0.16	–	–
ZrO_2	–	–	–	–	0.85	–	–
H_2O+	0.64	2.86	0.89	0.98	3.31	1.51	1.38
H_2O-	0.65	0.71	0.66	0.41	0.45	1.45	–
Total	99.88	100.43	100.20	99.87	100.35	99.74	100.15
C	–	–	–	–	–	1.22	–
Or	26.29	36.14	29.47	34.37	29.47	35.58	23.35
Ab	28.82	15.07	33.54	31.23	33.01	44.54	54.88
An	6.12	10.29	2.22	3.89	–	5.00	8.06
NaCl	–	–	–	–	0.35	–	–
Ne	25.84	15.12	22.15	18.29	21.58	6.53	7.74
Na_2SO_4	0.28	–	–	0.28	–	–	–
Wo	4.18	6.38	4.06	2.78	3.94	–	0.70
Ac	–	–	–	–	2.77	–	–
En	1.10	4.20	1.60	2.20	0.20	0.20	0.60
Fesil	1.85	–	0.79	0.26	1.72	–	–
For	–	–	–	–	–	–	0.21
Fai	–	–	–	0.30	–	–	–
Mt	2.32	3.02	2.78	3.02	0.70	1.62	1.86
Il	1.22	5.02	1.67	1.67	0.61	0.91	1.22
Hm	–	0.80	–	–	–	1.12	0.16
Zi	–	–	–	–	1.28	–	–
Ap	0.34	1.01	0.34	0.34	0.34	–	–
Pir	–	–	–	–	0.72	–	–
H_2O total	1.29	3.57	1.55	1.39	3.76	2.96	1.38
Total	100.05	100.62	100.17	100.12	100.37	99.68	100.16

1. Tahitite, Cao Pequeno, S. Tomé.
2. Tautirite, S. Tomé.
3. Analcitic Phonolite, S. Tomé.
4. Analcitic Phonolite, S. Tomé.
5. Nephelinitic Phonolite, Cao Grande, S. Tomé.
6. Alkaline Trachyte, Roça Monte Café. S. Tomé.
7. Nepheliniferous Trachyte, Monte Café, S. Tomé.

Potassic lavas, equivalent to feldspathoidal monzonites (No. 2) show potassium predominating over soda. The magmatic parameters of the specimen classify it as a microlitic heteromorph of borolonite, a rock rich in leucite. This specimen is akin to kivites, a leucitic lava common in the African Rift area and other regions, also to tautirite, a nephelinitic lava from Tautira, Tahiti, and chemically identical with tahitite but lacking hauyne. In the S. Tomé specimen, leucite is not discernible, the feldspathoids being nepheline associated with larger quantities of analcite. The first phase minerals are andesine and oligoclase-andesine as small phenocrysts and nuclei of angesine-labradorite and andesine with rims of the former minerals, the whole forming larger phenocrysts. Hornblende, aegerine-augite and augite, with relatively abundant sphene are other first phase minerals. The groundmass is microlitic, comprising potassic feldspar, along with aegerine-augite, augite, iron oxides and apatite. Feldspathoids form interstitial material. The rock is classed as a tautirite, the nepheline being largely associated with analcite.

It is well known that the lavas of the Atlantic islands are predominantly sodic, not potassic, and this feature also holds for the alkaline rocks of the African Rift region, except for the majority of the lavas in the area N of Lake Kivu, where potassium predominates slightly over sodium. We are also aware that the etindites of the Cameroun volcanoes are basic lavas rich in potassium but with soda dominant. As ASSUNCAO remarks (op. cit.) it would be interesting to study further potassic specimens from both S. Tomé and along the trend of the Cameroun fracture.

Nos. 3 and 4 serve to illustrate non-basaltic rocks deficient in silica, with virtual nepheline but containing feldspathoids, principally analcite, occurring interstitially. These more or less compact rocks have a trachytic structure, with phenocrysts of barilites and feldspar. The groundmass is of microlitic structure, with numerous well-formed, large, medium and small phenocrysts of albite, barkevitic hornblende and, more rarely, sphene. The albite and hornblende crystals combine to form huge aggregates, thus giving to the specimen (No. 4) a glomero-porphyritic structure. ASSUNCAO terms No. 3 an analcitic phonolite, and No. 4, an ancilitic phonolite with hauyne.

As in Principe (q. v.), feldspathoidal phonolites are present in S. Tomé. Here they appear to be either hauynic phonolites or then nephelinitic phonolites without hauyne. No. 5 is a specimen from Cao Grande, the most impressive plug in the island.

Alkaline trachytes are represented by No. 6. Under the microscope great masses of barilites can be observed, which reduce themselves into grains of an iron mineral. Large microlites of orthoclase, quasi-pure albite, magnetite and some hematite are present. A parametral study led ASSUNCAO to classify the rock as an alkaline-trachyte, with a certain tendency towards phonolite.

BOESE gave an analysis of a rock he classed as a 'sodalitic trachyte', but was later reclassified as a nephilionous trachyte by JÉRÉMINE (No. 7) who noted its similarity to rocks of this type from Mt. Roumpi, Cameroun.

A study of the parameters of these non-basaltic feldspathoidal rocks shows a distinct differentiation thus: 1. Nos. 1, 2, 3 and 4 are clearly feldspathoidal rocks equivalent to feldspathoidal monzonites or rocks with a certain chemical affinity. More leucocratic lavas of the series are trachytes, with a phonolitic tendency, as shown by No. 6. 2. The variation in the relation K_2O/Na_2O shows a potassic character (No. 2), or a sodic-potassic character as in No. 6.

Assuncao concluded that the volcanics of S. Tomé are principally alkaline (sodic), but also include calc-alkaline types similar to rocks studied in Western Cameroun by Jérémine.

Neiva (1954b) described six rocks from S. Tomé, two of which are given in Table 21, (Nos. 9 and 11) and two in Table 22 (Nos. 5 and 7) which he named respectively basanite, basalt, sodalitic phonolite and sodalitic trachyte. Two further analyses of a dolerite with olivine and a phonolite respectively are shown in Table 23.

Table 23 Chemical Analyses of Dolerite with olivine and Phonolite (Neiva, 1954b)

1. Rio Abade, above bridge, beside road leading to Agua Izé.
2. Base of Maria Fernandes.

	1.	2.
SiO_2	42.26	54.61
Al_2O_3	15.55	21.41
Fe_2O_3	1.63	2.38
FeO	9.92	0.90
CaO	10.40	1.32
MgO	4.93	tr.
K_2O	0.84	5.55
Na_2O	6.25	10.73
TiO_2	2.17	0.38
MnO_2	0.09	0.54
P_2O_5	0.68	0.22
Cl	–	0.68
S	–	0.04
SO_3	–	0.06
CO_2	4.23	–
H_2O+	0.44	1.33
H_2O-	0.81	0.33
Total	100.40	100.48

At the time of Fuster's publication (1954), he had only four analyses of S. Tomé rocks available, a trachydolerite (vd. No. 9, Table 21), a basalt and an aegerine-trachyte, all from Monte Café, quoted by Boese, and a nephelinitic phonolite from Cao Grande, quoted by Cavaco (1921). He was therefore aware that any comparisons made with other islands could only be provisional, but attempted to establish the following correlation between the lavas of S. Tomé and Fernando Poo:

Fernando Poo	S. Tomé
Basalts (with very little olivine). Alkaline basalts in part.	Basalts with feldspathoids (nephelinitic and leucitic)
Basalts with olivine (in part, alkaline basalts and subbasalts)	Basalts with little olivine.
—	Basalts with olivine
Picritic basalts, ankaramites, oceanites. Ultrabasic subbasalts.	—
	Phonolites.

Fuster & Neiva (1954a) both remarked that the alkaline basalts of S. Tomé have chemical characteristics very like those of gabbroic theralitic magmas of the soda series. The Monte Café basalt from S. Tomé (Boese) is very similar to the basalt from the Tom Yalla finca in Fernando Poo, which FUSTER classified as an ordinary basalt with neither nepheline nor olivine. The chemical analysis of this rock (cf. Tables 35, 36, No. 16, chapter on Fernando Poo) is intermediate between the alkaline (sodic) and the calc-alkaline series, the NIGGLI vaules placing it between the normal gabbroic-dioritic and the essexitic gabbroic-dioritic series.

According to FUSTER, the aegerine-trachyte of BOESE should be classed as a phonolite. As per the magmatic series of NIGGLI, these S. Tomé phonolites would be in the alkaline (sodic) series, analogous to a normal foyaitic magma for the nephelinitic phonolite from Cao Grande, to a bostonitic magma for the aegerine-trachyte (= phonolite) from Monte Café.

NEIVA (1954b) claimed that the dolerites and basanites of S. Tomé resembled in chemical character the glenmuirites and tephrites of Principe, the basalts of Fernando Poo, St. Helena and the Canary Islands, the limburgites and dolerites of Cabo Verde, with some trachydolerites of Madeira and with some trachydolerites of Ascension. The basalts of Monte Café, S. Tomé were akin chemically to some alkaline basalts of Tristan da Cunha and doleritic gabbros of S. Vicente (Cabo Verde). The S. Tomé phonolites resembled those of the Canary Islands, the phonolites and phonolitic syenites of Cabo Verde and to some phonolites of St. Helena. The sodalitic trachyte of S. Tomé resembled trachytes of Tristan da Cunha, St. Helena, Ascension, the Canary Islands and Madeira. Most of the phonolites of both S. Tomé and Principe belong to the normal foyaitic magma series, though the sodalitic trachyte belongs to the monzonite-foyaite magma series. The olivine-dolerites, basalts and analcimitic tephrites show a theralitic chemistry. Basanites are affiliated with a gabbroic-theralitic magma. Both NEIVA and FUSTER draw attention to the similarity of the S. Tomé volcanics to those of the other Atlantic islands.

BARROS (1960) commented that in the Manengouba, Bambouto and Roumpi Mountains of Western Cameroun there occur trachytes and phonolites accompanied by basalts such as are found in S. Tomé. The alkaline trachyte of Monte Café referred to by ASSUNCAO (No. 6, Table 22) and the nepheliniferous trachyte of Monte Café (No. 7, Table 22) are closely analogous to the phonolitic trachyte of Weme, Mt. Roumpi (JÉRÉMINE, 1943). Although neither NEIVA (1954a) nor ASSUNCAO (1957) mention the presence of leucite in the S. Tomé rocks, this mineral is described by PEREIRA (1943) from a leucitic basanite from Roça da Granja, S. Tomé, and thus similar to rocks from the volcano Etinde in Cameroun.

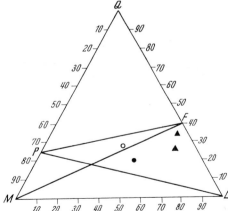

Fig. 38. QLM Diagram. (FUSTER, 1954)

● Alkaline Basaltic
○ Sub-Basaltic
▲ Phonolitic

Table 24

Jung and Rittmann Values for some S. Tomé Rocks (Barros, 1960)

Rock	Locality	Author	Jung Values SiO$_2$ (%)	Jung Values R	Rittmann Values Si°	Rittmann Values Az°
1. Limburgite	Porto Alegre (Praia Cléclé)	Assuncao, 1957	40.77	67.2	0.61	0.49
2. Alkaline Trachyte	Roça Monte Café	Assuncao, 1957	58.75	7.1	0.87	0.69
3. Neph.-Basanite	Praia do Ubabudo	Assuncao, 1957	41.59	70.4	0.63	0.45
4. Neph.-Basanite	S. Pedro, bridge at Rio M. Jorge	Assuncao, 1957	43.03	64.5	0.47	0.47
5. Neph.-Basanite	Roça Diogo Nunes	Assuncao, 1957	45.05	61.7	0.67	0.49
6. Anal.-Phonolite	Roça Diogo Vaz	Assuncao, 1957	55.82	15.9	0.81	0.67
7. Anal.-Phonolite	Roça Pontafigo	Assuncao, 1957	56.16	14.4	0.73	0.64
8. Tautirite	Roça D. Augusta	Assuncao, 1957	50.40	33.6	0.75	0.60
9. Limburgitic Basanitoid	Roça Porto Alegre	Assuncao, 1957	41.56	59.1	0.62	0.49
10. Tahitite	Cao Pequeno	Assuncao, 1957	54.24	19.8	0.67	0.62
11. Lab.- and Ol.-Basalt	Roça Agua Izé	Assuncao, 1957	41.69	75.0	0.70	0.47
12. Sodalitic Trachyte	Monte Café	Boese, 1912	60.34	14.0	0.88	0.68
13. Basnitoid	Monte Café	Boese, 1912	44.96	55.8	0.65	0.50
14. Andesite	Rapids below Monte Café	Boese, 1912	46.76	64.3	0.81	0.54
15. Basanitoid	Nr. lighthouse, Ilheu das Rolas	Neiva, 1954	41.22	71.8	0.64	0.45
16. Basanitoid	N part, Ilheu das Rolas	Neiva, 1954	41.69	68.8	0.63	0.46
17. Basanitoid	Cabeço Muquinqui (Roça Rio Ouro)	Neiva, 1954	45.56	52.6	0.65	0.52
18. Amphibolitic Basalt	Monte Macula (Roça Boa Esperança)	Neiva, 1954	47.52	54.6	0.79	0.55
19. Dolerite with Olivine	Rio Abada (Roça Agua Izé)	Neiva, 1954	42.46	59.4	0.60	0.49
20. Phonolite	Terreiro Maria Fernandes	Neiva, 1954	54.61	7.5	0.66	0.64
21. Phonolite	Cao Grande	Carvalho, 1921	55.61	12.4	0.77	0.65

Table 25

Niggli Values for S. Tomé Rocks (Barros, 1960)

Series	Group	Q	L	M	π	γ	μ	α	λ	β
Calk-Alkaline	Basalts	21.2	38.8	39.8	0.25	0.06	0.34	−0.35	1.94	−0.18
	Andesites	26.8	38.6	34.4	0.33	0.24	0.12	0.09	2.24	0.04
	Trachytes	34.3	60.9	4.8	0.04	—	—	−4.70	25.41	−0.15
Alkaline	Basanitoids Basanites Limburgites	17.1	36.6	46.1	0.21	0.24	0.41	−0.47	1.58	−0.29
	Phonolites	27.1	61.4	11.3	—	0.37	0.09	−3.67	10.86	−0.34

BARROS also presented diagrams for the S. Tomé rocks, those of JUNG, RITTMANN and the QLM triangle of NIGGLI. The data upon which these diagrams were constructed are given in Tables 24 and 25. In the JUNG table, diagrams are constructed using SiO_2 percentages as the abscissa and values of R for the ordinate. R represents the expression:

$$\frac{CaO}{CaO + Na_2O + K_2O} \times 100$$

As CaO, Na_2O and K_2O represent percentages, the two alkalis can be calculated from the analysed rock. Thus R is the percentage of CaO in relation to the total CaO in the alkalis.

In the RITTMANN table, the degree of silicity (Si°) is the abscissa, the degree of acidity (Az°), the ordinate.

$$Si° = \frac{si}{si + qz} \qquad Az° = \frac{si}{si + 100}$$

The well-known QLM triangle of NIGGLI needs no further comment.

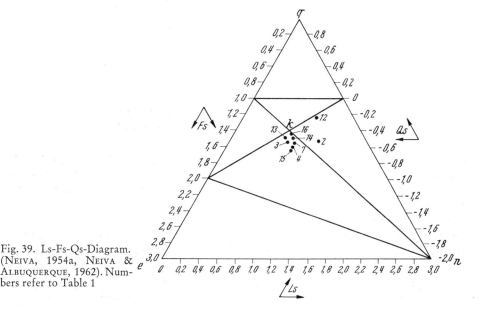

Fig. 39. Ls-Fs-Qs-Diagram. (NEIVA, 1954a, NEIVA & ALBUQUERQUE, 1962). Numbers refer to Table 1

Sedimentary Rocks

At Ubabudo, in the valley of the Abade river, and at Palmeiras, on the Agua Tomé, both localities some 7 km from the E coast, outcrops of whitish and yellowish, fine, compact sandstones with clay layers are exposed. These rocks, several metres in thickness, on occasion are so firmly consolidated that they have the appearance of quartzite (TEIXEIRA, 1949, BERTHOIS, 1955, TEIXEIRA, 1955, NEIVA, 1958, TENREIRO, 1961).

Quartz and quartzite are the dominant constituents, with fragments of the feldspars and muscovite characteristic. The cement is usually of argillaceous-siliceous nature, perhaps on occasion entirely siliceous. Now and again spheroidal jointing is present in the

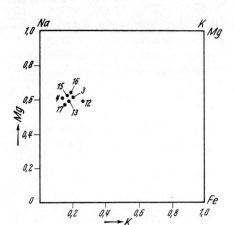

Fig. 40. K - Mg Diagram. (NEIVA, 1954a, NEIVA & ALBUQUERQUE, 1962). Numbers refer to Table 1

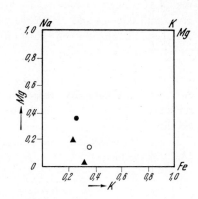

Fig. 41. K - Mg Diagram. (FUSTER, 1954)
● Alkaline Basaltic
○ Sub-Basaltic
▲ Phonolitic

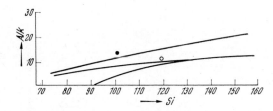

Fig. 42. Si - Alk Diagram of Basaltic Rocks. (FUSTER, 1954)

sandstones. The degree of rounding shown by most ingredients indicates that there has been a relatively high degree of transportation. The origin of the grains of quartz, quartzite, feldspars, muscovite, magnetite, etc. is to be found in the older granito-gneissic formations of the African littoral.

Associated with these sediments are asphaltic exudations and gaseous emanations. At Ubabudo as well as at Morro Peixe and Agua de Guadelupe, in the N part of the island, these oil shows issue through fissures in the tuffs, basalts and sandstones, at times under considerable pressure.

According to TEIXEIRA (1949) the Ubabudo sandstones belong to the 'Grès sublittoraux' of the Continental intercalaire, which occurs in coastal areas of Africa from Angola to the Ivory Coast.

At the time of writing by TEIXEIRA, microfossils had not yielded any indication as to the age of the sandstones, but by analogy with the littoral-laguno-marine deposits present along the western coasts of Africa, TEIXEIRA thought that these S. Tomé rocks might represent the Lower Cretaceous. (The 'Grès sublittoraux', whose age assignment has varied considerably since the term was first used by BARRAT in 1895, is now placed in the Lower Cretaceous. Vd. FURON, R. [1956].)

The beach sands of S. Tomé are overwhelmingly calcareous, comprising chiefly the detritus of shells and marine animals, associated with heavy mineral products of disintegration of the volcanics. (NEIVA and PUREZA, 1956, NEIVA & ALBUQUERQUE, 1962.)

Granulometric studies of the sands at Praia Sto. Antonio and Praia Sta. Joana, Ilheu das Rolas showed the following: For coarse sands, the valus of Md are between 1 mm and 0.5 mm; S_0 values are very well calibrated; values of Skqφ show that each one of the cumulative curves of samples studied are practically symmetrical.

The following heavy minerals were determined:

	Praia Sto. Antonio sands	Praia Sta. Joana sands
Augite	41.7 %	37.5 %
Much altered olivine	8.3	5.0
Magnetite	33.3	25.0
Basalt rock	16.7	12.5

The mineral compisition of the sands are as follows:

	Praia Sto. Antonio sands	Praia Sta. Joana sands
Calcite Dolomite	94.40 %	99.60 %
Augite	0.25	0.23
Much altered olivine	0.05	0.02
Magnetite	0.20	0.10
Basalt rocks	0.10	0.05

The calcareous material constitutes shell remains of Lammelibranchs, Brachiopods and Gastropods, also complete shells of younger forms of Gastropods, Echinoderm spicules, Ostracods *(Loxoconcha* and *Bairdia)* and Foraminifera tests, chiefly of *Amphistegina lessonii* and *Rotalia beccarii*, the commonest species in the sands of both S. Tomé and Principe.

Palaeontology

From S. Tomé, DINIZ (1959) described the following foraminiferal species, all collected from coastal localities on the N, E, SE and S sides of the island:

Praia da Baia de Ana Chaves, N side
  Rotalalia beccarii
  R. rosea
  Cribroelphidium vadescens
  Quinqueloculina cf. bicornis

Praia Calio
  Rotalia beccarii
  Cribroelphidium vadescens
  Siphoninoides glabra
  Virgula sp.

Praia de Fernao Dias, beside and N of quay
  Rotalia beccarii
  Cribroelphidium vadescens

Praia Pantufo, N side
  Amphistegina lessonii
  Rotalia beccarii
  Quinqueloculina cf. *seminulum*

Praia Lagarto, S of quay
  Amphistegina lessonii
  Rotalia beccarii
  R. rosea
  Cribroelphidium vadescens
  Quinqueloculina seminulum
  Q. sp.
  Spiroloculina cf. *ornata*

Praia Diogo Nunes
Rotalia beccarii
Cribroelphidium vadescens
Siphoninoides glabra
Cibicides refulgens
Quinqueloculina seminulum
Triloculina trigonula

Praia Rei, 20 m S of mouth of Agua S. Joao
Cribroelphidium vadescens
Rotalia beccarii
Amphistegina lessonii
Quinqueloculina cf. *bicornis*

Praia da Angra Furada, N of Pedra Furada
Amphistegina lessonii
Rotalia beccarii

Praia da Rib. Afonso
Amphistegina lessonii
Quinqueloculina viennensis

Praia da Baia da Praia Grande
Amphistegina lessonii
Rotalia rosea
Globigerinoides rubra

Praia da Cabana do Pai Tomas, or do Pesqueiro de Deus
Rotalia beccarii

Praia Zale
Amphistegina lessonii
Rotalia beccarii

Praia Lanca
Rotalia beccarii
Cibicides refulgens
Cribroelphidium vadescens

Praia Guegue
Amphistegina lessonii

Ilheu das Rolas:
Praia do Norte, 500 m SW of Sto. Anntonio bridge
Amphistegina lessonii

Praia do Sudoeste, 500 m NNW of Joana bridge
Amphistegina lessonii
Rotalia beccarii

The above specimens were obtained from cores 20 cms. in depth placed in the littoral sands. Outstanding as the commonest species are *Amphistegina lessonii* and *Rotalia beccarii*, typical littoral forms.

As per the study of the foraminiferal content of these sands, the age is placed as Aquitanian.

CHAPTER 8

Principe

General

Principe lies between lats. 1° 31′ and 1° 44′ N, longs. 7° 20′ and 7° 28′ E. The area is 110 km², with a maximum length and breadth of 19 km and 15 km respectively. Pico do Principe, 948 m, is the highest point.

As with S. Tomé there is much luxuriant vegetation right down to sea level, giving to Principe a rare beauty.

Cacao, coffee and cocnuts are the main items of export.

The island population is about 4500, giving a population density of about 40 inhabitants per km². Sto. Antonio, the capital, at the head of the deep embayment of Baia de Sto. Antonio on the NE coast, accounts for about one-quarter of the island population.

Physical Features

Geomorphologically, two distinct regions can be recognized: 1. the N, a relatively low, slightly undulating region, below 180 m in elevation, with small hillocks rising above the bevelled surface; 2. Central-S region, penetrating as a wedge into the former area, much more irregular topography, dotted with hills, peaks, having a more mountainous aspect.

In the former area, where basaltic-type rocks predominante, some fine examples of sea cliffs are to be found, e. g. at Precipicao no Norte. Considerable tracts of country are mantled with laterites.

In the Central-S part occur the three highest peaks, Pico, 948 m, Mencorne, 935 m and Cariote, 830 m, aligned E-W and thus forming a topographic barrier separating the northern part from the rest of the island. Phonolites and tephrites predominate in the Centre and S of Principe. It is thus possible to associate the two geomorphologic regions with corresponding areas of differing petrology.

Here and there in the island are boggy areas where slow deposition by streams tends to cause excessive silting. Such are to be found, e. g. at the mouths of the Rio Pagagaio and Agua Grande. By the seashore, e. g. near the beach at Burras and at Lapa, similar basins of silting occur. Common to all such marsh areas are mangrove plants.

As distinct from S. Tomé, the coastline displays more embayments, the larger and more penetrating ones occurring along the E and S coasts.

Climate

There is little difference between the principal climatic features of Principe and S. Tomé, q. v. The E-W barrier of high peaks shelters the northern two-thirds of Principe from the rain-bearing SE Trades, thus giving rise to a distinctively wet southern one-third, a much drier area in the N. As the land is considerably less high than S. Tomé, there is far less fog. According to some, Principe, an equatorial island, is held to have a remarkably fine climate, indeed almost bracing.

In both S. Tomé and Principe the summer is the dry season, yet on the western African mainland, summer invariably is the wet season.

As a summation, we may say that climatic extremes are not so great in Principe as in S. Tomé – average annual rainfalls are less, dryness less severe, temperatures somewhat more temperate.

Geology

In the northern half of the island, rocks of basaltic type predominante, whilst in the S, rocks of phonolitic-tephritic type are commoner.

Basalts show columnar and tabular jointing, sometimes spheroidal. Great columns of basalt are well observed on the S. Joaquim road near the beach at Caixao. In the NW, relatively extensive exposures of andesites are present. In places these also show jointing, usually horizontal, with either quadrilateral or pentagonal cross-sections. Near Abade, some 5 km ESE of Sto. Antonio, is an outcrop of trachyte which is related to a major dyke, striking NE-SW, cutting the vicinal basalts.

In the southern half of the island where rocks of phonolitic type predominate, columnar jointing also occurs, the cross-sections being of hexagonal shape, more rarely quadrilateral or octagonal, usually oriented vertically. Tabular jointing also is present here and there, spheroidal jointing very rarely. Volcanic necks are outstanding features, e. g. Focinhoa de Cao, Barriga Branca, Pico do Principe, Agulhas de Joao Dias Pai and Joao Dias Filho, Dois Irmaos, Morros Matemba, Caixao, Fundao, Papagaio, Oque Nazaré on the islet of Caraço and other localities. Such features testify to the powerful effects of erosion. Basalts also occur in the S, by Baia das Agulhas, also in the SE section. Essexites outcrop below Terreiro Velho and along the right bank of the Ribeira Agulhas near Anselmo de Andrade. Trachyte dykes cut basalts in the Rio Bibi, near the coast to the N of Cais General Fonseca, at Praia Cabinda lso. The dykes measure up to 1.3 m in thickness, with a dominant NW-SE strike. Trachytes also outcrop in small patches, e. g. between the beaches at Lemba-Lemba and Lapa, at the beach at Caixao, in the Ribeira Cangelolo etc.

Similar small trachyte outcrops occur in the N, e. g. at the mouth of the Ribeira da Forca, near to the capital town, near to S. José, as dykes cutting basalts at Praia das Burras and Paciencia, etc. At the beach at Sta. Rita, a trachyte dyke, 1.5 m thick, strikes NW-SE, dipping at 77° to the SW. Basalt dykes also occur, e. g. N of Cais General Fonseca, S of Praia Cabinda, N of Praia Caixao, on Ilheu Bombom, etc. These dykes generally strike NW-SE, and vary up to 1 m in thickness. A little SW of Praia Grande is a dolerite dyke, 0.8 m in thickness, striking N-S. By the bridge on the right bank of the Rio Bibi is a volcanic chimney of basaltic breccia, 3 m thick, cutting phonolites.

Geology 127

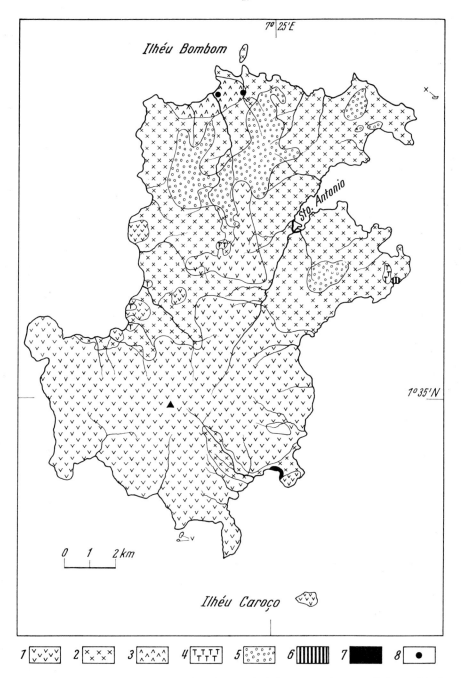

Fig. 43. Geologic Map of Principe. 1. Phonolite & Tephrite, 2. Basalt, 3. Andesite, 4. Trachyte, 5. Laterite, 6. Miocene Limestone, 7. Raised Beaches, 8. Oil Seepages. (After Neiva, 1956)

The commonest type of basalts are compact, dense rocks with large phenocrysts of olivine, augite and very often feldspars, most of which also have magnetite, some with large pyrite crystals. Textures are coarse and variable (TEIXEIRA, 1949).

Igneous Rocks

BARROS (1960), who made a petrological study of the volcanics of Principe, reviewed the various systems of igneous nomenclature and adopted to use that of RITTMANN for the eruptives of the island. Throughout he correlated this terminology with that of C. I. P. W.-LACROIX & VON WOLFE systems. Table 26 shows the correlations made by BARROS for the Principe rocks.

He classified the volcanics into the following principal petrographic types: Mafitic Nephelinites, Olivinitic Basalts, Phenobasalts, Nephelinitic Melanobasalts, Tephrites, Olivine-Trachyandesites, Alkaline-Trachytes, Phonolites and Leuco-Nephelinites.

Mafitic Nephelinites include rocks ranging from the basanitoid basalts of LACROIX to limburgites. They are poor in silica and rich in mafic minerals. Under this heading, two kinds of rocks are placed: 1. basanitoid basalts, well represented in the island (vd. Table 27, No. 27). The rocks are melanocratic, of porphyritic texture, phenocrysts of pyroxene and olivine being easily recognizable in an aphanitic groundmass. Under the microscope, two stages of crystallization are clearly evident, the first represented by titanaugite in large euhedral crystals, and olivine. The microlitic matrix comprises feldspars and much magnetite, titanaugite and sphene in lesser quantities. The olivine is densely fractured, with anhedric and subeuhedral shapes, altering into serpentine and a little iddingsite. BARROS remarked that the labradoritic basalt of ASSUNCAO (1957) (vd. Table 27, No. 1, Sao Tomé) resembled these rocks. 2. limburgites, of which specimen No. 29 is representative. It is a holomelanocratic rock with a porphyritic texture. The matrix is rich in iron minerals and possibly also analcite, with microlites of titaniferous augite, also magnetite grains. The phenocrysts include subeuhedral olivine, densely fractured and serpentinized, also titaniferous augite.

Olivinitic Basalts are similar to the foregoing rocks and are here represented by Nos. 23, 24. They are melanocratic, vesicular, of porphyritic or glomeroporphyritic texture. The former specimen shows micro-amygdales of natrolite and amygdales of presumed quartzitic material. Phenocrysts of pyroxene, appearing as greenish aggregates, give a curious arrangement like flowers. A 'lithophysaeic' texture is often present, with enormous number of oblong, subcircular cavities of quartz as filaments growing out of the walls of the cavities. This secondary silica is no doubt responsible for the quartz in the C. I. P. W.-LACROIX norms. The matrix in general is microlitic, containing microlites of labradorite, titaniferous augite and many grains of magnetite.

Phenobasalts include basaltic rocks of textural and compositional affinities, but excluding basalts with olivine. BARROS refers to six specimens in the collection of rocks made by MENDELSOHN during his petroleum investigations of the island (1942), (from the localities of Roça Sundy, bridge at Pedra Furada, bridge at Mae Marta and Ilheu Bombom), but omits to give chemical analyses or norms for these rocks. They show a microlitic texture, largely comprising labradorite, magnetite and vitreous interstitial material, with rare phenocrysts of labradorite, magnetite, augite largely titaniferous. Some speci-

Table 26   Correlations of the RITTMANN Systems of Nomenclature for Principe Volcanics with other Systems (BARROS, 1960)

Localities	VON WOLFE	C. I. P. W.-LACROIX	RITTMANN
1.	Phonolites, passing into Nephelinites	Nephelinitic Phonolite	Leuco-Nephelinite
2.	Phonolite	Nephelinitic Phonolite	Nephelinitic Phonolite
3.	Phonolite	Sodalitic Phonolite	Leuco-Nephelinite
4.	Nephelinite	Tephritic Phonolite	Leuco-Nephelinite
5.	Phonolite	Sodalitic Phonolite	Sodalitic Phonolite
6.	Nephelinite	Tephritic Phonolite	Nephelinitic Phonolite
7.	Nephelinite	Tephritic Phonolite	Leuco-Nephelinite with leucite
8.	Nephelinite	Tephritic-sodalite Phonolite	Sodalitic Phonolite
9.	Nephelinitic Basanite	Analcitic Essexite	Nephelinite
10.	Nephelinitic Basanite	Glenmuirite	Nephelinite
11.	Nephelinitic Basalt	Analcitic Basanite	Nephelinitic Melanobasanite
12.	Nephelinitic Tephrite	Analcitic Tephrite	Nephelinitic Tephrite
13.	Nephelinitic Tephrite	Nephelinitic Tephrite	Phonolitic-Nephelinitic Tephrite
14.	Trachydolerite	Basanitoid Basalt	Nephelinitic Melanobasanite
15.	Trapp Basalt	Basalt	Olivine-basalt
16.	Trachyphonolite	Alkaline-Trachyte	Alkaline-Trachyte
17.	Trachyandesite	Andesite	Olivine-trachyandesite
18.	Basanite	Basanitoid Basalt	Mafitic Nephelinite
19.	Basanite	Basanite	Nephelinitic Melanobasanite
20.	Limburgite	Limburgite	Mafitic Nephelinite

1. Cascade, Terreiro Velho, Forte de Santana.
2. Base of 'Os Dois Irmaos'; Roça Infante D. Henrique (Rio Cambamba-Bibi); Roça Esperança (Praia da Lapa).
3. Monte Papagaio; Escarpment, Portela do Fundao; Infante D. Henrique, by Mencorne.
4. SSW slope of Mencorne.
5. Praia da Lapa.
6. Beside 'Os Dois Irmaos'; ca. 3 km from railroad, Roça Infante D. Henrique.
7. Slope of Oque Pipi (Rib. Fria); Roça Esperança (Focinho de Cao).
8. Pico Charuto.
9. Right bank, Rib. das Agulhas (Roça Esperança).
10. Rio Bibi (Roça Infante D. Henrique).
11. Between Santana and Fortaleza.
12. Ilheu Caroço.
13. Roça Sundy (Praia Sundy); Praia da Campainha.
14. Roça Belmonte (Praia Uba).
15. Roça Paciencia (Praia das Burras, Rib. das Voltas); Ponta Capitao.
16. Near Abade.
17. Roça Sundy (ourtop in Rib. Izé).
18. Roça Sundy (Cajamanga, Castelo bridge).
19. Belmonte ca. 500 m to the S.
20. Roça Esperança (W of Terreiro de S. Joaquim, Horta Velha).

Table 27   Chemical Analyses (BARROS, 1960)

	1.	2.	3.	4.	5.	6.	7.	8.	9.	10.
SiO_2	55.33	55.47	56.20	55.10	55.28	55.81	57.78	53.41	53.58	53.65
TiO_2	0.05	tr.	0.14	0.20	tr.	0.10	0.40	0.79	0.54	0.20
Al_2O_3	22.38	22.03	22.08	21.05	23.45	21.55	19.57	19.29	20.94	21.34
Fe_2O_3	0.38	0.85	2.02	1.03	0.65	1.21	2.35	2.62	1.82	1.20
FeO	1.19	0.86	0.79	0.84	0.95	0.31	1.31	2.03	1.39	1.54
MnO	0.08	–	0.04	0.01	tr.	0.01	0.01	0.07	0.04	0.03
MgO	0.20	0.13	0.03	–	0.13	0.06	0.35	1.41	0.53	0.30
CaO	1.98	1.36	1.21	0.92	1.23	1.10	1.84	3.65	2.71	2.41
Na_2O	10.48	12.87	10.75	11.83	12.02	11.43	9.10	9.01	10.91	9.73
K_2O	5.88	4.68	5.69	5.57	4.55	5.57	5.60	4.63	4.71	6.21
H_2O+	1.12	0.74	0.72	2.70	1.80	2.06	0.96	2.13	2.02	2.71
H_2O-	0.55	0.31	0.36	0.40	0.21	0.77	0.23	1.06	1.03	0.47
P_2O_5	0.14	0.09	0.20	0.08	–	0.08	0.50	0.32	0.17	0.48
SO_3	–	0.07	0.02	0.01	0.14	0.03	0.08	0.01	–	0.02
Cl	0.19	0.25	0.28	0.37	0.40	0.33	0.40	0.05	0.02	0.19
Total	99.95 a	100.71	100.73	100.11	100.81	100.42	100.48	100.48	100.41	100.48

	11.	12.	13.	14.	15.	16.	17.	18.	19.	20.
SiO_2	53.74	55.76	41.51	42.01	44.52	45.46	53.60	54.74	58.08	44.30
TiO_2	0.35	0.20	4.70	0.26	0.90	1.80	0.55	0.40	0.85	3.49
Al_2O_3	17.77	20.99	12.23	13.33	20.92	18.80	24.25	21.45	19.81	14.80
Fe_2O_3	2.55	2.22	3.52	1.63	1.86	3.07	0.64	2.94	2.61	6.45
FeO	2.13	0.80	7.35	9.08	7.26	7.35	1.36	1.12	1.09	7.24
MnO	0.03	0.02	0.11	0.14	0.06	0.13	0.04	0.08	0.02	0.11
MgO	1.34	0.73	8.78	13.04	3.47	4.70	0.61	0.78	0.11	5.58
CaO	3.47	2.23	14.66	13.24	7.56	9.60	2.29	1.06	1.98	10.95
Na_2O	8.12	9.56	3.48	4.53	8.55	5.61	9.21	9.96	9.08	3.88
K_2O	7.09	5.59	0.80	0.99	2.45	2.21	6.08	6.41	5.05	1.33
H_2O+	2.92	1.18	1.75	0.35	0.82	0.19	0.70	0.31	1.46	1.30
H_2O-	0.49	0.27	0.66	0.65	0.61	0.93	0.72	0.87	0.61	0.70
P_2O_5	0.28	0.21	0.73	0.77	0.70	0.60	0.08	tr.	0.12	0.61
SO_3	0.07	–	–	–	–	–	–	–	–	–
Cl	0.09	0.43	–	–	–	–	–	–	–	–
Total	100.44	100.19	100.28	100.12a a	99.68	100.45	100.13	100.12	100.87	100.74

a) Includes CO_2, 0.10.

mens show a doleritic texture, some showing a texture similar to doleritic basalts. Under the heading of phenobasalts, BARROS places labradoritic basalts, andesinitic basalts and doleritic basalts – labradoritic or andesinitic.

Nephelinitic Melanobasalts are represented in the table by Nos. 20 and 28. The former is a compact, aphanitic, melanocratic rock with subconchoidal fracturing. The texture is typically microlitic with rare phenocrysts either of plagioclase or pyroxene. These pheno-

Table 27   Chemical Analyses (BARROS, 1960)

	21.	22.	23.	24.	25.	26.	27.	28.	29.
SiO_2	48.88	48.70	43.01	44.33	61.36	49.99	41.84	42.11	39.50
TiO_2	3.25	2.96	2.99	3.30	0.15	0.58	2.69	0.85	0.90
Al_2O_3	16.43	16.39	14.55	13.46	18.08	19.66	12.59	17.25	13.88
Fe_2O_3	4.88	9.18	5.60	8.47	3.41	5.92	5.33	5.32	5.68
FeO	5.21	1.15	8.55	4.78	0.64	3.73	7.25	7.89	9.56
MnO	0.12	0.24	0.01	0.21	0.15	0.46	0.49	0.29	0.36
MgO	2.86	3.48	8.45	8.81	0.32	3.62	11.43	7.12	11.46
CaO	8.32	5.66	10.21	9.90	0.65	7.13	12.29	10.24	12.04
Na_2O	6.79	6.34	1.82	2.03	5.82	3.49	2.06	3.06	3.17
K_2O	2.18	2.21	1.02	0.86	7.42	2.03	1.06	2.80	0.37
H_2O+	0.50	1.43	2.87	3.56	1.27	1.05	1.74	1.86	1.49
H_2O-	0.42	2.07	0.15	0.20	0.79	1.00	0.10	0.65	0.86
P_2O_5	0.97	0.95	0.53	0.45	0.49	1.20	0.89	1.04	0.95
SO_3	–	–	–	–	–	–	–	–	–
Cl	–	–	–	–	–	–	–	–	–
Total	100.81	100.76	99.76	100.36	100.55	99.86	99.76	100.48	100.22

1. Cascata do Terreiro, Velho.
2. Forte de Santana.
3. Base of 'Os Dois Irmaos'.
4. Monte Papagaio.
5. Escarpment, Portela do Fundao.
6. Infante D. Henrique, by Mencorne.
7. Praia da Lapa.
8. Near hill 'Os Dois Irmaos'.
9. SSW slope of Mencorne.
10. Slope of Oque Pipi (Ribeira Fria).
11. Ca. 3 km from railway (Roça Infante D. Henrique).
12. Pico Charuto.
13. Right bank, Ribeira das Agulhas (Roça Esperança).
14. Between Santana and Fortaleza.
15. Rio Bibi (Roça Infante D. Henrique).
16. Ilheu Caroço.
17. Roça Infante D. Henrique (Rio Cambamba-Bibi).
18. Roça Esperança (Morro Focinho de Cao).
19. Roça Sundy (Praia Sundy).
20. Roça Paciencia (Praia das Burras, Rib. das Voltas).
21. Praia da Campanha.
22. Roça Esperança (Praia da Lapa).
23. Roça Belmonte (Praia Uba).
24. Ponta Capitao.
25. Site near Abade.
26. Roça Sundy (outcrop in Ribeira Izé).
27. Roça Sundy (Cajamanga, Castelo bridge).
28. Belmonte (500 m to the S.).
29. Roça Esperança (W of Terreiro de S. Joaquim, Horta Velha).

crysts have a tendency to agglomerate, giving the rock an incipient glomeroporphyritic appearance. The plagioclase shows twinning of the Albite and Carlsbad-Albite varieties. The pyroxene is augite, tending towards pigeonite. Xenomorphic magnetite is abundant. Specimen No. 28 is also aphanitic and melanocratic. The groundmass is hemi-crystalline. Phenocrysts are mostly of olivine. Microlites of feldspar (andesine), titaniferous augite and olivine. Subeuhedral magnetite is abundant, of acicular shape, giving an arborescent appearance. There is an abundance of analcitic interstitial material and zeolitic cavities and filaments.

The major difference between these two specimens refers to the absence of olivine in the former and its richness in the latter. NEIVA (1954) described a caltonite or mesomelanocratic analcitic basanite which appriximates to No. 28.

Tephrites are of two kinds: 1. phonolitic nepheline-tephrites, compact, aphanitic, melanocratic rocks (Nos. 21, 22). Under the microscope they show a typical trachytic texture, with microlites of feldspar. Phenocrysts are rare in the first specimen but relatively abundant in the second. Feldspars are common (andesine, andesine-labradorite and labradorite, less commonoly, sanidine). Augite, tending towards pigeonite, aegerine-augite and magnetite are common. No. 21 is less rich in barilite than No. 22, but on the other hand contains more oxides of iron. No. 22 shows advanced zeolitization and in some cavities there are crustiforms of chalcedony. 2. nephelinite-tephrites. NEIVA (1954) described an analcitic tephrite from Ilheu Caroço which approximates texturally and in mineral composition to what BARROS (and RITTMANN) would term nephelinite-tephrite.

Olivine-Trachyandesites include typical andesites with olivine, or, when this is lacking, augite. Some rocks are transitional to labradoritic or olivine-basalts. 1. Andesinitic andesites are represented by No. 26. Megascopically the rocks are aphanitic, melanocratic and compact, the above specimen showing a porphyritic texture more evident than in other rocks of this kind. Under the microscope the groundmass is seen to be microlitic. Phenocrysts are of olivine, also a few augite crystals. Feldspars are typically andesine showing both Albite and Carlsbad twinning. Magnetite is abundant, usually much rounded. 2. Transitional andesites occur in the MENDELSOHN collection but no chemical analysis is given by BARROS. These show porphyritic texture and a microlitic matrix. Plentiful phenocrysts of titaniferous augite are present, and much fewer plagioclase – andesine and labradorite. Microlites of titaniferous augite and magnetite occur. Olivine is somewhat serpentinized. The rocks are said to have parameters similar to that of No. 26 above.

Alkaline-Trachytes are represented by No. 25. Microscopically the rock shows a pilotaxic trachytic texture. The matrix is formed of equidimensional microlites of sanidine and orthoclase in great quantity, acicular biotite and idiomorphs of magnetite. There are abundant phenocrysts of albite and some sodic orthoclase. Biotite also is plentiful, mostly transformed into ferruginous matter. A little isotropic material is taken to represent virtual nepheline.

Phonolites and Leuco-Nephelinites. Phonolites are well represented in the island and have been described in detail by NEIVA (1955a, 1955b). This author divided the phonolites into four groups:

1. Nephelinitic phonolites, where nepheline is the dominant feldspathoid.

2. Sodalitic phonolites, which besides nepheline also have sodalite as feldspathoids, giving a chemical analysis Cl > 0.3 %.

3. Tephritic phonolites, where orthoclase is dominant but with plagioclase also present.

4. Tephritic phonolites with sodalite, which, besides nepheline also have sodalite, giving a chemical analysis Cl > 0.3 %.

1. **Nephelinitic phonolites** (Nos. 1, 2, 3). Compact rocks with conchoidal fracture and holocrystalline porphyritic texture, with a matrix comprising orthoclase and sometimes aegerine-augite. The trachytic character of the microlites is more frequent around the phenocrysts. These latter comprise idomorphic and hypidiomorphic nepheline, with rare inclusions of sphene, sodic orthoclase with Baveno and Carlsbad twinning, idiomorphic and hypidiomorphic aegerine-augite, often having a zoned structure, prismatic aegerine and allotriomorphic barkevicite. Sphene sometimes occurs as phenocrysts. Magnetite, nepheline, apatite, natrolite and analcite can occur in the microlitic groundmass.

2. **Sodalitic phonolites** (Nos. 4, 5, 6, 7) are compact, light and dark grey rocks with conchoidal fracturing. Sometimes the phenocrysts are very large, at other times scarce visible. These comprise idiomorphs of nepheline, with inclusions of aegerine-augite, more rarely apatite and sodalite; idiomorphs of sodalite with typical inclusions; idiomorphs of orthoclase; hypidiomorphs of aegerine-augite; and in lesser proportion, aegerine, barkevicite, sphene and magnetite. The texture is holocrystalline and porphyritic. The groundmass is formed of abundant microlites of orthoclase, also aegerine-augite, nepheline, sodalite, sphene, magnetite and analcite. Rarely fibrous natrolite occurs.

3. **Tephritic phonolites** (Nos. 8, 9, 10, 11). Dark and light grey rocks, compact, with phenocrysts varying considerably both as regards size and quantity. These include: idiomorphic nepheline, some showing zeolitization and formation of analcite and natrolite, with inclusions of aegerine-augite and magnetite; idiomorphic orthoclase, sometimes with aegerine-augite and apatite inclusions; idiomorphic andesine and andesine-oligoclase, at times showing a little zeolitization; aegerine-augite, hypidiomorphic, with inclusions of magnetite; allotriomorphic barkevicite, with reaction crowns formed of magnetite and aegerine augite; hypidiomorphic sphene and a little magnetite. The groundmass includes abundant microlites of orthoclase, aegerine-augite, nepheline and sphene. A small quantity of apatite, analcite and natrolite may be present. The texture is holocrystalline, porphyritic.

4. **Tephritic phonolites with sodalite** (No. 12). Dark grey rock with phenocrysts of nepheline and feldspar cleary seen megascopically. The phenocrysts include: idiomorphic nepheline, often zeolitized into natrolite, with inclusions of aegerine-augite; idiomorphic orthoclase of sodic character; andesine; sodalite with numerous typical inclusions; aegerine-augite, often with zoned structure, and allotriomorphic barkevicite. The matrix comprises abundant microlites of orthoclase, aegerine-augite, nepheline, sodalite, magnetite and a little analcite.

BARROS grouped his phonolites and leuco-nephelinites into: 1. nephelinitic and sodalitic phonolites, 2. nephelinites. Under the former are placed Nos. 17 and 19. These show megascopically a porphyritic texture, with phenocrysts of nepheline, sanidine, orthoclase and aegerine-augite. Idiomorphic nepheline is often highly fractured, occasionally zoned. There is some hypidiomorphic hornblende present, with rather dark borders where grains of magnetite and aegerine-augite are concentrated. This magmatic corrosion of hornblende is very frequently seen in Principe phonolites. Sphene is very common, magnetite well represented, sometimes even as phenocrysts. The texture can vary considerably in such rocks. The sodalitic phonolites are megascopically porphyritic, melanocratic, of greenish hue. Phenocrysts include: orthoclase and anorthose in cross-form aggregates; euhedral nepheline, often highly fractured; abundant sodalite, often zoned, altering easily into stilbite, chabasite and natrolite; sphene occasionally present. The groundmass is composed of microlites of feldspar, nepheline, aegerine-augite, magnetite and ilmenite. Glassy interstitial material is probably analcite. The leuco-nephelinites (No. 18) have a porphyritic texture with abundant phenocrysts. These include nepheline, aegerine-augite and feldspars. The matrix is microlitic and glassy, inextricably mixed together. Under the microscope these rocks are somewhat difficult to diagnose. Accessory minerals include euhedral sphene and magnetite. The specimen here analysed appears to have considerable analcite, a hydrothermal alteration product of leucite.

Neiva summarized his account of the phonolitic rocks of Principe thus:

1. the rocks are holocrystalline, porphyritic, with a microlitic groundmass, the microlites being disposed in a trachytic character.
2. the phenocrysts which dominate are nepheline and sodic orthoclase, with aegerine-augite and barkevicite in smaller amounts, very small quantities of sphene and magnetite. According to the type of phonolite, there may be phenocrysts of sodalite and plagioclase (andesine or andesine-oligoclase).
3. in the groundmass, orthoclase, nepheline, aegerine-augite, with smaller quantities of sphene, magnetite, apatite, analcite, more rarely, sodalite, natrolite and stilbite, are present.
4. the rocks have an alkaline-sodic character, thus linking them to a normal foyaitic magma, sometimes showing relations to an urtitic magma, and less frequently, to an essexitic-foyaitic magma.

Neiva (1954) has described Nos. 13, 14, 15 and 16 in the table. Adopting his nomenclature, the following are the rock characteristics: No. 13, analcimitic essexite with olivine. Porphyritic texture with a doleritic base, showing tabular feldspars. Phenocrysts include: olivine passing into serpentinized and chloritized varieties; idiomorphic titanaugite, with zoned structure and inclusions of nepheline and natrolite; a little biotite; idiomorphic and hypidiomorphic magnetite. The matrix consists of: slivers of labradorite and some orthoclase; analcite resulting presumably from the alteration of feldspars: needles of natrolite in analcite; serpentinized olivine; augite; magnetite and idiomorphic nepheline. Very occasionally allotriomorphic calcite occurs. As per its chemistry, the rock approaches a gabbroic theralite and gabbroic essexite magma, and more particularly, to an ankaratritic magma.

No. 14, caltonite (analcimitic basanite). Meso-melanocratic rock, porphyritic texture and doleritic groundmass. Phenocrysts include: titaniferous augite predominates over olivine, also some labradorite. The augite is twinned and zoned, with inclusions of magnetite, rarely olivine. Idiomorphic olivine is slightly corroded, at times slightly serpentinized with magnetite inclusions. Labradorite has inclusions of augite. The matrix comprises prismatic microlites of labradorite and labradorite-andesine, idiomorphic and hypidiomorphic grains of titanaugite, idiomorphic crystals of magnetite, laminae and skeletons of ferrous hornblende, acicular apatite, some analcite with needles of magnetite, natrolite and hornblende, needles of natrolite, small amygdales of calcite, vitreous matter and isotropic palagonite. As per the parameters, the rock belongs to the ijolite family, approaching ankaratrites. The chemistry indicates a gabbroic theralite, more particularly, an ankaratritic magma.

No. 15, glenmuirite. Porphyritic texture, the groundmass of doleritic tendency. Phenocrysts include: hypidiomorphic olivine showing incipient serpentinization and inclusions of magnetite; zoned titanaugite with inclusions of olivine and magnetite; biotite, at times enclosing grains of olivine; magnetite. The doleritic matrix shows slivers of labradorite separated by analcite, many slivers of orthoclase, natrolite, needles of rutile and apatite crystals. As per the parameters, the rocks belongs to the theralite family, of essexite magma.

No. 16, analcimitic tephrite. Microlitic texture, tending towards trachytic. Many microlites of labradorite, idiomorphic and hypidiomorphic pigeonite, prisms of apatite, small needles of natrolite, analcite, fibrous chlorite, abundant granules of magnetite. As per the chemistry, the rock is classed as a theralitic-type magma, and by the parameters and mineralogical content, a tephrite.

As per the summation of NEIVA, the analcimitic essexite with olivine can be correlated with the ankaratrites of Annobon, the basalts, basanites and limburgites of Cabo Verde, the olivine-basalts of Madeira and the Azores. The glenmuirite and tephrite on a chemical basis, resemble the basalts of Fernando Poo, St. Helena and the Canary Islands, some trachydolerites of Ascension, the limburgites and dolerites of Cabo Verde, some trachydolerites of Madeira and some plagioclase-basalts of the Azores.

Table 28 shows the RITTMANN and NIGGLI parameters and coefficients for the Principe rocks listed in Table 27.

Table 28 — RITTMANN and NIGGLI Parameters and Coefficients (BARROS, 1960)

		1.	2.	3.	4.	5.	6.	7.	8.
RITTMANN Parameters	SiO_2	55.33	55.47	56.20	55.10	55.28	55.81	57.78	53.41
	alk	21.60	23.98	21.81	22.31	22.58	22.71	19.25	18.14
	k	0.26	0.19	0.26	0.23	0.20	0.24	0.29	0.25
	An	−0.03	−0.07	−0.04	−0.10	−0.03	−0.07	−0.04	−0.23
	fm	2.17	2.05	2.98	1.96	1.95	1.68	4.50	7.74
	ca"	10.14	3.31	2.37	3.54	2.23	3.09	2.82	7.89
	Group	D1	D1	C3	D1	D1	D1	C3	C3
NIGGLI Parameters and Coefficients	Q	23.14	21.27	23.80	23.36	22.35	24.10	27.50	23.11
	L	69.46	69.44	68.40	65.96	72.35	67.33	62.23	62.46
	M	7.39	9.28	7.75	10.64	5.28	8.57	10.40	14.52
	π	0	0	0	0	0	0	0	0
	γ	0.53	0.41	0.30	0.32	0.44	0.47	0.26	0.36
	μ	0.08	0.06	0.02	0	0.06	0.05	0.12	0.22
	α	−9.40	−8.08	−8.44	−5.81	−12.81	−7.27	−4.03	−3.82
	λ	18.79	14.96	17.65	12.40	27.40	15.71	11.95	8.60
	β	−0.50	−0.54	−0.48	−0.47	−0.46	−0.46	−0.33	−0.44
		9.	10.	11.	12.	13.	14.	15.	16.
RITTMANN Parameters	SiO_2	53.58	53.65	53.74	55.76	41.51	42.01	44.52	45.46
	alk	21.07	20.81	19.27	19.93	6.02	7.78	15.27	10.62
	k	0.22	0.29	0.36	0.28	0.13	0.12	0.16	0.20
	An	−0.05	−0.04	−0.09	−0.02	0.29	0.21	0.01	0.20
	fm	4.44	3.52	7.60	4.58	29.17	37.83	16.84	20.68
	ca"	4.04	3.37	5.43	2.85	11.68	10.71	9.69	5.82
	Group	D1	D2	C3	C3	D4	C8	D4	C8
NIGGLI Parameters and Coefficients	Q	21.89	22.19	23.36	24.40	16.12	11.25	14.19	19.74
	L	66.80	68.02	57.74	66.20	31.88	34.44	60.02	49.35
	M	11.27	9.77	18.89	8.84	51.98	54.29	25.77	30.89
	π	0	0	0	0	0.30	0.21	0.10	0.24
	γ	0.59	0.41	0.37	0.49	0.34	0.26	0.25	0.25
	μ	0.54	0.07	0.22	0.19	0.39	0.50	0.26	0.33
	α	−6.00	−7.11	−2.40	−6.69	−0.29	−0.64	−3.00	−1.27
	λ	11.85	13.92	6.11	14.97	1.22	1.26	4.65	3.19
	β	−0.51	−0.51	−0.39	−0.44	−0.23	−0.50	−0.64	−0.39

Table 28 — Rittmann and Niggli Parameters and Coefficients (Barros, 1960)

		17.	18.	19.	20.	21.	22.	23.	24.
Rittmann Parameters	SiO$_2$	53.60	54.74	48.88	43.01	48.70	58.08	44.30	44.33
	alk	19.89	21.35	12.36	3.75	11.72	18.67	7.15	3.90
	k	0.30	0.30	0.17	0.27	0.18	0.27	0.18	0.22
	An	0.04	−0.05	0.09	0.55	0.11	−0.02	0.30	0.51
	fm	3.39	5.81	16.56	31.90	17.45	4.03	25.57	31.35
	ca"	1.14	2.28	6.87	4.61	3.84	2.48	7.25	4.97
	Group	C3	D2	C7	B8	C7	C3	C8	B8
Niggli Parameters and Coefficients	Q	24.24	22.78	23.28	25.08	27.12	29.27	22.58	27.14
	L	71.65	66.83	49.63	32.89	50.85	63.22	38.83	31.55
	M	4.16	10.24	27.15	42.01	22.02	7.39	38.58	41.29
	π	0.05	0	0.09	0.55	0.12	0	0.30	0.51
	γ	0.29	0.21	0.36	0.16	0.23	0.51	0.28	0.17
	μ	0.25	0.21	0.64	0.48	0.42	0.03	0.33	0.52
	α	−16.96	−6.38	−1.07	0.22	−0.92	−5.22	−0.25	0.44
	λ	34.44	13.05	3.65	1.56	4.61	17.10	2.01	1.52
	β	−0.49	−0.48	−0.29	0.14	−0.19	−0.30	−0.12	0.28

		25.	26.	27.	28.	29.
Rittmann Parameters	SiO$_2$	61.36	49.99	41.84	42.11	39.50
	alk	16.27	5.52	4.15	7.39	4.75
	k	0.45	0.27	0.25	0.38	0.08
	An	–	0.41	0.46	0.35	0.44
	fm	4.90	17.72	36.65	28.56	39.51
	ca"	0.72	0.87	7.99	5.36	7.41
	Group	B2	B5	D8	C8	D8
Niggli Parameters and Coefficients	Q	35.79	32.89	17.74	18.30	13.80
	L	58.09	46.34	28.92	42.60	32.20
	M	5.96	20.77	53.27	39.10	54.00
	π	0.01	0.43	0.46	0.36	0.41
	γ	0.08	−0.05	0.21	0.58	0.18
	μ	0.12	0.41	0.48	0.39	0.46
	α	−0.13	0.28	−0.08	−0.77	−0.42
	λ	19.49	4.46	1.08	2.10	1.19
	β	−0.01	0.06	−0.07	−0.36	−0.35

Rittmann Classification

1. Light Nephelinite.
2. Light Nephelinite.
3. Nepheline-Phonolite.
4. Light Nephelinite.
5. Light Nephelinite.
6. Light Nephelinite.
7. Sodalitic-Phonolite.
8. Nepheline-Phonolite.
9. Light Nephelinite.
10. Light Leucitic-Nephelinite.
11. Nepheline-Phonolite.
12. Sodalitic-Phonolite.
13. Nephelinite.
14. Nepheline-Melano-Basanite.
15. Nephelinite.
16. Nepheline-Tephrite.
17. Nepheline-Phonolite.
18. Light Leucitic-Nephelinite.
19. Neph.-Phonolitic-Tephrite.
20. Olivine-Basalt.
21. Neph.-Phonolitic-Tephrite.
22. Nepheline-Phonolite.
23. Nepheline-Melano-Basanite.
24. Olivine-Basalt.
25. Alkali-Trachyte.
26. Olivine-Trachy-Andesite.
27. Mafic Nephelinite.
28. Nepheline-Melano-Basanite.
29. Mafic Nephelinite.

Table 29 is a summation of the NIGGLI values for the major groups of rocks discussed.
Table 30 presents data upon which JUNG & RITTMANN diagrams are constructed. (Vd. Sao Tomé for further remarks.)

Table 29   NIGGLI Values for Principe Rocks (BARROS, 1960)

Series	Group	Q	L	M	π	γ	μ	α	λ	β
Calc-Alkaline	Basalts	23.2	33.2	43.6	0.45	0.20	0.45	0.10	1.54	0.05
	Andesites	32.9	46.3	20.8	0.43	−0.05	0.41	0.28	4.46	0.06
	Trachytes	35.8	58.1	5.9	0.01	0.08	0.12	−0.13	19.50	−0.01
Alkaline	Basanitoids Basanites Limburgites	14.7	38.5	46.7	0.28	0.42	0.44	−0.70	1.68	−0.43
	Tephrites	23.3	49.9	26.6	0.15	0.28	0.46	−1.09	3.81	−0.29
	Phonolites	23.7	65.9	10.2	–	0.35	0.13	−6.50	14.40	−0.41

Table 30   JUNG and RITTMANN Values (BARROS, 1960)

Sample No.	Author	JUNG Values		RITTMANN Values	
		SiO_2 (%)	R	Si^0	Az^0
1.	NEIVA (1955)	55.33	10.7	0.65	0.64
2.	NEIVA (1955)	55.47	7.2	0.63	0.64
3.	NEIVA (1955)	53.41	21.1	0.68	0.62
4.	NEIVA (1955)	55.10	5.0	0.67	0.64
5.	NEIVA (1955)	55.28	6.9	0.63	0.64
6.	NEIVA (1955)	55.81	6.0	0.67	0.65
7.	NEIVA (1955)	57.78	11.1	0.75	0.65
8.	NEIVA (1955)	56.20	6.8	0.67	0.64
9.	NEIVA (1955)	53.58	14.7	0.65	0.62
10.	NEIVA (1955)	53.65	13.1	0.65	0.63
11.	NEIVA (1955)	53.74	18.5	0.69	0.62
12.	NEIVA (1955)	55.76	12.8	0.68	0.64
13.	NEIVA (1954)	41.51	77.4	0.64	0.46
14.	NEIVA (1954)	42.01	70.5	0.61	0.44
15.	NEIVA (1954)	44.52	40.7	0.53	0.50
16.	NEIVA (1954)	45.46	55.1	0.64	0.50
17.	BARROS (1960)	53.60	13.0	0.64	0.62
18.	BARROS (1960)	54.74	6.1	0.65	0.63
19.	BARROS (1960)	48.88	48.7	0.71	0.57
20.	BARROS (1960)	43.01	78.2	0.81	0.49
21.	BARROS (1960)	48.70	39.8	0.77	0.59
22.	BARROS (1960)	58.08	12.2	0.76	0.67
23.	BARROS (1960)	44.30	67.7	0.73	0.51
24.	BARROS (1960)	44.33	70.3	0.86	0.51
25.	BARROS (1960)	61.36	4.6	0.91	0.70
26.	BARROS (1960)	49.99	56.3	0.90	0.56
27.	BARROS (1960)	41.84	79.7	0.67	0.45
28.	BARROS (1960)	42.11	63.6	0.63	0.47
29.	BARROS (1960)	39.50	77.2	0.50	0.42

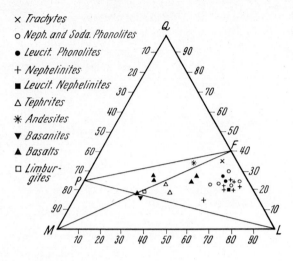

Fig. 44. QLM Diagram. (BARROS, 1960)

× Trachytes
○ Neph. and Soda. Phonolites
● Leucit. Phonolites
+ Nephelinites
■ Leucit. Nephelinites
△ Tephrites
✶ Andesites
▼ Basanites
▲ Basalts
□ Limburgites

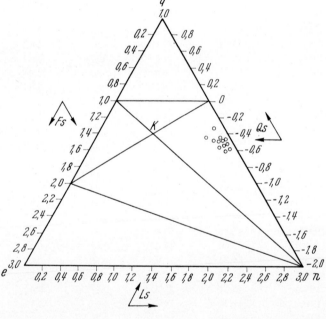

Fig. 45. Ls-Fs-Qs Diagram of Phonolites. (NEIVA, 1955b)

Laterites are well represented N of the parallel of the capital, Sto. Antonio, and SW thereof between Bella Vista and Abade (NEIVA & NEVES, 1957). In both areas, the landscape is mild, rather flat, basaltic rocks predominating in outcrop, and on laterite soils almost no cacao and inferior coffee production is found. In the superficial part of the laterites there is much fissuring into paralliped blocks, in some localities, and abundant development of fissuring so that the surface has a stoney appearance, e. g. between Oque Tres, Ponta do Sol and Sundy, at Pirimides, Sta. Rita and elsewhere. The laterite blocks,

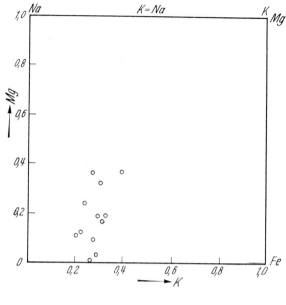

Fig. 46. K - Mg Diagram of Phonolites. (NEIVA, 1955b)

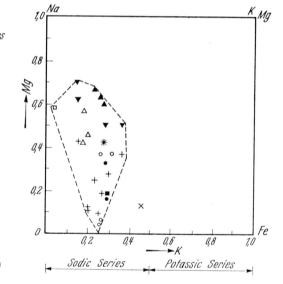

× Trachytes
○ Neph. and Soda. Phonolites
● Leucitic Phonolites
+ Nephelinites
■ Leucit. Nephelinites
△ Tephrites
✱ Andesites
▼ Basanites
▲ Basalts
□ Limburgites

Fig. 47. K - Mg Diagram. (BARROS, 1960)

immediately vesicular after their extraction, are fashioned easily with a machete into paralleliped shapes and used in the construction of single-storey houses, for example at Sto. Cristo, Nova Esperança and Belo Monte. In Principe laterites are known as 'Budo-Judeu'.

The laterites are ferruginous, chestnut-coloured, vesicular, or more exactly pluri-canalicular, colloformic and more rarely, pisolitic. The last-named type shows no structural

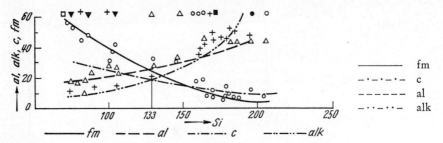

Fig. 48. Diagram of the Alkali Branch of Differentiation. (Barros, 1960)

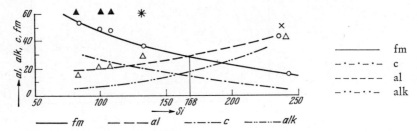

Fig. 49. Diagram of the Alkali-Calcic Branch of Differentiation. (Barros, 1960)

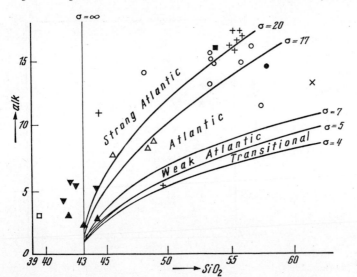

Fig. 50. Rittmann Serial Diagram. (Barros, 1960)

affinity with the mother-rock. Recently-fractured laterite has a chestmut colour on the walls of the canals or the vesicles, and in the interior, i. e. between the walls, the colour is reddish. Sometimes on the walls or then partially filling the vesicles and canalicules there is a whitish substance. The walls are composed of goethite and hematite, the former predominant, with which are associated limonite, gibbsite and kaolinite. Between the walls there is also goethite and hematite, but generally the latter is more abundant than the former, which former is associated with a little kaolinite.

Table 31 shows quantitative chemical analyses of four samples from the island.

Table 31 Chemical Analyses of Laterites (NEIVA & NEVES, 1957)

1. Vesicular and canalicular laterite, Uba-Chepique.
2. Vesicular laterite, Cascalheira.
3. Pisolitic laterite, road near Oca (S. Jorge).
4. Vesicular laterite, E of hill at Praia Grande.

	1.	2.	3.	4.
SiO_2	8.09	9.39	10.67	14.22
Al_2O_3	14.61	16.99	20.83	17.76
Fe_2O_3	51.73	51.54	42.77	40.55
FeO	0.52	2.06	0.16	1.13
CaO	–	0.12	0.09	0.23
MgO	0.06	0.14	0.07	0.13
K_2O	0.16	0.05	0.12	0.05
Na_2O	1.54	1.58	1.18	1.79
TiO_2	3.90	2.20	2.20	4.20
MnO	0.05	0.03	0.90	0.22
P_2O_5	2.28	1.05	1.69	1.31
H_2O+	16.85	14.59	18.89	18.02
H_2O-	0.04	0.04	0.05	0.06
Total	99.83	99.78	99.62	99.67

The origin of the laterites is related to the climatic and topographic conditions pertaining in areas where found. Lateritization generally occurs in more level regions under tropical environments. The genesis of autochthonous laterites has been explained by resorting to the influence of humic acids, ferrobacteria, climatic factors and the role of superficial waters. In Principe, the above authors believe that lateritization is a process of chemical degradation of the rocks by weather action in a tropical environment, essentially activated by the superficial imbibition of the rocks and by the slow movements of superficial waters over more flat terrain. The crust of the laterites has allowed the morphologic conservation of the surface, either plateau, plain or platform, but leads to sterilization of the soil.

Sedimentary Rocks

Sedimentary occurrences are very small. Shelly limestones are found at an elevation of some 130 m in the Forca valley, about 1.3 km SE of the capital, and at 900 m SSW of Oque Boi, in the Sta. Joana plantation. These outcrops intersect a water course which has its origin between Uba and the Ribeira Forca, the water emptying into the Rio Papagaio.

The limestones are highly fossiliferous, comprising abundant shells and internal casts of molluscs, gastropods and lamellibranchs, difficult to extract because of the calcareous cement which binds them together. Poorly-preserved micro-foraminifera, calcareous algae, spines and echinoid plates also are present.

The limestones, which rest on basalts, were first reported on by MENDELSOHN (1942), who also noted asphalt exudations and gaseous emanations. Further reference to the rocks has been given by TEIXEIRA (1949), BERTHOIS (1955), TEIXEIRA (1955), NEIVA (1958), SILVA (1956, 1958), SERRALHEIRO (1957) and DINIZ (1959). NEIVA (op. cit.) also mentioned crystalline limestones ca. 100 m SW of the trigonometrical mark at Abade, at an elevation of 70 m, which he considered likely contemporaneous with the calcareous limestones.

A chemical analysis of the limestone in the Forca valley (Table 32) clearly demonstrates the paucity in detrital quartz, seemingly confirmed in Table 33.

Table 32 — Chemical Analysis of Fossiliferous Limestone (BERTHOIS, 1955)

Soluble in HCl:	SiO_2	0.50	
	Al_2O_3	3.09	
	Fe_2O_3	1.15	
	P_2O_5	–	
	CaO	50.20	94.45
	MgO	0.10	
	CO_2	39.41	
	Cr	–	
Insoluble in HCl:	SiO_2	1.22	
	Al_2O_3	0.62	
	Fe_2O_3	0.21	
	CaO	0.19	2.29
	MgO	0.05	
	Na_2O	–	
	K_2O	–	
	H_2O+	2.40	3.01
	H_2O-	0.61	
	Total		99.75

Table 33 — Probable Mineralogical Composition of Fossiliferous Limestone (BERTHOIS, 1955)

Soluble Fraction:	Silica	0.50	
	Giobertite	0.17	
	Calcite	89.50	96.16
	Limonite	1.31	
	Gibbsite	4.68	
Insoluble Fraction	Quartz	0.12	
	$CaOSiO_2$	0.35	
	$MgOSiO_2$	0.10	2.38
	$FeOSiO_2$	0.26	
	Kaolinite	1.55	
	Water not used		1.19
	Total		99.73

According to TEIXEIRA (1955), these Burdigalian-Vindobonian limestones show facies characteristics very similar to the Miocene in the Lisbon and Algarve regions of Portugal, as well as having facies affinities with Miocene limestones in the Macaronesian islands. This opinion as to the age is identical with that of SILVA, but SERRALHEIRO though agreeing that the faunal content indicates a Miocene age, claimed it was not possible to be more precise.

SILVA believed that these Principe limestones were formed at depths ranging from 30 m to 70 m in a tropical environment.

On the right bank of the Ribeira Izé, near the mouth where it joins a small water course, at a place called Agua Petroleo, there is a small water well where gas bubbles through, and at times oil shows can be seen. In the valley of the Ribeira As-Duas-Aguas. Between Sta. Rita and Praia do Bombom, there are also oil shows. In both cases, the oil issues from fractures in the volcanics. NEIVA also reported a small outcrop of lignite in the upper reaches of a small stream where a swamp area occurs between Praia Grande do Norte and Santana. It is a thin exposure, intercalated with clays and dipping S at 40°.

At Praia Grande, beside the seashore, are beds of consolidated detritus, formed of volcanic fragments, especially basalts, and remains of bivalve shells and gastropods. These beds dip to the SSW at 9° and indicate a raised beach deposit. Conglomerates of basalt pebbles and cobbles and with a calcareous cement occur at Praia Cabinda and other localities.

Palaeontology

SILVA has listed the following fauna from the fossiliferous limestones of Principe:

Coelenterata
  Stylophora aff. *raristella* DEFRANCE

Echinodermata
  Clypeaster sp.

Lamellibranchiata
  Arca (Arca) turoniensis DUJARDIN 1837
  Arca (Senilia) aff. *senilis* LINNÉ 1758
  Pteria sp.
  Chlamys varia LINNÉ 1758
  Chlamys cf. *multistriata* POLI 1795
  Ostrea sp.
  Cardita (Venericardia) pinnula BASTÉROT 1825
  Cardita sp.
  Lucina (Loripinus) globulosa DESHAYES 1830
    cf. var. *hornea* DESMAREST 1901
  Codokia (Jagonia) aff. *pecten* LAMARCK 1818
  Laevicardium (Trachycardium) multicostatum BROCCHI 1814
  Dosinia exoleta LINNÉ 1758
  Tapes aff. *vetulus* BASTÉROT 1825
  Mactra subtruncata DA COSTA 1778 var. *triangula* RENIER 1804
  Tellina nitida POLI 1795
  Cuspidaria (Liomya) cf. *Dumasi* COSSMAN & PEYROT 1909

Gastropoda
  Turritella (Haustator) venus D'ORBIGNY 1852
  T. (Haustator) cf. *vermicularis* BROCCHI 1814
    var. *avermiculata* SACCO 1895
  T. (Zaria) cf. *pseudogradata* COSSMAN & PEYROT 1921
  T. (Turritella) eryna D'ORBIGNY 1825
  T. (Turritella) tricarinata BROCCHI 1814
  Tympanotonus margaritaceus BROCCHI 1814
    var. *Grateloupi* D'ORBIGNY 1852 race *Simplicior* VIGNAL 1910
  Strombus sp.
  Murex sp.
  Conus (Lithoconus) mercatii BROCCHI 1814
    var. *sharpeanus* COSTA 1866
  Terebra (Subula) aff. *plicaria* BASTÉROT 1825
  Actaeon sp.
  Cyclichna (Cyclichna) cf. *cylindracea* PENNANT 1777
  Scaphander sp.
  Umbraculum sp.

Fishes
  Sphyraena aff. *malembeensis* DARTEVELLE & CASIER 1943

Foraminifera
  Amphistegina lessonii D'ORBIGNY
  Textularia sp.
  Globigerina s. l. (*Globigerina* s. l. aff. *trilocularis* D'ORBIGNY)
  Triloculina gen?
  Quinqueloculina gen?
  Rotalia gen?

Algae
  Lithothamnium

SERRALHEIRO reported the following:

Lamellibranchiata
  Arca sp.
  Glycymeris sp.
  Ostrea sp.
  Cardita pinnula (BASTÉROT)
  Cardita sp.
  Lucina ornata AGASSIZ mut. simillima
    COSSMAN & PEYROT
  Phacoides columbella (LAMARCK)
  Cardium minimum PHILIPPI
  Cardium multicostatum BROCCHI
  Cardium sp.
  Dosinia lupinus (LINNÉ) var. nitens
    (DODERL) PANTANELLI
  Dosinia sp.
  Tellina nitida POLI
  Cordula sp.

Gastropoda
  Turritella aff. aspera SISMONDA
  T. tricarinata (BROCCHI)
  T. aff. tricarinata (BROCCHI)
  Turritella sp.
  Cerithium sp.
  Murex sp.
  (Tritonidae) sp.

Fish
  Tooth belonging to Sparidae

DINIZ described the following foraminiferal species from the localities referred to:

Praia do Abade
  Amphistegina lessonii
  Rotalia beccarii
  Quinqueloculina ungeriana
  Q. seminulum

Praia Neves Ferreira
  Amphistegina lessonii
  Rotalia beccarii
  R. rosea
  Textularia pseudorugosa

Praia Grande do Sul
  Amphistegina lessonii
  Rotalia beccarii
  R. rosea
  Quinqueloculina sp.
  Globigerinoides rubra
  Triloculina trigonula
  Cibicides refulgens

Praia Cabinda
  Rotalia beccarii
  R. rosea

Globigerina bulloides
Siphoninoides glabra
Cibicides refulgens
Quinqueloculina cf. bicornis
Triloculina gibba

Praia da Rib. Fria
  Amphistegina lessonii
  Rotalia beccarii
  Triloculina trigonula

Praia do Periquito, N area
  Amphistegina lessonii
  Rotalia beccarii
  Globigerina bulloides

Praia Caixao, N area
  Amphistegina lessonii
  Rotalia beccarii

Praia Rainha
  Rotalia beccarii
  Adeloaina laevigata
  Bolivina scalprata var. miocenica

Praia das Burras
  Rotalia beccarii

Praia Grande do Norte
  Neoalveolina philippinensis
  Quinqueloculina ungeriana
  Q. cf. seminulum

Praia de Sta. Rita
  Rotalia beccarii
  Quinqueloculina ungeriana

Praia da Ponto do Sol
  Rotalia beccarii
  Quinqueloculina viennensis

Praia Sundy
  Amphistegina lessonii
  Rotalia beccarii
  R. rosea

The two species of *Rotalia beccarii* and *Amphistegina lessonii* are the commonest, both in Principe and Sao Tomé. The foraminifera indicate a littoral facies, of Aquitanian age.

Economic Geology

The basalts, phonolites, tephrites and laterites are all used for construction purposes locally. Lime is made from the polyps which live in the marine waters close to the beaches, such as at Praia das Burras. Lime is also manufactured from the crystalline limestones, e. g. SW of Abade.

Cacao plantations are best developed on soils derived from basaltic rocks, the paraferralitic and ferrosialitic soils of CARDOSO & GARCIA (1962). Soils derived from phonolitic and tephritic rocks, the litho soils of these authors, are not nearly as good for cacao production. Soils derived on lateritic ground – paraferralitic soils with laterite – produce very little cacao but considerable coffee is grown instead. Coconut palms grow best on sandy soils of alluvial type, derived from phonolites and tephrites Forests are better developed in areas where the soils have developed from phonolitic and tephritic types of rocks, but in essence, forests and luxuriant dense vegetation is the natural vegetation of almost the whole island, the lateritic areas excepted.

Geological Evolution

The volcanic eruptions were formed before the Miocene, as the latter limestones rest on basalts. The time of vulcanism can be placed in the Cretaceous, probably at the end of Cretaceous times. In Principe, as in Sao Tomé, the basaltic occurrences were prior to the phonolitic outpourings. Already in the Miocene, much of the island was emergent. Evidence of this is seen in the outcrops of Miocene limestones occurring as high as 130 m above present sea level which were originally deposited from 30 m to 70 m below sea level, as per the ecological environments of the contained fauna. It can be postulated therefore that emergence of some 160 m to 200 m at least took place. Subsequently fracturing of the island took place, these fractures having in general an E-W, N-S and NW-SE trend, less often orientated NE-SW. The fractures formed channelways for the escaping magmas which, upon solidification and crystallization, formed basalts and trachytic dykes, the former being later. Following upon the phases of vulcanism and fracturing, gradual emersion occurred. In more recent times, modifications of the coastlines have developed, witnessed by the bevelled platforms, e. g. at Praia Grande do Sol, where consolidated detritals and older beach deposits, dipping to the SSW, are to be found. These beds are now being actively attacked by the waves, which leads us to presume, on the basis of their former position, that a slow immersion of the island has taken place.

THE SPANISH ISLANDS

CHAPTER 9

Fernando Poo

General

The island of Fernando Poo lies between lats. 3° 12' and 3° 48' N and longs. 8° 25' and 9° E, and at a distance of some 32 km from the Cameroun coast. Thus of all the South Atlantic islands treated here, Fernando Poo is situated closest to a mainland.

The area is 2009 km², being some 70 km in length and up to 37 km broad.

Pico Santa Isabel, the highest summit, rises to a height of 2972 m. The island is extremely mountainous, densely covered in vegetation, with usually abrupt and rocky coast-lines.

The volcanic soils are highly fertile; indeed, in spite of the ruggedness of the terrain, about one half of Fernando Poo is cultivated, and the island is quoted as being one of the most fertile areas of the entire West African coast. Much of the lower land in the N is given over to plantations, extending up to about 700 m. Cacao ist the principal agricultural product, but coffee, sugar cane, cotton, bananas, cassava, yams, tobacco, maize, copra, oil palms, cinchona bark and kola nuts all enter into the island economy. The lush forests support magnificent stands of ebony, oak, mahogany and other cabinet woods. At elevations varying from 1200–1500 m, pasturelands are found, the island being self-sufficient in dairy products. Higher yet is a semi-alpine vegetation, and the highest regions have moorlands. In the mountains around Moka European-type fruits and vegetables can be grown.

There are various small industries associated with the processing of agricultural products, also textile, paper, leather, soap and wood-carving industries.

The island population is about 62 000, of which only about one-third are indigenous Bubi, people of Bantu stock. Relatively large immigrations of Nigerians (22 000) and Camerounians are used in the harvesting of various crops, many of whom work only on a seasonal basis. Sta. Isabel, the capital, has a population of about 20 000, of whom 6000 are Europeans. San Carlos, the second town of the island, has about 13 000. It is thus evident that the island population is most unevenly distributed.

Physical Features

There is a high mountain range of volcanoes, craters, crater lakes and plateaux running down the axis of the island, and a smaller E-W trending range in the broader, southern part. From San Carlos to Concepcion extends a region of considerably lower terrain. In the major range, Pico Sta. Isabel is the highest point, whilst in the Gran Cordillera of the S, San Joaquin rises to 1951 m and San Carlos to 1882 m. Innumerable valleys, some many hundreds of metres deep, with abundant waterfalls, descend precipitously to the coasts, giving a highly dissected morphology and strong relief.

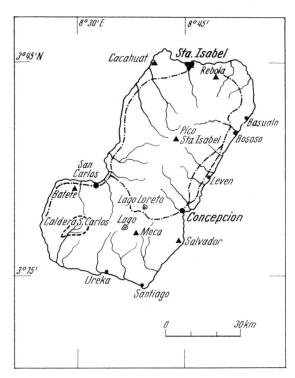

Fig. 51. Fernando Poo

Most of the coastline is cliffed and rocky. The coastal belt has a maximum altitude of 20 m, rising steeply inland. Only in the N, forming the hinterland of the capital, is there any appreciable extent of lower, milder topography.

Climate

Much of the lower areas of Fernando Poo experience a hot, humid, rather unhealthy climate, with the exception of the southern parts. Around Moka, in the highlands, where grasslands occur, the climate is indeed bracing and healthy, although much rain falls. The island experiences an equatorial-type climate, modified by island shape and maritime influences.

At. Sta. Isabel, the capital, the average maximum and minimum daily temperatures are 30° C and 21.7° C respectively, the average annual rainfall, 1798 mm, the average relatively humidity at all hours, 90 %. Along coastal stretches, the mean temperature is 27° C, with little variation during the year. Above an elevation of some 600 m, the climate is more healthy, and in some regions, e. g. around Moka, the climate is even temperate, although everywhere there is high rainfall and much more humid than in actual temperate climes.

July to October are the wettest months throughout the island, the rain-bearing winds coming from the SW, and hence the SW part of Fernando Poo receives most rain – Ureka has an average annual rainfall of 6000 mm, of which 1400 mm fall in the month of June! At Moka the average annual rainfall is 4000 mm, with a maximum of some 750 mm in September. December is the warmest month, August the coolest. The average annual temperature at Sta. Isabel, the capital, is 25° C, at Ureka, 23.7° C and at Moka, 18.5° C. The cooler, drier winter months in the more northerly highland areas, e. g. around Moka, are favoured by Europeans not only from the island itself but from neighbouring areas of West Africa as a place to enjoy more bracing, temperate conditions during holiday time. Everywhere during the summer months the climate is too wet, too humid, too utterly enervating to afford any incentive towards tourism.

Geology

Only volcanics are known in Fernando Poo, the island being a compound cone, a succession of flows, pyroclastics and agglomerates of essential basaltic composition.

Pico Sta. Isabel was the principal focus of eruption, the emissions from the centre occupying the northern part of the island. Rocks of similar composition but more recent in age occur in the southern part of the island, these being related to eruptive centres at Moka and Musola.

The eruption of highly fluid lavas from the three principal centres of Pico Sta. Isabel, Moka and Musola was interrupted by explosive phases. Small but very numerous craters, with vertical walls, were formed as a result of these explosions, the most notable of which are at Sta. Isabel and Moka. At the former site, a combination of marine erosion and subsidence has converted the breached crater into a small embayment, forming a natural harbour and thus favouring the siting of the capital with harbour facilities. The crater at Moka has been transformed into a lake which is fed by rains only.

As distinct from other islands in the Gulf of Guinea, trachytic rocks are unknown and the presence of phonolitic rocks is questionable.

The structural and textural variations amongst the basaltic rocks are indeed great, having little relation to the basic mineralogical composition of said rocks. FUSTER (1950, 1954) stated that every imaginable texture-structure between the compact, porous and vesicular basalts on the one hand and the porphyritic, coarse type and those totally aphanitic on the other hand, presented an extraordinary interest. Just as worthy of note is the remarkable constancy in mineral composition, the basalts having olivine, diopsidic or titaniferous augite, basic plagioclases, titaniferous magnetite and ilmenite, along with the normal accessories typical for such rocks.

FUSTER (1954), adopting the nomenclature of W. E. TRÖGER (1935, 1938), classified the basalts of Fernando Poo into:

1. Basalts, either without olivine or then very little. Have abundant phenocrysts of diopsidic augite, sometimes of plagioclase. Groundmass rich in plagioclase.
2. Olivine-basalts, rich in olivine and diopsidic or titaniferous augite. Varieties poor in plagioclase are transitional to the following groups.
3. Picritic basalts, very little plagioclase, and phenocrysts of olivine only.
4. Oceanites, very little plagioclase and abundant phenocrysts of olivine and augite, the latter sometimes titaniferous.
5. Ankaramites, very little plagioclase, phenocrysts of augite only.

Of these varieties, the olivine-basalts are by far the commonest.

Differences occur in mineral composition between the flow basalts and those forming pyroclastic material, the latter in general being more scoriaceous and poorer in phenocrysts, at times very vitreous. However the mineral composition of the phenocrysts is similar to that found in holocrystalline varieties.

Petrographic information on Fernando Poo rocks is given in the publications of SCHUSTER (1887), BOESE (1912) MACPHERSON (?) and FUSTER (1950, 1954). To date, we have a total of only 16 chemical analyses of such rocks, two from BOESE, four from FUSTER (1950) and the rest from FUSTER (1954).

Although the Fernando Poo rocks are overwhelmingly basaltic, indeed essentially so, there are chemical variations such that it is possible to place the rocks in groups. The principal variation tendency refers to the relative proportion of alkaline elements, allowing of a four-fold grouping (vd. Table 34). Tables 35 and 36 show the chemical analyses-norms and the mineralogical composition values of NIGGLI respectively.

Table 34 Petrochemical Grouping of the Basalts (FUSTER, 1954)

Petrochemical Types	Petrographic Types – Mineralogy
1. Alkaline Basalts	Basalts (without olivine or then very little.) Olivine-basalts, poor in the FE-Mg minerals.
2. Basic Sub-basalts	Typical olivine-basalts, rich in olivine or then augite.
3. Ultrabasic Sub-basalts	Picritic basalts, oceanites, ankaramites.
4. Ordinary Basalts	A basalt rich in plagioclase.

Alkaline Basalts. Specimens 1 to 4 have a strongly alkaline character, in which the alk parameter is high enough and the si parameter low enough so that in the computed mineral composition there is an appreciable proportion of nepheline. Accordingly, such rocks may be termed basanitoids, as per LACROIX, or then trachydolerites. FUSTER however preferred to name them alkaline olivine-basalts, being microscopically very similar to alkaline basalts without olivine. Nos. 1 and 2 have a high alk coefficient and a normal alumina proportion, whilst Nos. 3 and 4 have relatively less alumina yet very little alkaline variation. Such a variation in molecular values means a difference in the proportional of mineral constituents, so that whilst Nos. 3 and 4 have abundant Fe-Mg minerals (chiefly olivine), Nos. 1 and 2 have more plagioclase and less olivine.

Table 35 Chemical Analyses and Norms (Boese, 1912, Fuster, 1950, 1954)

	1.	2.	3.	4.	5.	6.	7.	8.
SiO_2	45.77	45.70	45.59	46.42	47.48	47.79	46.50	46.77
Al_2O_3	13.86	12.93	11.21	12.28	12.31	12.13	12.70	11.02
Fe_2O_3	8.97	5.74	6.58	1.56	3.97	4.12	4.88	6.41
FeO	5.95	9.68	8.37	9.89	9.29	8.57	8.49	7.88
MnO	0.36	0.13	0.24	0.10	0.18	0.27	0.26	0.32
MgO	5.54	5.53	8.31	11.03	8.34	7.33	8.30	10.71
CaO	7.71	10.31	8.45	11.63	9.41	11.42	9.92	7.28
Na_2O	3.83	3.44	3.66	2.59	2.92	2.67	2.38	2.88
K_2O	2.77	1.90	2.09	1.91	1.86	1.73	2.41	1.52
TiO_2	3.39	3.75	3.42	2.28	2.72	2.78	2.50	3.03
P_2O_5	1.26	0.19	1.42	0.29	1.42	0.68	0.74	1.10
H_2O+	0.29	0.44	0.46	0.31	0.49	0.51	0.70	0.78
H_2O-	0.09	a	0.14	a	0.03	0.07	0.38	0.38
Total	99.79	99.84	99.94	100.29	100.42	100.07	100.16	100.08

a) Dry at 110° C

	1.	2.	3.	4.	5.	6.	7.	8.
Or	16.5	11.5	12.5	11.0	11.0	10.7	14.5	9.0
Ab	30.0	19.9	27.3	6.7	26.5	21.3	19.6	26.3
An	13.0	14.7	8.2	16.2	15.2	16.5	17.4	12.9
Ne	3.1	7.1	3.7	9.9	–	2.0	1.3	–
Di	14.7	29.3	20.8	31.7	19.0	29.9	22.9	13.9
En-Hy	–	–	–	–	3.8	–	–	16.2
Ol	6.3	5.5	12.9	19.2	13.7	9.7	14.1	8.2
Mt	7.8	6.2	7.0	1.7	4.2	4.5	5.2	6.8
Il	4.8	5.4	4.8	3.0	3.8	4.0	3.6	4.4
Ap	2.6	0.4	2.8	0.6	2.8	1.4	1.4	2.3
Hm	1.2	–	–	–	–	–	–	–
Q	–	–	–	–	–	–	–	–

The alkaline basalts with a high alumina content and rich in the plagioclases are thus comparable to an essexitic gabbroic magma (No. 2) or then in those varieties having more alk, to a theralitic gabbroic magma (No. 1). Specimen No. 3, though being richer in olivine and having less feldspathoid, can also be compared to the former magma type. Specimen No. 4, being more femic, is to be related rather to a kaulaitic magma. This last-mentioned specimen has a lower alkali content than the other three, but on the other hand, a higher calcium content. This shows certain analogies with the ankaratritic or hornblenditic magmas, the specimen thus being more related to the alkaline basalts of Annobon than those of Fernando Poo.

Basic Sub-basalts. The olivine-basalts have not such a high alkali content as those mentioned above. Nos. 5–13, though showing some small individual variations, agree closely with rocks having a sub-basaltic chemictry, as established by C. Burri & P. Niggli (1945). Between the alkaline olivine-basalts and those olivine-basalts with a sub-basaltic chemistry there is no clear demarcation but rather a gradual transition from one to the other. However, Fuster takes a value of less than 7.5 alkali parameter, for

Table 35   Chemical Analyses and Norms (BOESE, 1912, FUSTER, 1950, 1954)

	9.	10.	11.	12.	13.	14.	15.	16.
SiO_2	48.65	47.60	45.21	46.85	45.73	46.01	45.44	47.80
Al_2O_3	11.13	12.16	12.76	10.25	11.20	7.86	7.16	13.70
Fe_2O_3	4.59	3.54	6.35	4.02	6.46	4.45	4.47	6.31
FeO	9.18	9.30	8.62	8.91	5.53	7.76	8.30	5.82
MnO	0.21	0.16	0.04	0.07	0.54	0.22	0.21	0.10
MgO	9.53	11.34	9.68	12.59	11.35	15.53	18.50	7.46
CaO	9.20	9.05	8.14	10.41	10.44	12.53	9.86	9.27
Na_2O	2.93	2.83	1.60	2.09	2.19	1.50	1.63	2.26
K_2O	1.03	0.82	1.42	1.10	1.35	1.15	0.78	1.62
TiO_2	2.46	2.28	2.91	2.57	3.22	2.30	2.27	2.87
P_2O_5	0.91	0.55	0.22	0.23	0.21	0.42	0.56	0.52
H_2O+	0.10	0.25	2.52	0.40	1.20	0.35	0.91	1.98
H_2O-	0.07	0.26	a	a	0.50	0.13	0.08	0.64
Total	99.99	100.14	99.47	99.49	100.13	100.01	100.20	100.42

a) Dry at 110° C

Or	6.2	4.7	8.7	6.5	8.2	6.7	4.5	10.2
Ab	26.5	25.3	14.8	18.7	20.2	12.5	14.3	21.1
An	14.5	18.2	24.5	15.5	17.3	11.3	10.0	23.2
Ne	–	–	–	–	–	0.5	–	–
Di	21.1	18.7	12.8	28.2	27.2	37.8	28.3	18.9
En-Hy	14.4	7.4	25.4	5.1	4.1	–	5.8	12.7
Ol	7.2	17.6	2.1	17.9	11.1	22.5	28.4	–
Mt	4.8	3.8	7.0	4.2	6.9	4.7	4.6	6.9
Il	3.4	3.2	4.2	3.6	4.6	3.2	3.0	4.2
Ap	11.9	1.1	0.5	0.5	0.4	0.8	1.1	0.6
Hm	–	–	–	–	–	–	–	–
Q	–	–	–	–	–	–	–	2.2

the silica parameter of 95, or an alkali parameter of 10 and a silica parameter of 110 as representing the range of the sub-basalts, which range would include all the olivine-basalts of Fernando Poo.

A study of Tables 35 and 36 shows interesting correlation between the degree of alkalinity and acidity within this group, namely, as the molecular proportion of silica decreases, the alkali content also becomes less. This of course is a normal variation occurrence in any basaltic series, but in the sub-basalts of Ferenando Poo, the alkali decrease is great enough to lead towards considerable basicity. This in turn has led FUSTER to establish the terms 'absolute alkalinity' and 'relative alkalinity'. The former will hold in a rock series becoming progressively more basic where the average diminution in the proportion of silica is accompanied by a similar diminution in the alkali content. 'Relative alkalinity' on the other hand will occur in a similar progressively more basic series where the diminution in alkali proportion is slow, so that the relative alkalinity on each occasion is greater, as the proportion of nepheline, or the ratios Ne/Ab or Ne/Feld will become larger in proportion as the deficit in silica becomes greater. In the same series it is therefore possible to envisage a progressive decrease in alkalis such that the previously-indicated relations remain constant, in which case the relative alkalinity of all the rocks in the series would remain the same. Should the alkali decrease be more accentuated in either instance, then the absolute alkalinity would then become a relative alkalinity in all the series. In other words, of two basalts which have the same silica parameters, that one is more alkaline which has a higher alkali value; of two basalts having the same alkali parameters, that one is more alkaline which has a lower alkali value.

The variation in the sub-basalts of Fernando Poo correspond to the latter condition mentioned, i. e. in general terms, the more basic are less alkaline, absolute as well as relative, than those of greater acidity where the silica values are exactly analogous to the alkaline basalts referred to above.

Of the analyses shown, Nos. 5, 6, 7 and 8 show an alkali-sodic chemistry more or less agreeing with an essexitic gabbroic magma – one of the least alkaline within the 'Atlantic' series. No. 8, which is somewhat more femic than the other three specimens, allies itself rather with a kaulaitic magma. No. 9, more plagioclasic and less femic than the above samples, is also less rich in alkalis than any of the samples Nos. 1–8, No. 4 excepted. The former feature allies the rock to an essexitic gabbroic magma, but the latter feature is more indicative of the normal gabbro magma. Nos. 10, 11 and 12 are relatively poor in alkalis yet show adequate femic characteristics, and can thus be considered as chemically intermediate between the alkaline-sodic series or kaulaitic magmas and the calc-alkaline series or hornblenditic magmas.

The calculated mineralogic composition of all these sub-basalts is in agreement with the chemical characteristics mentioned above. Only Nos. 6 and 7 contain nepheline; the rest have adequate silica to deny the formation of feldspathoids. In these basic sub-basalts, the total proportion of ferro-magnesian elements is high, on an average totalling 40 % by weight of the rock. Of such minerals, pyroxene is most abundant – about 30 % of the mineral content of the rocks being of pyroxene constitution. In this respect, these sub-basalts of Fernando Poo differ greatly from the typical alkaline basalts of the island, which in general have a much reduced ferro-magnesian content.

Ultrabasic Sub-Basalts. Common in the island are basaltic rocks very rich in phenocrysts of olivine and augite. According to the relative proportions of one or the other of these two minerals, these ultrabasics can be termed picritic-basalts (vd. reference to Tröger above) where the olivine proportion exceeds that of augite, oceanites where the two minerals are about equal in proportion, and ankaramites where pyroxene is much more abundant than olivine. Of these, oceanites and ankaramites are most common in Fernando Poo – vd. Nos. 14 and 15 respectively. The chemical features of these two specimens classify them as sub-basalts of extreme basic nature and unusually femic. Due to the small alkali content, these rocks show no relation to 'Atlantic' magmas, and are rather directly analogous to calk-alkaline magmas of strong basicity, especially to dialaguitic magmas.

The calculated mineral composition of these two rocks is worthy of note. In both there is a large olivine proportion and an even greater proportion of pyroxene, the pyroxene-olivine ratio being about 2 : 1. Very likely these ultrabasic sub-basalts are but a continuation of the basic sub-basalt series, the transition from one to the other series being gradational.

Table 36   Niggli Values (Boese, 1912, Fuster, 1950, 1954)

	1.	2.	3.	4.	5.	6.	7.	8.
Q	21.8	21.3	19.7	19.9	23.7	23.6	22.8	22.3
L	38.8	34.7	32.5	30.2	31.6	31.1	32.2	28.9
M	39.4	44.0	47.8	49.9	44.7	45.3	45.0	48.8
si	109	104	101	94	106	105	103	102
al	19.5	17.5	14.5	14.5	16.0	16.0	16.5	14.0
fm	48.0	47.5	54.5	52.5	52.5	48.0	51.5	60.5
c	19.5	25.0	20.0	25.5	22.5	27.5	23.5	17.0
alk	13.0	10.0	11.0	7.5	9.0	8.5	8.5	8.5
k	0.32	0.27	0.27	0.32	0.29	0.30	0.40	0.25
mg	0.39	0.40	0.51	0.63	0.53	0.51	0.53	0.58
π	0.20	0.25	0.15	0.32	0.38	0.32	0.32	0.27
γ	0.16	0.27	0.18	0.25	0.18	0.27	0.21	0.12
μ	0.33	0.29	0.41	0.47	0.44	0.37	0.42	0.51
α	−0.31	−0.14	−0.12	−0.01	0.18	0.19	0.09	0.19

	9.	10.	11.	12.	13.	14.	15.	16.
Q	24.5	23.1	24.2	21.6	21.5	18.5	17.0	27.4
L	28.3	28.9	28.8	24.5	27.4	18.8	17.3	32.7
M	47.2	48.0	47.0	53.9	51.1	62.7	65.7	39.9
si	101	100	98	97	96	87	83	113
al	19.5	15.0	16.5	12.5	14.0	8.5	7.5	19.5
fm	52.5	57.5	59.5	59.0	56.5	62.5	69.5	49.5
c	20.5	20.5	19.0	23.0	23.5	25.0	19.5	23.5
alk	7.5	7.0	5.5	5.5	6.0	4.0	3.5	7.5
k	0.19	0.16	0.36	0.26	0.29	0.33	0.24	0.32
mg	0.56	0.61	0.53	0.65	0.63	0.70	0.73	0.53
π	0.31	0.38	0.51	0.38	0.38	0.36	0.35	0.51
γ	0.18	0.24	0.11	0.20	0.21	0.24	0.17	0.19
μ	0.46	0.52	0.47	0.51	0.47	0.53	0.60	0.43
α	0.36	0.24	0.32	0.29	0.19	0.28	0.24	0.42

1. Basalt without olivine, San Carlos-Moka road, km 19.
2. Basalt poor in olivine, Vicinity of Basilé.
3. Basalt poor in olivine, Ilache cascades, Moka.
4. Olivine-Basalt, San Carlos Bay.
5. Olivine-Basalt, Beginning of road to Concepcion from Moka.
6. Olivine-Basalt, San Carlos-Batete road, 2 km from Batete.
7. Olivine-Basalt, Moka crater.
8. Olivine-Basalt, Sta. Isabel-Basakata road, km 20.
9. Olivine-Basalt, Basakata, near bridge.
10. Olivine-Basalt, Concepcion.
11. Olivine-Basalt, San Carlos Bay.
12. Olivine-Basalt, San Carlos Bay.
13. Olivine-Basalt, Sta. Isabel.
14. Ankaramite, San Carlos-Batete road, at Barcelonesa.
15. Picrite-Basalt, Sta. Isabel, Pta. Cristina.
16. Basalt, SW of Tom-Yalla farm.

Ordinary Basalts. An analysis by Boese (1912) of a rock from a large block occurring in basalt flows is given in No. 16. As the analysis shows it to be more acidic than others discussed above, it can be classed as a normal basalt. The feldspar content is high, more akin to that found in the alkaline basalts, the silicon content much greater than in any of the sub-basalts.

This basalt is intermediate between the essexitic gabbroic and normal gabbroic magma series. FUSTER was of the opinion that this type of basalt was somewhat rare in the island, for of all the specimens studied by him in person, none belonged to this category.

BOESE (op. cit.) claimed to have located a roughly-rounded phonolite pebble on the beach S of Basuola, but FUSTER was most sceptical about this, the more so as the description of the locality where the specimen was said to have been found is too vague.

BARROS claimed that the rocks which FUSTER classified as alkaline basalts were true basanitoids, and therefore the presence of virtual nepheline is not to be ignored, ranging from 3.1% (No. 1) to 9.9% (No. 4). BARROS thus claimed it was more correct to state that in the island we have fundamentally olivine-basalts diverging into opposite trends, one orientating towards picritic-basalts and ankaramites, the other towards basanitoids.

The rocks studied by BOESE and refered to by both JÉRÉMINE & FUSTER, verify that one of these (No. 16) contains free silica, approximating to a basalt occurring at Mt. Koupé in the Camerouns.

JÉRÉMINE summarized her findings of the Fernando Poo rocks by saying: "Ses laves sont du même type minéralogique que plusieurs de celles que nous avons étudiées dans les différents massifs du Cameroun". Further, BOESE also referred to a basalt with hypersthene in Fernando Poo, and a basalt with enstatite was mentioned by MESCH from Mont Grand in the Cameroun, from which JÉRÉMINE concluded that "la présence de ces pyroxènes ferro-magnésiens accentue la parenté des deux régions".

Table 37   JUNG and RITTMANN Values for Fernando Poo Rocks (BARROS, 1960)

Specimen Number	JUNG		RITTMANN	
	SiO_2 (%)	R	$Si°$	$Az°$
1.	45.77	53.8	0.71	0.52
2.	45.70	65.8	0.74	0.50
3.	45.59	59.5	0.70	0.50
4.	46.42	72.1	0.72	0.48
5.	47.48	66.3	0.77	0.51
6.	47.79	72.1	0.76	0.50
7.	46.50	67.4	0.76	0.50
8.	46.77	62.3	0.75	0.50
9.	48.65	69.9	0.76	0.50
10.	47.60	71.2	0.77	0.50
11.	45.21	72.9	0.80	0.49
12.	46.85	76.6	0.80	0.49
13.	45.73	74.6	0.77	0.48
14.	46.01	82.5	0.72	0.46
15.	46.44	80.3	0.72	0.45
16.	47.81	70.4	0.86	0.53

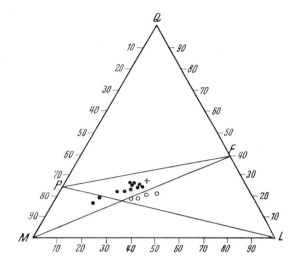

Fig. 52. QLM Diagram. (Fuster, 1954)

● Sub-Basaltic
○ Alkaline-Basaltic
+ Ordinary Basaltic

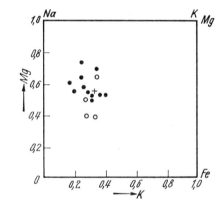

Fig. 53. K - Mg Diagram. (Fuster, 1954)

● Sub-Basaltic
○ Alkaline-Basaltic
+ Ordinary Basaltic

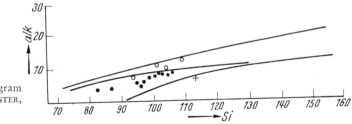

Fig. 54. Si - Alk Diagram of Basaltic Rocks. (Fuster, 1954)

CHAPTER 10

Annobon

General

Annobon, with an area of only 17 km², lies between lats. 1° 24′ and 1° 28′ S, longs. 5° 36′ and 5° 38′ E. The island measures some 6 km in length and 2 km broad. This Spanish island is under the administrative authority of Fernando Poo.

Like Fernando Poo, the island is remarkably beautiful, with an alternation of picturesque valleys and steep mountainous terrain. Santamina, the highest elevation, 700 m, in the southern part of the island, is an imposing height for such a small island.

The outline of Annobon shows several peninsulas projecting into the ocean. The S and W coastlines are more indented than elsewhere.

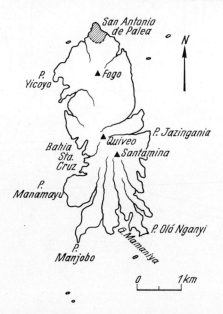

Fig. 55. Annobon. (LOBOCH, 1962)

Pico de Fogo, 450 m, in the N, is a bare volcanic cone with a small crater lake nestled within the summit, the surface of the lake being 200 m above sea level. During extended dry periods, the lake dries up. A rivulet drains northward from the lake to debouch into the sea W of the capital.

Of the many valleys only three include permanent streams.

There is a rich, varied and dense vegetal and forest growth. In earlier times, Dutch and British sailors would call at Annobon to fell the hard woods, e. g. oak, for various wooden constructions required on board ship. Agricultural land occurs as clearings in the larger valleys and along certain coastal stretches. Oranges, lemons, pineapples, bananas, mangos, papayas, cacao, coffee and cotton plants are all raised on Annobon. The surrounding ocean waters provide a plentiful supply of many kinds of fish, the chief diet of the people.

When first discovered by the Portuguese in 1471 there were no signs of human habitations, no four-footed animals but abundant bird and insect life.

Cotton weaving into items of clothings, mostly sent to Fernando Poo, is the chief industry.

The population of Annobon is about 1400. San Antonio de Palea, the capital, lies at the extreme N end.

Climate

Annobon enjoys a salubrious climate.

Three climatic seasons are recognized (LOBOCH, 1962): October to April is the wet time of year, May to July the dry season, August and September the transitional period. Stormy weather is very frequent in October, March and April. January and February are the warmest months.

In the dry season, strong winds are common, perhaps lasting five or six days at a time, raising large clouds of dust high into the skies over the island. The dry months can be most critical for the islanders, as then temperatures are lowest, fishing is often at a standstill because of the turbulent seas, crops may be jeopardized. This is the time of year when illness and deaths are most frequent, giving rise to the local expression "Epoca de muerte".

Geology

Like Fernando Poo, only volcanics are known in Annobon. According to SCHULTZE (1913) the island consists of basaltic flows along with some 'yellowish tuff'. The basaltic basement is cut by many basaltic dykes, measuring up to 20 m in thickness. The imposing crater of Quioveo, 640 m, constitutes the central part of the island, and the latest outpourings from the vent here flowed outwards and downwards from a gap in the NW rim of the crater. Although vulcanism appears to be a recent phenomenon, there is no historical record of such. Pico de Fogo, a trachytic plug, thrusts up through the northern rim of Quioveo, forming a monolithic hill, an easily recognizable landmark because of its whitish colour.

FUSTER (1950, 1954) stressed the strongly basic character of the Annobon rocks, indeed more ultrabasic than to be found in the other islands of the Gulf of Guinea. TYRRELL

(1934) studied closely five trachyte specimens, gave a chemical analysis of one and made appropriate comments on the trachytes of the island. FUSTER referred to the recognition by TYRRELL of trachytes but made scarce any further comment, and decided not to incorporate the chemical analysis of TYRRELL's biotite-trachyte as it showed an anomalous excess of alumina and was thought to be contaminated and had undergone secondary alterations. As far as the writer is aware, TYRRELL never replied in print to these opinions of FUSTER (TYRRELL died after the appearance of FUSTER's first paper), nor are there any publications other than those of these two geologists treating of the petrography of the Annobon volcanics, other than that of BARROS which latter merely quotes opinions of FUSTER. This perplexing problem of the trachytes is thus an open one, and all that can be said is that FUSTER had a better acquaintance with the island than TYRRELL, but it would be imprudent to reject the findings of the late Dr. TYRRELL, a petrologist of international repute.

Petrography

The strongly basic nature of the oceanites and picritic basalts referred to above, is only surpassed by analagous rocks in the Canary Islands, Cabo Verde and the Brazilian island of Fernando de Noronha (q. v.). Though showing extreme basicity, the content of alkalis is, in general, sufficiently high so that the rocks can be classed as alkali basalts of extremely basic type. It is to be noted, however, that the TYRRELL specimen (Table 38, No. 4) of oceanite has a considerably lower alkali content than the three FUSTER specimens. Adopting the magmatic series of NIGGLI, it can be said that the oceanites and picritic basalts of Annobon are transitional between hornblenditic and kaulaitic magmas, i. e. intermediate between alkaline and calc-alkaline magmas. However the relative alkalinity of the Annobon rocks is greater than in typical calc-alkaline magmas.

As regards the molecular norms, we note a high proportion of nepheline, feldspars less than 30%, but abundant ferro-magnesian content. Specimen No. 4 has a dominance of olivine over pyroxene, which thus lends itself to be classed rather as a picritic basalt. In No. 3 on the other hand, there is double the quantity of diopside over olivine. In Nos. 1 and 2, pyroxene predominates, but less so than in No. 3.

According to TYRRELL oceanites are represented in Annobon by their slaggy representatives, limburgites. The abundant oceanites are greyish-black in colour with many crystals of green olivine which are very fresh in thin section. The matrix comprises a mixture of microlites of labradorite, magnetite and dark glassy matter. When this glassy material becomes dominant we have limburgites. Olivine nodules are also common in the island, consisting either of pure olivine or then olivine, augite and magnetite rock.

The basic rocks of Annobon appear to be but an extension of the tendency towards variation of alkaline basalts established in Fernando Poo, and in all likelihood, the rocks of both islands issued from the same magma, modified by similar processes.

As per TYRRELL, the trachytes are compact, whitish rocks, rather chalky in appearance, and further, some are porous and friable. Under the microscope they consist almost entirely of sanidine, occurring as minute crystals, constituting the groundmass, and some phenocrysts of the same mineral. The feldspar is soda-orthoclase, rich in the albite molecule. The only coloured minerals present are flakes of biotite, nearly always altered to

Table 38   Data regarding the Volcanics (Tyrrell, 1934, Fuster, 1954)

	Chemical Analyses						Norms						Niggli Values			
	1.	2.	3.	4.	5.		1.	2.	3.	4.			1.	2.	3.	4.
SiO_2	42.24	43.31	43.21	39.20	61.55	Or	9.5	9.3	10.7	6.2		Si	80	82	85	70
Al_2O_3	10.63	8.89	9.79	12.49	21.15	Ab	1.8	3.8	7.2	0.8		Al	12.0	11.0	11.5	13.0
Fe_2O_3	4.30	4.40	6.12	3.47	1.24	An	13.6	12.6	12.5	22.2		Fm	60.0	60.5	57.5	63.5
FeO	8.78	9.48	9.13	9.48	0.84	Ne	13.6	12.6	12.5	10.0		c	21.0	21.5	22.5	19.0
MnO	0.09	0.09	0.13	0.15	–	Di	30.9	33.3	36.0	17.8		Alk	7.0	7.0	8.5	4.5
MgO	14.11	14.04	11.22	16.52	0.30	En-Hy	–	–	–	–		k	0.28	0.27	0.28	0.26
CaO	10.50	10.58	10.71	9.88	0.65	Ol	22.5	21.8	13.8	32.8		mg	0.66	0.65	0.58	0.70
Na_2O	2.74	2.76	3.06	1.98	5.31	Mt	4.5	4.5	6.5	3.6		Q	13.7	14.1	11.7	13.0
K_2O	1.58	1.62	1.82	1.05	5.55	Il	4.6	4.6	5.2	5.2		L	27.7	26.3	27.9	27.5
TiO_2	3.29	3.24	3.66	3.70	0.40	Ap	0.4	0.3	0.3	1.4		M	58.6	59.6	57.4	59.5
P_2O_5	0.33	0.17	0.17	0.71	0.28	Q	–	–	–	–		π	0.26	0.22	0.17	0.48
S	–	–	–	tr.	tr.							γ	0.21	0.22	0.25	0.22
(NiCo)O	–	–	–	tr.	–							μ	0.52	0.51	0.43	0.61
H_2O+	1.29	0.80	0.64	0.27	1.92							α	–0.24	–0.17	–0.22	–0.27
H_2O-	a	a	a	1.00	0.86											
Total	99.98	100.38	99.66	99.90	100.05											

a) Dry at 110° C.

1. Picritic Basalt, road from township to Laguna.
2. Picritic Basalt, vicinity of the township.
3. Oceanite, vicinity of the township.
4. Oceanite (Tyrrell specimen).
5. Biotite-trachyte (Tyrrell specimen).

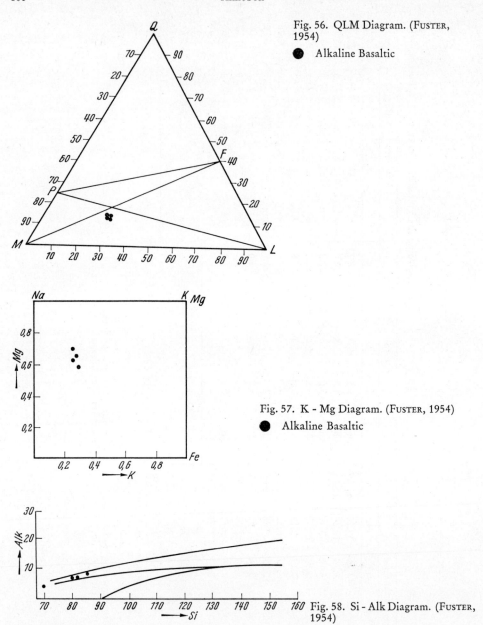

Fig. 56. QLM Diagram. (Fuster, 1954)

● Alkaline Basaltic

Fig. 57. K - Mg Diagram. (Fuster, 1954)

● Alkaline Basaltic

Fig. 58. Si - Alk Diagram. (Fuster, 1954)

some degree, a little magnetite and sphene. In some specimens, fragments of serpentinized olivine can be detected. The fine specimens studied microscopically by Tyrrell showed little variation in themselves, the only noteworthy change referring to the texture and the amount of biotite. One sample, however, was bauxitized, a common method of alteration for aluminous rocks such as trachytes occurring in tropical climes.

TYRRELL commented on the absence of trachybasalts from his collection. Such rocks, also trachytes, trachyandesites, augitites, limbutgites, and more richly alkaline types such as phonolites, nepheline-tephrites, nepheline-basanites, are common in other islands of the Gulf of Guinea and also in more distant St. Helena. From this, TYRRELL surmised that an investigation of larger collections from Annobon rocks than he studied – 13 from the island – would no doubt show the presence of some at least of these other rock types. Against this supposition we must note again that FUSTER is indeed sceptical of trachytes in Annobon and makes no mention of other rock types referred to by TYRRELL above.

BARROS (1960) gave data regarding JUNG & RITTMANN diagrams for the Annobon rocks, based upon figures presented in Table 39.

Table 39   JUNG & RITTMANN Values (BARROS, 1960)

Specimen No.	JUNG		RITTMANN	
	SiO_2 (%)	R	Si°	Az°
1	42.24	70.8	0.62	0.44
2	43.31	70.7	0.64	0.45
3	43.21	68.6	0.63	0.45
4	39.20	76.5	0.59	0.41

THE BRITISH ISLANDS

CHAPTER 11

Ascension

General

Ascension Island is centred in lat. 7° 57′ S, long. 14° 22′ W some 1130 km NW of St. Helena, the nearest land. The island lies distant from the nearest part of the South American continent by some 1960 km, and 1450 km from the African continent. Only one islet of any significant size, Boatswain Bird, lies off the coast, some 500 m distant from the main island.

Ascension has an area of 98 km^2, measuring 14 km W-E and 11.5 km N-S. The highest summit, Green Mountain, rises to a height of 859 m. DALY (1925) claimed that The Peak was the maximum height, "over 868 m", but later accounts invariably quote Green Mountain as the culminating point, elevation as shown.

Originally the island was bare and even today this is true of the lower elevations, except for a few shrubs that have seeded themselves. The flora is mainly South African, but many British plants are present. Green Mountain is so named because of its relatively profuse vegetation. There is no native fauna, but wild cats, wild goats (now extinct), wild donkeys (descended from animals brought in to work) and partridges were introduced by seamen. Sea birds are abundant and the surrounding waters abound in fish. Giant female turtles come to the island to lay their eggs but depart thereafter. No male giant turtles have ever been seen.

The island was discovered on Ascension Day, 1501, by the Portuguese navigator Joao da Nova, but remained uninhabited until the British took formal possession in 1815. In this year, Napoleon arrived on St. Helena, and the cautious British did not want a repetition of the Elba incident.

Georgetown the capital lies on the NW coast, with a relatively good anchorage on the leeward side. English Bay, NE of the capital, and Two Boats, 275 m up Green Mountain, are two very recent settlements began only a few years ago.

The Cable and Wireless Co. Ltd. have a station and farm on Ascension. During the last war, the Americans built an airfield to serve as a re-fueling base on trans-Atlantic flights. The Americans still maintain the airbase, also a missile-tracking station and a ground-station for Project Apollo – the mission to put a man on the Moon. In 1963 the British Broadcasting Corporation began installing a powerful relay transmitting station. It was because of the relatively great influx of labour – from the West Indies and St. Helena, plus British and American technicians, scientists, etc. – that the two new settlements of English Bay and Two Boats were founded.

The present population of Ascension is about 1400.

Ascension is becoming increasingly important as a communications centre, a strategic position between Africa and South America.

Physical Features

The volcanic island of Ascension rises from a depht of almost 3000 m, forming a magnificent gigantic cone. The island comprises a great number of individual vents of large and small size – more than 100 according to DALY (1925). (Throughout the chapter, all references to DALY relate to his 1925 publication, unless otherwise stated.) From the Green Mountain-Peak topographic complex, the land drops relatively steeply southward and eastward to sea level. The topographic asymmetry of Ascension, with highest elevations in the E and steeper slopes to S and E, is also reflected in the numerous vents, where erosion has accentuated the windward (SE) slopes and the drift of finer ejectamenta litter the gentler leeward slopes.

Tablelands, varying in elevation from 300–600 m surround the higher mountain areas. Ravines are very steep and narrow. The porous nature of most of the exposed rocks allow rain waters to sink underground, and essentially only after storms are the streams running. As a result, concrete catchments are required, located near the top of Green Mountain. Here also is a large dew pond which, according to DALY, is kept filled by the dripping of water from bamboo plants surrounding it. More recently, two desalinization plants have been installed in Ascension, assuring the increased population of adequate freshwater supplies.

Imposing sea cliffs occur in eastern Ascension and also on Boatswain Bird islet. The coastal shelf extends outwards to near the 100 m isobath, the sea-bottom descending more steeply down to the 200 m isobath.

The lee side of the island is pounded by great 'rollers' which smash on to the relatively weak rocks. These 'rollers' are the result of swells initiated by distant Atlantic storms. The SE coasts face the windward side, and here marine attack by turbulent seas is drastic.

Climate

The somewhat arid appearance of Ascension is due to the relatively high rock porosity which absorbs the rainfall, rather than to purely climatic reasons.

The average annual rainfall is 635 mm, with most falling in March and April.

The average summer temperature is 20° C, average winter temperature, 12.7° C. Maxinum daily temperature at sea level is 29° C, and at 610 m on Green Mountain, about 24° C.

There are strong and persistent winds throughout the year. Dominant winds are the SE Trades.

Cloudiness is rare except for frequent clouds resting on the summit of Green Mountain.

The weather is usually fine, with much warm sunshine, and the island is said to enjoy a most healthy climate.

Geology

The young volcanic island of Ascension is still in the process of construction, although no volcanic activity has occurred since its discovery in the 16th. century. Whilst the volcanic vents have often a very young appearance, it is somewhat strange that neither hot mineral springs nor fumaroles have been found.

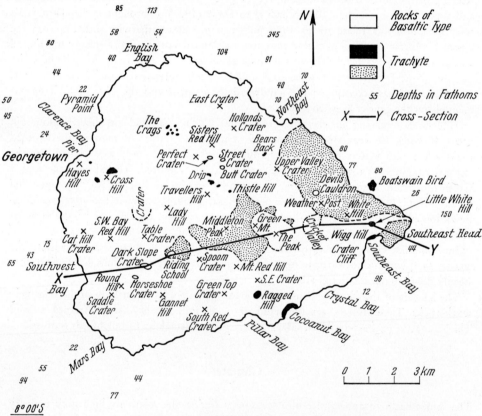

Fig. 59. Geological Sketch Map of Ascension Island. (After DALY, 1925)

Basalts and rocks of allied type comprise about 85 % of the island; rocks of trachytic habit constitute the remainder. Scoriaceous basaltic or trachydoleritic flows are exposed over more than half the area. As per ATKINS et al (1964) the great volume, wide distri-

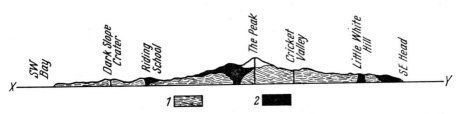

Fig. 60. Schematic W-E Profile across Ascension Island. 1. Basaltic-type rocks, 2. Trachyte. (DALY, 1925)

bution and variety of depositional structures of the pyroclastics are worthy of more detailed investigation than has hitherto been given. (These deposits appear to outcrop over some 35–40% of the island, according to the preliminary map of these authors.)

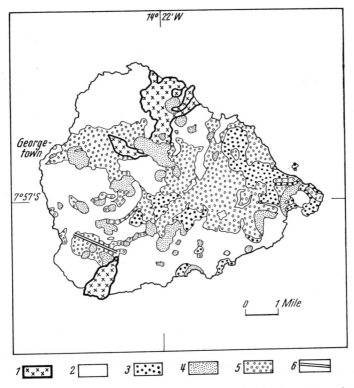

Fig. 61. Geological Map of Ascension Island. 1. Most recent basalt flows, 2. Basalt, 3. Trachyte, 4. Scoria and cinder cones, 5. Pyroclastics, 6. Wideawake airfield. (Based on Preliminary Map by ATKINS et al., 1964)

Trachytic masses occur principally in the east. DALY believed that the large composite cone of Ascension was chiefly of basaltic composition down to all depths below present sea level, the trachytic massifs being either intercalated with the basaltic flows and pyroclastics or then resting on them. In placing the trachytes in a minor role, DALY disagreed with DARWIN (1876) who had considered the trachytes as 'fundamental', the basalts overlying these. To DALY the trachytes were clearly younger eruptive phenomena than the basalts.

Basaltic flows have a rugged appearance, usually of blocky or aa type, larger flows having issued from the flanks of cones rather than from craters. Associated with basaltic flows are a wide variety of structural features, e. g. tumuli, hornitos, spatter cones, bocas, lava channels and lava tunnels. The channels or gutters, which DALY thought had been hollowed-out by hot lava streams, were interpreted by ATKINS et al as being the beds of such streams which built bordering levees as a result of periodic flooding, such as appears to be the case in Tristan da Cunha q. v.

Basaltic flows may display a series of transverse open fissures, separated by sharp ridges or spires, giving a crevasse-serac appearance, or then a more chaotic block aspect, blocks being the size of Vesuvian rather than Hawaiian dimensions. Smooth, ropy, pahoe-hoe type of flows are best seen in the vincinity of Southwest Bay, and are jointed in crude columnar fashion. DALY reported that basaltic flows seldom attained a thickness of 20 m, and 6 m is more the average. The thickest, 60 m at least, are to be found in the cliffs surrounding Cricket Valley. Hillocks of scoriae and bombs, up to 5 m in height, occur in linear fashion along some flows, e. g. E of Southwest Bay. Initially the colour of the flows is blackish-grey at the surface and floor, and although a weathered brownish tint is seen in older flows, nowhere are deep-red weathered colours seen in formerly deeply buried flows. However shorter, thinner flows show unmistakable red coloration, especially at Mountain Red Hill.

Table 40   Structural Classification of Basaltic and Trachytic Masses (DALY, 1925)

A. Basaltic Masses and other Femic Masses   (Masses other than basaltic specifically designated.)

1. Endogenous lava-dome (due to local eruption, without definite crater): Bears Back.
2. Fissure eruption: On Southeast Head (several small craters opened on the fissure). Trachyandesite.
3. Dominantly or wholly lava-formed cones:

    Dark Slope (breached)   Slag cones 800 m WSW of Lady Hill summit
    Driblet W of here   Three driblets, NE foot of Sisters Peak
    Table Crater (breached) Landing Pier cone, Georgetown (scoria cone) Trachydolerite
    Lady Hill   Hayes Hill. Trachydolerite
    Cat Hill (slag cone)   Twelve driblet cones, culminating at W end of series in Booby Hill

4. Composite cones, composed of flows and pyroclastics:

    Saddle Crater   Sisters Red Hill
    South Gannet Hill   Hollands Crater
    Round Hill   East Crater
    Horseshoe Crater   Southeast Crater
    Mountain Red Hill   Upper Valley Crater
    Butt Crater   Crater Cliff cone
    Street Crater   Three unmapped craters between Street and Holland craters
    Sisters Peak   1012 ft. cone NW of Sisters Peak
    Travellers Hill   1187 ft. cone E of Sisters Peak
    Thistle Hill   Breached cone, N foot of Bears Back

5. Dominantly or wholly pyroclastic cones:

    South Red Crater
    Southwest Red Hill   The Peak
    Perfect Crater   613 ft. cone, NW of Dark Slope. (Has effluent flows)

B. Trachytic Masses

1. Endogenous domes or crater-fillings, not visibly affected by axial subsidence:

Pillar Bay dome (probably)
Ragged Hill dome
Cocoanut Bay body, a relic, perhaps issued by explosion
Green Mountain dome, issued in a large basaltic caldera; thick overflows of trachyte. On the E, affected by second caldera explosion. The Peak cone of basaltic tuff-ash built in this cavity.
Weather Post dome, thick overflows of trachyte.
Affected by a major explosion at Devils Cauldron (caldera) and likely by a still greater explosion which formed the depression between Weather Post and ridge running N from White Hill.
White Hill dome, thick, stubby trachyte overflows.
Little White Hill dome, issued in centre of an older explosion crater, of which much of the basaltic rim is well preserved.
Wig Hill dome, veneered with basaltic scoriae.
Southeast Head dome, probably extended by thick overflows of trachyte. Fissured and flooded by a very young trachyandesite flow.
Boatswain Bird islet, likely an independent dome of monolithic trachyte.
Dome(?) at SE foot of Bears Back.
Stubby flow 500 m SE of 1187 ft. summit on Sisters Peak ridge.
"The Crags" dome, largely covered with younger basaltic ash and lavas.
Cross Hill dome, issued inside an older basaltic crater rim. At least one trachyte overflow. After solidification, covered by basaltic scoriaceous flows and cinders, which issued through the body of the dome.

2. Endogenous domes or crater-fillings, deformed by axial subsidence:

Riding School dome, issued in older basaltic caldera rim. Thick, stubby overflows.
"Drip" dome. At S foot of Sisters Peak ridge. Rose in older basaltic crater or caldera. Overflows of trachyte perhaps represented in outcrops of this rock E and NW of Thistle Hill summit.

DALY presented a list (Table 40) classifying the basaltic and trachytic masses on the basis of their essential features. Whilst the table cites many examples, we gather from DALY that it is not a complete tabulation, but it is of interest in a geological appraisal of the island. He elaborated in the text upon many examples listed, and a summation will be given here.

Basaltic cones show characteristic features. As already remarked asymmetry is prominent and breaching of both cones and craters is common, e. g. Hollands Crater, East Crater, Table Crater. Lava discharges are particularly located on the windward sides of Cross Hill, Spoon Crater cone and South Gannet Hill. The breaching of East Crater is an example of axial subsidence, a unique feature of the island amongst craters where basaltic lavas have welled upwards, and are possibly due to withdrawal of magma at depth, although the possibility that thermal contraction of a 'freezing' magma may be partially responsible cannot be excluded. Axial subsidence at crater sites is also believed to have taken place at Riding School and the "Drip" trachytic domes. Driblet basalt flows form many small, steep cones, the best examples of this being so aligned as to suggest that the vents mark the site of a single fissure. A good example of such occurs between Sisters Peak and Sisters Red Hill, where three of these conelets, 3–12 m in height, are arranged linearly for some 60 m. In one instance the cylindrical pipe at depht of 9 m broadens into a spacious domed chamber, suggesting that during activity molten lava in the pipe increased in volume downward – for some distance at least.

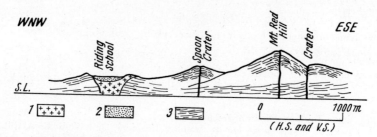

Fig. 62. Section through the basined dome of trachyte (1) at Riding School, the corresponding 'crater' filled with basined tuffs (2). Section continued through Spoon Crater and Mt. Red Hill, both of basaltic habit (3). (DALY, 1925)

Most trachytic masses are crater-domes or then flows from such. DALY expressed the view that rarely if ever could so many endogenous trachytic domes be found within such a small area as in Ascension. These domes are of both simple and complex type. Ragged

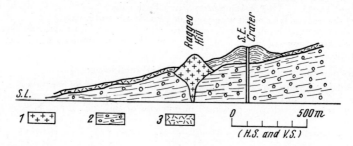

Fig. 63. Section through Ragged Hill trachyte dome (1), and basaltic cone of S. E. Crater, both resting on older flows (2). A young basalt flow from Green Mt. (3) has partly submerged both dome and cone. (DALY, 1925)

Hill dome is a perfect example of this type of volcanic structure. The base is concealed by younger flows from Green Mountain but a maximum of 80 m of the dome is visible on the S side of the latter, which shows it to be almost perfectly circular, with a diameter varying between 200–250 m. The trachyte, showing an abundance of large feldspar phenocrysts, displays marked concentric rifting, the plates so formed dipping outwards at angles up to 35° at the edges. Many vesicular lava inclusions are present, which probably are of basaltic type. The exterior of the dome is pitted, the silicified carapace giving a distinct ring when struck with a hammer. The steep slopes of the dome indicate a very high viscosity for the effluent trachyte.

Little White Hill dome has undergone scarce any erosional modification since formed. At the base it has a diameter of 150 m. Again the surface is pitted, and fluidal structure is more evident than in Ragged Hill.

Wig Hill dome towers over Southeast Bay in magnificent cliffed sections. The dome is draped with a layer, 5–20 m thick, of basaltic, scoriaceous agglomerate and driblet vesicular basalt flows, such that from a distance, the 'wig' looks like older basalts upraised and domed by the trachytes as they pressed upward to assume the domal shape. The lack of tensional effects in the rocks constituting the 'wig' led DALY to believe, however, that the 'wig' was a basaltic eruption through vent(s) which pierced the trachytic dome, i. e. the basaltic carapace is younger than, not older than, the underlying domal trachytes.

Cross Hill has the appearance of a typical cinder cone, the N, W and E slopes with lapilli several metres thick. Tuff and spatter bombs lie on the upper part, the tuffs having trachytic fragments, possibly derived from a trachytic dome some 75 m beneath the tuffaceous covering. The domal trachyte is a massive monolith, weathering at the summit into a highly rugose surface.

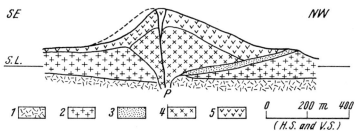

Fig. 64. Section through Cross Hill, showing older basalts (1) under younger basaltic (trachydoleritic) cone (2), partly destroyed by explosion. In the resulting caldera, tuff (3) was deposited. Later a monolithic dome of trachyte (4) formed in the caldera. A thick cap of scoria, tuff and breccia of basaltic habit (5) erupted on to the trachyte through one or more narrow vents, shown diagramatically at P. (DALY, 1925)

Riding School is an interesting feature, being a basaltic lava cone with a 500 m diameter crater which has been filled almost to the brim with monolithic trachyte. Possibly axial subsidence occurred here as the crater depression was already in existence before the next extrusive phase began, and it scarce seems likely that explosion was the cause of the crater as the outer slopes have few pyroclastics. Whether or not trachyte formed a true dome cannot be ascertained, but at least trachyte spilled over the crater rim to form a flow some 700 m in length on the E and NE sides. Within the crater hollow, ash, tuff and lapilli, probably derived from The Peak and Green Mountain area, are now preserved as well-stratified beds showing a distinct basinal structure, suggesting continued subsidence after the issuing trachyte had solidified. DARWIN had believed that some finer-grained beds within the crater were of lacustrine origin, but DALY surmised that they underlay basaltic lapilli. The bed of infusorial earth here noted by DARWIN was not found by either DALY or DOUGLAS (1923). Siliceous concretions are present within beds of acidic tuff. REINISCH (1912) gave an analysis of one of these concretions as follows:

SiO_2	90.07
$Al_2O_3 + Fe_2O_3$	2.85
MgO	0.04
CaO	0.09
Alkalis	0.03
Loss on ignition	7.31
	100.39

The highly pitted surface of the greyish trachyte at Riding School is quite remarkable, and no less the odd-shaped forms assumed, due to differential weathering of the trachytes and the so-called silicified 'veins'.

"Drip" dome is a small slaggy basaltic cone with a crater some 100 m in diameter, breached on the W side. Thick alluvial ash and lapilli obscure the form of this feature, but it appears to be a small dome having undergone axial subsidence. The name is taken because one of the few springs or 'drips' in Ascension is located here.

"The Crags" dome conprises many individual trachyte outcrops, separated by a thick layering of lapilli, bombs and wind-blown volcanic sand, the trachytes being all but smothered under a young basaltic flow issuing from vents along the northern base of Sisters Red Hill cone. It is assumed that the trachyte outcrops or 'crags' are merely part of a single, weathered dome centring near a point about 1 km NW of Sisters Peak. At the most southerly outcrop, a 10 cm. angular inclusion of hornblende granite was found by DALY, from which it may be suggested that granitic fragments in Ascension are not strictly confined to explosive breccias and tuffs.

Green Mountain dome shows the highest exposures of trachyte on the island, at an elevation of some 760 m. From this locus, viscous trachyte flows poured outwards to the N, NE, W and S. The eastern part of the dome was ripped open by one or several violent explosions in a caldera site measuring more than a kilometre in diameter. From here later explosions constructed the steeply-sloping Peak, which is chiefly composed of basal tuffs, ash and breccias. The Green Mountain trachyte shows distinct fluidal structure, and near the base of some flows, banding dipping at angles up to 50° to the S can be observed in northward-directed flows. Flows are often much brecciated, down to depths of 3 m. Some flows include many angular, vesicular fragments of basalt which were presumably derived from either the walls or the floor of the dome-vent.

Weather Post dome, White Hill and the large flow N of the Devils Cauldron constitute the largest trachytic body in the island. This composite mass is presumed to be an eruption product from a single centre, although the present-day topography would suggest that there were two principal sources of effluence, Weather Post and White Hill. At the northern edge of the rim forming Cricket Valley, the Weather Post trachyte has

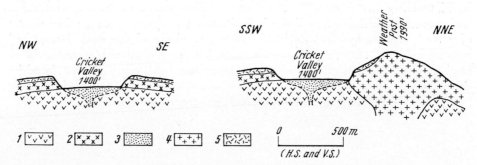

Fig. 65. Sections through Cricket Valley, showing caldera opened in basaltic flows of ordinary thickness (1), with one unusually thick flow of olivine-basalt (2), and filled to unknown depth with ash and tuff (3). The explosion also affected older trachytes (4) of Weather Post. Débris of the explosion shown on the surface (5). (DALY, 1925)

spilled over basaltic tuffs dipping to the SW, suggesting perhaps that these trachytes erupted inside the rim of an older basaltic crater or caldera. Thick, coarse breccias mantle much of the dome, and to the N there is a wide trachyte flow, at the head of which is

Devils Cauldron. This unusual explosion crater or caldera, elliptical in shape, with a length of 200 m and up to 60 m deep, is surrounded by very steep or vertical walls. DALY thought that the Cauldron was the locus of a separate dome-extrusion, the trachyte having issued through older, basaltic-type material. Well-bedded tuffs and agglomerates, up to 7 m thick, occur on the rim of the Cauldron, and on Weather Post pyroclastics up to 40 m thick occur. Tuffs dipping at angles up 70° on the SW side of the dome may owe this inclination to upturning by continuous rise of the viscous dome-magma.

White Hill dome shows trachytes petrographically very similar to those of Weather Post, but there is lacking any clear evidence of structural separation. On the walls of the large amphitheatre rising above Spire Beach, great thick tongues of trachyte, seemingly issuing from White Hill, are separated by trachytic tuffs and breccias. This flow appears very young, almost untouched by erosion. The high cliffs facing Boatswain Bird islet give an excellent view of the trachytes lying on a series of basaltic flows, the trachytes being heavily charged with angular inclusions of vesicular basalt.

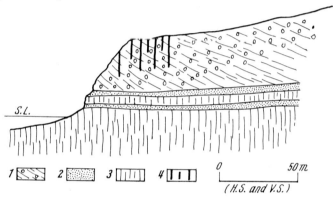

Fig. 66. Longitudinal section through a thick trachyte flow on the main island opposite Boatswain Bird islet, showing common upturning of trachyte flows near their lower extremities. Fractures due to tension. 1. Trachyte, 2. Tuff, 3. Basalt, 4. Fractures. (DALY, 1925)

Southeast Head dome is topographically quite level and was thought by DALY to be an endogenous dome of larger size than that of the adjacent Little White body. Upon solidification the trachytes of the Head dome were fractured by a vertical fissure extending for at least 1.2 km, and here and there along its length, scoriaceous trachyandesitic lavas spilled over on to the trachytes as narrow tongues. At one point the fissure was widened by explosion and a twin-crater formed.

The rock masses at Cocoanut Bay and Pillar Bay were assumed by DALY to represent a dome.

Boatswain Bird islet, on a topographic basis, does not seem associated with the flows of either the Weather Post or White Hill centres. It is more probable that the isle represents an independent dome issuing from below. This strongly cliffed islet, rising sheer to an altitude of ca. 98 m, is monolithic, shows a pitted surface, and the rock is petrographically similar to the trachytes of the mainland. Strong wave action has scalloped-out the cliffs to depths of 15 m below low tide. A fine example of a natural bridge, hollowed by the waves, is to be seen.

Pyroclastics are voluminous, of wide occurrence, and show considerable variation in their depositional structures. Basaltic, rhyolitic and trachytic tuffs, ash, lapilli, pumice, volcanic dust, volcanic bombs, scoria and agglomerates are widely known in Ascension. Seemingly nowhere has lateritization taken place, as not enough time has elapsed for the process. Whilst the greater preponderance of such ejectamenta are due to gravitational falls consequent upon explosive activity, on occasion the pumiceous dust and ash show excellent stratification and are so regular over considerable areas that Atkins et al thought it probable that these were deposited in "an extensive body of water scarcely subject to current action".

Xenoliths and ejected blocks have been recognized since the time of Darwin's visit. These commonly occur in basaltic and trachytic agglomerates of Green Mountain and Dark Slope Crater.

Dykes are not abundant and indeed Daly recorded a mere half dozen. To the SW of Cross Hill is a basaltic dyke cutting slaggy breccias, only 25 cm in width. A frothy basaltic glass dyke cuts a trappean flow at the N end of Southwest Bay, measuring about 15 cm thick. The best exposed dyke, up to 2 m in width and extending for at east 150 m, occurs on the W slope of Spoon Crater, transecting basaltic ash. Near the summit of Middleton Peak, a 2 m thick basaltic dyke protrudes some 3–4 m above a soft, thick, trachytic flow. This example and another one nearby seem to indicate that basaltic eruptions also took place subsequent to the trachyte phase.

Thin, localized, calcareous beach deposits are the sole sedimentary representatives. These are found in more sheltered bays on the lee side of the island.

Darwin and others commented upon phosphatic deposits on basaltic flows at several places. These encrustations are similar to those recorded on St. Paul's Rocks, q. v.

Faulting is rare, and such as was noted by Daly seemed restricted to tuffs and tuff-breccias. Normal faulting is likely due to the settling of igneous bodies, and reverse faults result from thrustings imposed by the more viscous trachytic magmas as they rose to form domes.

Petrography

The publications of Renard, Prior, Reinisch, Daly & Campbell Smith give information on the rocks of Ascension, later writings being either abstracts of what was published or then treating perhaps in greater detail some particular aspect of the rocks.

The "Challenger" collections were studied by Renard, Reinisch studied those gathered by the Deutsche Südpolar-Expedition, Campbell Smith investigated samples collected by the "Quest" expedition and Daly & Washington those collected by the former during his four week field study of the island.

Renard's publications of 1887 and 1889 gave to the world at large the first more detailed account of the petrography of the Ascension rocks. Daly regarded his major publication on the island as being supplementary to the work of those already published, and to date, Daly's is the most comprehensive report of the island geology to have appeared. The Oxford Expedition of 1964 is still engaged in research studies of the island, which will include the results of geological and geophysical surveys and geochemical investigations.

All and sundry pay homage to the initial observations of Darwin & Daly in particular confirms the correctness of so many of the renowned naturalist's findings. However Darwin's descriptions of outcrops and specimens suffer from a serious defect in that in so many instances he failed to indicate exactly the locality in question or then from where the specimen was found. Hence later workers were frequently not in a position to offer an opinion, one way or the other, as to descriptions given.

Renard classed the Ascension rocks into the major groups of augitic trachytes, obsidians, rocks transitional between trachytes and andesites-rhyolites, basaltic-type rocks, andesites, pyroclastics ("matières volcaniques incohérentes") and granitic, diabasic and gabbroic fragments within the eruptives. Renard, like Darwin, believed that trachytic rocks formed the fundament, a view not shared by Daly.

Darwin was at times hesitant to call some exposures either trachyte or tuffaceous sedimentary deposits. Because of the earthy appearance and friable nature of many trachytic outcrops, this was understandable. To rocks of trachytic habit Renard gave the general name augititic trachytes, characterized by more or less variable association of three features: monoclinic feldspar, augite and vitreous appearance, the microtexture being very similar in all these rocks. Table 41, No. 16 is an analysis of a specimen studied by Renard. The silica content is too high for a normal trachyte, but in this instance, this high percentage is attributed to the penetration of siliceous material into the rock after consolidation, such as occurs in the tridymite-trachytes of New Zealand. As frequently happens in trachytes, sodium shows a great preponderance over potassium.

Obsidians are closely associated with trachytic-type rocks, forming the "bancs zonaires" of Renard. The passage of zonal rocks into true obsidians is accomplished in several ways, Darwin having given detailed descriptions of the field relations of these obsidians. The only macroscopic element is sanidine. These typical black rocks may on occasion shows devitrification, and perlitic structure is quite common. No. 17 is an analysis of a Renard specimen from Green Mountain.

Renard stated that on Green Mountain there were many rocks forming a transition from augite trachytes to rocks of similar lithologic character. Under this grouping, he takes as examples an augitic trachyte passing into an amphibolitic andesite at the quarry near Georgetown, pyroxenitic trachyte passing into rhyolite at Red Hill. More clearly defined rhyolites were recorded and described from the crater at Riding School, rhyolitic tuffs at Dry Water Course, etc.

The basaltic rocks of Ascension were classed as of ordinary type, more rarely, of doleritic type. The former were considered commoner, of vesicular appearance and often scoriaceous.

Renard described several rocks which he said "se rapprochant beaucoup des basaltes et qu'on doit classer avec les andesites", but Daly found no true andesites during his investigations.

In the Green Mountain-Cricket Valley area, Renard remarked upon the occurrence of loose volcanic products, explosion material. Tuffs and bombs of various dimensions were given special attention.

Xenolithic fragments, first described by Darwin, were given considerable space by Renard, who mentioned in particular amphibolitic granites, granitites, gabbros and rocks of diabasic type.

174   Ascension

Mention was also made of the siliceous 'veins' and calcareous sands formed in coastal areas.

At times, DALY was somewhat critical of rock descriptions and analyses of REINISCH, noting REINISCH's rather impoverished samplings from the island.

CAMPBELL SMITH studied 60 thin-sections from Ascension, which included soda-rhyolites, pantelleritic trachytes, katophorite-trachytes, trachybasalts, olivine-basalts and xenoliths of plutonic rocks.

As DALY's studies are the most complete to date, we shall describe the Ascension rocks under headings listed by him.

Basalts. The most mafic types are olivine-basalts forming flows, either interbedded or then overlying olivine-free or then olivine-poor basalts, commoner in occurrence than those containing olivine. In character the olivine-basalts may occur as either rugged and irregular or then smooth and ropy in appearance. Scoria may be present in these flows, but less so than in basalts containing no olivine. Table 41, No. 1 is quite smooth but of hummocky appearance, of dark grey colour and shows quite numerous phenocrysts of feldspar within a compact, slightly vesicular matrix. The feldspars are bytownite and some labradorite, along with olivine. The groundmass shows much augite, iron oxide, possibly also magnetite and ilmenite, also apatite. Glass forms less than 5% by weight. Skeletal crystals of iron oxide are also present. In the petrographical nomenclature of DALY, he stated that the norm showed the specimen to be an ornose, in the salfemane class, of the order gallare. It is alkali-calcic and persodic, near camptonose, dosodic subrang, dosodic rang. No. 2 is a compact, somewhat vesicular basalt of greyish colour. The phenocrysts include bytownite which perhaps shows faint zoning, with labradorite on the exterior, and anhedra of olivine much less abundant. The matrix is a granular,

Table 41   Chemical Analyses (DALY, 1925)

	1.	2.	3.	4.	5.	6.	7.	8.	9.	10.
SiO_2	47.69	48.64	52.87	51.18	54.04	58.00	57.72	65.18	66.98	66.12
TiO_2	2.79	3.52	2.01	1.34	0.94	3.38	0.37	0.44	0.89	
Al_2O_3	16.23	15.54	16.68	21.41	19.58	14.92	17.64	15.91	14.30	15.51
Fe_2O_3	2.20	5.31	4.54	4.63	5.09	1.73	4.47	4.41	3.85	3.27
FeO	9.93	7.73	4.79	3.32	3.75	5.78	2.78	0.98	0.33	0.93
MnO	0.17	0.17	0.37			0.11	0.03	0.17	0.21	
MgO	7.15	4.96	3.92	1.75	1.99	2.23	1.01	0.10	0.30	0.17
CaO	10.02	9.03	7.32	6.56	5.54	4.50	4.36	0.81	0.83	1.05
Na_2O	2.87	3.60	4.63	4.72	4.70	5.88	5.50	6.24	6.76	6.31
K_2O	0.64	1.24	2.06	3.53	3.78	2.76	3.90	4.60	4.34	5.40
P_2O_5	0.59	0.64	0.52	0.48	0.31	0.71	0.57	0.08	0.22	
CO_2	0.04	0.03	n.d.					0.09	Nil	
H_2O-	0.09	0.16	0.40	1.08	1.16	0.09		0.45	0.08	1.98
H_2O+	0.19	0.18	0.30			0.31	1.65	0.53	0.44	
Total	100.60	100.75	100.41	99.88	100.58	100.40	100.00	99.99	99.70 a	100.74
S. G.	2.99	2.97	2.84		2.68			2.64	2.54	

a) Includes: BaO 0.04, ZrO_2 0.13.

Table 41 Chemical Analyses (Daly, 1925)

	11.	12.	13.	14.	15.	16.	17.	18.	19.	20.
SiO_2	63.98	67.05	63.02	55.10	71.88	70.99	72.71	71.42	69.70	65.59
TiO_2	0.28	0.10			0.25					
Al_2O_3	16.00	15.43	15.75	18.56	12.85	14.84	12.80	14.09		
Fe_2O_3	2.57	3.25	0.52	6.80	3.60	3.76	2.64	1.41		
FeO	2.12	1.25	3.15	0.03	0.05	0.35	1.48	2.32		
MnO					0.29	Tr.	Tr.			
MgO	0.64	0.16	0.38	0.62	0.18	0.14	0.10	0.08		
CaO	1.58	1.06	1.49	0.70	0.60	0.60	0.58	0.80		
Na_2O	6.45	6.12	6.11	3.17	5.32	5.94	6.50	6.01		
K_2O	5.18	5.32	5.21	4.00	4.78	2.40	3.87	3.52		
P_2O_5		0.04			0.05					
CO_2										
H_2O-	0.61	0.56	4.83	8.30	0.18	0.40	0.48	0.85	0.94	0.87
H_2O+					0.17					
Total	99.41	100.34	100.46	100.45 b	100.20	99.42	100.16	100.50		
S. G.					2.58					

b) Includes: Sol. in H_2O 3.17.

1. Olivine-Basalt. Surface flow of SW group.
2. Olivine-poor Basalt. Flow, N wall of Cricket Valley.
3. Trachydoleritic Basalt. Flow, Mountain Red Hill.
4. Trachydolerite. Hayes Hill (Reinisch).
5. Trachydolerite. Landing Pier cone (Reinisch).
6. Trachyandesite. Flow from fissure, Southeast Head.
7. Average of nine Trachyandesites, reduced to 100%.
8. Trachyte. Ragged Hill dome.
9. Trachyte. Southeast Head dome. (Special tests for S and Cr_2O_3 gave negative results.)
10. Trachyte. From a point "N of Dark Slope Crater" (Reinisch).
11. Trachyte. Cross Hill quarry (Reinisch).
12. Trachyte. From a point "half-way up Green Mountain" (Reinisch).
13. Trachytic Pumice. Riding School (Reinisch).
14. Trachytic Tuff. Riding School (Reinisch).
15. Quarz-Trachyte. Outflow, White Hill dome.
16. Augite. Trachyte. Weather Post Hill (Renard).
17. Obsidian. Green Mountain.
18. "Rhyolithobsidian". N of Riding School. (Reinisch).
19. Spherulitic "Rhyolithobsidian". Outer mantle, Riding School (Reinisch).
20. "Obsidianknollen". From "Kegelmantel", Green Mountain (Reinisch).

fluidal composite of labradorite, diopsidic augite, augite, iron oxide and apatite. The norm shows it to be camptonose. No. 3 is thought to be representative of the most abundant basalts on the island, and is from a flow issuing from Mountain Red Hill and wrapping around the cone of South Red Crater. This somewhat vesicular, grey, compact rock has a few phenocrysts of bytownite and stubby prisms of augite. The matrix is a composite of labradorite, augite, magnetite, possibly ilmenite and apatite. A few small granules of olivine occur. No glass was noted. Daly claimed that the lower specific gravity of the sample correlated it with the less mafic Red Hill basalts. The norm indicates it belongs to the dosodic subrang andose, ot the alkali-calcic rang andase, of the

Table 42   Molecular Norms (Daly, 1925, and Fuster (A), 1954)

	1	1A	2	2A	3	3A	4	4A	5	5A
Or	3.33	4.0	7.23	7.5	11.68	12.0	20.57	21.0	20.57	20.7
Ab	24.63	25.8	30.39	33.0	40.30	41.7	32.49	34.4	39.82	42.2
An	29.43	29.7	22.52	23.3	18.90	18.8	26.97	27.0	21.96	22.1
Ac										
Ne							3.41	4.8		
Di	13.60	13.3	15.14	14.9	11.28	11.7	2.16	2.2	2.87	3.2
Wo										
En		8.9		8.6		6.9				3.6
Hy	8.83		9.14		6.05				3.43	
Ol	10.48	10.9	0.15	2.3			2.38	2.8	0.90	0.9
Mt	3.25	2.3	7.66	5.6	6.50	4.8	6.73	4.8	7.42	5.3
Il	5.32	4.0	6.69	3.6	3.80	2.8	2.58	2.0	1.82	1.4
Ap	1.24	1.1	1.24	1.2	1.24	1.0	1.01	1.0	0.67	0.6
Hm										
H₂O	0.28		0.37		0.70		1.08		1.16	
Q						0.3				
Total	100.53 a	100.0	100.53	100.0	100.45	100.0	99.38	100.0	100.62	100.0

a) Includes: CaCO₃ 0.10.

	6	6A	8	8A	9	9A	10	10A	11	11A
Or	16.68	16.2	27.24	27.4	25.58	25.7	31.69	33.2	30.58	30.5
Ab	49.78	52.7	53.92	56.0	49.26	52.3	49.78	54.5	53.45	56.5
An	5.84	6.1	1.67	1.8						
Ac					6.93	6.7	3.23	3.2	0.92	0.8
Ne										
Di	9.98		0.43		1.73		0.86		6.18	
Wo		5.2	0.12	0.4		1.1	1.74	2.3	0.23	3.1
En		6.2		0.4		0.8		0.4		1.7
Hy	4.55	3.3								1.3
Ol										
Mt	2.55	1.8	2.55	1.8			3.02	2.3	3.25	2.4
Il	6.38	4.6	0.76	0.6	1.06	0.9			0.61	0.4
Ap	1.55	1.4	0.31	0.1	0.31	0.4				
Hm			2.72	1.8	1.44	1.0				
H₂O	0.40		0.98		0.52		1.98		0.61	
Q	2.76	2.5	10.14	9.5	11.70	10.9	8.34	4.1	3.60	3.3
Total	100.47	100.0	100.04 b	100.0 c	99.49 d	100.0 e	100.64	100.0	99.43	100.0

b) Includes: CaCO₃ 0.20.   d) Includes: CaO.TiO.SiO₂ 0.78, ZrSiO₄ 0.18.
c) Includes: Co 0.2.   e) Includes: ZrO₂ 0.2.

Table 42　　　　　　　　　　　　　　　Molecular Norms (DALY, 1925, and FUSTER (A), 1954)

	12	12A	15	15A	16	17	18	18A
Or	31.14	31.4	28.36	28.5	13.90	22.80	20.57	20.5
Ab	49.78	52.3	38.78	41.8	50.30	44.02	50.83	54.0
An					3.01		1.11	1.0
Ac	1.39	1.2	5.08	4.8		7.39		
Ne								
Di	0.86		0.86			2.45	2.45	
Wo	1.74	2.1		0.9				1.0
En		0.4		0.3				0.3
Hy		0.1			0.40	1.28	1.95	2.6
Ol								
Mt	3.94	2.9	0.46	0.2	1.16		2.09	1.5
Il	0.15		0.46	0.4				
Ap			0.09	0.2				
Hm				1.1	3.04			
H_2O	0.56		0.35		0.40	0.48	0.85	
Q	10.62	9.6	23.40	21.8	28.56	21.54	20.58	19.1
	100.18	100.0	100.20	100.0	99.40	101.07	100.43	100.0
			f		g	h		

f) Includes: $CaSiO_4$ 0.70.　　　g) Includes: Co 1.33.　　　h) Includes: Na metasilicate 0.61.

order germanare, of the dosalane class. The specimen is obviously intermediate between a true basalt and a trachydolerite and was classed as a trachydoleritic basalt by DALY.

The ordinary type basalts which RENARD described from Red Hill are permeated throughout by iron oxide, with phenocrysts of plagioclase, rarely olivine, more rarely augite. Olivine is altered into hematite, giving the mineral a reddish hue, and penetrated by parallel, curved filaments of trichite. The feldspar is often corroded or them fractured, the particles separated one from the other. Frequently undulating extinction traces can be detected.

CAMPBELL SMITH noted olivine-basalts similar to those from Red Hill mentioned by RENARD, and also basaltic rocks from flows near Dark Slope Crater. Those occurring on the N side of Devils Riding School are poorer in olivine, and may be correlated with the olivine-poor basalt of DALY from the N wall of Cricket Valley (No. 2).

FUSTER (1954) termed Nos. 1 and 2 subbasalts. No. 1 is much more basic and femic than the trachydoleritic basalt or the trachydolerites (Nos. 4 and 5). This olivine-basalt, like the majority of rocks of this type, is intermediate between the alkali-soda and calc-alkaline series, with more affinities to the former, and hence is of magma-type between essexitic gabbroic and normal gabbroic. No. 2 is poorer in olivine and higher in absolute alkalinity, being intermediate between an essexitic gabbroic and a gabbro-dioritic magma. The calculated mineralogical composition shows a higher albite proportion than in typical basalts with olivine.

Trachydolerites. REINISCH described specimens from Hayes Hill, a scoriaceous cone at the water front at Georgetown and from a smaller scoria cone just N of the Landing Pier here as being trachydolerites (Table 41, Nos. 4, 5). The rocks were strongly vesicular, brownish-black in colour, representing slag or bomb fragments. In places the rocks showed superficial layers of glass and could comprise small phenocrysts of plagioclase when viewed microscopically. The groundmass consisted of dimly transparent, dark brown glass with abundant opaque specks and minute grains of iron oxide, abundant granules of magnetite and large crystals of labradorite and andesite. DALY claimed that both analyses classed the specimens as being of the dosodic subrang andose, of the alkali-calcic rang andase, of the perfelic order germanare, of the dosalane class. He was doubtful regarding the accuracy of the alumina and magnesia determinations. REINISCH also described basaltic trachydolerite occurring in two separate beds at Cricket Valley in which were present phenocrysts of labradorite-andesine, augite and olivine, set in a matrix comprising andesine, augite, olivine and magnetite and cemented by alkali-feldspar and anorthoclase. Mica, apatite and a barkevikitic hornblende are accessories. Half-way up Green Mountain he mentioned a trachydolerite flow enclosing fragments of trachyte, the latter partially melted. Trachydoleritic lapilli were also noted on The Farm, on the NE slope of the Mountain. However DALY was sceptical of these Green Mountain occurrences, and believed the rocks to be ordinary basalts. He conceded that perhaps trachydoleritic basalts were present N of Cross Hill, but accurate determination of the alkali-feldspar and its amount was not possible.

RENARD spoke of "des roches basaltiques du type des dolérites", greyish in colour, almost saccharoidal in appearance, coarse-grained, with plagioclase crystals visible to the naked eye. Under the microscope the plagioclase was seen to be labradorite, augite is intercalated between feldspathic lamellae in the matrix. Olivine is corroded, the greenish alteration products often being fibrous, rather resembling hornblende. Unfortunately RENARD does not mention where his doleritic type rocks occur in the island.

FUSTER termed the rocks named trachydolerites by REINISCH as being acidic, alkaline basalts, including also in this category the trachydoleritic basalt of DALY (No. 3). He believed that the relative proportion of the alkalis was sufficiently high to include them in such a terminology, admitting however that the acidity was very high for such types, but on the other hand, the ferromagnesian elements were lower than usual. In all three specimens, the proportion of feldspar, principally albite, is very high, more than three-quarters of the total weight of the rocks in question. NIGGLI values for No. 3 place it as belonging to a beringitic or sodic gabbroic magma; No. 4 is more alkaline and also more aluminous, comparable to a rouvillitic magma; No. 5 is considered as a normal essexitic magma.

Andesites ?. RENARD spoke of "Certaines roches, se rapprochant beaucoup des basaltes et qu'on doit classer avec les andésites, se rencontrent en plusieurs points de l'île, en particulier à Red Hill". The exterior appearance resembled basalt in some cases, but there also occurred more earthy-looking rocks of reddish colour with thick crusts of sublimated hematite. Microscopic study led RENARD to class the rock as pyroxenitic andesites, bronzite being the pyroxene. The essential feldspar was microlites of andesine or oligoclase. The mineral he identified as bronzite was always altered, giving it a reddish hue. Some sections showed an octagonal form recalling that of augite sections, whereas the reddish coloration of the prisms gave a resemblance to some olivines. However, the optical properties distinguished the mineral from augite, and the outlines of the sections and elongated form of the prisms ruled out olivine. Only rarely were the bronzites large enough so as to render a microporphyritic appearance, but where some prismatic sections attained this dimension, they were frequently profoundly hollowed-out. Around the borders of larger bronzites there was a pronounced fluidal tendency.

RENARD also classed as andesites vein rocks in trachytes forming an elevation described by DARWIN as 'Crater of an old volcano'. DARWIN gave a description of these 'veins', attributing them to local silicification or normal trachytes (vide infra).

DALY found in several thin sections of basaltic rocks examples where the olivine was partially altered to a deep reddish-brown material which he thought might be iron-stained serpentine, as well as olivine crystals so elongated as to resemble an orthorhombic pyroxene such as bronzite. He did not find any true andesites in Ascension and believed that the pseudomorphs described by RENARD were in reality olivine.

CAMPBELL SMITH re-examined the RENARD specimen and concluded that it was a trachybasalt, perhaps a trachyandesite.

It would therefore appear that it is very doubtful if true andesites occur in the island, although such cannot be dogmatically ruled out.

Trachyandesites. From the tongue of a ragged flow on the western slope of Southeast Head, some 200 m from the shore at Southeast Bay, DALY collected a dark-grey, vesicular rock containing some xenoliths of labradorite-augite. (As the xenolithic character of the rock was only made manifest after microscopic examination, in the field the rock was thought to be an olivine-free basalt, as was incorrectly reported in DALY's 1922 publication.) A few glitters from feldspar micro-phenocrysts are seen in a dense groundmass, as viewed with the naked eye. The phenocrysts are mostly andesine, sometimes with external carapaces of oligoclase-andesine. Sodiferous orthoclase is also presumed, and spare, small, automorphic augite phenocrysts. The matrix, microcrystalline to cryptocrystalline, comprises an aggregate of oligoclase-andesine, orthoclase, diopside, dust-like ilmenite or titaniferous magnetite, some apatite and interstitial glass. The chemical analysis is shown in No. 6. On the basis of its norm characteristics, DALY placed the specimen in the dosodic subrang akerose, of the domalkalic rang monzonase, of the order germanare, of the dosalane class. He listed the average analyses of nine trachyandesites quoted by ROSENBUSCH (No. 7), from which he concluded that the specimen was a trachyandesite. The mineralogical composition of the DALY's specimen shows the predominance of feldspars over other components (75 %). The NIGGLI values show features of the alkaline-sodic series, and FUSTER claimed the rock was analogous to a sodic syenitic magma.

Trachytes. These rocks were studied in detail by DALY who examined some 50 thin-sections from various trachytic bodies. Before him, DARWIN, RENARD, PRIOR & REINISCH had given descriptions of these rocks, and later CAMPBELL SMITH commented further, but to date, DALY's account is the most complete.

The macroscopic descriptions given by DARWIN cannot be improved upon, even although he was merely giving generalities and not related to specific examples. The weaker, earthy, whitish types of trachytes resemble, en masse, tuffaceous sediments, and indeed DARWIN was at first inclined to consider them so, but finally ruled-out such a classification because of the feldspar crystals and grains of a black mineral in the same quantities and proportions as in trachytic rocks with which these varieties were associated.

RENARD gave the general name augitic trachytes to the samples he studied, characterized by the variable association of monoclinic feldspar, augite and vitreous material.

Mineralogical composition was remarkably constant, the mineral character was distinct, and variation in the rocks referred rather to texture and the greater or lesser importance of the role of glassy material. Varieties ranging from holocrystalline to vitreous showing only rarely tiny crystals of sanidine and augite were encountered, and when the vitreous matter developed further at the expense of these minerals, the rocks passed into obsidians. The microtexture of most rocks studied showed the dominance of microliths of sanidine and augite. No. 16 is an analysis of a RENARD specimen from Weather Post Hill, and can be taken as typical of these augitic trachytes.

According to DALY, the trachytes, without exception, show a predominance – often up to 80 % – of alkali-feldspar, chiefly soda-orthoclase, with variable proportions of anorthoclase. The nominative minerals are aegerine, diopsidic augite, riebeckite, cossyrite and other brown amphiboles of which aegerine is by far the commonest. Unquestioned riebeckite appears to be restricted to Green Mountain dome and to the trachytic tuffs and projectiles in the area. Mica was not observed in any thin-section. Like RENARD, DALY remarked upon the comparative uniformity, qualitative, mineralogical and chemical, of the trachytes of Ascension.

The trachyte of Ragged Hill dome (No. 8) is a normal representative, with porosity features, lack of bubble-vesicles, typical for these rocks. Feldspar phenocrysts are so abundant that the specimen has the appearance of syenite. Microscopically the rock shows soda-orthoclase, anorthoclase, in a matrix of these same minerals, some diopside, much aegerine magmatically altered into an opaque material resembling magnetite, a greenish-brown to opaque material suggesting riebeckite or allied amphibole, and anhedra of presumed titaniferous magnetite. Although both quartz and apatite occur in the rock norm neither were seen under the microscope, nor was any glass identified. The specimen is of the dosodic subrang nordmarkose, of peralkalic rang nordmarkase, of perfelic order canadare of persalane class. The specimen from Southeast Head dome, (No. 9) is non-vesicular, compact with rare flashes of light betraying the presence of phenocrysts, of anorthoclase. The matrix shows micro-crystalline mass of soda-orthoclase, anorthoclase may be present, micropoikilitic quartz and colorless glass, and sprinkled throughout are minute, deeply-coloured aegerines of corroded or skeletal form, some magnetite, a few grains of colourless diopside, with apatite rare and zircon very rare. It is uncertain whether or not riebeckite is present. The micropoikilitic quartz, which comprises 10 % by weight of the rock, has a habit similar to the free quartz in the siliceous 'veins'. Such quartz was found in Samoan trachytes by DALY, PRIOR found it in British East Africa, Aden and Ascension, and RENARD also noted its presence in Ascension. PRIOR thought that such quartz was not the result of ordinary weathering, and DALY claimed it was more likely of late-magmatic origin, deposited by residual solutions of the magma after crystallization of other constituents. The norm places the specimen in the dosodic subrang kallerudose, peralkalic rang liparase, quardofelic order britannare, persalane class. It was classed as a somewhat vitrophyric aegerine-trachyte by DALY.

REINISCH gave a chemical analysis (No. 10) of a glass-rich diopside-trachyte collected from the northern base of Dark Slope cone. DALY found no trachyte in situ here and supposed that the REINISCH sample was a fragment which had been carried from the Riding School trachytic body. The rocks at the latter locality are glass-free, non-vesicular, minutely porous, having rare phenocrysts of soda-orthoclase, anorthoclase and aegerine.

The matrix, microcrystalline to crypto-crystalline, comprises essentially the same minerals along with aegerine-augite. A sample from a flow on the NE side showed no quartz, but in the vicinity quartz, amounting up to 10% by weight, occurs in micropoikilitic form. REINISCH also referred to "alkali-trachyt-perlit" in a stream bed on the N side of Riding School, and gave two analyses (Nos. 13 and 14) of a trachytic pumice at the foot of the Riding School cone, on the N side, and a reddish trachytic tuff from the SW side of the cone.

At the "Drip" dome and at the NW base of Thistle Hill occur aegerine-trachytes almost identical with those of Riding School. The microphenocrysts include anorthoclase and extremely little aegerine, set in a groundmass composed essentially of soda-orthoclase, aegerine and diopsidic pyroxene. No quartz was identified in three thin-sections from here by DALY, but at Thistle Hill a little micropoikilitic quartz was noticed.

The greenish-grey trachytes of "The Crags" dome may exhibit phenocrysts of soda-orthoclase and anorthoclase. The groundmass comprises essentially a microcrystalline aggregate of soda-orthoclase and aegerine, much less diopsidic augite and magnetite, and a relatively high content of green-greyish, highly pleochroic amphibole of moss-like appearance which may be riebeckite.

The trachytes of Cross Hill show great variety in colour. Weathering was considered as an unlikely cause for this by DALY and he was unable to account for this feature. REINISCH gave an analysis of a sample from the quarry at the base of the western side of the Hill (No. 11), where the rocks show a powdering of dust-like grains of amphibole and irregular needles of aegerine-augite and aegerine. Microphenocrysts of soda-orthoclase and anorthoclase occur in a matrix of the same minerals, also aegerine-augite, micropoikilitic quartz and magnetite. The darker varieties of the trachytes have no amphiboles and the groundmass shows a few deeply-coloured anhedra of a vague, brown mineral which may be cossyrite.

The trachytes of Green Mountain are of two principal types, those in which aegerine is the only essential mafic constituent and those where both aegerine and alkaline amphibole are present. The aegerine-trachytes, light-grey in colour, are free of quartz, and occur as thick flows which outcrop on the road up the Mountain at the 1400 foot contour. The amphibole-bearing types occur quite commonly, and from Valley Tank, where the trachytic rim of the caldera in which The Peak cone was constructed, DALY made special studies. The compact, grey rock shows a few, small phenocrysts of soda-orthoclase and anorthoclase in a groundmass of the same minerals, aegerine-augite, augite and some riebeckite. Cossyrite was not identified in thin-sections.

REINISCH had reported cossyrite-arfedsonite-trachytes on Green Mountain at the elevation where the sanitorium is and also from a stream N of Donkey's Plain. No. 12 is a chemical analysis by REINISCH of a specimen collected half-way up the Mountain, which DALY classed in the dosodic subrang, nordmarkose.

DALY remarked that fragments, both large and small, of riebeckite-trachytes were quite plentiful in breccias encountered on the road up Green Mountain, in breccias on Middleton Peak ridge, scattered over the valley bottom followed by the Green Mountain road and also to the W and NE of Travellers Hill. Microscopic study made of some specimens showed that apart from riebeckite, at least three other varieties of amphibole occurred. These all had a ragged, mosslike habit similar to riebeckite, the same single and double refraction. All these moss-like amphiboles had a corroded appearance suggesting thex had been subject to reactions due to residual water-gas or then other liquids of a late-magmatic phase. Thes dark-coloured varieties were seemingly affected by water-vapour or then water-gas which removed the silica and concentrated iron oxides, resulting in a more or less opaque solid.

It is to be noted that REINISCH reported the occurrence of cossyrite in all his specimens of amphibole-trachytes, but DALY invariably found it difficult to prove the presence of the mineral, although he frequently suspected its presence.

CAMPBELL SMITH recognized several varieties of trachytes from the "Quest" collection. The cossyrite-arfedsonite-trachyte of REINISCH (No. 12) he re-named a pantelleritic-trachyte. His specimens of such came from the W slope of Green Mountain and from Riding School. A trachyte very rich in minute, pale-yellow augite (very similar to those occurring in sodalite-bearing trachytes of Gough Island, q. v.) was collected from the southern rim of Riding School. The texture and general mineral composition showed close analogies to the Gibele type of trachyte from Pantelleria described by H. S. WASHINGTON (1913), a type also well represented amongst the soda-trachytes of Kenya. Aegerine-diopside-trachytes were represented by a compact, olive-grey rock from E of Devil's Ashpit, overlooking Southeast Head. These rocks were so named by DALY and closely resembled RENARD's augitic-trachytes from Weather Post Hill. A loose boulder from Riding School was found to be a katophorite-trachyte. RENARD described similar rocks from the quarry at Georgetown. Another rock, presumably from the same quarry, was described by REINISCH as "katoforittrachyt" and DALY likewise described the same rock – No. 4.

FUSTER again remarked upon the unusual chemical similarities of the trachytes of Ascension. A study of the NIGGLI values (Table 43) shows that variation in the rocks is almost exclusively conditioned by modifications in the corresponding value of silica; in fact, except for si, the other parameters are practically equal throughout. The alkaline character of the rocks is evidenced by the high alk parameter. This factor, in combination with the high acidity, shows a chemistry comparable to the more acid sodic magmas of NIGGLI, especially to those magmas of alkaline-granitic group. These Ascension trachytes appear to fall between nordmarkitic and gibelitic magma types. The calculated mineralogical composition of specimens Nos. 8 to 12 and No. 13 show free quartz in proportions varying from 3.3% to 10.9% and thus some of these, as per FUSTER, should perhaps rather be classed as rhyolitic, i. e. rocks of rhyolitic composition without actual quartz. The proportions of feldspars, especially albite, are very high, varying between 75% and 85%, and between 52% and 56% for sodic feldspar. In those specimens with al-alk negative differences, acmite is present between the pyroxene molecules.

Quartz-Trachytes (Rhyolites). The rock forming White Hill dome shows an alternation of bluish-grey and white layers and is highly fluidal or platy. The eutaxitic structure is associated with the degree of crystallization of the original glass. Macroscopically aphanitic, in thin-section the rocks show microphenocrysts of soda-orthoclase in a groundmass of layered, colourless glass in which occur tiny, ragged needles of aegerine, skeletal crystals of alkaline feldspar, magnetite and micropoikilitic quartz. DALY's analysis of a specimen (No. 15) places it in the sodipotassic subrang liparose, of the peralkalic rang liparase, of the order britannare, of the persalane class. The chemi-

Table 43   Niggli Parameters (Fuster, 1954, Barros, 1960)

	1.	2.	3.	4.	5.	6.	8.
Si	108	119	143	146	158	190	276
Al	21.5	22.5	26.5	36.0	34.0	28.5	39.5
Fm	47.0	43.5	37.0	25.0	29.0	31.5	19.0
C	24.0	23.5	21.0	20.0	17.5	15.5	3.5
Alk	7.0	10.5	15.5	19.0	20.0	24.5	38.0
K	0.13	0.18	0.23	0.33	0.33	0.23	0.33
Mg	0.51	0.42	0.43	0.30	0.30	0.34	0.04
Q	27.2	27.7	30.8	30.6	32.7	34.1	41.7
L	35.7	38.3	43.5	54.3	51.0	45.0	51.1
M	37.1	34.0	25.7	15.1	16.3	29.0	7.2
	0.51	0.37	0.26	0.30	0.26	0.08	0.02
	0.17	0.18	0.19	0.06	0.08	0.23	0.05
	0.44	0.54	0.35	0.28	0.27	0.27	0.05
	0.30	0.19	0.21	−1.11	−0.24	1.06	3.18
Q_7	−20	−23	−15	−30	−22	−8	24
Ti	4.7	6.4	4.1	2.7	1.8	8.4	1.2
P	0.54	0.58	0.66	0.68	0.35	0.98	

	9.	10.	11.	12.	13.	15.	18.
Si	299	266	253	291	268	365	352
Al	37.5	39.0	37.5	39.5	39.5	38.5	41.0
Fm	17.0	15.0	18.5	16.0	15.0	16.5	15.5
C	4.0	5.0	6.5	5.0	6.5	3.5	4.0
Alk	41.5	41.0	37.5	40.0	39.0	41.5	39.5
K	0.30	0.36	0.35	0.37	0.36	0.37	0.28
Mg	0.11	0.07	0.21	0.07	0.15	0.07	0.04
Q	43.5	39.5	38.5	42.7	40.3	50.4	49.5
L	46.8	52.6	52.2	50.2	52.7	42.2	45.3
M	9.7	7.9	9.3	7.1	7.0	7.4	5.2
					0.01		0.01
	0.09	0.21	0.25	0.23	0.28	0.11	0.15
	0.07	0.04	0.14	0.04	0.10	0.04	0.04
	3.80	1.68	1.2	3.9	2.20	9.03	1.1
Qz	45	82	3	32	12	108	94
Ti	2.9		0.9	0.3		1.2	
P	0.28						

cal analysis of the rock would show it to be a rhyolite, but Daly believed that the silica percentage had probably been increased as a result of infiltration during a late-magmatic phase, the magma being nearer to typical trachyte rather than typical rhyolite. In consequence, he termed the rock a quartz-trachyte. The trachytes of Little White Hill and Wig Hill were believed by him to be very similar to the above specimen.

At Weather Post Hill a large flow extending to the N of Devils Cauldron was stated by Daly to be a very homogeneous aegerine(-diopsidic)-trachyte with many irregular grains and short needles of black, opaque material in the groundmass, which he thought might be magmatically altered cossyrite, likely aegerine also. Soda-orthoclase and less abundant anorthoclase occurs both as phenocrysts and in the matrix. In three thin-sections studied from here, no quartz was found.

Renard had classified the rocks here as augitic trachytes (Vd. No. 16) and thought that the high silica content was due to infiltration of quartz likely of secondary origin. Daly classed this specimen as of the dosodic subrang kallerudose, of peralkalic rang liparase, of quadrofelic order britannare, of persalane class.

The Boatswain Bird islet monolith was termed an aegerine-diopside-trachyte, poor in diopside. Micropoikilitic quartz was quite prevalent. Small amounts of moss-like material may be altered riebeckite or then a closely similar amphibole.

Campbell Smith noted the "Quest" collection soda-rhyolites of Comende type (Comendite), taken from the NW slope and on the cinder cone of Green Mountain. As noted above, Daly, and also Renard believed that infiltration had occurred in the trachytes during a late-magmatic period, the quartz being of secondary origin. Prior was of a different opinion, claiming that all the quartz was not of secondary origin and believed that some of the quartz-bearing rocks which he described should be termed rather soda-rhyolites. Campbell Smith thought it most unlikely that the abundant quartz in his comendite specimen could be explained by infiltration, but left the matter there. (He noted that a hyaline soda-rhyolite had been collected in 1843 by members of the Ross Antarctic Expedition from the base of Green Mountain.)

Fuster included in his rhyolites those acid rocks with actual quartz, viz. Daly's quartz-trachyte from White Hill (No. 15) and Reinisch's "rhyolithobsidian" from the Riding School area (No. 18). As one would presume the acidity of these rocks is much greater than in trachytes, but the alkaline character is still present, since the alk parameter is as high as in the trachytes. The close relation between both groups of rocks is evident by inspection of the other Niggli parameters, with the exception of si. It follows therefore that the rhyolites are also comparable to the same group of granitic magmas of the alkaline series, though being richer in silica, Fuster would ally them to normal alkaline granitic magmas. A study of the mineralogical composition of the rhyolites likewise shows similarity to the trachytes with the exception of quartz, in which the value is as high as 20 % for rhyolites.

Obsidians. Darwin gave detailed descriptions of the massive, zoned, and spherulitic obsidians of the island, descriptions which were praised by both Renard & Daly, the former quoting lengthy excerpts from Darwin's publication. Darwin mentioned occurrences in the western part of Green Mountain where there are passage or transition zones between trachytes and obsidians and where also obsidians are present in the floors of trachytic flows. The transition to obsidian is accomplished in several ways, involving angular, concretionary obsidian and schistoidal feldspathic rock.

Renard found the specimens of Green Mountain in no way different from the characteristics well enough known of obsidians. Weathered rock has a greyish colour and a somewhat earthy appearance, and when such rocks decompose they have a pronounced greasy appearance like pechstein. The only mineral which can be observed by naked eye is sanidine which at times can be detached from the rock as quite large, glassy grains. The vitreous base is never homogeneous, being pock-marked by tiny, lamellar crystals of sanidine and augite. Some obsidians show devitrification, and perlitic structure may be quite common.

DALY found obsidians as projectiles, as chilled surface and floor phases of trachytic flows, but saw no large, independent obsidian masses. The non-pumiceous glass is restricted to areas where larger trachyte domes are present, such as Riding School, Green Mountain, Weather Post Hill, Devils Cauldron and White Hill. Obsidians are very common in trachytic breccias and tuffs which floor trachytic flows. The abundant slivers of greenish augite present are of diopsidic type, with microphenocrysts of soda-orthoclase very much rarer.

Specimen No. 17 is an obsidian from Green Mountain reported by RENARD. No. 18, as noted above, is a rhyolitic obsidian from the Riding School region quoted by REINISCH. Nos. 19 and 20 show silica and water analyses, after REINISCH, of a spherulitic rhyolite-obsidian from the outer mantle at Riding School, and of an "obsidianknollen" from the Green Mountain dome respectively. DALY classified Nos. 17 and 18 as falling in the dosodic subrang kallerudose.

CAMPBELL SMITH mentioned obsidians from Wideawake Valley, associated with pumice.

BELL (1967) has commented upon obsidians and natural glasses occurring in the island. He classified the natural glasses present as belonging to four types, described below.

1. Lavas, either flows or fragments. The glass is associated with salic lavas, including trachyte, quartz-trachyte or rhyolite, pantellerite and comendite varieties, as well as hyaline facies of pantellerite and as obsidian. He noted salic lava bodies with interbanded obsidian within crystalline 'trachyte' rock near the base of the flows, the two having differences in chemical composition. BELL surmised that this dintinction chemically and petrographically might be due to a too viscous fraction which did not crystallize, and also that the obsidian might represent a denser phase which sank during eruption down throught a trachytic phase which, at that time, was less dense due to high gas content.

2. Pumice. Unwelded and welded pumice is associated with pyroclastic flows deposits at several localities, e. g. on Green Mountain road, on the ridge at Grazing Valley, etc. In the welded pumice at the former locality, obsidian lenticles are streaked out to lengths up to 45 cm. Pumice also is found as independent bedding formations which may show size grading. These are believed to be deposits resulting from subaerial pyroclastics falls, in some instances later sorted in an aquatic enviroment. Pumice has also been observed at the base, more rarely the top, of trachytic bodies, being thought to represent a highly vesiculated variety of lava.

3. Mesostasis in plutonic xenoliths. BELL refers to two distinct types here. One is a coarse-grained feldspathic xenolith from a lava at East Crater which contains interstitial glass. The mesostasis is colourless in the vicinity of the feldspar but brown in colour where near sodic pyroxene and/or opaque oxide. Reaction is evident along the crystal edges in contact with glass, the latter being thought to be a product of partial melting ot the xenolith minerals due to their incorporation within the lava. The other instance refers to a plagioclase-augite-olivine cumulate xenolith which occurs in redistributed tuffs on Middleton Ridge. Here the mesostasis is dark, and various extents of crystallization of the opaque minerals and plagioclase crystals is obvious. Cumulus crystals contacting the glass are completely euhedral, with no reaction visible. In this instance, the mesostasis is thought to be a chilled intercumulus liquid, with quench crystals within it.

4. Glass also occurs as selvages on volcanic bombs, but Bell does not elaborate upon the topic.

Siliceous 'Veins' in Trachytes. Darwin wrote an excellent account of rib-like ridges occurring on the weathered surface of trachytic bodies which he noted especially at Riding School. These 'veins' are of white colour, on occasion reddish or yellowish, showing either conchoidal or angular fracturing, enclosing white, powdery material in cavities. The contained crystals of glassy feldspar and tiny black specks and stains are similar to those in the country-rock of altered trachytes. These 'veins' were also noted in scoriaceous basalts. They vary greatly in size and change their thicknesses very suddenly, from 3–4 mm in with up to 25 mm broad. They are inclined at all angles to the horizontal and may lie horizontally. Usually they display a curved form and frequently anastomosing is common. The hard ribs stand up to weathering better than the enclosing trachytes or basalts, and may project a metre above the country-rock surfaces. Darwin was greatly astonished at the sight of these 'veins', saying he had never seen or read of any quite like those of Ascension. They looked to him like ferruginous seams resulting from some kind of segregation such as often occurs in sandstones. Many ochre- and red-coloured jasper and siliceous sinter 'veins' and irregular masses indicated an abundant source of silica in the environment. As these 'veins' differed from the trachytes only in being harder, more brittle but less fusible, Darwin assumed that they were due to segregation or infiltration of siliceous material, such as occurs with oxides of iron in many sedimentary rocks.

Renard noted amongst samples collected by the "Challenger" expedition from Riding School and Red Hill, clear evidences of siliceous infiltration. One of the rocks from Red Hill he classed as a true siliceous tuff, showing under the microscope that the matrix comprised entirely small, angular quartz grains strongly pressed together, and between crossed nicols had the appearance of the matrix of some quartz-porphyries. At Riding School, a white, very hard volcanic glass showed almost complete transformation into silica, microscopically showing glass in the pores and interstices of which chalcedonic quartz was present. A specimen from Southwest Bay Renard thought might be a piece of clinker, but Daly stated that here there was only clinkery phases of basaltic lava, and the sample showed no relation to the 'veins' in trachytes.

Daly agreed with Darwin in attributing these 'veins' or silicified interfaces as Daly preferred to call them, to local silicification of normal trachyte. The silica so introduced is always micropoikilitic quartz, such as occurs in normal trachytes. The proportion of quartz by weight in these 'veins' was estimated to be between 15% and 25%, five to ten times the amount found in trachytic country-rock. Daly did not think that the free silica was introduced in the process of ordinary weathering of trachytes but rather resulted from late-magmatic action. He hypothesized that when motion of the effluent trachyte ceased, its glass cooled, crystallized and underwent tensional stress, similar to that causing columnar jointing in lava. Via actual or potential partings, magmatic steam escaped, and from the volatile solution, quartz was precipitated. Seemingly a little hematite was likewise formed, giving rise to typical reddish coloration. These precipitates are not true veins, according to Daly, but simply trachytic portions which were modified by magmatic fluids – gaseous solutions – which migrated along the interfaces of the trachytes. Daly believed that similar gaseous solutions affected the pyroxenes and amphiboles of the normal trachytes, in which leaching-out of silica from the original minerals was accompanied by transformation of much or all of the aegerine or riebeckite into pseudomorphs of iron oxides, etc.

Xenoliths and Ejected Blocks. The occurrence of plutonics in volcanic projectiles was evidently first reported upon by WEBSTER (1834) who mentioned the presence of granite, syenite, graywacke and argillaceous schist. Neither of the last two mentioned exotic rocks have been recognized by later students of the island geology.

DARWIN was the first to give a more detail description of these ejected blocks. From the Green Mountain region he described granitic fragments embedded in scoriaceous volcanics, grouping them as follows: 1. a syenitic rock, striated and dotted with reddish parts. Potassic feldspar in good crystal form, small quartz grains, hornblende all recognizable. 2. a brick-red fragment composed of feldspar, quartz and dark particles of hornblende. 3. a whitish feldspar mass containing dark-coloured hornblende. 4. a rock comprising an aggregate of large, dark-tinted crystals of labradorite, between which occur granules of whitish feldspar, lamellae of mica and altered hornblende, but no quartz. DARWIN also mentioned finding elsewhere a conglomerate containing small granite fragments, jaspoid material and porphyry, embedded in a pebble cut by innumerable thin veinlets of concretionary pechstein passing into obsidian. These conglomerate beds are parallel, slightly undulating and continue for only a short distance, thinning at their extremities and resembling lentiles of quartz in gneiss. DARWIN thought that these fragments could not have been projected as isolated pieces from a volcano, but originated from a fluidal mass like obsidian.

RENARD examined specimens which he claimed were crystallines expelled from great depth by volcanic action. From the Green Mountain area he recognized amphibolitic granites, friable, the grains having a vitreous appearance, the feldspathic mass being whitish and sprinkled with black specks which were like tiny hornblende crystals. The texture is granitoid. The feldspar was identified as orthoclase, including fine lamellae of albite, i.e. microperthite. The typical cleavage of hornblende identifies the mineral. Quartz, the last mineral to crystallize, shows sections remarkably, the crystals are relatively large and sometimes show gas bubbles. Only occasionally can fine needles of tourmaline be seen in the quartz. Zircon also is present. Another specimen showed a porphyritic texture with large orthoclase crystals – up to 3 cms in length – in a biotite groundmass which had a somewhat gneissic appearance. Another granitic sample from the same area showed quartz crystals severely shattered into many pieces. In the same general area, RENARD also found rocks of diabasic type, though lacking the microstructure of true diabase. The specimens were deeply altered, with feldspar and biotite visible to the naked eye in a granitoid structure. Microscopically the rock comprises an aggregate of labradorite, augite, hornblende and biotite. The augite has more the appearance of augites found in diorites than diabases. Cordierite may be present, although this could be an altered feldspar. CAMPBELL SMITH concluded that neither orthoclase nor olivine were present in RENARD's diabasic specimen, that the rock was rich in both biotite and barkevikitic hornblende and was probably related rather to the essexites. From Red Hill, RENARD recognized olivine-gabbros, showing microscopically lengthy pericline crystals enclosing and moulded by augite, a mélange near to bytownite and anorthite (RENARD gave a chemical analysis of the plagioclase which showed 30% albite, 70% anorthite) and augite often showing no evidence of cleavage but many grooves caused by regular fractures. Another specimen from here showed no olivine but instead rare sections of a colourless, rhombic mineral of strong relief, identified as enstatite.

From Green Mountain-The Peak area, DALY, on the basis of microscopic study, recognized the following types of xenoliths: 1. alkali hornblende-biotite-granite, rich in microperthite, 2. strongly miarolitic hornblende-syenite, with accessory quartz – a quartz-syenite, 3. augite-hornblende-quartz-syenite, transitional to granite, 4. salic diorite, transitional to monzonite, 5. mottled, sugary, miarolitic augite-hornblende-diorite, 6. typical olivine-gabbro, 7. typical olivine-free gabbro, 8. unusual olivine-free gabbro, of highly automorphic augite and labradorite embedded in opaque, black cement, likely ilmenite, 9. typical coarse-grained wehrlite. Granitic boulders of the above kinds are common in the streams and main valleys to the N of Green Mountain. In a trachyte at "The Crags" dome, DALY found a single angular inclusion very like the alkali hornblende-granites occurring in basaltic tuffs at Green Mountain. From Dark Slope Crater many somewhat angular xenoliths of gabbro occur in scoriaceous basaltic flows. Most are of olivine type but olivine-free varieties are not lacking. In the latter types, some have essential diallage, some show diopsidic augite without diallage partings, and still others contain both pyroxenes. As DALY found no quartz in specimens from this locality, he doubted if granitic xenoliths were present here.

A complete list of ejected blocks and xenoliths, without micro-pegmatite, known to DALY is given below:

> Alkali amphibole-granite, with/without micropegmatite
> Alkali hornblende-biotite granite
> Pyroxene-hornblende-quartz-syenite
> Hornblende-syenite, verging on quartz-syenite
> Monzonitic diorite
> Typical augite-biotite-diorite
> Augite-gabbro
> Abnormal augite-gabbro, with ilmenitic mesotasis (no olivine)
> Typical olivine-gabbro
> Typical wehrlite (peridotite)

To the above can be added the types recorded by RENARD:

> Biotite-granite, bearing some micropegmatite
> Biotite-bearing 'diabase' (essexite as per CAMPBELL SMITH)
> Enstatite-bearing, olivine-free gabbro.

All xenoliths tend to be fritted, are usually brittle and show well-developed cleavages in the essential minerals. Stringers and small bubbles of brown glass show incipient fusion, and in general the essential minerals show a dull appearance due to fluidal and glassy inclusions.

DALY expressed the opinion that as these xenoliths showed strain features in the minerals such as would be expected if shot-up through hot volcanic vents, it was no easy matter to decide whether or not these rocks represented deep-seated magma phases of the Ascension cone. The gabbro, diorite and wehrlite fragments are no doubt differentiates of a deep-seated magma, but the occurrence of more salic representatives is more problematical. It was DALY's view that these consanguinous rocks, granite and syenite, originated from a single mass or syngenetic mass, of plutonic rock. As both in chemical constitution and mineralogical content these salic rocks differ greatly from the trachytes and rhyolites of the island, they scarce can be deep-seated differentiates, syngenetic with the exposed lavas. DALY was thus led to conclude that the granitic and syenitic fragments were representatives of an older basement on which the island cone rested. This announcement, made in 1925 just after the appearance of the more widely circulated English translation of WEGENER's controversial tome on continental drift, was indeed significant. Here was a young oceanic island, probably formed in the Pleistocene or not much earlier, built up of a series of basaltic and trachytic flows into the form of a massive cone rising 3000 m from the Atlantic depths, which, amongst its xenoliths resulting from explosive action, were 'continental' rocks, seemingly suggesting a 'continental' fundament far removed from present continental coasts.

Petrology

DALY presented tables showing comparisons of average chemical analyses of the Ascension basalts with those of other regions in the world, also Ascension basalts, trachydolerites, trachytes and rhyolites with those of Tutuila Island, Samoa, which he had previously studied. The tables demonstrated the relative uniformity of basalts on a worldwide scale. He disagreed with REINISCH when the latter claimed, on the basis of a study of far fewer Ascension specimens, that the Ascension rocks were throughout special alkaline basalts: to DALY they were common basalts, representing a primary magma, belonging to the calc-alkaline or sub-alkaline group of rocks. As already noted, FUSTER classed the Ascension basalts as belonging to the acid alkaline and subbasalt groups.

DALY went into considerable detail as to the origin of trachyte, so common in the island. It was his opinion that common basalt was the source magma, trachyte a derivative of such, Ascension being an excellent area to show the intimacy between the two rock types. He based his contention on the following facts, gathered from Ascension, Samoa and elsewhere; 1. the close relations, both in time and space, of common basaltic and trachytic eruptions, 2. the significance of trachytic volumes compared to the accompanying basalt, 3. presence of trachydolerite and other volcanics transitional from basalt to trachyte, 4. the alternation of trachytic and basaltic eruption at central vents, 5. the absence of large trachytic fissure eruptions, 6. trachytic eruptions from central vents usually preceded by long periods of quiescence, 7. important explosive outbursts preceding the issuing of trachytes, 8. scarceness of ordinary bubble-vesicles in trachytes, except in minor pumiceous phases, 9. trachytes show high viscosity, hence tendency to form endogenous domes, crater-fillings, short, thick flows, 10. innumerable cases of transitions from trachytic to rhyolitic lavas, sometimes to phonolitic lavas, 11. quite common occurrence of micropoikilitic quartz in trachytes, and 12, quite frequently trachytes are associated with limburgites and such highly mafic types. Not all of the above observations are to be noted in Ascension, e. g. neither phonolitic nepheline-bearing nor ultra-mafic lavas have been found in the island, but Ascension does show many of those features which can be ascertained in other trachytic regions of the world where more detailed studies have been made.

The Ascension trachydolerites and trachyandesites are but steps in the transition from basalt to trachyte. Rhyolitic obsidians and quartz-trachytes represent small-volume segregations of magma rich in silica taking place at the summits of trachyte columns in a magma condition, indicative of gravitational separation of differentiation units. To DALY the trachytes are the result of pure fractional crystallization, the change from basalt being effected through the intermediary trachydoleritic and trachyandesitic stages, the quartz-trachytes and rhyolitic obsidians being smaller, localized, gravitational separations of silica-rich magma at the apices of trachytic cylindrical masses. This differentiation required the upward movement of both liquid and gaseous fluids.

DALY noted many instances where trachytes either lay upon or then penetrated basaltic bodies, and further, basaltic xenoliths occurred in places in trachytes. This would therefore argue for an older age for the basalts. However, trachytes can be observed, e. g. at Green Mountain, Cross Hill, etc., which are overlain by rocks of basaltic type, and here this would argue for a younger age for the basalts. These latter basalts are relatively small in volume as compared to the trachytes, whereas trachytic rocks have a far smaller volume than basaltic rocks when they lie on, or are penetrated by, the latter. To account for the basalts resting on trachytes, DALY surmised that if the magma chamber in which the trachytes developed has a larger cross-section at depth than in the higher conduit filled with trachyte, then the mother-magma directly below the conduit would experience little contamination from gravitational sinking of crystals formed. The expulsion of the trachyte would also entail the rising upward of this original basaltic magma, which could follow-up behind the trachytic extrusion in a relatively short time. Hence such a mechanism could explain the basalt-trachyte-basalt sequence such as occurs at Green Mountain and Cross Hill. DALY was of the opinion that the absolute lengths of time between each trachytic eruption were not great, that the visible trachytic bodies were likely all generated during one, relatively short period of igneous history.

ATKINS et al saw no justification for postulating an earlier trachytic phase of eruption and a later basaltic one; to these authors both types were produced during various periods in the history of the island.

Some of the trachytic domes and basaltic flows have a much younger appearance than others, indicated by the degree of weathering to which they have been subjected. Although there is no historical record of vulcanism in Ascension, ATKINS et al thought that some basaltic flows were "perhaps only a few hundred years old". It is indeed quite possible that the building of the visible part of the great composite cone which is Ascension is a phenomenon of the Pleistocene.

BELL (1965) presented a short note concerning the method of eruption in Ascension. In general, whatever the type of lava extruded, has been of the central-vent type. He pointed to three important characteristics associated with the trachytic bodies which had a bearing on their eruptive mechanisms: 1. early explosion action, witnessed by pumice beds lying below trachytic flows, 2. instances of updoming by trachytic bodies of earlier basaltic flows, 3. expulsion of large trachytic bodies in the vicinity might possibly have caused the caldera of Cricket Valley, though here explosive action may also have been a cause.

DALY had remarked upon the absence of fissure eruptions in Ascension, but BELL refers to a discontinuous fissure measuring 1.6 km striking E-W and intersecting Southeast Head trachytic dome. Explosive action and outpourings of trachyandesite can be seen from several of the 25 small craters aligned along this fissure. It was BELL's opinion that trachytes at the eastern end of the island were built up through emplacement of domes, expulsion from small craters and also from fissure eruptions.

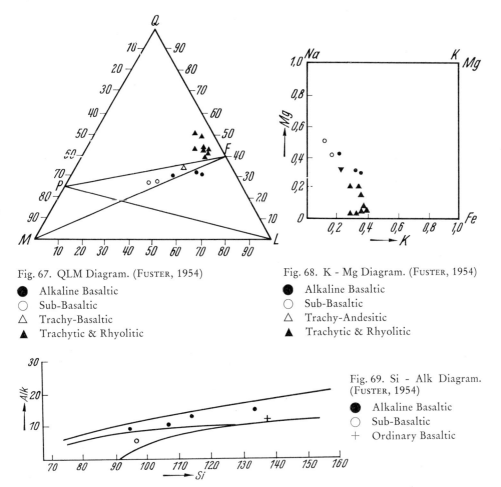

Fig. 67. QLM Diagram. (FUSTER, 1954)
● Alkaline Basaltic
○ Sub-Basaltic
△ Trachy-Basaltic
▲ Trachytic & Rhyolitic

Fig. 68. K - Mg Diagram. (FUSTER, 1954)
● Alkaline Basaltic
○ Sub-Basaltic
△ Trachy-Andesitic
▲ Trachytic & Rhyolitic

Fig. 69. Si - Alk Diagram. (FUSTER, 1954)
● Alkaline Basaltic
○ Sub-Basaltic
+ Ordinary Basaltic

Diagrams presented by FUSTER illustrate further chemical characteristics of the rocks listed in Table 42. With the exception of Nos. 4 and 5, the basalts, trachytes and rhyolites group themselves around the M-F line of the QLM diagram. The k-mg diagram shows a normal position for any alkaline-soda series, there being a complete transition from trachytes and rhyolites to subbasalts, passing through trachyandesites and alkaline basalts. From the locus of the rocks in the si-alk diagram, we can conclude that the alkaline basalts occupy an extreme position, corresponding to excessive acidity, whereas the subbasalts are more centrally placed, one at the boundary of the alkaline basalts, the other at the limit of ordinary basalts.

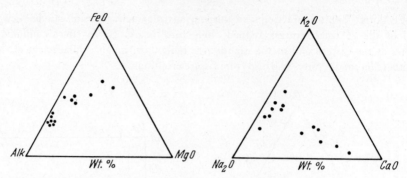

Fig. 70. MgO-FeO-Alk Triangular variation Diagram. (LE MAITRE, 1962)
CaO-Na$_2$O-K$_2$O Triangular variation Diagram. (LE MAITRE, 1962)

Two interesting features of Ascension which distinguishes it from other islands in the Gulf of Guinea are: 1. the acid volcanic associations, and 2., the occurrence of holocrystalline xenoliths in the lavas.

BARROS (1960) presents tables giving further information. Table 43 essentially based upon FUSTER but with further additions by BARROS gives NIGGLI parameters for 14 of the specimens listed in Table 41. Table 44 gives JUNG and RITTMANN values. (We would

Table 44   JUNG and RITTMANN Values (BARROS, 1960)

Specimen No.	JUNG		RITTMANN	
	SiO$_2$ (%)	R	Si°	Az°
1.	47.69	74.0	0.84	0.51
2.	48.64	65.1	0.83	0.54
3.	52.87	52.2	0.90	0.59
4.	51.18	44.2	0.82	0.59
5.	54.04	40.3	0.87	0.62
6.	58.00	34.2	0.95	0.65
8.	65.18	7.0	1.09	0.73
9.	66.98	6.9	1.17	0.74
10.	62.12	8.2	1.44	0.72
11.	63.98	11.9	1.01	0.71
12.	67.05	8.4	1.12	0.74
13.	63.02	11.6	1.05	0.72
15.	71.88	5.6	1.49	0.78
18.	71.42	7.7	1.36	0.77

note that specimen No. 4, named a trachydolerite by REINISCH and accepted as such by DALY and FUSTER, was classed by BARROS as a basanitoid basalt, on the basis of its 4.8% nepheline. Similarly No. 6, named trachyandesite by WASHINGTON, DALY & FUSTER, BARROS would prefer to call a basalt, with an acidity approaching that of trachyte, containing 2.5% virtual quartz.)

Isotope Geochemistry

To date we have two publications yielding a small amount of data on the isotopic composition of lead and strontium from Ascension rocks, those of TILTON et al and of GAST et al. Table 45 is compiled from tables presented in these two papers. Both reports refer to interisland and intraisland variations concerning Ascension and Gough. (Vd. chapter on Gough for details of that island.)

Table 45   Isotopic Composition of Pb and Sr (TILTON et al, 1964, GAST et al, 1964)

Rock	Daly Spec. No.*	Locality	$\frac{Pb\,206}{Pb\,204}$	$\frac{Pb\,206}{Pb\,207}$	$\frac{Pb\,206}{Pb\,208}$	$\frac{Pb\,207}{Pb\,204}$	$\frac{Pb\,208}{Pb\,204}$	$\frac{Sr\,87**}{Sr\,86}$	$\frac{Sr}{Rb}$
Olivine-basalt	2765	Travellers Hill	19.43	1.240	0.4958	15.67	39.20	0.7025 0.7028	15
Olivine-poor basalt	2740	Portland Point	19.55	1.247	0.5008	15.68	39.04	0.7025 0.7028	–
Obsidian bomb	2775	–	19.50	1.2471	0.4973	15.64	39.21	–	–
Aeg.-riebeckite trachyte	2716	Crags dome	19.72	1.255	0.5000	15.71	39.44	0.7073	0.25
Trachyandesite	2864	Southeast Head	–	–	–	–	–	0.7025	–
Trachyte	2863	Southeast Head	–	–	–	–	–	0.7045	–

* not listed in Table 41.   ** Normalized to Sr 86 / Sr 88 = 0.11940.

As regards the interisland aspect, there is a significant difference in the isotopic composition of lead in basalts, due essentially to the variation in the abundance of Pb 206, with higher Pb 206/Pb 204 ratios in Ascension. This would suggest that in this island, the lead is associated with a higher uranium-lead ratio environment. TILTON et al did not believe one could assume from these ratios that the basalt leads in both islands are uncontaminated. Regarding variations within Ascension, these authors pointed out that the Pb 206/Pb 204 ratios are higher in the trachytes than in the basalts or trachyandesites, and the same holds for the Pb 208/Pb 204 ratios. These trachyte-basalt variations cannot readily be explained from contamination by surface or oceanic lead. The authors suggest that either the trachytes originated from distinct reservoirs which had higher uranium-lead ratios than the basalt reservoirs, or then that both rocks had a common reservoir but incorporated varying parts of a mineral assemblage in which there occurred different uranium-lead ratios from mineral to mineral. However each rock type is taken to have originated in the outer mantle.

As regards strontium isotope ratios, GAST et al point out that the two trachyte samples from Ascension are distinctly more radiogenic than in the more basic samples, a feature also noted for lead above. The lead and stontium variations show the usual behaviour geochemically of uranium, thorium, lead, rubidium and strontium, i. e. the enrichment in uranium and thorium in relation to lead is also accompanied by an enrichment in rubidium with respect to strontium. The authors are opposed to a magma-contamination theory to accunt for the above lead isotope ratio variations in the basalts and trachytes of Ascension, nor does it seem feasible that the Sr 87/Sr 86 ratio variations in the basalts-trachytes can be explained by the admixture of pelagic sediments with a primary magma.

The above opinions of TILTON et al and GAST et al regarding a possible role of contamination or mixing of magma, assumes the absence of older crystallines in the immediate depth vicinity of the island. But as has already been noted, earlier workers have commented liberally upon holocrystalline xenoliths in the flows of the island, possibly representatives of an "older terrane of granitic rocks", as per DALY. Beside these older views on the xenoliths, we must place the more recent study of TILLEY who, as noted previously, believed that these inclusions were the plutonic equivalents of some of the more acidic extrusives. If indeed DALY was correct in theorizing upon an older terrane, perhaps several hundred million years in age, then of course a convenient source is available for the radiogenic lead and strontium of the volcanics. But even assuming such a source, GAST et al find it difficult to visualize how the isotopic lead composition in the basalts could be affected without at the same time also altering the chemical composition and the isotope strontium composition. They thought it possible that the noted high uranium-thorium ratio in the Ascension rocks was related to the source of the rocks rather than to magmatic differentiation processes, or then perhaps was due to similar high ratios in some parts of the mantle, observing that the uranium-thorium ratio of the mantle, inferred from isotopic data from Ascension, is analagous to that inferred from chemical data in the circum-Pacific area. Regional chemical heterogeneities in the upper mantle were thought by GAST et al as the most acceptable explanation for the noted variations in isotopic composition of lead and strontium between Gough and Ascension rocks.

Economic Geology

DARWIN noted enamelled surfaces mantling some rock outcrops in Ascension. The material was similar to that observed by him on St. Paul Rocks (q. v.) and was thought to be an impure phosphate of lime. RENARD later studied samples of this material and concurred with DARWIN, its microscopic features being very similar to the St. Paul Rocks occurences, in both instances being due to the decomposition of bird excrements.

DALY noted that thin phosphate deposits on basaltic flows were quite numerous in the island of Ascension. A specimen from the roof of a pressure cavern in a young flow was submitted to RICHARDS (1928), who chemically analysed the sample, as shown below. The whitish substance had a specific gravity of 2.30, and was classed as newberyite.

It would appear that no effort has been made to exploit commercially the phosphate, doubtless because of its sporadic occurrence and relatively small quantity.

Table 46   Chemical Analysis of Newberyite (RICHARDS, 1928)

SiO_2	3.76	Na_2O	1.34
FeO	0.62	K_2O	0.94
MnO	0.39	H_2O+	7.39
MgO	0.21	H_2O-	0.83
CaO	40.50	P_2O_5	44.19

CHAPTER 12

St. Helena

General

St. Helena, forever associated with the exile of Napoleon Bonaparte, lies between lats. 15° 54′ and 16° 01′ S, longs. 5° 37′ and 5° 47′ W, some 1900 km and 2900 km respectively from the nearest coasts of Africa and South America. DALY (1927) made the arresting observation that: "With the exception of the still smaller Ascension Island, the 115 square kilometres of Saint Helena form the only dry land in an area of 15 000 000 square kilometres, or three per cent of the earth's surface".

The island measures 17 km in length and 10.5 km broad, with an area of 118 km² (HIRST, 1951), rising to a maximum elevation of 823 m in Diana Peak.

Discovered in 1502 by the Portuguese navigator Joao de Nova Castella, at which time it was uninhabited and densely forested, the island of St. Helena was to experience a varied history. Portuguese, Spaniards, Dutch, French and British have all vied for its possession, its location as a watering and victualling station on the Cape of Good Hope route to India and the Far East rendering it of prized importance. The East India Company, that unique concern which blazed the path of British colonialism, held a Charter for the island from 1673 until 1836. During the time of Napoleon's melancholy confinement on St. Helena from 1815 till his death in 1821 therefore, the island was actually privately owned, although during this period the Crown assumed control.

Ascension and Tristand da Cunha are dependencies of St. Helena.

The indigenous forests, of which gumwood, ebony, redwood, white cedar and tree-ferns are the most prominent, are no longer seen. Most of the green areas comprise New Zealand flax *(Phormium tenax)* and pastureland. In the grounds of private dwellings there are to be found trees and shrubs flourishing from many parts of the world in this cool, mild and equable climatic environment. Today there is very little timber and the obtaining of even twigs for firewood is a problem. Because of the extensive areas given over to the flax plantations, sheep grazing and cattle ranching, there is relatively little land left over for the growing of foodstuffs. Fish is the principal item of diet.

The only indigenous bird is a plover *(Aegialitis Sanctae)*, but canaries, pheasants, pigeons, turkeys, ducks, geese and fowl have been introduced. Snakes are absent but lizards are plentiful. Goats and sheep are responsible for much of the erosional damage. Donkeys are beasts of burden.

The 1961 population of some 4600 inhabitants comprises mostly Islanders – the native population who are descendants of European, Indian, Chinese, other Far Eastern and African origin. These English-speaking peoples have lived a quiet, respectable life, friendly and mannerly, many of whom during the years have sought wider horizons in Britain and South Africa.

Jamestown, 1568 inhabitants in 1961, the capital and only town, lying on the NW coast, is partially built on an old delta of the James Valley river.

Physical Features

St. Helena, a composite volcanic cone deeply eroded at the summit, rises from the seafloor at a depth of 4400 m. The base of this cone is ca. 130 km mean diameter, with an area of some 12 000 km^2 – more than ten times the area of the base of Etna, and a volume at least twenty times that of Etna (DALY). As the highest point on the island rises to 823 m, the total height of this immense cone is some 5000 m, a truly gigantic phenomenon.

The axis of the island extends NE-SW, parallel to the island trend, and here are aligned the major peaks, Diana Peak 823 m, Actaeon Mountain, 818 m, High Peak, 798 m, Hooper Rock, 691 m, The Bairn, 615 m and several other unnamed peaks above 600 m. Diana Peak, Actaeon Mountain, High Peak, Hooper Rock and White Hill, 542 m, have an amphitheatre arrangement, a 12 km long ridge open towards the S with streams converging into the main one and emptying into Sandy Bay. DARWIN & OLIVER were of the opinion that this amphitheatre was a vast explosion crater, but DALY opposed this view, arguing that there is no widespread and superficial layer of fragmental material over the slopes leading downward on the outside of the half-bowl, the shallowness of the explosion crater being inconsistent with its other dimensions, and lastly, the ridge extending from Diana Peak via White Hill to Sandy Bay Barn is convex and not concave to the Great Basin or Great Hollow, such as one would expect in the case of an explosion crater. HIRST was in agreement with DALY in considering the basin an erosional feature.

The valleys draining the highlands, more or less centrifugal in pattern, are very steep and deep (some +300 m and more) as well as narrow and are only active after heavy rains. HIRST suggested that probably in inter-glacial pluvial periods, these streams were then perennial and with much greater erosive power than at present. Almost all streams reach sea-level at grade, but some shorter valleys where more resistant lavas overlie tuffs have been truncated by backward erosion of sea cliffs, such that they debouch at elevations from 100–200 m above present sea-level.

Marine erosion has resulted in awesome destruction of the island, with consequent development of truly imposing cliffs. Precipices, many of which are unscalable, rise almost sheer up to heights as great as 670 m. Sea stacks, both large and small, likewise bear testimony to the receding strand-line. Apparently coral reefs have never retarded cliffing, even although the mean temperature of the waters around St. Helena in the coldest month is about 20° C. At many localities along the NW or leeward side of the island, very steep cliffs rise upward from the inner sides of inclined rock benches. The elevation of the benches at the landward side are about 3.3 m above mean sea-level. At Munden Point it is unusually high, 5.8 m above the same datum. Many of these benches are cut out of massive lava flows, but some from coarse talus strongly cemented together.

The average slope of these benches is 1 : 5 or 1 : 6, in reasonable agreement with the average slope of the present inshore sea bottom. There also occur lower benches which, according to Daly, merely represent the surfaces of outwardly dipping lava flows. All evidences of the upper bench on the windward side of the island have been destroyed by marine denudation. Several dry sea caves occur at the inner edges of the higher benches, giving, as per Daly, "unequivocal evidence of a sensibly uniform emergence of at least 4 metres but not greater than 6 metres". Daly also noted recent cliffing of gravel deltas at the James and Rupert streams, suggesting that here sea waves developed scarps varying from 3–6 m in height.

A graded shelf entirely surrounds St. Helena. Out to the 50-fathom line, this shelf varies from 1.3 km in width to 3 km. From the 50-fathom to the 200-fathom line, the slope varies from 1 : 1.2 to 1 : 2.9, this slope presumably being related to the edge of the detrital embankment. The inner shelf likely has been formed by marine denudation, whereas the outer one is due to deposition of material so eroded.

Climate

The cool, pleasant climate results from the SE Trades, blowing steadily during the year. Possibly because the warm South Atlantic currents do not come near the island, St. Helena is cooler than might be expected. Summer temperatures at sea level range from 20° C to 29° C; in winter, from 14° C to 21° C. In the higher interior, it is 5–10° C cooler.

Mean annual rainfall varies likewise according to altitude. On the exposed central highland ridge, it is ca. 1525 mm; at the elevation of Longwood (Napoleon's residence), 533 m, it varies between 1140 mm and 1270 mm; at Jamestown it is about 250 mm. Summer rains are heavier but less persistent than those of winter, at which time mists are forever rolling down from the central highlands, causing a somewhat damp, chilly climate. The coastal belt in general has low rainfall; at middle altitudes, rainfall amount is greater and also more consistent, whereas on either side of the central ridge at higher elevations not only is the amount of rainfall greater than elsewhere but there is also much mist and cloud, notably in winter.

Both Daly and Hirst mentioned the agreeable climatic conditions for field studies.

Geology

Probably the earliest account of the geology is to be found in an anonymous publication of 1805, but the initial geological reports of value are those of Darwin (1876, 1901) and Oliver (1869). The most detailed account of the geology, though chiefly concerned with petrology and structure, is that of Daly (1927), who spent some 5–6 weeks on the island compiling a reconnaissance map and field investigations. Hirst (1951) has written a good, brief report, based on a 3-week stay on St. Helena. Baker has spent two field seasons on the island, published two papers to date, and is working on a doctoreal thesis to be published later.

Petrographic data are to be found in the publications of Prior (1903), Reinisch (1912), Daly (1922, 1927), Campbell Smith (1930), Fuster (1954), Barros (1960), Baker et al (1967), Baker (1968).

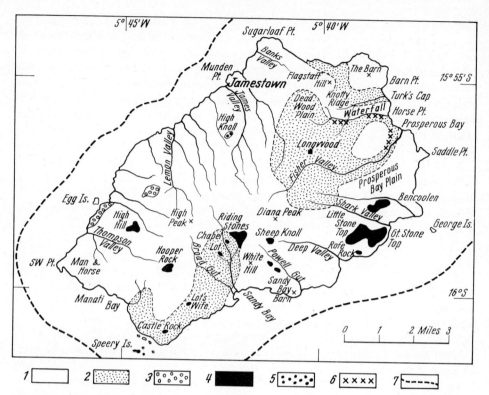

Fig. 71. Geological Sketch Map of St. Helena Island. 1. Mainly lavas of Upper Complex, 2. Mainly tuffs of Upper Complex with Dykes Swarms, 3. Tuffs of Upper Complex where prominently exposed, 4. Phonolites and Soda-Trachytes, 5. Trachydoleritic Basalt, 6. Manganese ore, 7. 50-Fathom Isobath. (HIRST, 1951, based largely on DALY, 1927)

Igneous Rocks

DALY discussed the igneous rock formations under two headings, those of the Main Massif and those of the older NE Massif, these constituting "essentially a basaltic doublet" of the emerged part of the great volcanic cone. By far the major part of the visible mass erupted from the exogenous dome in the southwestern part of St. Helena, constituting the Main Massif with an original presumed height of some 1200 m. Here bedded lavas have outward dips from the focal area represented by the great amphitheatre – the Grat Basin of DALY or the Great Hollow of HIRST. Although several necks are present in this bowl-like region, it is not considered that any or all of these were the vents through which the major part of the Massif was erupted, for they are of phonolitic or trachytic composition and appear to be younger than even the youngest basalts. Contrariwise, necks are absent in the NE Massif, yet similar basaltic flows mantles are found, from which it is deduced that the flows issued from numerous fissures, i.e. the island seems to be a composite volcano of central type, the eruptivity taking place over a considerable area and not through a simple, single pipe. This localized fissure-eruption mechanism of St. Helena was a dominant feature of other oceanic volcanoes during their later stages, for the island was already old before it poked its summit above the waves.

The flows dip outward at the rim of the amphitheatre at an average of 7°, whereas the average mid-slope outward dips are nearly 10°, and along the NW shores, cliff sections show flows dipping seawards at angles up to 15°. The flows vary in thickness from a metre to more than 30 m, and at Cole's Rock, in the Great Basin, a flow has a minimum thickness of 55 m. Commonly the basalts are amygdaloidal and scoriaceous at the surface. Thicker flows have a trapp habit. Columnar structure is rare, block-jointing more common. Only occasionally are ropy lavas of pa-hoe-hoe characteristics to be seen, doubtless due to active altering agencies, and aa or block lava was very seldom erupted. Xenoliths are very rare in the flows. Pyroclastics are not too evident in either Massif. On the Main Massif these comprise brown tuffs, 1–2 m thick, interbedded amongst the flows, but thicker local deposits are found where flank eruptions have taken place, e. g. at the old quarry at Jamestown, along the shore S of Egg Island, etc.

In Thompson valley occurs the only intrusive sheet in the Main Massif. This slightly curved, lenticular mass has a maximum thickness of some 12 m and transects at low angles several basaltic flows. This feature, showing excellent columnar structure, is presumed to be basaltic also, but DALY did not study the feature in detail.

Dykes are basaltic, the partial later fillings of fissures. They are best seen in the sea cliffs, where it is noticed that they rapidly decrease upwards. They have strikes of NE-SW and NNW-SSE, the former trend being commoner. DALY was of the opinion that probably a significant erosional unconformity lies between the basalt complexes and their mantles of flows, and DARWIN had suggested that some of the older dykes may have been truncated by erosion during the relatively slow growth of the island, although DALY doubted if there were lengthy intervals between the outpouring of the flows. The basal complex appears to be a relatively homogeneous, poorly stratified mass of reddish and brown basaltic tuff, agglomerate and thin slaggy flows, cut by innumerable dykes of basaltic composition. The dykes, mostly only a few centimetres thick, are highly irregular, forming "local networks of unmappable intricacy" (DALY, 1927). In the case of these thinner dykes, there is no clear evidence of erosional control by any lithological distinction between dykes and tuff. On the other hand, larger dykes have a pronounced topographic importance, standing-up above the surface of the complex. DALY thought that the imposing neck-like masses of Castle Rock and Lot's Wife might be local enlargements of master dykes rather than pure necks. The great development of dykes on St. Helena is strikingly different from their extreme rarity on Ascension, doubtless because the latter younger island never experienced such powerful tensional strains as St. Helena.

Pipe-like masses of alkaline rocks, representative of volcanic necks, penetrate the basalts of the Main Massif. These include such prominent elevations as Lot, Sheep Knoll, Chapel, Castle Rock, Hooper Rock and the remarkable plug, Lot's Wife. Each of the phonolitic and trachytic bodies appear to have erupted as monolithic masses, and because of their strength, project well above the basaltic mantles. Actual contacts of these bodies with the basalts are difficult to come by, and it is rather their pronounced topographical expression which advertises their presence. Great Stone Top and Little Stone Top constitute a significant phonolite dome, perhaps as much as 300 m thick originally. High Hill is an exogenous trachytic dome, with a relatively large lower mantling of pyroclastics. Lot and Castle Rock, two phonolite necks, display excellent vertical jointing, with marked columnar structure. High Knoll, in the James Valley, is a vertical trachydoleritic basalt pipe, 180 m in length and 60 m maximum width, narrowing to 15 m wide where it is in contact with the underlying basalt flows.

BAKER (1968) discusses in some detail (vd. sub) the trachytic and phonolitic dykes and parasitic bodies in the Main Shield Volcano, which are considered to be the latest vulcanism in the island. Three major highly alkaline dykes and some smaller ones here strike NE-SW. The larger dykes can be traced for 6 km, with individual exposures extending up to a kilometre. Thicknesses generally vary between 15 and 25 m, though occasionally up to 100 m. Compactness characterizes these bodies, though some dykes may show central vesicular structure. Parasitic bodies are often altered whereas the dykes show unusual freshness.

The NE Massif is also a composite exogenous basaltic dome which had probably completed its growth before the NE flows of the Main Massif were erupted. Its initial height is considered to have been less than that of the Main Massif, about 800–1100 m. Most of what has been said regarding the Main Massif applies here also. Though at first glance seemingly a homogeneous complex, in reality it comprises a maze of dykes in a milieu of soft pyroclastics. Two principal dyke swarms occur, one vertical and trending NE-SW, like those of the Main Massif; the other dipping at an average angle of 70° and striking N 10–20° W. All are narrow and show chilled contacts. Between The Barn and Flagstaff Hill is a multiple dyke comprising some 200 single dykes, suggesting a very lengthy endurance of tension.

The Knotty Ridge complex shows tuffs and breccias dipping at angles less than 15°; at The Barn, dips are 25°–30° to the NNE; at Flagstaff Hill, 22° to the SW; between Banks and Rupert Valleys, dips vary from 15° to 35°, always to the W; on the ridge E of Jamestown, dips are 5°–15° to the NW. These younger flows of the NE Massif are thinner than those of the Main Massif, varying from 1–6 m in thickness, though one of 30 m occurs in The Barn and even 60 m thick at the Waterfall. Thicker dykes show a tendency towards columnar structure.

The abnormally high dips of the flows of the NE Massif were attributed by DARWIN to anticlinal uplift exerted by upward pressure of the magma injected into or beneath the complex, a view shared by OLIVER. DALY likewise inclined to such an opinion and further suggested that the 70° dipping dykes here were initially vertical and underwent rotatory displacement associated with this uplift.

HIRST, in his geological map, has two major geological divisions of the island, named "Mainly lavas of Upper Complex" and "Mainly tuffs of Lower Complex with dyke swarms", thus following the original concept of DARWIN. He recognized tuffs of the former Complex where these are prominently exposed. DALY, on the other hand, has two major divisions on his map, labelled "Basaltic flows and interbedded pyroclastics", largely corresponding to HIRST's Upper Complex, and "Knotty Ridge and Sandy Bay Complexes", corresponding to HIRST's Lower Complex. However the latter author maps a large area, centred on Longwood, as belonging to his Lower Complex, whereas DALY included this region in his basaltic flows-interbedded pyroclastics unit. The basaltic Main Massif is older than the NE Massif, both of which constitute the Lower Complex.

Doubtless I. BAKER will add much to our general knowledge of St. Helena when his thesis is completed. To date, BAKER has presented only two publications, of which one, published with colleagues of Imperial College, London (1967) deals primarily with radiometric datings of samples collected, along with the briefest of petrographic descriptions of samples analyzed. A map is presented (Fig. 72) along with a very short account of the major divisions of the volcanics on a temporal basis. As per these authors, two broad basaltic shield volcanoes comprise some two-thirds of the island, named the Main Shield Volcano and the Lower Shield Volcano. In the NE is a smaller volcano, partially buried beneath the flanks of the Main Shield Volcano. A third centre of eruptivity occurs in the centre and E, lava outpourings here being essentially through fissures with dyke concentrations common.

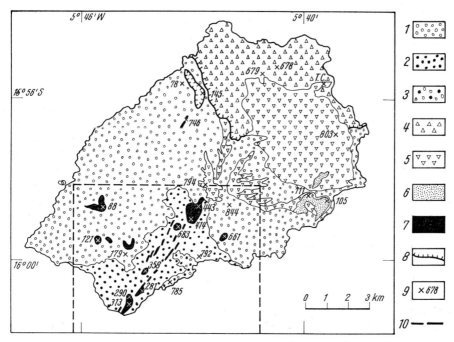

Fig. 72. Major Stratigraphic Divisions of St. Helena Volcanics. 1. Main Shield Volcano (SW Volcano), 2. Lower Shield Volcano (SW Volcano), 3. Oldest part of Lower Shield Volcano (SW Volcano), 4. NE Volcano, 5. Third Volcanic Centre – Late Lavas, 6. Late Trachytic Flows & equivalent Extrusives, 7. Alkaline Intrusives, 8. Unconformity, 9. Samples petrographically & radiometrically analyzed. (BAKER et al., 1967) – – – – – – – – Area shown in Fig. 77

A comparison of the geological sketch maps of DALY, HIRST & BAKER et al would suggest that much of DALY's "Basaltic flows and interbedded pyroclastics", which corresponds to HIRST's "Mainly lavas of Upper Complex", would be placed by BAKER et al in their "Main Shield Volcano". DALY's "Knotty Ridge" unit corresponds to HIRST's "Mainly Tuffs of Lower Complex", placed by BAKER et al in their "Northeast Volcano" unit. DALY's "Sandy Bay Complex" corresponds to HIRST's "Mainly Tuffs of Upper Complex" and to BAKER et al's "Lower Shield Volcano". In the centre and E of St. Helena, where HIRST mapped areas in both his "Mainly Tuffs of Upper Complex" and "Mainly Lavas of Upper Complex", BAKER et al have distinguished their "Third Volcanic Centre" with late lavas.

The conclusions of BAKER et al regarding the sequence of vulcanisms and ages will be given later.

Sedimentary Rocks

DALY claimed that more than three-quarters of the basaltic rocks of St. Helena were deeply altered to a depth of several metres, lateritization resulting in a mantle of laterites. HIRST expressed surprise that DALY should have considered the weathered, variegated tuffs of the Lower Complex as laterite, stating: "They bear not the slightest resemblance to the residual laterite so common in tropical countries... There is no true laterite on the island, but there are some deep red-weathering ferruginous basalts... If true laterite were present, there would not be the difficulty there is in finding a suitable self-binding top-dressing for the roads".

At diverse localities along the shores, calcareous sand dunes of moderate size are present. These comprise small shell fragments, mostly marine, along with varying proportions of basaltic grains. MELLISS (1875) mentioned their occurrence at Lot's Wife Ponds, Banks, Ridge and Rupert Beach, and DALY studied the sands at Potato Bay, E of Sandy Bay, lying at an elevation of 61 m, and at 213 m some 300 m E of the summit of Sugarloaf Hill. At the former locality the sands are poorly cemented and cross-bedded, totalling about 10 m in thickness, but originally thought to be 30 m thick at least. This somewhat impure limestone lies on basaltic flows which had been eroded before their deposition. Much of the original carbonate has been dissolved by rain-water, now forming a tenaceous travertine cement for scree and gravel lower down in the deep valley. At Sugarloaf Hill the sand deposit measures some 5 m in thickness, but aeolian and rain-wash action has greatly reduced its original thickness. The sand is indifferently cemented with calcium carbonate, which forms travertine on the slope immediately below. Here DARWIN found land shells, bird bones and some large eggs (water-fowl?). The land shells had one living representative and two extinct species.

The presence of such coarse calcareous sands at present elevations of 60 m and 213 m is attributed by DALY to unusually strong winds in former times, rather than invoking an assumed uplift of the island, in whole or in part. DARWIN had commented upon the absence around St. Helena of shelly beaches, whereas Ascension Island has well-developed beaches of calcareous detritus. DALY claimed that these old St. Helena dune deposits were formed long after the time when the island had a shelving coast, such as occurs at Ascension; rather they were formed after the deep valley and high cliffs had already been developed. The St. Helena beaches comprise only boulders and pebbles of volcanic rocks, but at certain places offshore, in depths of 5–10 fathoms, patches of white sand are present. It was DALY's belief that these sands comprised broken shells and algal remains. A lowering of sea-level with moderate speed would cause some of this sand to be thrown landward by waves, and waves would drive it further upwards and inland. These growing and migrating dunes would continue to do so until a wave-cut bench at the new sea-level once more attained an equilibrium profile, or then until sea-level rose again. The former alternative is believed by DALY to be exemplified by the post-Glacial 5 m eustatic lowering of sea-level. It is also possible that the more significant eustatic changes in sea-level associated with Pleistocene glaciation and deglaciation are responsible for the present location of these old dunes, but at this time the true explanation evades us.

REINISCH (1912) had reported a graywacke from St. Helena which contained angular particles of quartzitic schist, tourmaline-bearing sandstone, phyllite, gneiss and granite. Neither WEBSTER, DARWIN, OLIVER nor DALY found any evidence on the island for quartz-bearing gneissic, granitic or sedimentary rocks, and it must be assumed that REINISCH's specimen was 'imported' into the island, perhaps as dumped ship ballast.

Petrography of the Igneous Rocks

During the visits of several expeditions to austral regions, e. g. "Challenger" expedition, 1873–76, the German South Polar expedition of 1901–03, the "Quest" expedition of 1921–22, specimens were collected and studied, but the most detailed petrographic descriptions resulted from DALY's visit of 1921–22 (1927), and such later works as those of FUSTER (1954) and BARROS (1960) rely chiefly on the findings of DALY.

According to DALY, about 99 % of the volume of the visible island comprises material of femic, basaltic nature; the rest is salic, phonolitic character. As with many other oceanic islands, lavas intermediate between basaltic and highly salic types are most rare, if not entirely absent.

Basalts

REINISCH described many St. Helena specimens as being trachydolerites, but DALY argued that as all except one were deeply altered by weathering, it was extremely doubtful if such a classification was appropriate, and preferred instead to consider the lava flows as olivine-rich basalts. REINISCH's one "tolerably fresh" specimen, as per DALY (Table 47, No. 1), from Ladder Hill, showed, according to REINISCH, alkali-feldspar and a trace of nepheline in the matrix, the pyroxene being 'titanaugite'. Otherwise the specimen shows

Table 47 — Chemical Analyses of Basaltic Rocks (DALY, 1927)

	1.	2.	3.	4.	5.	6.
SiO_2	43.72	45.50	46.12	47.10	49.99	50.02
Al_2O_3	17.32	11.87	15.24	18.56	18.81	18.37
Fe_2O_3	7.21	3.09	5.70	1.94	4.62	4.25
FeO	6.03	9.25	6.42	9.55	7.16	6.78
MnO	–	0.11	0.11	0.09	0.12	0.05
MgO	6.01	10.40	7.21	4.64	2.82	3.26
CaO	12.00	10.69	8.85	8.01	6.72	6.75
Na_2O	3.40	2.52	3.77	4.98	3.91	4.81
K_2O	1.57	0.84	1.37	1.25	1.04	2.00
TiO_2	0.81	4.47	4.72	2.18	2.80	1.82
P_2O_5	0.32	0.14	0.54	0.52	0.58	0.34
Cr_2O_3	–	0.12	–	–	–	–
SO_3	–	–	–	0.20	–	–
H_2O+	1.80	0.98	0.20	0.37	0.69	0.94
H_2O-	–	0.37	0.05	0.36	0.65	0.23
Total	100.19	100.35	100.30	99.75	99.91	99.62
S. G.	–	2.962	2.922	2.902	2.831	2.813

mineral constituents similar to ordinary olivine-basalts. Specimen No. 2 is a dark greyish-green, porphyritic basalt, poor in vesicles, with olivine and augite phenocrysts. There are thin films of serpentine around some of the olivine but otherwise the rock has undergone little alteration. The groundmass comprises basic labradorite, augite, olivine, apatite and much magnetite – likely titaniferous. Glass cannot be detected. The texture is doleritic to fluidal. As per the analysis of the norms of this specimen (No. 2, Table 48) there is a decided leaning towards picritic composition. Probably more abundant than olivine-rich basalts are normal olivine-poor basalts, of which No. 3, Table 47 is representative. In hand specimen this compact, non-vesicular rock has a greyish-green colour. Small phenocrysts include augite, olivine, faintly zoned plagioclase ranging from acid bytownite to acid labradorite. In the groundmass the feldspars appear more acidic than in specimen No. 2. A representative olivine-poor basalt from the Main Massif is shown in No. 4. It is

Table 48   Molecular Norms of Basaltic Rocks (DALY, 1927)

	1.	2.	3.	4.	5.	6.
Q	–	–	–	–	3.96	–
Or	9.45	5.00	8.34	7.23	6.12	11.68
Ab	8.91	19.65	27.77	26.20	33.01	36.16
An	27.24	18.90	20.29	24.46	30.02	22.80
Ne	10.79	0.71	2.27	8.80	–	2.27
Co	–	–	–	–	0.20	–
Di	24.08	26.56	15.77	9.68	–	7.51
Hy	–	–	–	–	11.88	–
Ol	5.18	14.87	7.49	13.95	–	7.67
Mt	10.44	4.41	7.19	2.78	6.73	6.03
He	–	–	0.80	–	–	–
Il	1.52	8.51	8.97	4.26	5.32	3.50
Ap	0.67	0.34	1.34	1.34	1.34	0.67
H_2O, etc.	1.80	1.35	0.25	0.93	1.34	1.17
Total	100.08	100.30	100.48	99.63	99.92	99.46

1. 'Trachydolerite' flow, Ladder Hill. (R. REINISCH).
2. Olivine-rich basalt flow, road near cable station.
3. Olivine-poor basalt flow, path from Knotty Ridge along SE slope of The Barn.
4. Olivine-poor basalt flow, turn of road 900m SE of top of Diana Peak.
5. Olivine-poor to olivine-free basalt flow, road 800 m due W of summit of High Knoll.
6. Trachydoleritic basalt, pipe or crater-filling, at High Knoll.

a dark greenish-grey rock, compact, non-vesicular and of trapp habit. Microphenocrysts include a few minute crystals and laths of basic labradorite, with a dense matrix of augite, labradorite, much iron ore and apatite, and a strongly fluidal texture. It can be classed as a common basalt, somewhat enriched in soda. No. 5, taken from the middle portion of a normal flow, has a similar petrography to the previous specimen. No. 6, taken from a pipe penetrating tuffs at High Knoll, is dark grey in colour, compact, showing a few pores filled with zeolites. There are small anhedra of olivine phenocrysts and more abundant zoned feldspar, basic labradorite at the cores, probably oligoclase at the periphery. The confused groundmass comprises aggregates of augite, much magnetite and feldspar laths which may include potash feldspar as well as plagioclase more acidic than labradorite. The rock is classed as a trachydoleritic basalt.

As per C. BURRI & P. NIGGLI (1945) the major part of these specimens of DALY would be placed in the alkaline group of basalts, viz. Nos. 1, 3, 4 and 6, which are the most representative types of Ascension Island (q. v.). A consideration of the NIGGLI values (Table 49) led FUSTER to classify Nos. 1 and 3 as basic magmas of the soda series, between

Table 49 NIGGLI Values for Basaltic Rocks (FUSTER, 1954)

	1.	2.	3.	4.	5.	6.
Q	21.0	21.7	23.1	24.6	32.7	28.6
L	41.1	28.2	37.9	46.2	44.4	47.3
M	37.9	51.1	39.0	29.0	22.9	24.1
si	94	97	106	113	137	133
al	22.0	15.0	20.5	26.5	30.5	29.0
fm	41.5	54.5	47.5	39.5	38.0	36.5
c	27.5	24.5	21.5	20.5	20.0	19.0
alk	9.0	6.5	10.5	13.5	12.0	15.5
k	0.24	0.18	0.20	0.14	0.15	0.21
mg	0.46	0.61	0.53	0.43	0.31	0.36
π	0.41	0.39	0.33	0.33	0.43	0.29
γ	0.23	0.22	0.17	0.13	0.00	0.14
μ	0.32	0.46	0.43	0.37	0.44	0.30
α	0.51	0.17	0.17	0.66	0.41	0.36

those of gabbroic essexite and gabbro-doleritic essexite type, somewhat more acidic than the alkaline basalts of Fernando Poo (q. v.). No. 4, the commonest type of basalt on St. Helena, can be correlated with a gabbroic theralitic magma very rich in alkalis. No. 6 is intermediate between a dioritic and an essexitic magma. These alkaline basalts show high proportions of nepheline and total feldspars – between 50% and 70%. In general, the proportion of pyroxene is greater than olivine.

FUSTER would classify No. 2 as a subbasalt, of moderate alkali content, less aliminium but more femic, which would place the specimen between kaulaitic and hornblenditic magmas. Likewise No. 5 would be classed as a normal basalt by FUSTER. DALY claimed that in mineral content it was similar to No. 4, but FUSTER argued that the relative proportions of its distinctive minerals would be much different and hence the chemical nature would differ. It shows a higher acidity than all other basalts, whilst the relative alkalinity is less, and thus, as per BURRI and NIGGLI, it is classed as an ordinary basalt with reduced basicity, to be correlated with an orbitic magma type.

CAMPBELL SMITH (1930) referred to a nepheline-basanite from near sea-level at Sandy Bay. This compact, aphanitic rock shows no phenocrysts in hand specimen. It has a strongly fluidal texture and shows ca. 10% Na-bearing silicate. Another weathered specimen of nepheline-basanite from near the same locality resembled REINISCH's "basaltic trachydolerite" from Ladder Hill.

Table 50 gives molecular norm values for these six specimens as presented by FUSTER.

Table 50   Molecular Norms of Basaltic Rocks (Fuster, 1954)

	1.	2.	3.	4.	5.	6.
Or	9.3	5.3	8.1	7.3	6.3	11.8
Ab	10.0	21.3	29.9	27.9	36.2	38.8
An	28.0	18.5	21.0	25.5	31.5	23.2
Ne	12.7	1.1	2.5	9.3*	–	3.0
Di	24.8	27.5	16.0	9.3	–	8.0
En–Hy	–	–	–	–	12.4	–
Ol	5.8	16.2	9.0	14.1	–	7.6
Mt	7.7	3.2**	5.5	2.1	5.0	4.5
Il	1.2	6.4	6.6	3.0	4.0	2.6
Ap	0.5	0.3	1.1	1.0	1.1	0.5
Hm	–	–	0.3	–	–	–
Q	–	–	–	–	3.5	–

* 0.5 % Th.   ** 0.2 % Cr.

Phonolites

Daly stated that the salic masses in St. Helena – domes, necks, crater-fillings, some thick flows, chonoliths, two powerful dykes – were true phonolites and closely allied to soda-trachytes. In true phonolites, the proportion of nepheline does not exceed 12 % by weight, and is absent or in very minor amounts in soda-trachytes.

Specimen No. 7 (Table 51), similar to one described by Prior (1903), is a fine-grained, light greenish-grey, slightly porphyritic rock with a peculiar shimmer due to the fluidal arrangement of the principal feldspar, anorthoclase, which constitutes the bulk of the rock. As per Prior the feldspar yields "curious undulose extinctions", showing also "very conspicuous wavy lines of parting". After anorthoclase, aegerine-augite, somewhat pleochroic and zoned, is the next commonest mineral. Prior claimed that the chief pyroxene was aegerine and he also detected cossyrite, but the latter was not recognized by Daly. Small aggregates of iron ore, minute euhedra of nepheline, magnetite and rare apatite comprise the accessories. No. 8, like specimens collected from Lot, Sheep Knoll and Hooper Rock by Daly as well as a specimen studied by Campbell Smith from the "Quest" collection in a dyke half-way between Sandy Bay and Lot's Wife, shows the same unusual shimmer and fluidal structure. Campbell Smith stated that the only specimen in the British Museum coming actually from Lot's Wife, collected in 1815, was an aphyric trachytoid phonolite. A specimen collected by Douglas (1923) from "opposite Lot, on the same strike at 720 feet" contains a few phenocrysts showing very fine but quite definite twin lamellation, which, as per Campbell Smith, "may be referred to as potash-oligoclase. Except for these phenocrysts and the coarseness of its texture, this rock is quite similar to the trachytoid phonolite of Lot's Wife". No. 9, a fine-grained, compact, greenish-grey rock, microscopically shows ragged prisms of aegerine-augite in a groundmass of sanidine-like feldspar and probably anorthoclase. Aegerine-augite, magnetite, apatite and some cossyrite (?) are present. Although no nepheline occurs in this specimen, it is present in rocks at Sheep Knoll, the mineral having a blurred appearance, likely altered to hydro-nepheline. Daly classified the specimen as a soda-trachyte or a phonolitic trachyte, a rock probably transitional to true phonolite found on the NW

Table 51   Chemical Analyses of Phonolitic Rocks (DALY, 1927)

	7.	8.	9.	10.	11.
SiO_2	60.90	56.94	59.98	60.92	62.34
Al_2O_3	19.47	16.89	16.90	18.01	17.35
Fe_2O_3	1.62	3.73	2.73	2.66	1.70
FeO	1.41	3.36	4.13	2.88	2.53
MnO	0.17	0.17	0.16	0.12	0.19
MgO	0.06	0.41	0.24	0.05	0.09
CaO	0.85	3.11	2.62	1.68	1.31
Na_2O	7.60	8.06	7.12	7.70	6.88
K_2O	6.23	3.86	4.04	4.70	5.79
TiO_2	0.30	1.47	1.30	0.37	0.66
P_2O_5	0.24	1.02	0.15	0.34	0.27
ZrO_2	–	–	–	–	0.14
S	–	–	–	–	0.02
H_2O+	1.19	0.03	1.22	0.43	1.23
H_2O-	0.20	0.70	0.38	0.18	0.28
Total	100.24	99.75	100.57	100.04	100.78
S. G.	2.514	2.671	2.635	2.671	2.600

7. Phonolite of Lot pipe or neck.
8. Phonolite of Little Stone Top.
9. Phonolite of Sheep Knoll dome or crater-filling.
10. Phonolite of "Chapel" pipe.
11. Trachytic phonolite of Hooper Rock pipe or crater-filling.

side of the Knoll. No. 10 is a shimmering, trachytic phonolite, similar to No. 9 in field habit and microscopic character. The feldspars, aegerine-augite, nepheline and accessories are all like those of No. 7. No. 11 also is much like the specimens from Lot but show a somewhat darker bluer colour. Anorthoclase, soda-sanidine (?), small nepheline euhedrals and aegerine-augite are present, along with some ragged grains of cossyrite (?), colourless diopsidic pyroxene, and magnetite and apatite as accessories.

Without having carried out chemical analyses, DALY presumed, on the basis of study of thin sections, that phonolites also occur on Great Stone Top dome, Bencoolen flow, the dyke alongside this flow and the Riding Stones pipe. The two small pipes between Sheep Knoll and Sandy Bay Barn, and also Speery Isle, are believed to be phonolitic trachytes with traces of nepheline, and the High Hill crater-dome, a soda-trachyte without visible nepheline. Neither a chemical nor microscopic study suggested that the chonolithic intrusion at Rofe Rock and at Castle Rock are phonolitic trachytes, only field habit suggesting such a classification.

A phonolitic trachyte (?) specimen from Sandy Bay studied by CAMPBELL SMITH has a surface shimmer similar to the Lot specimens, but the texture is neither trachytic nor fluidal.

In Table 52 the norms for these phonolitic rocks described by DALY are shown.

FUSTER (Tables 53 and 54) remarked that these phonolites form a series of increasing acidity, with a progressive though irregular augmentation in the absolute alkalinity. The

Table 52   Molecular Norms of Phonolitic Rocks (DALY, 1927)

	7.	8.	9.	10.	11.
Or	36.70	22.80	23.91	27.80	34.47
Ab	46.11	49.78	55.54	55.28	52.92
An	–	–	2.22	0.56	–
Ne	9.94	8.52	2.56	5.25	1.99
Co	0.20	–	–	–	–
Ac	–	2.31	–	–	1.39
Wo	–	1.16	0.35	0.12	–
Di	2.46	5.13	7.74	4.93	4.15
Ol	–	–	–	–	0.61
Mt	1.62	4.18	3.94	3.94	1.85
Il	0.61	2.89	2.43	0.76	1.21
He	0.48	–	–	–	–
Ap	0.67	2.35	0.34	0.67	0.67
H_2O, etc.	1.39	0.73	1.60	0.61	1.67
Total	100.18	99.85	100.63	99.92	100.93

Table 53   NIGGLI Values for Phonolitic Rocks (FUSTER, 1954)

	7.	8.	9.	10.	11.
Q	33.2	29.8	33.8	33.7	36.1
L	61.9	54.8	53.9	57.4	56.1
M	4.9	15.4	12.3	8.9	7.8
si	231	190	212	222	261
al	43.5	33.5	35.5	39.0	43.0
fm	10.0	21.5	21.0	16.5	8.0
c	3.5	11.0	10.0	6.5	6.0
alk	43.0	34.5	33.5	38.0	43.5
k	0.35	0.24	0.27	0.29	0.36
mg	0.04	0.09	0.05	0.01	0.07
π	0.01	–	0.24	0.01	–
γ	0.16	0.64	0.27	0.20	0.20
μ	0.05	0.06	0.04	0.01	0.03
α	4.94	1.31	0.52	1.54	0.50

norm shows nepheline in variable proportions. The richest in feldspathoids are Nos. 7 and 9, more than 10%; the other samples vary between 2.7% and 6.3% nepheline. All the rocks are highly feldspathic, comprising more than 85% in mineral content. These St. Helena phonolites show a deficiency in silica and the aluminium and alkali content varies between negative and very slightly positive. They thus may be compared to sodic magmas of greater acidity, namely, foyaitic magmas. No. 7 is of umptekitic-foyaitic magma type; Nos. 8 and 10 are of umptekitic type and Nos. 9 and 11, of albititic-umptekitic magma type.

DALY, in his petrographic summary of the samples presented, called attention to the following points: 1. the similarity between the average chemical composition of four

Table 54 Molecular Norms of Phonolitic Rocks (FUSTER, 1954)

	7.	8.	9.	10.	11.
Or	35.8	22.6	23.8	27.3	33.8
Ab	47.4	51.4	58.5	57.2	55.1
An	0.7	–	2.2	0.7	–
Ne	11.6	10.4	3.2	6.3	2.7
Ac	–	2.4	–	–	1.1
Wo	0.9	3.7	4.0	2.1	1.9
En	0.3	1.1	0.5	0.1	0.3
Hy	0.9	1.5	2.9	2.3	2.4
Mt	1.6	3.0	2.8	2.7	1.4
Il	0.4	2.0	1.8	0.6	0.8
Ap	0.4	1.9	0.3	0.7	0.5

Table 55 Comparison of Average Analyses of St. Helena Basalts (computed as water-free) and Phonolites, reduced to 100.0 (DALY, 1927)

	Basalts		Phonolites		
	A	B	C	D	E
No. of Analyses	4	198	5	11	25
SiO_2	47.62	49.87	60.23	59.96	57.45
Al_2O_3	16.27	15.96	17.74	19.74	20.60
Fe_2O_3	3.87	5.47	2.49	2.11	2.35
FeO	8.17	6.47	2.87	1.37	1.03
MnO	0.11	0.32	0.16	0.20	0.13
MgO	6.33	6.27	0.17	0.56	0.30
CaO	8.65	9.09	1.92	2.75	1.50
Na_2O	3.82	3.16	7.50	6.21	8.84
K_2O	1.14	1.55	4.93	5.15	5.23
TiO_2	3.57	1.38	0.42	0.61	0.41
P_2O_5	0.45	0.46	0.40	0.12	0.12
H_2O+			0.82	1.02	2.04
H_2O-			0.35	0.20	
Total	100.00	100.00	100.00	100.00	100.00

A St. Helena, four flows.
B Average of 198 analyses of world-wide distribution.
C St. Helena, five rock bodies.
D Auvergne, France.
E Average of 25 analyses, world-wide distribution.

studied St. Helena basalt samples with 198 analyses of basalts of world-wide distribution (Table 55). 2. the general field habit of the phonolites is remarkably similar to that of the classic phonolite occurrence in the Auvergne, France. Both the Auvergne 11 analyses and the 25 world-wide analyses of phonolites show a lower silica content whereas the soda content of the world-wide analyses is higher than in either the St. Helena or Auvergne results. 3. although trachydolerite types occur on the island, the preponderant lava body is a common basalt. 4. chemical varieties between the trachydolerite basalt and phonolites have not so far been identified on the island. However, structural relations, volume ratios and sequence of eruption – basalts first, then phonolites – suggest that the latter have been derived from a common basaltic magma.

DALY pointed out the similarity in basaltic character between St. Helena and Ascension Island, and both FUSTER and BARROS remarked upon the similarity between St. Helena and other islands of the Gulf of Guinea and hence with the nearby continent.

To complete petrographic data, Table 56 gives JUNG and RITTMANN values for St. Helena rocks, according to BARROS.

Table 56   JUNG and RITTMANN Values (BARROS, 1960)

St. Helena Samples	JUNG		RITTMANN	
	SiO (%)	R	Si°	Az°
1.	43.72	70.7	0.69	0.48
2.	45.50	76.1	0.76	0.49
3.	46.12	63.2	0.74	0.51
4.	47.10	56.2	0.74	0.53
5.	49.99	57.5	0.93	0.60
6.	50.02	49.7	0.87	0.57
7.	60.90	5.8	0.84	0.69
8.	56.94	20.6	0.79	0.65
9.	59.58	19.0	0.83	0.67
10.	60.92	11.9	0.88	0.68
11.	62.34	9.3	0.95	0.75

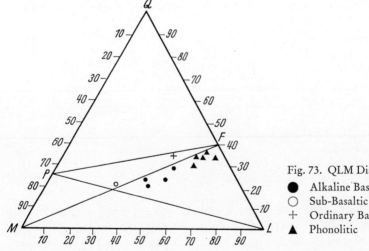

Fig. 73. QLM Diagram (FUSTER, 1954)
● Alkaline Basaltic
○ Sub-Basaltic
+ Ordinary Basaltic
▲ Phonolitic

As previously remarked, BAKER et al selected samples of rocks collected by BAKER from St. Helena for radiometric dating. The same specimens were examined microscopically,

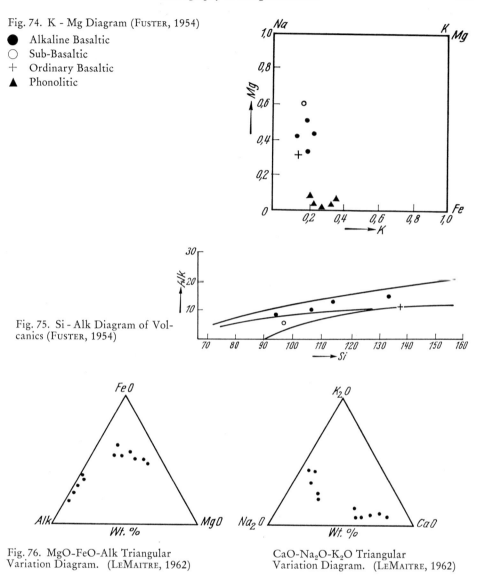

Fig. 74. K - Mg Diagram (FUSTER, 1954)
● Alkaline Basaltic
○ Sub-Basaltic
+ Ordinary Basaltic
▲ Phonolitic

Fig. 75. Si - Alk Diagram of Volcanics (FUSTER, 1954)

Fig. 76. MgO-FeO-Alk Triangular Variation Diagram. (LEMAITRE, 1962)

CaO-Na_2O-K_2O Triangular Variation Diagram. (LEMAITRE, 1962)

and in their 1967 paper brief petrographic descriptions are given. (The localities from where the samples were taken is shown in Fig. 72.) The essential features of these 24 samples are as follows:

TC: A slightly altered, fine-grained trachyandesite. Fayalitic olivine microphenocrysts in good trachytic texture.
78: Very fine-grained trachybasalt. 'Iddingsitized' matrix olivines. Apatite and titanomagnetite phenocrysts.
88: Medium-grained trachyte. Rare fayalitic olivine.
105: Highly oxidized, slightly altered trachyte. Perfectly unaltered alkali feldspars.

111: Slightly altered trachyte, perfect fluidal texture. Fayalitic olivines partially altered to green phyllosilicates.
121: Phonolitic trachyte, with perfect trachytic texture.
145: Porphyritic olivine-pyroxene basalt.
179: Ditto.
281: Phonolite. Phenocrysts of alkali feldspar, euhedral nepheline, anhedral aegerine-augite.
290: Phonolite. Matrix shows poikilitic aegerine, rare cossyrite.
313: Phonolite. Alkali feldspar prisms in fine mass of feldspar laths.
358: Phonolitic-trachyte, fine-grained. Rare phenocrysts of alkali feldspar in very fine groundmass.
414: Ditto.
443: Ditto. Perfect trachytic texture.
661: Phonolitic-trachyte, microphenocrysts of alkali feldspar and nepheline.
678: Porphyritic pyroxene-olivine basalt. Slight phyllosilicate alteration of olivine along cracks.
679: Porphyritic pyroxene-feldspar-olivine-titanomagnetite basalt.
683: Phonolitic trachyte, medium-grained.
746: Highly oxidized trachybasalt with olivine. Plagioclase laths with weak fluidal texture.
785: Porphyritic olivine-pyroxene basalt.
792: Basalt, medium-grained. Late interstitial biotite. Very slight traces of green phyllosilicates in groundmass.
794: Basalt. Microphenocrysts of olivine and plagioclase in matrix of pyroxene, titanomagnetite, plagioclase and olivine.
803: Basalt, fine-grained. Some alteration of olivine along cracks to phyllosilicates.
844: Porphyritic olivine-pyroxene-basalt.

BAKER (1968) subjected 33 specimens of the late intrusives to chemical analysis, of which 25 were analyzed for all major elements and 8 for only Na_2O and K_2O (Table 57). Of these 25 analyses, he lists full analysis for five samples (Table 58, p. 214).

Table 57 Analyses of Alkalis of the Late Intrusives (BAKER, 1968)

Specimen No.	Na_2O	K_2O	Total Alkalis	Zone
281	9.02	5.15	14.17	14%
403	8.93	5.10	14.03	
628	8.22	5.51	13.73	
638	8.34	5.00	13.34	
319	8.63	4.98	13.61	
11	8.94	4.93	13.87	
290	8.37	5.12	13.49	
359	8.45	5.11	13.56	
337	6.94	5.45	12.39*	13%
347	6.86	5.62	12.48*	
369	8.20	5.09	13.29	
379	8.91	4.86	13.77	
406	8.19	5.11	13.30	
765	8.35	5.13	13.48	
167	7.25	5.04	12.29	
682	7.41	4.81	12.22	12%
789	7.33	5.14	12.47	
395	7.29	5.14	12.43	

Table 57   Analyses of Alkalis of the Late Intrusives (BAKER, 1968)

Specimen No.	Na_2O	K_2O	Total Alkalis	Zone
121	6.65	5.07	11.72	
232	7.25	4.58	11.83	
415	7.04	4.71	11.75	
421	6.79	4.26	11.05	11%
452	7.64	4.96	12.60*	
846	7.44	4.26	11.70	
671	5.97	3.53	9.50*	
662	7.03	4.38	11.41	
259	6.80	3.71	10.51	
87	6.88	3.44	10.32	10%
468	7.03	3.79	10.82	
751	6.36	3.72	10.08	
323	6.37	3.33	9.70	
438	4.60	2.09	6.69	
85	5.18	2.33	7.51	

* Anomalous values

281 125 m dyke, NE of Asses Ears
403 2.5 m dyke, E of Lot's Wife
628 85 m dyke, Castle Rock
638 Sill-like dyke extension, SW of Asses Ears
319 Dyke extension, W side of Asses Ears
11 Ditto, N side of Asses Ears
290 Loose block, 250 m N of Asses Ears
359 40 m dyke, N of Lot's Wife
337 75 m dyke, W of Lot's Wife
347 Dyke extension, Lot's Wife
369 30 m dyke, White Rocks
379 40 m sill-like dyke extension, White Rocks
406 30 m dyke, E of Lot's Wife
765 7 m dyke, W of Lot's Wife Ponds
167 Parasitic intrusion, Hooper's Rock
682 Volcanic pipe, Lot
789 8 m dyke, S side Sandy Bay
395 9 m dyke, Broad Gut
121 Intrusive mass, Wild Cattle Pound
232 Intrusive dome, Thompson's Valley
415 Arcuate feeder dyke, Riding Stones Hill
421 Collapse cavity infilling, Riding Stones Hill
452 Ditto
846 Volcanic pipe?, N of Sheep Knoll
671 Intrusive dome?, Sheep Knoll
662 Intrusive dome? E of White Hill
259, 87, 468 Parasitic intrusion High Hill
751 15 m dyke, Powell's Valley
323 Small intrusive mass, N of Tripe Bay
438 8 m dyke, offshoot from irregular intrusion, High Knoll
85 Flow from High Knoll intrusion

Form of the SW Magma Chamber

BAKER (1968) attempted to estimate the form of the SW magma chamber based upon a consideration of the variation in composition of the late, highly-alkaline intrusives occurring within this shield volcano. Here there occurs a concentration of feeder dykes and parasitic bodies whose petrographic characteristics have been mentioned above.

From Table 57 and Fig. 77 it is noted that, with the exception of four anomalous samples, there is a systematic variation away from the centre in all the major oxides, although the total-alkali content shows this variation clearest. Contouring these changes, BAKER was able to demonstrate this variation cartographically, from phonolites in the

Table 58   Analyses of some Late Intrusives (BAKER, 1969)

	281.	628.	167.	662.	751.
SiO_2	59.34	61.80	62.09	60.29	58.94
Al_2O_3	19.78	18.05	18.52	18.57	18.11
Fe_2O_3	2.22	3.29	1.86	1.11	2.87
FeO	1.24	0.82	1.76	3.81	4.08
MgO	0.18	0.08	0.07	0.32	0.59
CaO	0.85	1.10	1.26	2.43	2.66
Na_2O	9.02	8.22	7.25	7.03	6.36
K_2O	5.15	5.51	5.04	4.38	3.72
H_2O+	1.48	1.04	1.32	1.34	1.83
H_2O-	0.24	0.17	0.37	0.17	0.26
TiO_2	0.06	0.13	0.14	0.23	0.44
P_2O_5	0.01	–	0.06	0.08	0.08
MnO	0.20	0.15	0.16	0.25	0.25
Total	99.77	100.36	99.90	100.01	100.19

Petrography

281: Medium-fine grained, trachytic texture. Alk-feldspar laths and sub-poikilitic zoned aegerine-augite/aegerine. Finer matrix of small euhedral nephelines, alk-feldspar, aegerine, augite. Opaque minerals rare. Modally nepheline just over 12%.

628: Very fine-grained, trachytic texture. Alk.-feldspar laths, minute euhedral nepheline, poikilitic to sub-poikilitic aegerine-augite. Some opaque minerals and rare poikilitic cossyrite.

167: Medium-fine grained, trachytic texture. Alk.-feldspar laths, rare subhedral nepheline, strongly pleochroic aegerine-augite. Opaque minerals and very rare fayalitic-olivine.

662: Medium-fine grained, trachytic texture. Alk.-feldspar laths in felted mass of feldspar, aegerine-augite, fayalitic-olivine and rare subhedral nepheline. Opaque minerals of two generations abundantly scattered troughout.

751: Fine-grained, perfect trachytic texture. Elongate alk.-feldspar laths, strongly poikilitic, yellow-green aegerine-augite, rare fayalitic-olivine. Opaque minerals abundant.

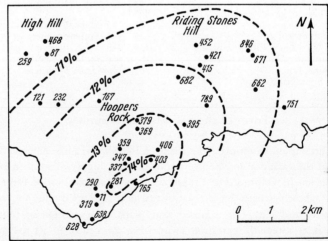

Fig. 77. Analyzed Specimen Locations of the Alkaline Late Intrusives and contours of Total-Alkali content plotted from Table 57. (BAKER, 1968)

central area outwards through trachytes to trachybasalts. The pronounced trachytic texture of these late intrusives indicates a partially crystallized magma when the intrusions took place. The earlier basaltic and intermediate feeder dykes show unmistakable distribution patterns similar to the variation in total-alkalis of the later intrusives.

BAKER believed that this zonal variation in the composition of these late alkaline intrusions resulted from withdrawal at varying levels of a differentiated, high-level magma chamber by means of dykes and parasitic centres. Those rocks having the highest total-alkali content, hence the most highly differentiated, are thought to indicate the culmination of the magma chamber, outwards from here the intrusives showing increasing basicity. A progressive decrease in both modal and normative nepheline away from the centre is to be noted.

It was his belief that the magma chamber was perfectly differentiated and that 10% of its volume comprised magma having a silica content higher than 57%. Thus 10% of the volume of this chamber lies above the locus of the 10% total-alkalis level. Half of the chamber is assumed to contain magma showing more than 7% total-alkalis (50% silica), and thus the 7% level divides the chamber into two equal volumes, the top half of which contains the alkali-rich magma.

From the above assumptions, combined with the distribution pattern of the total-alkali variations, BAKER attempted to assess the shape of the elongate chamber in cross-section, as well as its depth below the surface. Citing various evidences, he concluded that the chamber likely had a major diameter of 18–19 km, minor diameter 13–14 km, thickness somewhat more than 2 km, thus giving a volume of 300 km^3 approx. for the magma chamber (Fig. 78, p. 216).

Structure

The relatively large amphitheatre of the Great Basin or Hollow may, at first sight, suggest circumferential faulting, but DALY stated that there was no field evidence to support this view. Only one fault was seen, 600 m W of Diana Peak, trending parallel to the NE-SW striking dyke system of the Basin. This great half-bowl is due to differential erosion, neither the result of explosion nor of sinking along circumferential faults. Indeed, according to DALY, faulting is extremely rare in St. Helena and only one fault was seen excellently exposed in sea cliffs and canyon scarps in the course of circum-navigating the island in a small boat. This fault is at Crown Point, NW of the summit of Sugarloaf Hill, with a nearly vertical fault surface, striking about N 15° E, downthrow to the E with a displacement of about 10 m. In thus denying a role to faulting in the construction of the island, DALY disagreed with DARWIN, the latter claiming that during the slow elevation of the island, the borders were elevated more than the central area, the former being thus upraised and not sloping gently into the lower central region but separated from it by curvilinear faults. To DALY, from ridge to shore, the mantling flows of the Main Massif conform in dip, their inclinations being only what one might expect in basaltic cones which have been constructed by successive flows and explosions without concurrent or subsequent updoming of said flows.

As already remarked, the high dips of the flows at The Barn and Knotty Ridge in the NE Massif are believed to result from tilting due to the upward pressure of the injecting

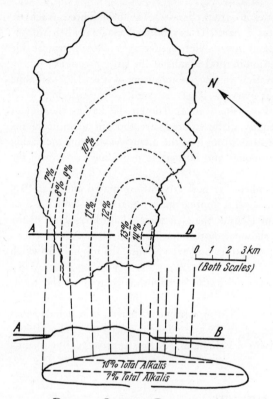

Fig. 78. Contours of Total-Alkali content of the Late Intrusives, and reconstruction of the underlying Magma Chamber. (BAKER, 1968)

magma. DARWIN, OLIVER & DALY were in general agreement on this point. However, the view of MELLISS (1875) that there was an anticlinal uplift of the basaltic rocks along the axis of the island from The Barn to the SW end of St. Helena is contested by DALY. Local changes in dip here result from variations in the initial attitudes of the flows, and such features as High Knoll and High Hill acquired their character as a result of powerful and massive local extrusions of magma.

DALY devoted several pages relating to the island and isostasy. W. BOWIE (1917) listed the following data for the Jamestown station:

Theoretical gravity	978.418 dynes
Correction for elevation	−.003 dyne
Correction for topography and compensation	+.177 dyne
Observed gravity	978.712 dynes
New Method anomaly, based on formula used in 1912	+.112 dyne
New Method anomaly, based on formula used in 1917	+.120 dyne

Values are based on the absolute value of gravity at Potsdam. The New Method anomaly was calculated on the assumption that isostasy is complete at the station. This would mean that the lavas exposed and the base above the sea bottom were simply moved parts of the 'crust' which were once originally parts of the layer above the assumed level of isostatic compensation. (The New Method of calculation of anomalies was based upon the Pratt concept of isostasy.)

DALY doubted if the topographic and isostatic compensation correction was a high as 0.177 dyne, for although the island cone is porous, the density from exposure to sea bottom is likely higher than that of a continental layer at the same level, and it was conceived as unlikely that below the cone base the density differed essentially from that of the surrounding 'crust'. He contended, therefore, that the island excess mass was much greater than the computed excess, on the assumption of uniform isostatic compensation to a depht of 113.7 km. The 1917 BOWIE anomaly would mean excess matter in the cone below the Jamestown station equal to layer of rock having a density of 2.7 and a thickness of almost 1200 m.

A. BORN (1923) listed the following data for St. Helena:

	Heigh of station (m)	Bouguer anom. (dynes)	Free-air anom. (dynes)
Longwood	533	+0.263	+0.314
Jamestown	10	+0.296	+0.297

With an average Bouguer anomaly for 20 islands listed by BORN of +0.243 dyne, and assuming no compensation, then the excess mass of St. Helena would be equal to a layer of 2.7 density rock having a thickness of some 2800 m. BORN was of the opinion that oceanic volcanic cones such as St. Helena were unstable, but DALY found no evidence to confirm this in St. Helena. The magnificent sea cliffs indicate constancy of position since probably pre-Glacial times. According to MELLISS: "Earthquakes happen so rarely, and when they do are so slight, that they scarcely need be noticed as occurring at all. Four only have been recorded during the last 370 years...". That regional compensation under such a coneload as St. Helena, resulting in some degree of downward bending of the crust, occurs is indeed highly feasible, but that the sub-oceanic crust is able to support such a cone as St. Helena is accepted.

Economic Geology

As the prime object of HIRST's visit to the island was to investigate manganese deposits, his paper treats more fully of the economic mineral occurrences and prospects.

The manganese deposits were, of course, reported on by earlier workers, e. g. OLIVER, MELLISS. These deposits are spread fairly widely throughout the island, e. g. near Sandy Bay Beach, Gumwood, near Francis Plain, near Sidepath, but the only significant deposits are in the NE.

The variegated tuffs underlying the Longwood and Deadwood plains are cut by several gulleys. Just below the junction of Netley Gut anf Sheep's Pound Gut, draining from the latter plain below Flagstaff Hill, occur anastomosing veins of psilomelane, at times forming nodules, on the S bank of the stream. Weathering action here has caused manganese ore to be re-deposited lower down the slope in the form of soft ore or wad, occassionally as definite bands about 60 cms. thick, which lie in and partially hide the friable brownish 'sand' washed down the slope. Both the above varieties of deposit can be traced intermittently along the S side of the valley, but are absent on the N side where lavas outcrop. In Springs Valley, between Turk's Gap and Horse Point, some 6 m of white, decayed tuff overlain by grey lava forms a clearly exposed cliff face. Where two small streams join to form Springs Valley, at an elevation of about 120 m, this 'claystone' band shows thin veins and irregular clumps of hard, botryoidal ore. Below is found the brownish 'sand' with soft manganese ore weathered from the manganiferous bed. Bluish-white decayed tuff can be traced southwards in the cliff behind Prosperous Bay. At Horse Point, at Holdfast Tom and in Fisher's Valley and intermediate inaccessible places between these localities, veins and clumps of similar ore are found. In the last-mentioned locality, the ore occurs in agglomerate. Between Prosperous Point and Stonetop Gut manganese as thin coatings on lava are found in Dry Gut.

Although the tuffs are continuous from Flagstaff to Dry Gut, concentrations of manganese are sporadic, and were not observed on the S side of Longwood and Deadwood plains, although similar tuff series are present.

Table 59 — Analyses of Manganese Ores, in percentages (Hirst, 1951)

	1.	2.	3.		1.	2.	3.
MnO_2	74.48	55.98	56.16	P_2O_5	0.91	0.63	0.82
MnO	6.61	4.12	6.96	Mn	52.18	38.56	40.88
SiO_2	3.12	14.49	6.55	Fe	0.56	3.08	2.83
Fe_2O_3	0.80	4.40	4.04	P	0.40	0.28	0.36

1. Hard, botryoidal ore, Springs Valley.   2. Hard, botryoidal ore, Holdfast Tom.
3. Soft ore, near junction Netley Gut and Sheep's Pound Gut.

The masses and veins of botryoidal ore protrude clearly where present, and could easily be separated from its matrix. Analyses of these ores are shown in Table 59. No. 1 is typical of the hard, botryoidal ore when cleaned. The high silica content of No. 2 is due to a higher proportion of tuff. No. 3 occurs as 'boulders' and an uncertain layer comprising manganese washed out and re-deposited from the nodules and veins occurring in botryoidal form in tuff. Hirst ruled out the manganese deposits of St. Helena as an economic proposition, chiefly because of inadequate quantity.

Nodules of hematite and limonite are profusely strewn in places in Longwood and Deadwood plains. As crushed material it has been tried as top dressing for roads, but it lacks binding qualities.

Small, sporadic occurrences of iron sulphide in the lavas have been mistaken for gold or copper ore in past times. Quartz veins are lacking on the island, nor is there any other geological environment conducive to the occurrence of gold.

Fertilizers are sorely needed in the island as the soils are deficient in lime. Some calcareous 'sandstones' and tuffs might prove suitable, as also phosphatic rock from old bird rookeries. An analysis of the phosphate by Hirst was as follows:

	Total P_2O_5 (%)	Available P_2O_5 (%)
Solid material from Prosperous Bay plain	25.25	11.09
Phosphatic 'dust', same locality	18.47	9.58

Present water supplies are adequate, presuming proper usage and additional storage.

Hot springs are no longer present. Melliss reported there was a warm spring at Longwood in 1810, but he himself failed to locate the site.

Age of St. Helena

Both OLIVER & MELLISS reported the occurrence of tree-trunk casts in basalts, especially at Friars Ridge, near Jamestown, which suggested a moister climate than nowadays. However no fossils of diagnostic value have been found in the island. That St. Helena is relatively of great age is however witnessed by the depths and widths of the valleys and by the heights of the sea cliffs, very often consisting of very hard rocks.

The flora and fauna of the island were commented upon by A. R. WALLACE (1895), who referred to the studies of WOLLASTON, HOOKER, MELLISS and others, as well as his own observations. Of the beetles, 129 species are "truly aborigines" and 128 species are found "nowhere else on the globe". It was his opinion that the St. Helena insect fauna dated back to the Miocene or perhaps even earlier. Of the land shells, WALLACE indicated immigrations from unknown lands which certainly dated back to the Miocene, perhaps even Eocene times. WALLACE's conclusion that both the flora and fauna are ancient, Miocene or earlier, was echoed by DALY who also claimed that a Miocene age was "sufficiently conservative". As seen below, however, these opinions may require some amending.

Geochronology

BAKER et al carried out radiometric datings on 24 samples collected from the island, adopting the K/Ar method of analysis. For each of the specimens, potassium analyses

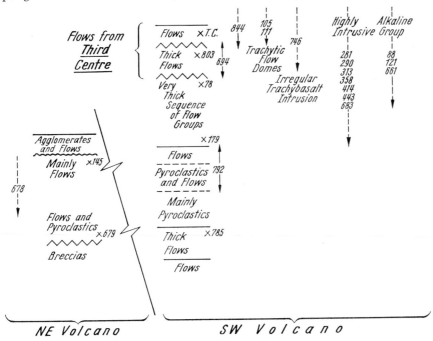

Fig. 79. Locations of dated specimens in simplified stratigraphic chart. (Arrow heads indicate minimum & maximum ages based on field evidence.) (BAKER et al., 1967)

were made in duplicate and argon analyses duplicated for six specimens. Replicate potassium values agreed to within 1 %; analysis of inter-laboratory standards gave estimates of absolute accuracy of ± 2 %.

Sample No.	Age (x 10⁶ yrs.)	Stratigraphic Sequence
281	6.8 ± 0.4	Intrusives
88	7.3 ± 0.4	Intrusives
661	7.3 ± 0.4	Intrusives
78	7.3 ± 0.4	Lower-Main Shield flows
443	7.5 ± 0.4	Intrusives
121	7.5 ± 0.4	Intrusives
358	7.6 ± 0.4	Intrusives
683	7.7 ± 0.4	Intrusives
290	7.7 ± 1.2	Intrusives
414	8.1 ± 0.4	Intrusives
111	8.1 ± 1.0	Late lava flows
313	8.2 ± 1.0	Intrusives
746	8.4 ± 0.6	Intrusives
TC	8.5 ± 0.4	Late lava flows
105	8.7 ± 1.0	Late lava flows
803	9.2 ± 1.0	Late lava flows
844	9.3 ± 0.6	Late lava flows
794	9.6 ± 0.8	Late lava flows
179	10.1 ± 0.8	Lower-Main Shield flows
792	10.9 ± 0.8	Lower-Main Shield flows
785	11.3 ± 1.0	Southwest Volcano
678	11.4 ± 1.0	Northeast Volcano
145	11.4 ± 1.0	Northeast Volcano
679	14.0 ± 1.0	Northeast Volcano

Table 60 Radiometric Datings and Stratigraphic Sequence (Modified after BAKER et al, 1967)

Table 60, modified after BAKER et al, gives the radiometric ages and stratigraphic positions of specimens studied.

These authors stated that these age determinations were in close agreement with the stratigraphic positions of the various groups, except that specimen No. 78 was believed to be anomalous, possibly because argon was lost as a result of re-heating of the overlying flow.

The NE and SW volcanoes show an overlap of isotopic ages. Specimen No. 785 is from the lowest levels of the SW volcano, No. 145 from the top part of the NE volcano, whereas a stratigraphically low specimen from here (No. 679) gives an age of 14.0 ± 1.0 x 10⁶ yrs. The authors surmise that because of the absence of late intrusives in the NE, there was a shift of eruptive centres of some 12 km to the SW, indicating a total transfer of major extrusive activity.

The radiometric ages would suggest that the SW basaltic shield volcano was constructed during approximately 3 million years ($8.5-11.3 \times 10^6$ yrs). Late flows above the unconformity (vd. Fig. 72) seem to be decidedly later ($8.5-9.6 \times 10^6$ yrs) than the mass of the shield comprising extrusives ($10.1-11.3 \times 10^6$ yrs), the flows comprising less than one-quarter of the volume of the volcano.

The highly alkaline intrusives of the western part of St. Helena, totalling less than 5% of the total volcanic products, are in general notably younger than the shield, varying from $6.8 \pm 0.4 \times 10^6$ yrs to $8.4 \pm 0.6 \times 10^6$ yrs. BAKER et al thought that these late intrusives might represent products of a late, highly differentiated magma chamber occurring at a high level within the volcanic structure.

Attention to the age/volume relationships within the island show that there has been deceleration in activity in the formation of the principal volcano. The late flows and secondary flank activities seem to have taken place within a period almost the same as that required for the construction of the main shield, and the volumetrically insignificant late alkaline intrusives likewise represent approximately a similar time-interval.

The authors believed that the total subaerial activity of the island involved a minimum of some 7.5 million years, that the single main volcano had an active minimum life of 4 million years. Such time periods show good correlation with preliminary age determinations in the Canary Islands (A. ABDEL-MONEM, N. D. WATKINS & P. W. GAST, 1967), but on the other hand do not agree with the findings of I. G. GASS for the Tristan da Cunha group (vd. reference in Tristan chapter), who postulated a million years of volcanic activity followed by some 20 million years of erosional activity. BAKER et al claimed that the geochronological evidence for Nightingale Island, in the Tristan group, equally well supported the contention that this island was built up over a period of some 18 million years.

For St. Helena, BAKER et al believed that the visible island represented about only 5% of the total volume of the volcanic pile and involved a period of some 7.5 million years of its history. However, ages cannot be extrapolated down into the volcanic mass as the rate of submarine extrusion is not known.

From the radiometric datings presented by these authors, however, it can be stated that the formation of St. Helena took place more than 14 million years ago, and thus much older than the age of vulcanism on the North Atlantic Ridge. The island is a product of Neogene times, possibly Miocene.

Geological Evolution

DARWIN distinguished an older 'basaltic' series and a younger 'grey feldspathic lavas' and tuff series, the former constituting a "much-broken ring or rather a horse-shoe of basalt" at the circumference of the island, the later lavas breaching this, particularly on the eastern side, and emanating from a great central crater of general oval shape. The circumferential basalt mountains once had their steep escarpment facing inwards, the strata facing outwards, it being the opinion of DARWIN that these dips were greater than could be accounted for solely by flowing down a sloping surface. These basaltic mountains, which "rest on older and probably submarine beds of different composition" thus constitute "craters of elevation", being raised and pushed outward by emanations from the central area which relieved the submarine forces.

DALY disagreed that any such distinction into the above two series could be made, as microscopic and chemical studies revealed no rock differences for localities specifically mentioned by DARWIN. DALY claimed that mantling flows, e. g. those of the Main Massif, showed dips which are commonly to be seen in basaltic cones, and further, he saw no evidence of curvilinear faults, such as proposed by DARWIN, separating the inward-facing basalt scarps from the central region of 'feldspathic lavas' and tuffs of DARWIN.

DARWIN thought that the lavas of the Knotty Ridge and Sandy Bay complexes, lying beneath his basaltic and feldspathic series, were formed under the sea, because of the occurrence of gypsum and salt, rounded pebbles in the tuffs and abundant amygdales to be found in these basal volcanic strata. DALY found no evidence here to support DARWIN's thesis, though he admitted that the tuffs and agglomerates most certainly did not have the features of being subaerially formed, and inclined to accept DARWIN's suggestion as a working hypothesis. This hypothesis would imply an emergence of some 500 m since the lower complex tuffs and agglomerates were covered by the mantling flows of 'basaltic' and 'feldspathic lava' and tuff series. Such a post-Miocene emergence would not be out of keeping with that postulated for some other Atlantic islands, e. g. Madeira.

Following upon the formation of the basal complexes, further tensions resulted in fissuring and the outpouring of basalts in the Knotty Ridge mass. These fissures were dominated by a NNW-SSE pattern which formed the chief loci for the flows and pyroclastic outbursts which ended volcanic events in the NE Massif. Fracturing accompanied or followed these eruptions, likely resulting in tilting of blocks and steepening of flow dips in The Barn and between Knotty Ridge and Rupert Valley. DALY was of the opinion that nowhere in the NE Massif is there a clearly defined central orifice, the eruptions having occurred through fissures only.

The Sandy Bay area marks the principal volcanic centre of the island, where long-continued vulcanism resulted in tuffs and smaller flows at first, then thicker and more numerous flows. Flows from here spread E and NE, unconformably covering the SW flank of the NE Massif to a height of almost 600 m above present sea-level. It is possible that volcanic activity began here during the later period of activity in the NE Massif. Here also fissure eruptions predominated, accompanied by explosive action from subsidiary cones. Two principal fissure patterns were NE-SW and NW-SE, the former of much greater proportion. DALY believed that basaltic flooding continued until the Main Massif achieved an elevation of 1000–1300 m above sea-level.

When the basaltic phase had ended, alkaline rocks were extruded from several pipes, it being presumed that the alkalines in general are later than the basalts, although some might be almost contemporaneous with younger basalt flows. Nowhere have basaltic dykes been observed cutting the alkalines.

Since the phenomenon of alkaline vulcanism, erosion has been the dominant feature in the development of St. Helena. The stream regimes and the points of application of wave action were obviously influenced by eustatic oscillations of sea-level which accompanied the fluctuations of Pleistocene ice-caps. When sea-level was lower than at present, streams could not deepen their valleys greatly, as hard basalts formed rock benches then exposed in lower stream courses, and waves must have swept away much of the more friable material on the ancient shelves. But during the relatively short periods of lowered sea-level, wave action had little effect in cliffing the hard basalts.

The last significant event was eustatic lowering of sea-level in post-Glacial times. This was estimated by DALY to be some 5 m in St. Helena, and to have occured in late-Neolithic times, perhaps some 3500 years ago and no later than the beginning of the Christian era. As far as is known, St. Helena has shown great stability since first discovered by Europeans some 460 years ago.

CHAPTER 13

Tristan da Cunha Group

General

The Tristan da Cunha group, lying in latitude 37° 05′ S, longitude 12° 17′ W, comprises the islands of Tristan da Cunha, Inaccessible, Nightingale, Stoltenhoff and Middle. The nearest land to the group is Gough Island, some 370 km to the SSE. Inaccessible is some 35 km WSW of Tristan da Cunha; Nightingale about 34 km to the SSW of the main island, these two smaller islands being some 20 km distant from each other. Stoltenhoff and Middle Islands lie off the N coast of Nightingale Island. It is usual to refer to the whole group by the simple name of 'Tristan'.

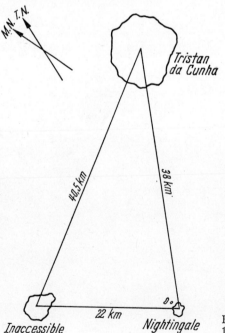

Fig. 80. The Tristan da Cunha Group. (DUNNE, 1941)

The main island has an approximate area of 95 km², is roughly circular in shape, measuring some 11 km in length by 9.5 km in breadth. The highest summit, The Peak (Queen Mary's Peak) reaches 2062 m. Inaccessible has an area of some 10 km² and rises to 548 m. Nightingale covers an area of 2.6 km², highest point, 396 m. The islets are small indeed, Stoltenhoff with an area of some 25 000 m² rises to 105 m, Middle island with an area of 26 300 m² reaches an elevation of 65 m.

The island group, along with Ascension and Gough Islands, are dependencies of St. Helena, but all are administered by the British Colonial Office.

The flora and fauna are of great interest, chiefly because of the isolated position of the group and the loci of origin of many species – South America, Africa, Australasia, South Atlantic and Indian Ocean islands.

From a distance, the main island has a bleak and barren appearance. However there is an altitudinal zoning of the native vegetation, chiefly determined by climate but also by topography and soils. Soils and vegetation practically cease at an elevation of some 1350 m. The publication of N. M. WACE & M. W. HOLDGATE (1958) gives a detailed account of the botany of the main island, and U. HAFSTEN on the basis of pollen analyses of peat samples, concluded that no major vegetational changes had occurred during the last 5000 years, except during the last few hundred years when the island became inhabited. (The preliminary reports by M. W. HOLDGATE on the flora, and D. E. BAIRD on the fauna, as appendices in BAKER, GASS et al, indicate that the 1961 volcanic activity on Tristan da Cunha had only local effects on the plant and animal life, and overall, demage was not critical.)

The main crop is the potato, but a considerable number of apples and peaches are grown, all restricted to the main centre of habitation. Sheep, cattle, donkeys, geese and fowl are raised for domestic needs. Crayfishing is of some importance.

Tristan was discovered in 1506 by the Portuguese navigator Tristao d'Acunha. The Dutch took an early interest in the island, first landing there in 1643. The first settlement was made in 1810 by the British, and a British garrison arrived in 1816, probably aimed more at keeping an eye on American privateers than any attempted rescue of Napolean from St. Helena. The British left the following year, but Corporal Glass and his family decided to remain from which stems the real founding of the colony.

The Tristanians are of mixed origin. Male ancestors are chiefly of European and American descent, whereas the ancestors of the greater part of the females came from Europe, Africa and Malaya.

The population of Tristan da Cunha (the other islands are uninhabited) is about 300. The main centre is Edinburgh (often merely referred to as the 'Settlement'), near the NW coast, lying at an elevation of some 30 m. Only in this vicinity is there cultivated land, known as Potato Patches.

During the tremors and eruptions of late 1961, the entire population was first evacuated to Nightingale Island, then to South Africa, and eventually to Britain. But the people just could not adjust to the tempo of modern living and in such strange environments, and almost all returned back in 1963.

Physical Features

Tristan da Cunha

The main island rises out of the ocean like a steep cone, and, as DUNNE remarked, from a photograph or view from out at sea, presents a typical textbook example of a volcano.

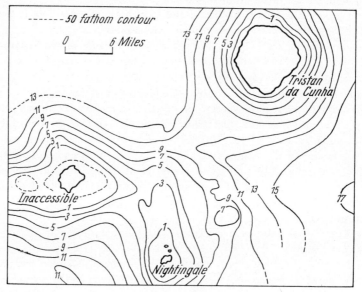

Fig. 81. Bathymetric contours at 200 fathom intervals. (BAKER, GASS, et al., 1964)

From most of the coastal area cliffs rise, often almost sheer, as high as 900 m above S. L. Fringing areas of flatter land, sloping seaward, are extremely scarce, and indeed only four are present. The largest one is where the settlement of Edinburgh is located in the NNW. Slopes here vary from 5° to 7° and occupy a region some 6 km in length and 1.5 km broad, bordered seawards by cliffs ranging from 5–15 m high. In the S are two smaller areas, around Cliff Pt. and Stony Hill, and at Sandy Pt. on the E coast is the smallest. In places sea caves 4 m above S. L. may be seen.

Whether or not faulting had anything to do with the formation of these restricted more level areas cannot be definitely ascertained. Alluvial wash and coalescing alluvial fans comprise the surface.

Between 600 m and 900 m, forming what is locally known as the Base, slopes are more gentle, forming an inclined plateau, with inclinations as low as 8°. From about 900 m, the land, known collectively as the Peak, rises more steeply at angles from 20° to 30° up to the central summit at 2062 m. The cinder cone begins at ca. 1740 m and continues to the top where there is a small crater lake with walls rising 20 m to 80 m above its surface.

It is to be noted, however, that on the E side of the island, this clear distinction between the slopes of the Base and the Peak is lacking, slopes here flattening-out much more gradually. On the W coast, where the distinction is marked, BAKER, GASS et al claimed that the steeper Peak slopes were determined by the angle of rest of the pyroclastics, the

Base slopes by fluid lavas, and they were not in agreement with DUNNE who invoked structure as the cause of slope differences. Nor would the above writers agree with DUNNE that the upper steeper slopes were comprised mostly of more viscous lavas.

The most impressive cliffs occur in the NW of the island – rising to 900 m – but on the E side, they have half this height, and in the SE, about one-fifth this height. Of interest is the fact that development and height of the cliffs bears no relation to present fringing coastal strips. Neither these latter nor the cliffs are the result of faulting; the former are due to flows from minor eruptive centres whereas the latter are the result of marine erosion.

The drainage is radial, their thalwegs reflecting the topographic slopes of the Peak, Base, cliff edges. When rains are copious and some of the streams are full, they plunge over cliff edges and thicker lava flows, some as high as 100 m as magnificent waterfalls. The greater part of the drainage is underground, and only for a few hours after rain do the gulches and gutters carry water. The larger gulches rise on the Peak, are narrow, V-shaped canyons, up to 60 m in depth, with step-like longitudinal sections, gravelly floors and vegetation. The gutters rise on the Base, are less long, less deep and have no gravelly beds but only vegetation. In times of heavy rainfalls, the bare Peak area allows of rapid runoff, quickly filling the channels and hence active fluvial erosion takes place. The gutters have gentler slopes, the terrain is carpeted in vegetation, scouring is less active, erosion greatly reduced. But as a considerable proportion of the surface water becomes lost to underground flow, fluvial erosion is actually a somewhat slow process.

Small alluvial fans develope below the cliffs, best observed on the small coastal strips, where they have been incised by the streams.

Springs are important, the chief ones occurring at the base of the cliffs, but they also are found high up on the slopes of the Peak. The Settlement derives its water supply from such a spring nearby at the foot of the cliffs.

Erosion is most active on the Peak and along the base of the sea cliffs. On the former, steep slopes, lack of vegetation, large amount of scoriaceous material allows of easy denudation. Marine erosion, with undercutting by the waves, with resultant landslides, is most active. BAKER, GASS et al refer to wave attack eating back some 10 m of the new lava field during a two month period in early 1962. By the same token, long-shore currents during the same period created a bar 150 m long, 10 m broad, cutting off the sea and forming a lagoon.

Between elevations of about 950 m to 1750 m, hogback ridges have been eroded out. The crest of these ridges are seldom broader than 6 m at the top, with slopes of 45° on either side. At ca. 1050 m elevation, the hogback crests begin to develop drainage systems which deepen into trenches, some 7 m deep, which continue for ca. 180 m gradually widening and splitting the hogbacks into divergent branches. Between 975 m and 950 m the ridges fan-out and form a continuous slope – the Base.

Many small secondary cones are scattered over the surface of the island. Seldom rising higher than 200 m, of almost perfect shape, they comprise explosion vents, scoria mounds, breached scoria cones with lava fields and effusive centres.

Between elevations of some 600 m and 900 m, in the NE part of the island, are three explosion craters containing miniature crater lakes. The largest measures 600 m from rim to rim, lake level 120 m below the rim, the water only a metre or two deep.

In no place does folding or faulting assume major proportions. Slumping, rock-falls, landslides do occur, small adjustment fractures, especially in tuffaceous beds, are present here and there. Dykes are radial in pattern and intrusive necks are common.

Inaccessible Island

The name given the island is appropriate, for almost everywhere near vertical cliffs rise to elevations varying between 150 m and 548 m, and landing has always been a difficult and hazardous matter.

Maximum elevations occur in the extreme W. Most of the interior shows rather level, undulating topography, sloping down gradually eastwards. Over the abrupt slopes forming the central interior region, streams tumble in cascades down to the shores. In the N and NW prominent landslide topography occurs. The NE central area has undergone greatest stream dissection, the principal drainage being in this direction down to the shore at the usual landing site. Along the southern coast are one or two abrupt conical masses, and the offshore islet of Pyramid Rock rises nearly vertical to over 75 m. A shallow water platform, depths less than 180 m, extends for as far as 9 km out from the island, leading BAKER, GASS et al to postulate that Inaccessible is only a small emergent part of an original very much larger volcanic island.

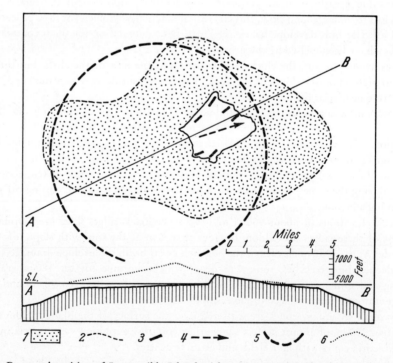

Fig. 82. Presumed position of Inaccessible Island with respect to original volcanic Island. 1. Submarine platform, 2. 100 Fathom contour, 3. Major dyke trend, 4. General dip of Main Sequence, 5. Postulated margin of original Island, 6. Postulated profile of original volcanic cone. (BAKER, GASS, et al., 1964)

Nightingale Group

In Nightingale Island a prominent ridge runs N-S in the eastern extremity, culminating at 396 m, highest point on the island. This ridge is joined by a saddle to more subdued, irregular topography of the central and western areas, with depressions occupied by swampy ponds. S of the ponds, in the central area, is the second highest elevation, 289 m. There are no permanent streams. In the N, NE, E and SE, sea cliffs vary in height between 10 m and 46 m, but along the NW, W and SW coasts, cliffs rise sheer as high as 183 m.

Separated from Nightingale by a channel only some 300 m wide lies Middle Island, with many off-shore rocks. Near-vertical cliffs drop down to the shores throughout. The interior is somewhat flattopped in appearance, with minor depressions.

NW of Middle Island and separated by a 600 m deep water channel, lies Stoltenhoff Island. The major islet is rimmed by 100 m high vertical cliffs. The relatively flat interior slopes gradually to the NW. At the SW end of the islet are two smaller masses, separated by defiles weathered along major joints.

Climate

The climate is equable but wet, with violent gales common in winter, and most of the year strong winds from the NW and SW sectors are the rule. November to March are generally the best months.

The average summer temperature on Tristan da Cunha is 20° C, average winter temperature, 12° C. February is the warmest month, 18° C, August the coldest, 11° C.

High relative humidity, average 80 %, is characteristic throughout the year.

Rains are frequent, the average annual at the Settlement being 1650 mm, about 5200 mm on the Peak. During much of the year, skies are cloudy.

Some reports say frost is unknown, hail showers are rare and snow even more rare. On the other hand, it is reported that snow often lies above 900 m from June to October. Whether snow or not, opinions appear unanimous that the higher slopes are clouded in dense mist for much of the year.

On the basis of pollen analyses of peat samples from the main island, HAFSTEN believed there had been no distinct revertent climatic development during the past 5000 odd years.

Geology

For ease of treatment, the islands comprising the Tristan da Cunha group will be considered separately.

Tristan da Cunha

The island consists essentially of a composite volcanic cone, formed chiefly by a central vent from which alternate layers of basaltic flows and pyroclastics were erupted. Although there was a principal central conduit, smaller parasitic centres emitted scoria and some thick flow-banded trachybasalts.

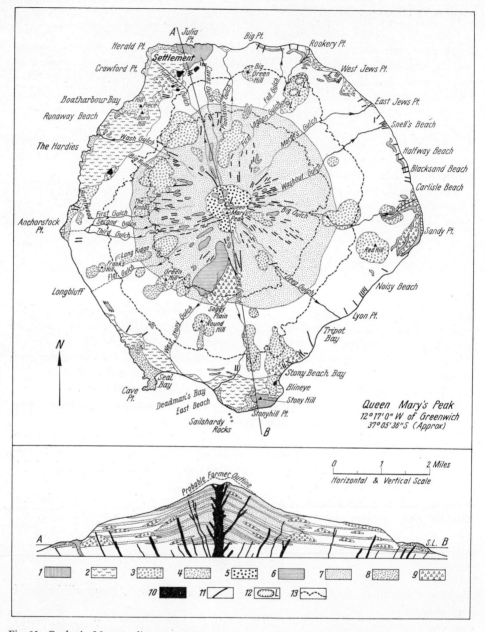

Fig. 83. Geologic Map & diagrammatic cross-section of Tristan da Cunha. 1. Recent Trachyandesite eruptions of Stony Hill & 1961, 2. Alluvium-mostly outwash deposits, 3. Surface cinder cones, 4. Lavas from surface cinder cones, 5. The Peak cinder cone, 6. Prominent lava flows, 7. Chiefly pyroclastics (Main volcanic Sequence), 8. Chiefly lavas (Main volcanic Sequence), 9. Pyroclastic centres (Main volcanic Sequence), 10. Intrusive masses, 11. Dykes, 12. Crater lakes, 13. Approx. position of 600 m & 900 m contours. (BAKER, GASS et al., 1964)

The inclination of the flows from the original central vent in the Base region varies from 5° to 10° and thus the topographic slopes here agree rather well with the dip of the flows. Higher up, in the Peak region, alternating flows and pyroclastics have a radial inclination up to about 25°, presumably due to the greater development of fragmental material as one approaches the central vent. (BAKER, GASS et al mention that here the proportion of lava to pyroclastics is much lower than further down on the Base, lava

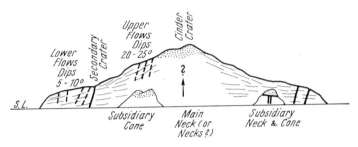

Fig. 84. Schematic section of Tristan da Cunha, showing types of volcanics units. (DUNNE, 1941)
– – – – Smaller dykes ——— Larger dykes

sometimes constituting only 25 % of the sequence.) The pyroclastics of the Peak consist mostly of reddish and blackish scoria, sometimes with bombs and intercalated lavas. S of the highest elevation is a earlier secondary vent in the form of noteworthy plug of trachyte and thus the upper steeper slopes of the island result from two centres of extrusion.

Because of dense vegetal covering, rock exposures on the Base are best observed in gulches and gutters, where basic lavas and pyroclastics dip radially outwards at approximately the same angle as the surface slopes. Outcroppings of the lavas across the stream channels result in waterfalls. Going upwards from the edge of the cliffs towards the steeper Peak area, the proportion of pyroclastics to lavas increases.

There are many secondary centres of eruption – over thirty have been mapped. BAKER, GASS et al have classified such into four main types:

a) Explosion centres, comprising a crater with little or no eruptive material in the vicinity, represented by three small radially-directed crater lakes in the NE part of the island. These ponds have been scalloped out of the seaward dipping lavas. There are three separate vents at depth, broadening out funnelwise nearer the surface so that their rims coalesce, forming V-shaped outlets.

b) Scoria mounds, with almost all material pyroclastic. Most comprise scoria and cinders, on occasion partly welded into agglomerates. The pyroclastics may be one or two hundred metres thick, and the mounds may measure several hundred metres across. Twelve such mounds are described in detail by the above authors.

c) Breached scoria cones, lava flows descending from breached crater. These are pronounced, well-preserved features. Invariably the breached part is on the downhill side, with the lava extending below as far as 800 m. The two best examples of such features are Hillpiece and Burnt Hill, lying between Potato Patches and the Settlement. It is surmised that effusive and pyroclastic action went on contemporaneously at these cones, the lavas being distinctly more basaltic than where effusive action only was taking place. Thirteen such cones are described by the above authors.

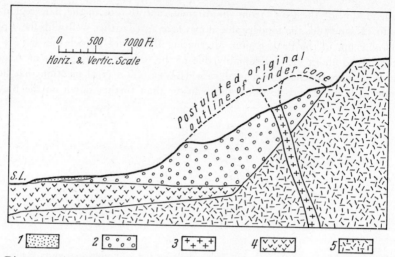

Fig. 85. Diagrammatic NW-SE section through the Burntwood parasitic centre. 1. Alluvium, 2. Pyroclastic debris (Burntwood centre), 3. Volcanic neck (Burntwood centre), 4. Leucitic Trachybasalts of Settlement coastal plain, 5. Interbedded lavas and fragmental horizons of the main sequence. (BAKER, GASS, et al., 1964)

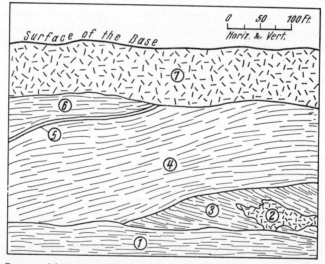

Fig. 86. Sandy Pt. parasitic centre. Sketch of S bank of Big Gulch. 1. Detritus in bed of Big Gulch, 2. Irregular intrusion of Trachybasalt, 3. Red ash and cinder with thin lava flows dipping SW at 10°, 4. Loosely cemented bright red ash and scoria dipping E at 15° to 20°, 5. Ft. layer of bright orange ash, 6. Dark red ash and cinder dipping S at 10°, 7. Thick flow of columnar pointed Trachybasalt. (BAKER, GASS, et al., 1964)

d) Loci of only effusive activity. These are represented by Stony Hill and the 1961 volcanic outburst, with thick flows of block lava. Stony Hill shows close similarity with the recent vulcanism, especially in the petrography of the rocks in question. The 1961 eruption will be mentioned later.

Baker, Gass et al, on the basis of radiocarbon datings of soil horizons under the Big Green Hill cinder cone, a scoria mound, suggest an age of 10 000 years for this Hill and several breached cones, whereas other breached cones and scoria mounds indicate a somewhat older age.

The great circle of cliffs which almost everywhere rise precipitously from the shore up to the seaward extent of the Base show a general horizontality of greyish lavas and red-brown pyroclastics, intersected by vertical dykes. As the island was formed from a central vent with many secondary centres of eruptivity material derived from both origins is decipherable in these immense cliffs, which, however present serious abstacles to the field geologist, due chiefly to their extreme steepness. Baker, Gass et al estimated that in every 600 m vertical of cliff there were some 60 individual lava flows with subordinate pyroclastic beds. The parasitic centres have contributed chiefly pyroclastics, but flows as thick as 100 m and up to 350 m broad are also encountered.

The above authors present views as to how the cliffing might have occurred: the island might once have been much larger, the upper slopes descending right to sea level and continuing in depth, in which case cliffing would be a subsequent phenomenon. On the other hand, perhaps the island was never much larger than at present but grew higher by the addition of further vulcanism at the same time as it was trimmed back by marine erosion. A marine abrasion platform offshore seems to be lacking, the water deepening regularly. If the cliffs are to be interpreted as a step, rather than a nick, in the profile, then this would suggest that the latter hypothesis is the more likely.

Dunne remarked that the only important fault on the island occurred "in the main cliffs S of Anchorstock Gulch. It is practically vertical with a throw of about 60 m and strikes 70° W of N". His map does not show the location of this Gulch, nor that of Baker, Gass et al, who in turn make no mention of any such fault.

The very small total area of relatively low flat land near the coasts comprise lava flows, pyroclastics and at the surface coalescing fans backed by the high cliffs and fronted by small sea cliffs thus actually constitute small piedmont alluvial plains. In some places, e. g. the Settlement plain, alluvium is seen to be some 100 m thick in places. Baker, Gass et al were quite certain that faulting had nothing to do with the formation of these small plains, but on the other hand, Dunne thought it unlikely that their presence could be accounted for other than by faulting, though he presents no concrete evidence of such. The small plain in the lee of the island at Sandy Point suggests that here subaerial erosion has been of greater importance than marine erosion, the Base area being embayed, allowing of deposition to form this small area of low ground.

Lava flows average between 2 m and 3 m in thickness, breadths of 1 km and lengths about the same can be observed. Flows are usually sandwiched between reddish-brownish rubble or pyroclastics. As seen in the major cliffs, pyroclastics appear to constitute between 15 % and 20 % of the rock sequence. The lower surface of the flows is uneven and brecciated, often showing chilled rims up to 5 cms. thick in non-porphyritic rock, up to 15 cms. thick in highly porphyritic rock. New flows spreading out over rubble and/or fragmental material have often picked up this and incorporated such as xenoliths in the basal half metre of the flow. The lower half of flows are usually compact and crystalline with well developed columnar jointing in thicker flows. The upper surface of flows have a decidedly scoriaceous appearance. In general, the degree of vesicularity increases upward in the flows, and indeed some even have the appearance of pumice. The upper metre or so of individual flows are of broken lava and fragmented parts of pahoehoe lava. On the other hand, highly porphyritic basic lavas have quite smooth upper surfaces with pronounced vesicularity.

On the Peak, basaltic flows occur but are not so prominent as those of trachyandesite. In the main cliffs flows can be observed comprising trachybasalts, basalts, olivine-basalts, feldspar-phyric-basalts, porphyritic ankaramites, with trachybasalts as the commonest.

Black, brownish and reddish cinders are abundant everywhere and mantle the highest slopes of the island, as well as forming the principal material of the secondary vents and craters. Tuffs are not common, the largest and most important occurrence being behind the Settlement. Whilst flows maintain a close degree of thickness throughout their lengths, the pyroclastics vary rapidly in thickness. Usually finer-grained pyroclastics are well bedded, and display vertical grading. Where the grains are more even in size, stratification is lacking.

The radial pattern of the many dykes is striking. They are most prolific on the Peak, several can be seen on the major cliffs but as the Base is a constructional feature, they are very scarce here. Generally dykes are vertical, seldom broader than 2 m, and may protrude above the surrounding terrain as much as 20 m, and on the Peak, can be traced for several hundred metres. Chilled rims are invariably present. Baking of the host rock only occurs in the case of pyroclastics, affecting merely a few centimetres.

DUNNE remarked that all dykes were olivine-basalts, with one exception. BAKER, GASS et al, on the other hand, state that petrographically the dykes show the same range as the lavas of the main sequence.

DOUGLAS (1930) had mentioned tunnels in the lava into which streams disappear, to emerge lower down where a more impervious, resistant layer causes the water to become visible again.

BAKER & HARRIS have written of somewhat similar features, namely, lava channels or trenches. These may be as much as some 850 m in length, up to 25 m broad and as deep as 8 m. They can be well observed at Stony Hill, at Long Ridge and at the latest eruptive centre. Frequently the sides of the channels rise well above the surrounding terrain, forming parallel ridges or levees. These authors deem it necessary to distinguish lava levees from lava morains, the former term being taken to mean constructional features (of solid lava, not loose blocks) built up by the solidification of lava overflowing from a lava stream, whereas lava moraines apply rather to blocks and cinders falling and rolling down from ridges of the lava streams and forming block walls similar to lateral and terminal ice moraines.

BAKER & HARRIS envisaged the formation of the lava channels and levees as somewhat similar in development to a stream which rises in flood and pours over its banks – forming lava levees, and when the quantity of lava diminishes in the 'stream', the channel aspect is acquired. They did not believe the lava channels were collapsed tunnels, even although the origins are similar.

DUNNE remarked that in the southern part of the island (he did not specifiy more particularly), wind-blown sand dunes had been piled up to elevations as high as 200 m above S.L. and extending inland for as much as 1 km, with older, lower beds semiconsolidated. This may be so, but BAKER, GASS et al make no mention of such, though their map shows considerable areas of outwash deposits behind Seal Bay and Stony Hill.

Inaccessible Island

Some 90% of the island comprises interbedded pyroclastics and basaltic lava flows, all gently dipping eastwards. BAKER, GASS et al presented a tentative table showing the age sequence as follows:

Surface pyroclastic centres
Trachybasalt lavas
Dykes
Parasitic pyroclastic centres ⎱ Approximately
Main basaltic sequence   ⎰ contemporaneous
Trachytic domes and lavas

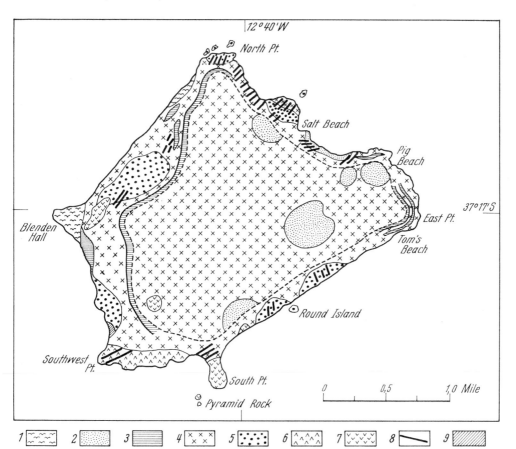

Fig. 87. Geologic Sketch Map of Inaccessible Island. 1. Detritus, 2. Surface cinder cones, 3. Thick Trachybasalt flows, 4. Main sequence – basaltic lavas and pyroclastics, 5. Pyroclastic centres with main sequence, 6. Trachyte lava, 7. Trachyte domes, 8. Dykes, 9. Landslide area. (DUNNE, 1941, and BAKER, GASS, et al., 1964)

Trachytic domes occur in the SW. As these represent the presumed oldest rocks, and, as previously remarked, present-day Inaccessible Island might possibly represent part of a once much larger volcanic cone, the eastern undissected part thereof, in which case these trachytic lavas may be representative of the more central parts of the original volcano, though all proof of such a contention is lacking as of now. On the other hand, DUNNE,

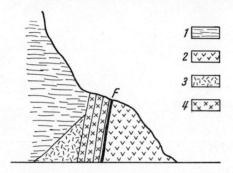

Fig. 88. Section of large dykes and trachyte dome, at Blenden Hall, Inaccessible Island. 1. Lava flows of main cliffs, 2. Trachyte dome, 3. Hornblende-bearing tuff neck, 4. Dykes, F = Possible fault plane. (DUNNE, 1941)

the first to observe these trachytic domes, did not evidently regard them as the oldest rocks, for he speaks of a 'basal complex of the West', comprising a great number of dykes which have replaced, up to 75 %, the country rock which is a completely serpentinized olivine-basalt. (Admittedly rough seas prohibited DUNNE drastically in the executing of field work on this island, his observations being made chiefly from a boat at sea.) These domes are striking features, with smooth surfaces with outward curving columnar jointing, presenting rock faces 100 m in height.

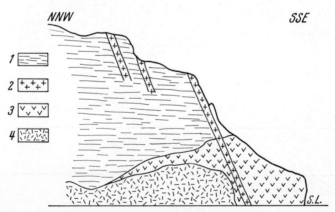

Fig. 89. Section at the bluff, SW Point, Inaccessible Island. 1. Flows of the main cliffs, 2. Trachy-Andesite dykes (?), 3. Trachyte intrusion, 4. Basal complex. (DUNNE, 1941)

Trachytic lavas form a tongue between the 'basal complex' and the main basaltic flows at SW Point. BAKER, GASS et al postulated that as these lavas are dipping eastward, their locus of origin must have been to the W of the present island.

The main basaltic sequence (olivine-basalts, as per DUNNE) comprise something like 130 individual flows, average less than 2 m in thickness, with interbedded tuffs and cinders comprising some 10 %. The dip varies between 3° and 5° to the E. These flows are similar petrographically to those on Tristan da Cunha, which, according to BAKER, GASS et al, range from basalts, through olivine-basalts to ankaramites.

The above authors noted 10 parasitic pyroclastic centres, the cones varying between 400 m and 800 m in diameter, from 60 m to 245 m in height. Flows associated with these pyroclastics are thin and more steeply dipping – between 10° and 25°.

The dykes, of olivine-basalt or ankaramites in general, are well jointed, both horizontally, vertically and cross-columnar. Usually dykes are thin, tending to occur in swarms, but where not occurring as swarms, dykes are much thicker – up to 10 m in width. Dykes are relatively rare in the eastern end of the island. On the other hand, those of the western coastal region are not only numerous but well-defined and petrologically different from the others, for here they are mugearites, but mineralogically identical with mugearite dykes in the E.

Along the N coast, the dykes strike SW-NE; along the W coast the strike is WSW-ENE, whereas along the S coastal stretches it is presumed the dykes strike NW-SE.

Principally in the E, constituting what DUNNE called 'The Thick Flows', are a system of flows, average thickness 4 m, but attaining thicknesses of 45 m, of trachybasalt or olivine-trachyandesite composition (mugearites). Flows of such composition in the island appear to be associated with secondary pyroclastic centres, and DUNNE postulated that those of the E may have originated from a large breached crater cone ca. a kilometre to the W thereof. The geological sketch map of the island given by BAKER, GASS et al shows similar flows paralleling the NW and SW coasts and about 1 km inland.

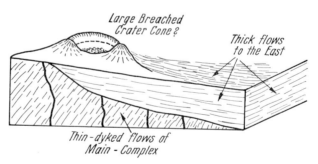

Fig. 90. Section of eastern part, Inaccessible Island showing position of schematic crater. (DUNNE, 1941)

Several surface cinder cones are present, the largest being that above referred to by DUNNE. This one comprises red scoriaceous basalts intermixed with cinders, measures some 320 m across the rim and has its steepest slope facing W although it is breached on the E side. BAKER, GASS et al believed that these parasitic cones on Inaccessible were essentially the same in all respects to those of the Base in Tristan da Cunha, although it should be noted that these authors also made many of their observations of the island's geology from a boat.

It was the opinion of DUNNE, and appears to be echoed by the above authors, that Inaccessible Island is much more complicated structurally than either Tristan da Cunha or Nightingale, and the latter held the view that as regards the extent of erosion undergone, the island was intermediary between that of the other two islands.

Nightingale Group

Most of the island comprises trachytes and trachyandesites, mostly porphyritic, and volcanic ash and agglomerate. The 'basal agglomerate' of DUNNE, or the 'older pyroclastic sequence' of BAKER, GASS et al outcrops along the northern, eastern and southeastern sea cliffs. This comprises a basal agglomerate composed of fragments of trachyte, trachyandesite, trachybasalt, sometimes olivine-basalt, cemented together by brownish glass, ash and cinder. Lapilli have excellent banding striking S 70° E, frequently dipping at 40° to the N but sometimes at 35° to the NW. DUNNE was uncertain whether such were primary dips or then the result of tilting ... perhaps due to both.

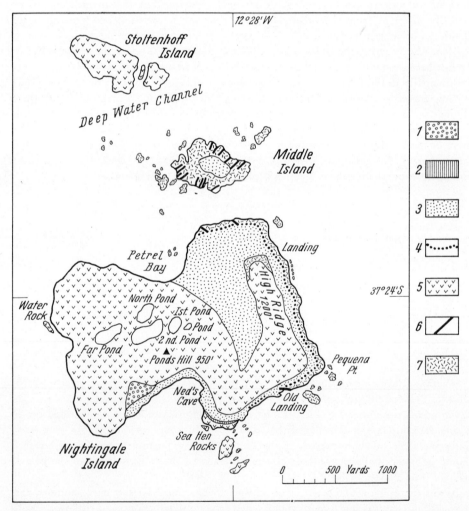

Fig. 91. Geologic Sketch Map of the Nightingale Group. 1. Talus, 2. Raised beach deposit, 3. Younger Pyroclastics, 4. Boulder bed, 5. Trachyte, 6. Trachybasalt, 7. Older Pyroclastics. (BAKER, GASS, et al., 1964)

These agglomerates form the base of terraces lying 12 m above present S. L. (vd. below), which, according to Dunne, represents an old eroded surface brought down to sea level before eustatic change took place.

Rocks of trachytic character constitute about three-quarters of the island. Though the mode of intrusion is presumed to have been of intrusive origin, this cannot be corroborated. Occasionally vertical columnar jointing is seen, especially in isolated outcrops. Lying within these trachytic exposures are circular depressions which Douglas (1930) thought were explosion craters. In spite of the morphology of these lakes being indeed very similar to depressions of explosive origin, there are no indications of explosive products in the vicinity, and both Dunne & Baker, Gass et al agreed that these ponds were erosional features within an original irregular topography.

Lying between 5 m and 33 m above S. L. and varying in thickness between 0.3 m and 4.5 m is a boulder bed comprising pebbles and boulders of porphyritic trachyte. This bed, occurring in the cliffs along the eastern side of the island, lies unconformably on the volcanic ash and agglomerates, the upper surface having a slight dip seawards, and being overlain by younger pyroclastics. Baker, Gass et al interpreted this boulder bed as representing a fossil beach deposit, indicative of a quieter period during the vulcanism of the island when earlier formed rocks were being eroded.

Tuffs and agglomerates occur lying unconformably upon all the older rocks. The lowest unit is a sandy tuff, slightly cross-bedded, and this fact, along with the excellent stratification, suggest it was deposited in shallow water, and plant remains may suggest a lagoonal environment. Dunne thought that these tuffs, as seen for example at the Landing on the NE coast, were at least older than some trachytic flows, and if all the trachytes and trachyandesites were more or less contemporaneous and so younger than this sandy tuff layer, then the major portion of the island would represent a relatively late eruption. On the other hand, Baker, Gass et al state that this tuff lies on coarse agglomerates of their 'older pyroclastic sequence', above which occur the trachyte masses. However they make no comment on the above statement of Dunne, and hence the problem rests as is. (Dunne included this sandy tuff in his 'basal agglomerate', equivalent to the 'older pyroclastic sequence' of Baker, Gass et al.)

At the Sea Hen rocks off the S coast, a good sequence is exposed of these younger pyroclastics. The following is given by Baker, Gass et al:

Sea Hen tuff	25 ft.
Trachytic lava	18 ft.
Fine tuff with abundant plant remains and numerous ash partings	10 ft.
Raised beach deposits	15 ft.
Weathered trachyte	30 ft.

Of interest here are the abundant carbonaceous plant remains, which, on the basis of radiocarbon dating, gave ages of 39.160 (+ 6090, − 3410) years B. P. The topmost tuffs are similar to those at the Landing. On rocky islets offshore here a 3 m thick sandy tuff has so many plant remains it could be regarded, in part, as lignite.

The raised beach deposit will be mentioned below.

BAKER, GASS et al presented the following table of events for Nightingale Island:

Raised beaches   Period of erosion

Younger pyroclastic sequence } Fine ash to coarse agglomerate, with local horizons rich in plant remains

Period of erosion } Formation of boulder bed, comprised chiefly of trachytic material

Intrusion and extrusion of trachyte masses

Older pyroclastic sequence } Chiefly yellow agglomerates issuing from various centres and cut by numerous basic dykes

Middle Island is composed essentially of the same type of pyroclastics as form the basal beds in Nightingale Island, the geological sequence being the same as in the NE part of the latter. Along the N coast of Middle Island, the pyroclastics (older) have a dip to the N of between 20° and 30°, but in the S of the island, the dip is to the S at an angle of 60° maximum.

Sandy tuffs similar to those above described on Nightingale Island occur on the S coast of Middle Island, lying some 50 m above S.L. in unconformable relation to highly dipping older tuffs and agglomerates. These sandy tuffs also appear to be of lagoonal origin.

Innumerable trachybasalt dykes are seen along the northern and southern coasts, with a general NE-SW strike, and unconformably overlain by the above sandy tuffs.

Stoltenhoff Island is a monolithic structure of biotite-trachyte. The larger islet is completely surrounded by cliffs some 100 m high. The major joint pattern is orientated NNE-SSW, with a secondary pattern aligned almost E-W. Marine erosion along the major pattern has separated two smaller islets at the E end of Stoltenhoff. BAKER, GASS et al thought that the island was a monolithic dome-like intrusion, similar to that forming the eastern N-S ridge in Nightingale Island and South Hill on Inaccessible Island.

DUNNE devoted a section to the discussion of eustatic change in sea level of the Tristan group. Sloping rock benches, abandoned terraces and sea-caves are common around the coasts of Tristan da Cunha which all testify to a relative emergence of some 5 m. These are well expressed at Sandy Point and between Stony Beach and Seal Bay. Similar features can be seen along the W and SE coastal areas of Inaccessible Island. Benches, terraces and sea-caves are even more strikingly developed on Nightingale Island, but here and in neighbouring Middle Island, relative emergence amounts to some 12 m instead of 5 m. (The above section quoted from BAKER, GASS et al at Sea Hen rocks, Nightingale Island, indicates a raised beach deposit lying above some 10 m of weathered trachyte, but unfortunately we do not know if the base of the trachyte was measured from sea level or higher up.) DUNNE was in partial agreement with DALY's postulation of a 5 m post-glacial eustatic shift as deduced from many regions throughout the world, including oceanic volcanic islands. To account for the 12 m change in Nightingale Island, DUNNE proposed that either this eustatic change was a separate and older one than the 5 m shift, or then Nightingale Island underwent a further 7 m change after the 5 m rise. He admitted it was difficult to decide, on the basis of evidence available, which hypothesis might be correct, but admitted it was somewhat difficult to account why and how Nightingale-Middle Islands should have risen a further 7 m when these islands in essence formed but a small part of an immense composite volcanic unit. (It should be noted that DOUGLAS (1930) claimed that Middle Island was once connected with Nightingale Island, but he doubted if Stoltenhoff Island was ever united with Nightingale, and DUNNE agreed with these views.)

Petrography

Early geological information on the Tristan group is mostly concerned with petrographic descriptions of specimens collected by various expeditions. RENARD (1885, 1887, 1889) described samples collected by the H. M. S. Challenger expedition; SCHWARZ (1905) described those of the H. M. S. Odin expedition; DOUGLAS (1930) described those of the 'Quest' expedition, as also did CAMPBELL SMITH (1930).

The first systematic study of the Tristan group was made by DUNNE (1941), whose publication is chiefly concerned with petrography. The Royal Society expedition (BAKER, GASS, HARRIS & LE MAITRE, 1964) members were dispatched in late 1961-early 1962 to study the recent alarming volcanic activity, and the report is not only the latest but also the most comprehensive dealing with all aspects of the geology of these islands.

In compiling the chapter on the Tristan da Cunha group, the author has relied chiefly on the publications of DUNNE and BAKER, GASS et al.

Nomenclature

DUNNE classified the rocks of the island group as follows:

    Olivine-alkali-basalts
    Hornblende-alkali-to-trachybasalts
    Trachyandesite-basalts   a) Mugearites
                                    b) Trachy-andesite-basalts (tephritic)
    Trachyandesites
    Biotite-soda-trachytes
    Feldspathoid-bearing trachytes
    Essexitic Gabbros
    Augitic-Hornblenditic-Biotitic-Anorthositic coarse-grained Xenoliths

BAKER, GASS et al thought that DUNNE's classification was somewhat cumbersome and so simplified this by classifying the rocks as

    Alkali-basalts   Trachyandesites
    Trachybasalts   Trachytes

along with appropriate prefixes, to cover the alkali-basalt – trachyte range of rocks occurring. They note that the rocks do indeed contain fair amounts of normative nepheline, but because this mineral never occurs in crystalline form (except specimen No. 30) the adoption of a terminology to indicate undersaturation was thought to be unnecessary. The divisions of the rocks are arbitrarily chosen, based upon their chemistry, with boundaries as shown in Fig. 92.

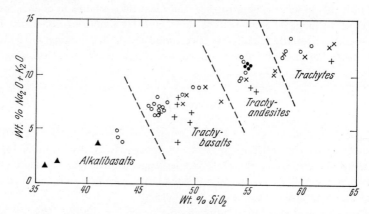

Fig. 92. Petrographic nomenclature: Total Alkali/Silica Diagram. (BAKER, GASS, et. al., 1964)
▲ Tristan Xenoliths   ○ Tristan lavas   ● Tristan new lava
× Nightingale Group lavas   + Inaccessible lavas

Seldom do these authors specify the types of pyroclastics present, but DUNNE is more specific.

Rock Types

The rocks listed and described by DUNNE (which includes collections studied by earlier workers) and BAKER, GASS et al will be treated separately, the chemical analyses being shown in Tables 61 and 62 respectively.

Olivine-basalts and Olivine-alkali-basalts

These rocks vary in texture from strongly porphyritic to aphyric, they may be vesicular to dense if occurring as flows, or then anamesitic to doleritic if occurring as dykes. The melanocratic nature may vary from 40% to 80%.

Porphyritic types are the commonest, in which olivine and augite or plagioclase phenocrysts may be as large as 3 cms. In the more compact, black lavas, the phenocrysts may be either totally of ilmenite, magnetite, or then these two minerals. Most porphyritic types show phenocrysts of olivine, augite and plagioclase. The matrix is fine-grained, comprising pigeonitic augite, subordinate olivine, plagioclase laths with some interstitial alkali feldspar, 'anemousite and frequently some glass. The glassy nature becomes more pronounced in more scoriaceous parts of the rocks and may indeed comprise the entire groundmass. Iron oxides, serpentine, chlorite and saussurite occur as alteration products; apatite, ilmenite and magnetite occur as accessories.

As representative of these types of rocks occurring as flows, Table 61, No. 1 shows the chemical characteristics. Phenocrysts of augite, some olivine, plagioclase and ore comprise 42% by volume of the rock, the groundmass being largely of plagioclase, alkali feldspar, 'anemousite', glass and augite. On a mineralogic basis, the rock resembles basalts found in volcanic oceanic islands, and chemically it is similar to many rocks analysed from Atlantic, Pacific and Indian oceanic island provinces. Specimen No. 1 can be termed a basic olivine-basalt tending towards alkalinity; a melanocratic increase would change the specimen into one of the ankaratrite group, and an increase in olivine, into an average olivine-basalt.

Olivine-basalts occurring as dykes comprise the same minerals as those present as flows, but are lighter coloured and of different texture. No. 2 is representative of the dyke rocks, showing plagioclase laths and interstitial orthoclase, anemousite and glass, these phenocrysts constituting some 60% of the rock. Augite is prominent in the groundmass, and euhedral apatite and biotite flakes with ore comprise the accessories. No. 3 is an olivine-basalt from Inaccessible Island, whilst Nos. 4 and 5 are trachybasalts studied by CAMPBELL SMITH from the 'Quest' collection, which DUNNE preferred to call hornblende-alkali-basalts. Similarities between specimens 4 and 5 and a basalt from Madagascar and a trachydolerite from Kilimanjaro respectively were pointed out by CAMPBELL SMITH & DUNNE showed similarities chemically between the doleritic olivine-basalts as dykes in Tristan da Cunha with rocks from Kerguelen Island, Maui (Hawaii Islands), Mont Doré (Auvergne) and St. Helena, pointing out however that the chief difference between the Tristan rocks and the others referred to the ratio between the alkalis.

Hornblende-alkali to trachybasalts

Where the olivine-basalts show a smaller proportion of olivine – not exceeding 5%, a greater abundance of leucratic minerals – usually more than 55%, where the interstitial amount of alkali feldspar and nephelinitic material is greater – from 10% to 20%, and where titanium-rich basaltic hornblende occurs, then the rocks are named as above. Transitions occur between these rocks and the olivine-basalts on the one hand, trachybasalts on the other hand.

Most rocks have a greyish colour, are porphyritic, aphanitic to sub-fluidal matrix. Phenocrysts comprise: zoned augite, often showing hour-glass structure, brown basaltic hornblende showing all stages of resorption, plagioclase usually strongly zoned and commonly resorbed. Amongst the accessories, tiny phenocrysts of ore and apatite are most frequent, with occasional flakes of biotite. Transitional varieties to olivine-basalt show anhedra of olivine and magnetite. The groundmass comprises confused patches and laths of plagioclase, tabular pigeonitic augite, ore and a few scattered grains of olivine, usually altered. Alkali feldspar and anemousite occur interstitially, and a small proportion of brownish isotropic glass sometimes can be observed. Nos. 4 and 5 are representative of this class of rock. No. 4 was collected at the edge of the crater lake at an elevation of 1950 m, whereas No. 5 was got near sea level at Herald Point.

Trachy-Andesite-Basalts

These rocks are best developed on Inaccessible Island but also occur on other islands. Under this heading can be grouped rocks which by others have been named olivine-poor basalts, trachydolerites, andesine-andesites, andesine-basalts, trachybasalts and trachyandesite. The greyish rocks occur either as flows or dykes, and have textures which may be either porphyritic, aphanitic, doleritic or fluidal. They may be considered as feldspathic rocks with a ratio of plagioclase to alkali feldspar of 5 : 1–2. Generally they show about 30 % melanocratic constituents, but as this may be as low as 5 %, then melanocratically some are trachyandesites, others trachybasalts.

DUNNE divided the group into two sub-divisions, the first of which was typified by the predominance of olivine over augite and the absence of an undersaturated leucocratic mineral. Such rocks were named mugearites by CAMPBELL SMITH, and indeed they are closely analagous to those from the type locality in Mull, Scotland. The other subdivision has very little olivine, perhaps none at all, along with interstitial undersaturated leucocratic constituents, and are here referred to as tepheritic trachy-andesite-basalts.

a) Mugearites. Many flows in the central and eastern parts of Tristan da Cunha and dykes in the western coastal area of Inaccessible Island comprise rocks of this type. These alkali-feldspar (oligoclase-andesine) rocks, 30–40 % melanocratic, may be well crystallized, of porphyro-doleritic appearance, ot then a fluidal arrangement of plagioclase laths with scattered phenocrysts may be evident. Specimen No. 6 is typical of the former, consisting of euhedral olivine, irregular augite, cubic ore and laths of andesine, with no interstitial material. Biotite flakes form the essential accessory. No. 7 is a fluidal-type mugearite, but as some pneumatolysis had occured in the specimen, it is scarcely typical. At Blenden Hall, Inaccessible Island, the fluidal variety is well represented, the rocks showing well-twinned laths of plagioclase enclosed in alkali feldspar, the whole displaying a marked fluidal arrangement when viewed microscopically. Phenocrysts of plagioclase and olivine, if present, are orientated parallel to the general flow direction. Biotite which frequently is the only phenocryst, occurs in some specimens. Apatite and basaltic hornblende occasionally are present as inclusion-filled phenocrystic euhedra. No. 8 is from a likely dyketype intrusion at South Point, Inaccessible Island. Phenocrysts comprise olivine and augite, and considerable less interstitial alkali feldspar is present. Ore is quite abundant, and some flakes of biotite. Under the microscope, the rock is very similar to some essexite boulders at Blenden Hall.

Rocks similar to the above are present in many places in the islands of the Pacific, Auvergne region, etc., though described under different names, and the trachydolerites of St. Helena and Madeira are chemically comparable. The mugearites of Mull, however, are slightly richer in alkalis and magnesia.

b) Tephritic Trachy-Andesite-Basalts. No. 9, collected at 1400 m on Tristan da Cunha, at First Lagoon Gulch, shows glomeroporphyritic clusters of plagioclase and some slightly resorbed hornblende, a few euhedra of olivine, magnetite and apatite, all in a strongly fluidal matrix of plagioclase laths, small pyroxene grains, elongated skeletal olivine, ore, and rare hornblende flakes. Interstitial matter comprises sodic plagioclase, alkali feldspar and glass. Nos. 10, 11 and 12 are from dykes on Middle Island, which can traced into Nightingale Island. The rocks here differ considerably from the Tristan da Cunha occurrences, although considered contemporaneous in age. Here the groundmass may approach that of the olivine-basalts and hornblende-alkali-basalts, or then to that of the trachy-andesite-basalts and trachyandesites. These Middle Islnd rocks are about 60 % melanocratic by volume, all are porphyritic, with the following commonest phenocryst associations: augite, augite-olivine, augite-olivine-hornblende, augite-hornblende. Subordinate to the above, or then absent perhaps, are strongly corroded plagioclase phenocrysts. The augites are strongly zoned, of titaniferous variety; olivine is partially altered to serpentine and the brown basaltic hornblende is unresorbed to an unusual extent.

Leucocratic varieties of these rocks may vary from aphyric to dopatic. Aphyric types have a groundmass similar to porphyritic types, but the amount of phenocrysts – chiefly hornblende – vary in abundance from dyke to dyke.

Nos. 11 and 12, studied by CAMPBELL SMITH, are representative of the aphanitic type of rocks and correspond most closely with the magma giving rise to these class of rocks. A few phenocrysts of plagioclase, partly resorbed hornblende, and magma cubes are present in a fluidal matrix of tabular and lath-shaped plagioclase, pyroxene, ore, serpentinized grains of olivine, tiny flakes of either biotite or hornblende. Interstitial material is partly orthoclase, but most is probably anemousite. (Note: DUNNE believed that the chemical analysis of No. 12 by CAMPBELL SMITH was in error, for in thin-section this specimen showed no great difference from No. 11, and DUNNE attributed the mistake to an over-estimate in the silica content.)

Trachyandesites

DUNNE considered that almost the entire island of Nightingale was composed of rocks of this type, occurring also on Tristan da Cunha and Inaccessible Islands.

The Nightingale rocks were termed 'hybrid' trachyandesites, resulting from heterogeneous contamination of a trachytic, partly crystallized melt with 'foreign' phenocrysts of basaltic hornblende, plagioclase, augite, ore and apatite. According to the amount of xenocrysts which are admixed, a range of rocks from trachyandesites to biotite-trachytes outcrop. The degree of corrosion and solution of foreign material varies greatly, sometimes very slightly affected to complete disappearance. No. 13 is considered quite representative of the Nightingale rocks, which is greyish in colour, porphyritic, with phenocrysts of corroded basaltic hornblende, brownish biotite with magnetite rims, plagioclase often rimmed with a sodic border grading into orthoclase, colourless, zonal augite and greenish diopsidic augite. Accessories include sphene with resorption borders of ilmenite grains, apatite, and a few zircon grains. The matrix is a fluidal compound of alkali feldspar and plagioclase laths, grains and needles of colourless pyroxene, and grains of ore. Reddish varieties of this type of rock show a more trachytic groundmass, and biotite is the dominant dark phenocryst. Hornblende may have disappeared entirely, the olivine shows resorption, resulting in an augite rim around an iddingsite nucleus, the augites are of diopsidic and aegerine-augite varieties. The accessories are similar but sphene is not resorbed.

Interbedded with the lavas on Nightingale Island are thick layers of clastic material of similar mineralogical composition. In these brecciated tuffaceous layers, parts may be strongly vesicular, showing inclusions of large hornblende, plagioclase and ore. Field inspection shows that the hornblende was introduced during highly gaseous eruptive intervals, and spread through the lavas. Inclusion-filled apatites also occur in the hornblendes and masses of similar apatite and ore.

On Tristan da Cunha at an elevation of ca. 1300 m, on the NE side of the island, are somewhat vesicular flows, of uniform porphyritic appearance with well-developed flow structure. No. 14 shows phenocrysts of augite with features similar to those in the hornblende-alkali-basalts, partly resorbed hornblende and tabular plagioclase. Interstitially there is alkali feldspar, some anemousite and some grains and microlites of augite. Of accessories, there are anhedra of ore, very small phenocrysts of apatite and flakes of biotite.

On Inaccessible Island, at Blenden Hall, many remnant boulders and a few isolated outcrops of trachyandesites are found, and also as thick flows in the cliffs in the NW, W and SW of the island. These are closely associated and grade into mugearites, distinction being based on the more feldspathic groundmass and hornblende as the dominant dark phenocrysts in the trachyandesites. The rocks are light grey in colour, weathering to still lighter colours, are compact, with only a few phenocrysts of black hornblende and glomeroporphyritic plagioclase. Aphyric and vesicular types are also present. No. 15 is a dense, grey rock, with a few phenocrysts of basaltic hornblende slightly resorbed around the edges, glomeroporphyritic clusters of plagioclase in a fluidal, turbid matrix, diopsidic pyroxene and serpentinized skeletal olivine with interstitial alkali feldspar. Microphenocrysts of magnetite and apatite are accessories. No. 16, a dark grey, weathered porphyritic rock, shows phenocrysts of resorbed basaltic hornblende, some prismoids of augite, a flake or two or biotite and tabular plagioclase. The matrix is similar to the above specimen except that plagioclase occurs as tabular, zoned crystals. Calcite, serpentine and chlorite are present in small quantities.

High up in Cave Gulch, Tristan da Cunha, several flows are present which are presumed to have originated from the neck forming the second peak. These flows can also be termed trachyandesites, an analysis being given in No. 17. The light grey, compact, porphyritic rocks have evenly distributed phenocrysts of barkevitic hornblende, glomeroporphyritic clear zonal plagioclase, biotite flakes and zonal dioposidic augite. Accessories include small phenocrysts of magnetite, apatite, larger phenocrysts of sphene. The groundmass is a fluidal aggregate of plagioclase laths, alkali feldspar, grains and needles of diopsidic augite and abundant ore. Some boulders seen at West Point, Inaccessible Island, are very similar to these rocks and likely indicate a late phase of the trachyandesites.

CAMPBELL SMITH described a rock collected from the summit of Middle Island which he named a tephritic trachyte (No. 18). The rock forms a hard, compact matrix of breccia in a volcanic neck, and close by are rocks of chemical similarity to which RENARD (1889) gave the name 'phonolitic tufa'. Few phenocrysts occur of plagioclase, augite, ore and resorbed hornblende, in a very fine-grained liquidous groundmass of plagioclase which is mostly enclosed in alkali feldspar, along with tabular augites, ore grains and often incipient mica. Chemically, these rocks are close associates of the trachyandesite-basalts dykes, trachyandesites and alkali-trachytes which also occur on Middle Island, as well as Nightingale Island.

Biotite-Trachytes

These rocks are represented as domes and monolithic masses on Inaccessible Island, as flows intermixed with 'hybrid' trachyandesites on Nightingale Island, as lapilli in the agglomerates of Middle Island and constitute the whole of Stoltenhoff Island. Those in the last-named locality are dense, greyish or reddish brown when weathered, somewhat porphyritic. No. 19, collected some 10 m above S. L. in the NW part of the islet, shows a few phenocrysts of plagioclase, biotite, diopsidic augite and rarely anorthoclase. The plagioclase is well twinned and of zonal character often tending towards a very sodic

For comparative purposes, the analysis of the RENARD specimen is given below:

SiO_2	48.09
Al_2O_3	19.05
Fe_2O_3	3.44
FeO	5.59
MnO	0.00
MgO	3.50
CaO	9.42
Na_2O	5.06
K_2O	2.88
H_2O+	0.67
H_2O-	0.00
TiO_2	4.38
P_2O_5	0.00
Total	102.08

Specimen from NE Nightingale Island, opposite Middle Island.

rim bordered by alkali feldspar or then more frequently penetrated and replaced by the latter. Biotite is strongly pleochroic, but mostly is totally resorbed to yield a patchwork of iron oxides and other materials. Augites show good hour-glass structure and are often zoned. The matrix comprises alkali feldspar, diopsidic augite, magnetite cubes and biotite flakes, and, on occasion, sphene; magnetite and apatite are accessories. Seemingly no feldspathoids have been noted in these Stoltenhoff rocks but on Middle Island, rocks otherwise identical, do show some yellowish sodalites. The biotite-trachytes of Inaccessible Island are all quite similar and mineralogically analagous to the specimen described above. No. 20 is from the dome at Blenden Hall, Inaccessible Island. This reddish-white, porphyritic rock has phenocrysts of plagioclase, anorthoclase, biotite and barkevikitic basaltic hornblende, also inter-penetrating replacement growths of plagioclase and alkali feldspar. The matrix comprises laths of orthoclase, squares of strongly zoned plagioclase, short, tabular diopsidic augite, biotite flakes and ore. Cristobalite occurs in some vesicular interstices. Tiny euhedral phenocrysts of magnetite and apatite constitute the accessories. From SW Point, same island, trachytes similar to the above are found, but the phenocrysts of anorthoclase-plagioclase intergrowths are larger, the biotite is mostly resorbed, there are remnants of resorbed hornblende, and there are inclusion-filled, quite large phenocrysts of apatite. Specimens collected at S Point are identical to the above, but the plagioclase is somewhat fresher and is only partially replaced by alkali feldspar. The dome forming the summit of Inaccessible Island conprises deeply weathered trachyte, composed alkali feldspar, plagioclase and diopsidic augite.

Feldspathoid-bearing Trachytes

The most leucocratic rocks in the Tristan group are feldspathoid-bearing trachytes. DUNNE was not sure whether or not nepheline was present in any of these rocks, but sodalite seemed more definite. Such rocks are exposed on Tristan da Cunha, Nightingale and Middle Island, and in each, the rocks are distinctive. DUNNE classed them as of two types:

a) *Sodalite-plagioclase-trachytes.* No. 21, from the Peak crater, main island, is grey in colour when fresh, whitish when weathered, and has a rather drusy and somewhat porphyritic character. Small blue phenocrysts of sodalite and pyrite cubes are visible in hand specimen. Microscopically the rock is composed of glomeroporphyritic clusters of plagioclase, some being untwinned and strongly zoned and rimmed by alkali feldspar, whilst others do show twinning. There occur also resorbed biotite flakes, aegerine-augite, euhedra of sphene, some bluish sodalite grains. The groundmass is trachytic, comprising alkali feldspar and diopsidic augite, in which zircon, apatite and magnetite occurs as accessories. Chemically the rock is similar to the biotite-trachyte of Stoltenhoff Island.

b) *Sodalite-trachytes.* In eastern Nightingale Island, several large flows occur of compact, light greyish rocks with a pronounced flow structure. Where the rocks have been subjected to weathering, they acquire a strange mottled appearance, for which Dunne could offer no explanation. Specimen No. 22, in thin-section, shows long prismoids of anorthoclase-orthoclase, aligned parallel in a fluidal aggregate of alkali feldspar, with cubes and hexagons of very small sodalites occurring interstitially. On a chemical basis, the specimen is analagous to many aegerine-augite trachytes and pulaskites.

On Middle Island some lapilli and the agglomerate at the Landing on Nightingale Island are essentially variations of the above in which anorthoclase, orthoclase phenocrysts are larger and perhaps include a lath or two of plagioclase. In the agglomerate of the flow E of the Landing, the trachytes have been contaminated with basaltic hornblende and plagioclase, but the matrix seems to have been little affected.

Specimen No. 23 is somewhat porphyritic and of brownish colour. Chief phenocrysts include irregular large sodalites, an occasional prismoid of anorthoclase, skeletal aegerine-augite and magnetite associations, some euhedra of sphene and zircon. The non-fluidal matrix comprises alkali feldspar, prisms of aegerine-augite, tiny magnetite cubes and minute, clear sodalites. Chemically this specimen is very similar to many phonolites and phonolitic trachytes, but the fm values of the latter are almost double those of the sodalite trachytes.

Essexitic Gabbros

On Inaccessible Island these rocks outcrop at Blenden Hall, and at W Point boulders of these rocks are found. No. 24, from the former locality, has a hypidiomorphic, granular texture, and consists of euhedral-subhedral olivine altered to serpentine, with iron oxides along cracks and cleavage planes, augite and some laths of plagioclase enclosed in a little alkali feldspar of anorthoclasic character in which are irregular intergrowths of sodic plagioclase. The accessories include grains and skeletal forms of ore, brown biotite and small apatites. The W Point boulders have the same minerals, but biotite is more plentiful and melanocratic minerals are less abundant.

These essexitic gabbros are chemically more rich in MgO than average rocks of this type, and indeed are more akin to some olivine-basalts, e. g. those from Juan Fernandez island group off the coast of Chile. (Baker, Gass et al make no reference to essexitic gabbros in Inaccessible Island, doubtless regarding such occurrences as olivine-basalts.)

Table 61

	Olivine-Basalts Ol.-Alk. Basalts			Horn.-Alkali to Trachy-Basalts			Trachy-Andesite-Basalts					
	1.	2.	3.	4.	5.	6.	7.	8.	9.	10.	11.	12.
SiO_2	42.51	46.24	47.76	47.44	46.31	49.30	48.10	49.30	50.02	48.76	50.64	52.1
Al_2O_3	12.82	15.75	15.68	17.17	17.36	18.85	19.39	16.36	19.58	15.10	21.75	20.48
Fe_2O_3	4.89	3.04	3.34	3.10	3.27	4.37	8.02	3.69	2.29	4.23	0.19	3.8
FeO	8.64	7.39	6.55	7.07	8.12	5.80	3.03	5.94	5.24	5.27	5.93	2.2
MnO	0.13	0.14	0.07	0.14	0.17	tr.	0.13	0.16	0.12	0.07	0.07	0.0
MgO	9.63	4.92	6.05	4.65	4.64	3.95	2.91	5.31	3.15	3.54	2.56	1.96
CaO	12.22	9.57	9.02	9.78	9.74	7.66	4.61	7.40	6.32	7.45	6.88	5.55
Na_2O	2.83	4.12	4.22	3.71	3.67	3.89	4.81	4.09	5.15	4.43	5.27	3.92
K_2O	1.40	2.60	1.94	2.95	2.79	2.65	2.56	1.71	3.71	3.96	3.73	3.75
H_2O+	0.41	0.29	1.87	0.28	0.21	0.45	1.87	1.67	0.21	3.11	0.44	2.21
H_2O-	0.05	0.05	0.00	0.11	0.08	0.30	2.08	0.49	0.00	0.43	0.31	1.96
TiO_2	4.29	4.48	3.05	4.02	3.64	2.02	2.60	3.18	3.39	2.93	2.50	2.00
P_2O_5	0.38	1.20	0.49	n.d.	n.d.	0.64	–	0.49	1.02	0.83	–	–
Total	100.20	99.79	100.04	100.42	100.00	99.96	100.11	99.95	100.20	100.11	100.27	100.04
S. G.				2.85	2.83	2.68					2.72	2.41

a) Includes ZrO_2 0.10.   b) Includes Cl_2 0.06.   c) Includes Cl_2 0.32.   d) Includes Cl_2 0.32.

1. Olivine-Basalt, Sandy Point, TdaC.
2. Dolearitic Olivine-Basalt, Behind Settlement, TdaC.
3. Olivine-Basalt, West Pt., Inaccessible.
*4. Trachy-Basalt, TdaC.
*5. Trachy-Basalt, TdaC.
*6. Mugearite, Inaccessible Is.
*7. Mugearite, Inaccessible Is.
8. Mugearite, South Pt. Inaccessible Is.
9. Trachy-Basalt, First Lagoon Gulch, TdaC.
10. Trachy-Basalt, Middle Is.
*11. Trachy-Basalt, NE Peak, Middle Is.
*12. Trachy-Basalt, dyke, Middle Is.
13. Trachy-Andesite, Eastern section, Nightingale Is.
14. Trachy-Andesite, First Lagoon Gulch, TdaC.
15. Trachy-And
16. Trachy-And
17. Trachy-And
*18. Tephritic Tr
19. Biotite-Trac
20. Biotite-Trac
21. Sodalite-pla

Chemical Analyses of Rocks (Campbell Smith, 1930, Dunne, 1941)

	Trachy-Andesites					Biotite-Trachytes			Feld. bear. Trachytes			Essex. Gabbro	Xenoliths		
	13.	14.	15.	16.	17.	18.	19.	20.	21.	22.	23.	24.	25.	26.	27.
	54.40	54.43	55.51	54.94	54.35	57.10	59.91	62.36	58.20	62.23	58.71	48.11	35.70	36.76	40.76
	17.34	18.23	18.20	17.69	19.97	21.05	18.44	18.99	19.10	18.43	20.03	14.29	10.45	16.18	19.19
	3.04	3.24	3.04	2.97	1.52	1.76	2.88	1.76	2.24	1.68	2.16	2.01	10.88	13.13	7.62
	3.69	2.66	4.03	3.29	3.15	2.01	1.20	0.87	1.28	0.60	0.45	7.63	9.54	3.87	2.29
	0.07	0.09	0.13	0.12	0.24	0.11	0.07	0.04	0.08	0.12	0.10	0.15	0.12	0.06	0.03
	2.50	1.52	2.05	2.87	1.35	1.20	0.97	0.78	0.81	0.11	0.00	10.13	9.75	7.50	6.05
	4.94	5.83	4.67	4.73	4.52	4.13	2.73	1.78	3.58	1.16	0.98	9.71	16.32	14.98	14.07
	5.11	5.64	5.43	5.05	6.71	5.01	6.05	6.30	6.30	6.77	7.56	2.92	1.53	1.59	2.42
	4.73	4.71	3.15	3.76	4.53	5.33	5.83	5.01	5.94	6.28	6.03	0.98	0.14	0.33	1.36
	1.11	1.55	0.56	1.03	0.38	0.97	0.30	0.63	0.90	1.60	3.19	0.46	0.61	0.83	0.70
	0.00	0.00	0.25	0.47	0.04	0.43	0.16	0.24	0.13	0.06	0.22	0.09	0.00	0.00	0.00
	2.38	1.82	2.12	2.18	2.18	1.11	1.21	0.72	1.33	0.65	0.39	3.27	5.13	4.88	3.56
	0.69	0.42	0.56	0.62	0.89	n.d.	0.22	0.21	0.21	0.07	0.06	0.31	0.05	0.02	1.99
	100.10	100.14	99.70	99.72	99.83	100.21	99.97	99.69	100.16	100.01	100.13	100.06	100.22	100.13	100.04
	a					2.61			b	c	d				

13. ...desite, Western section, Inaccessible Is.
14. ...desite, Western section, Inaccessible Is.
15. ...esite, Cave Gulch, TdaC.
16. ...achyte, Middle Is.
17. ...chyte, Stoltenhoff Is.
18. ...hyte, Inaccessible Is.
19. ...ioclase-Trachyte, Crater Lake, TdaC.
22. Sodalite-Trachyte, E side, Nightingale.
23. Sodalite-Trachyte, Stony Beach, TdaC.
24. Essexitic Gabbro, Inaccessible Is.
25. Augititic endogenic inclusion in Basalt lava, Sandy Pt., TdaC.
26. Hornblenditic endogenic inclusion in Basalt, Sandy Pt, TdaC.
27. Biotite-bearing, plagioclase-rich endogenic inclusion in red cinder, Half Way Beach, TdaC.

(* = Campbell Smith specimen)

Beiträge zur Regionalen Geologie der Erde Band 10: Mitchell-Thomé, S. Atlantic Islands
Gebrüder Borntraeger, Berlin · Stuttgart

Augititic-Hornblenditic-Biotitic-Anorthositic Coarse-grained Xenoliths

CAMPBELL SMITH referred to hornblendic types of coasre-grained xenolithic inclusions and bombs, and DUNNE noted that they were associated with almost every secondary cinder cone of the main extrusive mass on Tristan da Cunha. They form approximately-rounded masses varying up to 30 cm in diameter and are 'cemented' along with basaltic blocks in the cinders or basaltic obsidian. They are to be interpreted generally as included cognate xenoliths rather than volcanic bombs. At Sandy Point, Tristan da Cunha, they occur scattered throughout the thick, fine-grained basaltic flow underlying the porphyritic olivine-basalt analysed in No. 1. At Stony Beach, a 20 m wide neck is filled with black, fragmented lava in which about 5 % by volume comprises xenoliths. A dyke on Nightingale Island, and basaltic flows on Inaccessible Island show much fewer similar inclusions.

Both as regards texture and mineral content, the xenoliths show great variation, which might suggest that they represent sporadic, heterogeneous, crystal agglomeration within a crystallizing magma rather than a 'truly' consolidated rock from such a magma. Textures are often hypidiomorphic to xenomorphic granular, and frequently coarse pegmatitic and fluidal schlieren-like varieties can be seen. On a volumetric basis, augite, hornblende and plagioclase are the dominant minerals; other minerals may include olivine, biotite, ore, titanite and apatite.

No. 25 is from a basaltic flow at Sandy Point, and is classed as an augititic endogenic xenolith, augite comprising 60.7 % by volume, plagioclase, 11.6 %, hornblende 7.9 %, titanite, apatite and interstitial matter, 19.8 %. These minerals seem to have crystallized almost simultaneously, the ore likely being slightly earlier. With an increase in the amount of hornblende, this type of rock passes over into a hornblenditic xenolith variety, and is here represented by No. 26. This specimen shows 32.4 % plagioclase by volume, 27.8 % hornblende, 27.6 % augite and 12.2 % ore. Here the crystallization order appears to have been plagioclase-ore-augite-hornblende. Specimen No. 27 is a friable rock showing recognizable fluidal character. Biotite in large flakes is present, the augite is non-zoned and the hornblende is often occurring parallel as thin blades, and frequently forms reaction rims around augite. The nature of the hornblende and large patches of biotite which have been changed to the so-called rubellanitic variety, indicate that the specimen underwent heating of different intensity. The mineral composition, on a volumatric percentage basis is: plagioclase 52.2, hornblende 24.6, biotite 14.0, augite 4.2, apatite 4.0, ore and titanite, 1 % each.

Xenoliths are indeed of great petrologic, genetic and volcanologic importance, but throughout the world not a great deal of rock analyses have been made of such endogenic inclusions, although of course many collections and descriptions have been given by various authors from many localities. The Tristan specimens appear to show a close analogy with those studied by LACROIX from Mont Doré, Auvergne, France, which he named mareugites, differing chiefly in that the former have no feldspathoids, although such are not always present in the LACROIX specimens. Chemically one can note some similarity between Nos. 25 and 26 and certain issites, avezacites and yamaskites, of the pyroxenite and hornblendite families, whilst No. 27 is akin to such types as berondites, melilite-fasanites, etc. It is also to be noted that chemically, the Tristan rocks are quite like many calcic basic volcanites such as basanites, melilite-ankaratrites, biotite-melilites, etc.

As already remarked, BAKER, GASS et al treated of a somewhat more simplified classification of the rocks. They pointed out the dominance of basalts, especially trachybasalts, and the more acidic character of the Peak material compared to that of the Base and the Main Cliffs.

Of distinct interest in the lavas is interstitial leucite (LEMAITRE & GASS, 1963) and the following points are noteworthy: it occurs only interstitially, it is a very rare mineral in oceanic islands, and apparently it has substituted for nepheline in an essentially silica-undersaturated magma series.

The following descriptions of rocks refer to Table 62.

Table 62   Chemical Analyses of Rocks (BAKER, GASS et al, 1964)

	28.	29.	30.	31.	32.	33.	34.	35.	36.	37.
SiO_2	42.93	42.43	45.5	45.70	46.48	46.00	46.01	46.07	45.96	46.36
TiO_2	3.73	4.11	3.3	3.65	3.10	2.83	2.19	3.08	3.44	3.54
Al_2O_3	12.05	14.15	18.3	16.70	16.68	17.03	16.84	17.06	17.84	16.19
Fe_2O_3	5.58	5.84	2.5	3.73	4.12	3.79	7.61	2.59	4.53	3.66
FeO	8.27	8.48	8.4	7.28	7.30	7.47	5.37	8.32	6.21	6.94
MnO	0.16	0.17	0.1	0.17	0.18	0.23	0.18	0.18	0.20	0.18
MgO	10.28	6.71	4.6	4.89	4.65	4.80	4.75	4.72	4.13	4.57
CaO	12.58	11.91	10.0	9.91	9.40	9.54	9.36	9.35	9.61	9.45
Na_2O	2.36	2.77	4.2	3.96	3.80	4.04	3.74	4.01	4.27	3.97
K_2O	1.47	2.04	3.0	3.10	3.07	3.11	2.72	3.16	3.16	3.15
H_2O+	0.15	0.34	0.1	0.09	0.57	0.09	0.01	0.12	0.17	0.29
H_2O-	0.11	0.44	0.2	0.12	0.07	0.02	0.08	0.06	0.05	0.19
P_2O_5	0.59	0.58	0.3	0.84	0.90	1.06	1.18	1.22	0.52	1.42
Total	100.26	99.97	100.5	100.14	100.32	100.01	100.04	99.94	100.09	99.91

	38.	39.	40.	41.	42.	43.	44.	45.	46.	47.
SiO_2	46.2	47.06	48.54	49.52	53.90	54.04	54.95	54.53	54.76	54.1
TiO_2	3.5	3.44	2.98	3.18	1.77	1.81	1.58	1.62	1.62	1.7
Al_2O_3	18.1	17.14	18.00	17.72	19.00	19.54	19.63	19.35	19.06	20.0
Fe_2O_3	3.1	3.29	3.78	2.55	3.37	2.60	1.62	4.85	3.15	2.0
FeO	6.7	6.65	5.18	5.66	3.05	3.35	3.31	1.20	2.95	3.3
MnO	0.2	0.18	0.18	0.18	0.18	0.19	0.18	0.18	0.18	0.2
MgO	4.6	4.35	3.32	3.42	1.68	1.66	1.42	1.50	1.51	1.8
CaO	9.4	9.00	8.49	7.58	6.25	6.22	5.73	5.76	5.60	5.2
Na_2O	4.7	4.08	4.74	4.94	5.04	5.26	5.89	5.84	5.87	6.1
K_2O	3.3	3.40	3.38	3.88	4.53	4.53	4.95	4.83	4.89	5.6
H_2O+	0.2	0.37	0.14	0.29	0.19	0.11	0.00	0.00	0.03	0.2
H_2O-	Tr.	0.27	0.03	0.15	0.28	0.19	0.01	0.02	0.03	0.3
P_2O_5	0.5	0.75	1.18	1.09	0.74	0.75	0.43	0.38	0.36	0.3
Total	100.5	99.98	99.94	100.16	99.98	100.25	99.96 a	100.22 b	100.17 c	100.8

a) Includes 0.27 Cl, 0.08 F.   b) Includes 0.11 Cl, 0.12 F.   c) Includes 0.13 Cl, 0.10 F.

Table 62   Chemical Analyses of Rocks (BAKER, GASS et al, 1964)

	48.	49.	50.	51.	52.	53.	54.	55.	56.
SiO_2	54.66	58.0	59.6	60.7	48.59	57.31	58.07	62.50	46.4
TiO_2	1.60	1.2	0.5	0.5	2.61	1.59	1.25	0.38	2.6
Al_2O_3	19.91	19.5	19.6	20.5	17.18	17.01	18.01	20.39	17.7
Fe_2O_3	3.07	1.7	2.4	2.3	2.64	6.47	2.92	0.31	3.3
FeO	2.73	2.2	0.1	0.4	5.12	0.28	2.30	0.84	7.2
MnO	0.18	0.1	0.2	0.2	0.17	0.12	0.13	0.09	0.2
MgO	1.10	1.0	0.4	0.2	3.53	2.05	1.15	0.02	5.8
CaO	5.56	3.3	1.3	1.4	8.08	3.71	3.15	1.28	9.3
Na_2O	5.85	6.5	5.7	6.2	4.13	5.38	6.00	5.93	3.9
K_2O	5.03	5.3	6.6	6.7	3.27	5.17	6.02	7.08	2.6
H_2O+	0.00	0.2	2.3	1.0	1.86	0.10	0.28	0.64	0.8
H_2O-	0.00	0.1	1.3	0.4	1.30	0.28	0.38	0.69	0.9
P_2O_5	0.29	0.2	0.05	0.03	0.87	0.52	0.43	0.10	0.2
Total	100.22 d	99.3	100.05	100.53	99.95	99.99	100.09	100.25	100.9

d) Includes 0.23 Cl, 0.10 F.

28. Base, Main Cliffs, E end, Sandy Pt. — Ankaramites
29. 300 yds. W of Caves Gulch. 100 ft. above SL. — Olivine-Basalt
30. East Molly Gulch. 4000 ft. — Trachybasalt
31. Hottentot Gulch. 5400 ft. — Trachybasalt
32. Small cinder cone, just N Big Sandy Gulch. — Trachybasalt
33. W headland, Boat Harbour Bay. — Trachybasalt
34. N end, Stony Beach. 1800 ft. — Trachybasalt
35. 'Pillow Lavas' foreshore, immediately W of 1961 lava. — Trachybasalt
36. Sandy Pt., 100 yds N of East End Gulch. — Trachybasalt (Flow-banded)
37. Pigbite Gulch. 60 ft. — Trachybasalt
38. Inside crater wall, Frank's Hill. — Trachybasalt
39. 100 yds. E of summit crater lake. 6550 ft. — Leucite-bearing Trachybasalt
40. Blineye parasitic cone. — Leucite-bearing Trachybasalt
41. Noisy Beach, 0.75 mile N Lyon Pt. 40 ft. — Leucite-bearing Trachybasalt
42. Stony Hill. SE lava. — Trachyandesite
43. Stony Hill. Prominent spine near summit. — Trachyandesite
44. 1961 eruption. Glassy bomb, 100 yds. W of tholoid. — Trachyandesite
45. 1961 eruption. W flank of tholoid, 40 yds. from summit. — Trachyandesite
46. 1961 eruption. Central flow. — Trachyandesite
47. W side of 'Ridge-where-the-goat-jumped-off'. 2500 ft. — Porphy. Trachyandesite
48. 1961 eruption. NE extremity of lava field. — Trachyandesite
49. SE flank of Peak. — Plagioclase Trachyte
50. 400 yds. W of Settlement Quarry, foot of Main Cliffs. — Phonolitic Alkali Feldspar Trachyte
51. 10 yds. E of Specimen 50. — Alk. Feldspar Trachyte
52. Cutting agglomerate N end of High Ridge, Nightingale Is. — Trachybasalt
53. 200 yds. NE of North Pond. Nightingale Is. — Trachyandesite
54. N end, High Ridge. Nightingale Is. — Plagioclase Trachyte
55. Hardies off Sea Hen Rocks, Nightingale Is. — Alk. Feldspar Trachyte
56. Headland, E of waterfall, Salt Beach, Inaccessible Is. — Trachybasalt

A. Picrite Basalts

Ankaramites. In hand specimen this rock (No. 28) is highly porphyritic with many phenocrysts of black pyroxene and less olivine, together constituting some 40% to 50% of the rock. The groundmass is aphanitic. Microscopically the phenocrysts are seen to be pyroxene, olivine, plagioclase and iron ore. The slightly zoned diopsidic-augite phenocrysts are usually euhedral and inclusions of iron ore are common. Colourless olivine is perfectly fresh, euhedral to subhedral and often has good cleavage. Plagioclase is progressively zoned from An_{85} in the centre to An_{45} at the borders, occurs as complex, twinned subhedral crystals, frequently as aggregates. More or less equidimensional iron ore is abundant. The matrix shows an intergranular texture, comprising essentially plagioclase, pyroxene and iron ore. The plagioclase is present as a felted mass of laths, with much granular iron ore and greenish grains of pyroxene, with olivine in minor quantity. Sometimes abundant needles of apatite are present.

B. Alkali Basalts

These are less common in Tristan than in many other oceanic islands. They are either equigranular or porphyritic, and with increase of olivine and augite phenocrysts, they pass into picrite basalts.

Alkali-Olivine-Basalt (No. 29). To the naked eye this is a greyish rock with a few small vesicles and a few phenocrysts of black pyroxene and red-brown olivine. In thin-section, the olivine phenocrysts are seen to be to euhedral to subhedral, often with good cleavage and no zoning. Iddingsitization is only slight, restricted to cracks and narrow border areas. The pyroxene phenocrysts are purple-brown titaniferous augite, with slight zoning, the rims sometimes packed with granular iron ore inclusions. The groundmass comprises pyroxene feldspar, iron ore, very small amount of amphibole and no olivine. The plagioclase has slight zoning, occurring as complexly twinned laths. In some interstices are areas of isotropic material, likely leucite, often containing apatite needles. Some flakes of brown amphibole and reddish biotite also occur in the matrix.

C. Trachybasalts

These are the predominant rocks of the islands and show considerable range in texture, occurring as flows, dykes, intrusive bodies and scoriaceous material in cinder cones. Many types contain interstitial leucite, and are here named leucite-bearing trachybasalts.

Trachybasalt. No. 30, a compact, greyish, aphanitic rock which, under the microscope is somewhat microporphyritic, having small phenocrysts of brownish titaniferous augite in a groundmass comprising plagioclase pyroxene, iron ore and some olivine. Plagioclase occurs as multiple-twinned laths, showing imperfect flow structure. The pyroxene is of same type as the small phenocrysts, occurring as rounded, prismatic grains, sometimes with iron ore inclusions. Iron ore is abundant as euhedral grains. Euhedral, fresh olivine very scarce, and some alkali feldspar occurs in some interstices. The specimen is taken from a dyke, but No. 31 is from a prominent flow. Megascopically this is a dense, somewhat porphyritic rock with pyroxene and plagioclase phenocrysts set in an aphanitic groundmass. In thin-section the euhedral to subhedral purple-brown pyroxene is likely a titaniferous augite, usually containing inclusions of iron ore. The plagioclase phenocrysts are slightly zoned with a tendency towards corroded outlines and may contain dust-like inclusions of iron ore. The matrix is a felted mass of ill-defined plagioclase laths and abundant granules of iron ore, along with pyroxene and olivine (?). The interstices are filled with alkali feldspar. No. 32, a highly vesicular trachybasalt, is a piece of a scoriaceous pyroclastic block so vesicular that in the centre it has a 'spongy' appearence. To the naked eye it is a dark grey, glassy rock with rare phenocrysts, and in thin-section is identical with the next specimen, No. 33. The latter is a vesicular trachybasalt, from the ropy surface of a lava at the base of the cliff section. The rock is dark grey in colour, aphyric, with many small elongated vesicles. Microscopically it comprises plagioclase, pyroxene, some olivine and iron ore set in interstitial glass. Plagioclase laths show some alignment and flow round the vesicles, being composed of albite, with Carlsbad twinning, the cores being approximately An_{65}. Pyroxene, euhedral to subhedral, shows some evidence of zoning, is unusually free of inclusions, as also is the perfectly fresh small olivine crystals. Iron ore occurs as clusters of somewhat equidimensional grains. Cloudy brown glass occurs interstitially. Specimens Nos. 32, 33 and 35, the last-named a trachybasalt from the 'pillow' lavas just N of the Settlement, are all identical in thin-section, and all came from parts of the Hillpiece parasitic centre. Trachybasalt from a columnar jointed lava hill (No. 34) is compact and aphanitic, with a few small phenocrysts of pyroxene and plagioclase. Microscopically there are seen a few phenocrysts of pyroxene, plagioclase and iron ore. Both pyroxene and plagioclase are anhedral. Resorbed basaltic hornblende crystals are probably xenocrysts. The matrix is intergranular, very fine-grained, comprising plagioclase, pyroxene, iron ore and amphibole. The plagioclase consists of minute laths, iron ore comprises small granules, the pyroxene is rounded and elongated. Pale-brown, pleochroic irregular flakes of amphibole also are present. The spotty appearance of the groundmass is due to the concentration of femic constituents along with a certain amount of isotropic material (leucite?). The flow-banded trachybasalt (No. 36) is an aphanitic rock with prominent 'flow' structures on weathered surfaces. In thin-section are seen a few microphenocrysts of iron ore. The matrix is intergranular, comprising plagioclase, iron ore, pyroxene and apatite. Viewed microscopically, the flow bands become vague. Plagioclase occurs as felted masses of laths, showing some zoning and multiple twinning. Interstitially occurs granular iron ore and pyroxene. Some flakes of slightly pleochroic amphibole can be detected. Hexagonal prisms of turbid apatite are fairly common. No. 37 is a greyish, compact rock with a fair number of small phenocrysts of black pyroxene and plagioclase. Microscopically the pyroxene is seen to be euhedral brownish titaniferous augite. Plagioclase also is euhedral and somewhat zoned. Equidimensional iron ore and small subhedral olivine crystals are also present. The matrix shows an intergranular texture, comprising plagioclase, pyroxene and iron ore. The plagioclase present as laths shows distinct parallel arrangement. Much granular iron ore occurs, also pyroxene between the feldspar laths. The interstices are filled with small amounts of alkali feldspar and leucite. No. 38 is a scoriaceous trachybasalt, a highly vesiculated dark-grey, aphanitic rock, forming part of a pyroclastic block. Under the microscope are seen a few small scattered phenocrysts of titaniferous augite. The matrix is partly crystalline, comprising slender laths of plagioclase, minute granules of iron ore and elongated grains of pyroxene, all set in a purplish-brown glass. Small grains of olivine (?) also occur in the groundmass. No. 39 is from a small plug at the summit crater lake, is massive, slightly porphyritic, with phenocrysts of black amphibole, black pyroxene and plagioclase laths in an aphanitic groundmass. Microscopically the phenocrysts comprise partly resorbed amphibole, pyroxene and plagioclase. The amphibole is similar to basaltic hornblende present in plutonic xenoliths (see below). It has a reaction rim consisting of granular iron ore, feldspar and pyroxene (?). The pyroxene is thought to be a slightly titaniferous augite, often showing simple twinning. The complexly twinned plagioclase is strongly zoned from An_{80} to An_{70}, is subhedral to euhedral and often packed with dust-like iron ore in the outer zones of the crystals. The matrix has an inter-granular texture, comprising plagioclase, pyroxene, iron ore, leucite and alkali feldspar. The plagioclase forms a felted mass of laths which often show simple twinning. Pyroxene and iron ore tend to be granular. In parts of the matrix the interstices are filled with alkali feldspar, often full of small apatite needles: in other areas, leucite fills the interstices, also often full apatite needles. No. 40 comes from a volcanic neck in the Blineye parasitic centre. It is a compact, greyish, aphanitic rock, some phenocrysts of plagioclase and amphibole visible megascopically. In thin-section the rock is slightly 'porphyritic', with 'phenocrysts' of plagioclase, resorbed amphibole, pyroxene, accessory iron ore and apatite. Very likely many of the 'phenocrysts' are actually xenocrysts derived from the plutonic xenoliths. Plagioclase occurs as euhedral crystals, complexly twinned and slightly zoned at margins. The amphibole is all but entirely resorbed and is now represented by iron ore, pyroxene and feldspathic material. It is deduced that the original amphibole was anhedral basaltic hornblende. The pyroxene is anhedral titanaugite, often with dust-like inclusion of iron ore. The matrix has a very fine-grained intergranular texture, consisting of plagioclase, iron ore, pyroxene, leucite and sodalite. Minute flakes of biotite (?) also present in small quantities. The 'spotty' appearance of the groundmass is due to concentrations of femic constituents and leucite. The latter occurs interstitially and also fills small cavities. Sodalite was only positively identified in X-ray powder-photographs. No. 41 is a greyish, aphanitic rock with a few small irregular vesicles. Pale-grey leucite-rich spots cover ca. 10% of the surface, the spots often being arranged in roughly parallel lines, thus giving a flow-banded appearance. In thin-section there occur some microphenocrysts of faintly zoned plagioclase and rare pyroxene. The groundmass has a fine-grained, intergranular texture, comprising plagioclase, pyroxene, iron ore, some olivine and leucite, as also a few minute flakes of amphibole and biotite. Pale-grey spots on the weathered surfaces are seen as rounded areas in the matrix where the interstices are entirely filled with leucite. If these areas become larger, they are often crowded with acicular needles of apatite.

D. Trachyandesites

In the secondary centres these are the chief rock types, also occurring as flows in the main sequence, probably also as dykes. No. 42 is a greyish, finely porphyritic rock with some phenocrysts of feldspar and elongate hornblende. In thin-section shows hyalopilitic and somewhat trachytic texture, with subparallel phenocrysts of plagioclase, pyroxene and amphibole in a microcrystalline to glassy matrix. Phenocrysts of plagioclase display large unzoned cores of labradorite and finely-zoned borders ranging to andesine. Most plagioclase is free of inclusions and alternation products. Pyroxenes have a prismatic habit, show well-defined zoning. It is possible that the core represents soda-augite, the wide rim, titaniferous soda-augite and the thin outermost border, titaniferous aegerine. Amphibole phenocrysts always show some degree of resorption, and is at times completely replaced by ore. Two varieties, both likely basaltic hornblende, were recognized. The hyalopilitic matrix comprises a network of small plagioclase laths, aegirine microlites and ore in a turbid aggregate of alkali feldspar and abundant glass. The tholoid from the same locality, very similar to the above specimen, is analysed in No. 43.

Nos. 44, 45, 46 and 48 are taken from the region of the 1961 eruption. The material of the initial tholoid (No. 45), the central flow (46) and the dome show a uniformity, being greyish, finely-vesicular rocks with occasional stumpy prismatic plagioclase crystals and elongate amphibole in an aphanitic base. In thin-section, the rocks show fluidal texture, with here and there large phenocrysts of plagioclase, clinopyroxene and basaltic hornblende, along with microphenocrysts of plagioclase and pyroxene in a microcrystalline to glassy matrix. The large phenocrysts of plagioclase have basic labradorite cores surrounded by a zone ranging to sodic andesine. Microphenocrysts of plagioclase show similar cores with borders ranging outwards to An_{40}. Large phenocrysts of clinopyroxene sometimes occur, and frequently enclose poikilitically small plagioclase laths and ore granules. The core of these phenocrysts is evidently aegerine-augite, with borders probably of a titanaugite. Strongly coloured amphibole phenocrysts may be basaltic hornblende, but because of titanium present in analyses, also sphene and ilmenite, may be these amphiboles are akin to kaersutite in composition. The abundant ore is predominantly ilmenite. A most interesting constituent, occurring as scattered, discrete grains, is the mineral haüyne. Nos. 47 and 48 are from lava fields. No. 47 is a porphyritic rock with many phenocrysts of white plagioclase, black, elongated amphiboles and black plates of biotite, all set in a greyish aphanitic groundmass. Microscopically the phenocrysts comprise plagioclase, amphibole, biotite, iron ore, some pyroxene and apatite. The plagioclase shows normal zoning, often bordered by a thin layer of alkali feldspar, and frequently occurs as glomerocrysts. The amphibole may be barkevitic hornblende, is euhedral to subhedral, very fresh and almost unzoned. The biotite is strongly pleochroic and shows very slight indication of resorption. Apatite and pyroxene also occur. The matrix comprises a felted mass of alkali feldspar laths. Minute needles of colourless pyroxene and euhedral grains of iron ore are abundant in the matrix. (Note that No. 47 is NOT related to the 1961 eruption.) No. 44 is a volcanic bomb associated with the 1961 tholoid. The bombs in general have highly vesicular, scoriaceous crusts. A microscopic study of one of these (not the above specimen) showed phenocrysts of plagioclase, clinopyroxene and basaltic hornblende in a predominantly glassy matrix. The few large plagioclase crystals show a network of glassy material in the core. Plagioclase microphenocrysts are somewhat zoned. The basaltic hornblende shows no resorbed rim or corona of iron ore. Amphibole seems rather more abundant in bombs than in specimens from the initial tholoid, lava field and dome. Sometimes phenocrysts of aegerine-augite are present. The pyroxene microphenocrysts have a prismatic habit. Ilmenite crystals are common, but no haüyne has been detected. The matrix comprises plagioclase and pyroxene microlites.

E. Trachytes

Usually these rocks occur as intrusives, more rarely as lava flows, Mineralogically they can be divided into two groups: plagioclase trachytes and alkali-feldspar trachytes. In some there is a tendency towards phonolite.

Plagioclase Trachyte. No. 49 is from a volcanic plug. The specimen is compact, porphyritic, having lath-like phenocrysts of plagioclase and black, elongated pyroxenes and amphiboles, set in an aphanitic matrix. Foliation is very clear in the phenocrysts. Microscopically the phenocrysts are seen to incorporate plagioclase, amphibole, pyroxene, iron one and sphene. Plagioclase occurs as complexly twinned laths, with some oscillatory zoning and often a thin border of alkali-feldspar. Basaltic hornblende is euhedral to subhedral, always partly resorbed to an aggregate of granular iron ore, often with apatite inclusions. Aegerine-augite is euhedral to subhedral showing slight zoning and inclusions of iron ore and apatite needles. Euhedral sphene and prismatic apatite also occur. The groundmass comprises blocky laths of alkali-feldspar, pyroxene needles and euhedral grains of iron ore. Foliation can also be observed in the matrix.

Phonolitic Alkali-Feldspar Trachyte. To the naked eye, No. 50 shows many small phenocrysts of nepheline, with one or two phenocrysts of alkali-feldspar. In thin-section the phenocrysts are nepheline, alkali-feldspar and minor sphene. Nepheline is present as euhedral, hexagonal prisms with a yellow alteration product border, having many inclusions of pyroxene. The alkali-feldspar occurs as subhedral laths and there are also some exsolution blebs of a more sodic phase. The matrix is very cloudy, comprising ill-defined alkali-feldspar laths, elongated grains of aegerine-augite and iron ore.

Alkali-Feldspar Trachyte. Megascopically No. 51 is a compact rock of somewhat speckled appearance. In thin-section, phenocrysts are alkali-feldspar, iron ore and sphene. The alkali-feldspar is euhedral to subhedral, occurring as laths and often clusters, showing signs of exsolution to a more sodic phase. Sphene occurs euhedrally. The groundmass has a very fine trachytic texture, consisting of alkali-feldspar, aegerine-augite and iron ore. The pyroxene is oxidized and is present as very ragged crystals frequently poikilitic. Iron ore is scattered throughout the matrix abundantly. A colourless isotropic mineral (sodalite?) often occurs as a coating to the alkali-feldspar laths and also as discrete crystals where sometimes it is associated with small nepheline crystals.

F. Plutonic Xenoliths

BAKER, GASS et al examined some 250 specimens of these rocks from Tristan da Cunha and in general confirmed the observations of CAMPBELL SMITH & DUNNE regarding these rocks. Mineralogically they are relatively simple, comprising essentially plagioclase, clinopyroxene, amphibole and iron ore, in all proportions. As accessories there are apatite, biotite and sphene. Olivine is very rare, hypersthene entirely absent, thus quite unlike the gabbroic xenoliths of Gough and Ascension Island. Plagioclase is usually quite fresh, has a composition in the labradorite-bytownite range, and sometimes is slightly zoned. Clinopyroxene appears similar to titanaugites in the lavas. Often it shows irregular alteration to a brown amphibole. Iron ore sometimes occurs as exsolved rods and blebs. The amphibole is similar in appearance to that seen in many trachybasalts and trachyandesites. DUNNE has named the amphibole kaersutite, which commonly occurs as discrete blade-shaped crystals. In xenoliths found in effusives, the amphibole is usually altered to a granular mixture of ilmenite, pyroxene (referred to by DUNNE as rhoenite) and plagioclase. Frequently such alteration is complete and at such a stage the xenolith is ready to split-up into xenocrysts. Iron ore occurs usually as large irregular masses, more rarely as discrete grains. Apatite might be abundant. Biotite is present as a secondary mineral, with a tendency to be present in xenoliths derived from tuffs and cinder cones.

Specimens 52–55 are from Nightingale Island, and No. 56 is from Inaccessible Island.

Trachybasalt. No. 52 is from a thin basic dyke intruding volcanic agglomerates on High Ridge. The rock is dark-grey, aphanitic, with occasional phenocrysts of pyroxene. Microscopically the rock is fine-grained with flow texture. Plagioclase laths often show only simple twins. Pyroxene is present interstitially as tiny granules, also as rare crystals. Iron ore granules occur in the interstices also, some unidentifiable ferromagnesian material and some residual patches of leucite. Rare xenocrysts of resorbed basaltic hornblende and occasional fragments of trachyte also are present. No. 56, a dyke rock from Inaccessible Island, is classed in the table by Baker, Gass et al as a trachybasalt from the Salt Bay area. They give general petrographic descriptions of ankaramites, olivine-basalt and basalt dykes in this region, but none referring to this actual specimen.

Trachyandesites. No. 53 has a porphyritic appearance, with phenocrysts of white plagioclase, dark-brown amphibole and pyroxene, in a greyish aphanitic groundmass. In thin-section, the phenocrysts are seen to include plagioclase, pyroxene, highly resorbed basaltic hornblende and apatite. The subhedral plagioclase is complexly twinned, slightly zoned and at times has a rim of alkali-feldspar. Pyroxene is zoned from purplish titaniferous augite to greenish aegerine-augite. Basaltic hornblende is frequently completely resorbed. The matrix consists of felted material of ill-defined alkali-feldspar laths, granules of aegerine-augite and deep-red hematite (?).

Trachytes. No. 54 is a compact, porphyritic rock with phenocrysts of black amphibole and pyroxene, and plagioclase, the matrix being bluish and aphanitic. Microscopically the phenocrysts were identified as plagioclase, pyroxene, resorbed basaltic hornblende and apatite. Plagioclase is slightly zoned, is complexly twinned, euhedral to subhedral in form. The euhedral pyroxene is aegerine-augite, a second generation of elongated pyroxene crystals also being present. Basaltic hornblende is usually completely resorbed. Elongated, turbid apatite is present but not common. The groundmass comprises blocky crystals of zoned alkali-feldspar, small grains of euhedral iron ore and minute, rod-like crystals of another pyroxene. No. 55 was from a columnar jointed lava underlying tuff horizons. To the naked eye the rock is coarse-grained, but actually is porphyritic, with many phenocrysts of white feldspar and some black elongated pyroxenes, set in a greyish, crystalline groundmass. In thin-section the phenocrysts comprise alkali-feldspar, aegerine-augite, iron ore, sphene and zircon. The alkali-feldspar is present as tabular crystals, twinning is not common, and there are blebs of an exsolved sodic feldspar present. Iron ore, sphene and zircon occur as accessory phenocrystal minerals. The matrix comprises a mass of blocky and felted lath-like crystals of alkali-feldspar, with granules of aegerine-augite and iron ore.

Norms and Modal Composition

Tables 63 to 67 refer to statistics provided by Dunne. In Table 63 it is seen that in the olivine- and hornblende-alkali-basalts, the normative molecular quantities of potash feldspar vary between 8.5 and 17.3, nepheline between 5.3 and 10.7. The normative nepheline is not ascribed to the presence of large quantities of augite by Dunne but rather to the fact that it is chiefly suppressed in the matrix, mostly in the form of anemousite, lesser so in the form of glass. The ratio of normative amounts of An to Ab varies between 1.4 and 0.6.

Table 63   Norms (Dunne, 1941)

Sample Number	Q	Or	Ab	An	Sod	Ne	Wo	En	Hy	Fo	Fa	Mt	Hm	Ru	Ap
1.	—	8.5	12.0	18.3	—	8.2	16.1	—	—	20.1	7.7	5.2	—	3.1	0.8
2.	—	15.8	22.5	16.8	—	8.8	9.9	—	—	10.3	7.3	3.2	—	3.2	2.2
3.	—	11.8	29.5	18.3	—	5.3	9.6	—	—	12.7	6.0	3.6	—	3.2	1.0
4.	—	17.3	15.8	21.5	—	10.7	12.2	—	—	9.7	6.8	3.2	—	2.8	n.d.
5.	—	16.8	16.1	22.7	—	10.2	10.4	—	—	9.7	8.0	3.5	—	2.6	n.d.
6.	—	15.8	34.0	26.5	—	0.9	2.3	—	—	8.3	4.5	4.6	—	1.4	1.7
7.	.8	15.7	41.8	23.8	—	1.8	0.4	—	—	6.3	—	7.4	0.9	1.9	n.d.
8.	—	10.3	37.7	21.6	—	—	4.8	5.6	—	7.1	5.4	3.9	—	2.3	1.3
9.	—	21.5	28.3	19.3	—	10.5	2.5	—	—	6.5	4.9	2.3	—	2.3	1.9
10.	—	24.2	29.0	9.8	—	7.6	9.3	—	—	7.6	4.0	4.7	—	2.1	1.7
11.	—	22.0	20.2	24.2	—	15.8	3.9	—	—	5.2	6.8	0.2	—	1.7	n.d.
12.	—	23.2	36.5	28.3	—	—	0.1	5.7	—	—	0.6	4.2	—	1.4	n.d.
13.	—	27.8	39.5	10.3	—	4.1	4.0	—	—	5.2	2.8	3.2	—	1.7	1.4
14.	—	27.8	36.5	10.5	—	8.4	6.4	—	—	3.2	1.6	3.2	—	1.5	0.8
15.	—	18.5	49.2	16.0	—	—	1.6	5.6	—	0.1	3.3	3.2	—	1.4	1.1
16.	—	22.5	45.5	14.8	—	—	2.0	3.6	—	3.3	2.4	3.2	—	1.5	1.2
17.	—	26.2	35.5	11.0	1.0	13.0	2.5	—	—	2.7	3.1	1.6	—	1.8	1.6
18.	—	—	—	—	—	—	—	—	—	—	—	—	—	—	—
19.	—	33.8	48.8	5.8	—	2.9	2.5	—	—	2.0	—	3.0	—	0.8	0.4
20.	1.2	29.2	56.2	8.5	—	—	0.1	2.2	0.3	—	—	1.8	—	0.5	—
21.	—	34.5	40.5	6.3	1.0	8.0	4.0	—	—	1.7	0.4	2.3	—	0.9	0.4
22.	1.4	37.0	50.2	2.7	5.0	—	1.1	0.3	—	—	—	1.6	0.1	0.5	0.1
23.	—	35.5	45.5	4.2	5.0	7.4	0.1	—	—	—	—	1.4	0.5	0.3	0.1
24.	—	5.8	26.2	23.0	—	—	9.3	7.5	—	15.3	7.9	2.2	—	2.3	—
25.	—	0.8	2.1	22.0	—	7.4	25.1	—	—	22.1	5.9	11.9	—	3.7	—
26.	—	2.0	7.8	37.5	—	4.4	16.0	—	—	16.2	—	9.2	3.4	3.5	—
27.	—	8.2	13.5	38.2	—	5.2	8.4	—	—	12.7	—	5.6	1.7	2.5	4.0

Table 64

NIGGLI Values of Tristan Samples (DUNNE, 1941)

Sample Number	si	al	fm	c	alk	k	mg	c/fm	t	p	Magma Type
1.	85	15	51	26.5	7.5	0.25	0.56	0.50	6.5	0.3	Ankaratritic-Alk-Issitic
2.	113	22	39	25	14	0.30	0.46	0.65	8	1	Normal Theralite-Gabbroic
3.	116	22.5	41	23.5	13	0.24	0.53	0.57	5.5	0.5	Ditto
4.	115	25	37	25	13	0.34	0.45	0.68	7	–	Ditto
5.	110	24	38.5	25	12.5	0.34	0.42	0.64	6.5	–	Ditto
6.	127	29	36	21	14	0.31	0.42	0.58	4	0.7	Normal Dioritic to Mugearitic
7.	135	32	36	14	18	0.26	0.33	0.38	5.5	–	Ditto
8.	129	25	41	21	13	0.21	0.50	0.50	6	0.6	Ditto
9.	137	31	31	18	20	0.32	0.43	0.58	7	1.2	Normal Essexitic
10.	133	24	35	32	19	0.37	0.41	0.62	6	1	Mugearitic
11.	138	35	24.5	20	20.5	0.32	0.42	0.81	5	–	Rouvillitic
12.	164	38	24	19	19	0.39	0.38	0.79	5	–	Essexitic-Akeritic (?)
13.	167	31.5	28	16	24.5	0.38	0.41	0.57	5.5	1.0	Kassaitic to Nosykombitic
14.	168	33	22	19	26	0.35	0.33	0.90	5	0.5	Kassaitic
15.	175	33	28	16	23	0.27	0.35	0.56	5	0.8	Kassaitic to Nosykombitic
16.	171	32.5	29	16	22.5	0.33	0.45	0.54	5	0.8	Ditto
17.	170	37	19	15	29	0.31	0.33	0.81	6	1.0	Larvikitic-Essexitic-Foyaitic
18.	190	41	16	15	28	0.41	0.37	0.94	3	n.d.	Larvikitic-Leuco-Syenitic
19.	213	39	16.5	10.5	34	0.39	0.31	0.63	3	0.3	Pulaskitic
20.	243	43.5	13	7.5	36	0.34	0.36	0.58	2	0.4	Nordmakitic
21.	200	39	14	13	34	0.38	0.30	0.93	3	0.3	Pulaskitic
22.	252	44	8	5	43	0.38	0.09	0.58	2	0.1	Bostonitic
23.	225	45	8	4	43	0.34	0.00	0.50	1	0.1	Alkali-Syenite-Aplitic
24.	106	18.5	51	23	7.5	0.18	0.65	0.45	5.5	0.3	Normal Gabbroic to Essexite-Gabbroic
25.	64	11	55	31	3	0.06	0.47	0.57	7	0.04	Issitic
26.	71	18.5	47	31	3.5	0.12	0.46	0.66	7	0.01	Ditto
27.	89	24	36	33	7	0.27	0.54	0.90	6	2	Berondritic

Table 65 Average Niggli Values of the Main Groups of Rocks (Dunne, 1941)

Rock Group	si	al	fm	c	alk	k	mg	Magma Type
Pyroxenitic and Hornblenditic Xenoliths	67	15	51	31	3	0.09	0.47	Issitic
Olivine Alkali Basalts	85	15	51	26.5	7.5	0.25	0.56	Ankaratritic
Olivine-hornblende Alkali Basalts	113	23.5	39	24.5	13.0	0.31	0.47	Normal-Theralite-Gabbroic
Trachy-andesite-Basalts, including Mugearites	133	29	34	19.5	17.5	0.30	0.42	Normal Essexitic
Trachy-Andesites	170	33.5	25	16.5	25	0.33	0.37	Kassaitic
Plagioclase-rich Trachytes	206	39	15	12	34	0.38	0.30	Pulaskitic
Trachytes	240	44	10	5	41	0.35	0.15	Nordmarkitic

Table 66 Modal Composition, measured and estimated, in Volume Percentages (Dunne 1941)

	1.	2.	3.	4.	5.	6.	8.	9.
Plagioclase	2P 23G	16P 30G	35	45	45	53	50	5P 40G
Alk. Feldspar Anemousite Anorthoclase Sodalite	15[1]	17	15	10[2]	15[3]	20	16	20
Glass			10					10
Augite	28P 10G	3P 20G	21	30	30	8	11	6
Diop. Augite Aeg. Augite								
Hornblende				5				2
Olivine	8P 2G	5P 2G	9	p	p	12	13	10
Biotite			2			2	1	
Apatite Titanite Zircon	p	1		p	p	1	1	1
Ore	4P 8G	6	8	10	10	4	5	6
Serp. Calcite Chlorite.							3	

In the trachybasalts and trachyandesites of Inaccessible Island (Nos. 6, 7, 8, 15 and 16), olivine is more abundant than augite. Compared to the olivine-alkali-basalts and the other trachyandesite-basalts and trachyandesites, these rocks show much greater saturation and indicate a somewhat special group, with affinities to specific Pacific rocks rather than the pronounced Atlantic character of the others. The normative amount of potash feldspar varies between one-fifth and one-third of the total alkali-feldspar. Normative anorthite constitutes ca. 30 % of the total feldspar in the mugearites, and 20 % in

Table 66   Modal Composition, measured and estimated, in Volume Percentages (DUNNE 1941)

	10.	13.	14.	15.	16.	17.	19.	20.
Plagioclase	2.6P	9.3P	10.5P	4P	2P	8P	6P	8P
	10G	18G	20G	75G	75G	40G[8]		10G
Alk. Feldspar	50[4]	50	40				84	70
Anemousite								
Anorthoclase								4P
Sodalite								
Glass			10			35		
Augite	25	0.4P	3.5P	5[6]	10[7]	1P		
		11G	9G			6G		
Diop. Augite							4	0.5P
								3G
Aeg. Augite								
Hornblende	5.4	5.8	2.5	5	2	3.5	3[10]	3[11]
Olivine	2							
Biotite		0.5						
Apatite			2[5]	1	1	2[9]		
Titanite								
Zircon								
Ore	5	1P	1.5P	5	5	0.5P	3	1.5
		2G	3G			3.5G		
Serp. Calcite Chlorite.				5				

	21.	22.	23.	24.	25.	26.	27.
Plagioclase	10P			45	11.6	32.4	51.2
Alk. Feldspar	75	90	86				
Anemousite							
Anorthoclase				15			
Sodalite	3	5	10				
Glass							
Augite				14	60.7	27.6	4.2
Diop. Augite	1P						
	5G						
Aeg. Augite		3	2.5				
Hornblende					7.9	27.8	24.6
Olivine				23			
Biotite	1			3[13]			14.0
Apatite							4.0
Titanite							1.0
Zircon	1						
Ore	4	2[12]	1.5		19.8	12.2	1.0
Serp. Calcite Chlorite.							

P = Phenocrysts   G = Groundmass   p = Present

[1] Includes Anemousite and Glass.
[2] Includes Glass.
[3] Includes Glass.
[4] Includes Glass, etc.
[5] Includes Sphene and Zircon
[6] Includes Olivine.
[7] Includes Olivine.
[8] Includes Alk. Feldspar.
[9] Includes Sphene and Zircon.
[10] Includes Biotite.
[11] Includes Biotite.
[12] Includes Titanite.
[13] Includes Apatite and Ore.

Table 67 Molecular Base (DUNNE, 1941)

Sample Number	Rock Name	Si	L			
			Kp	Ne	Hl	Cal
25.	Pyroxenitic Xenolith	64	0.5	8.7	–	13.2
26.	Pyr.-Horn.-Plag. Xenolith	71	1.2	9.1	–	22.5
1.	Alk. Olivine-Basalt	85	5.1	15.4	–	11.0
27.	Horn.-Biot.-Plag. Xenolith	89	4.9	13.3	–	22.9
24.	Essexitic Gabbro	106	3.5	15.7	–	13.8
5.	Alkali Basalt	110	10.1	19.9	–	13.6
2.	Doleritic Alk. Ol.-Basalt	113	9.5	22.3	–	10.1
4.	Horn. Alk. Basalt	115	10.4	20.2	–	12.9
3.	Alk. Ol.-Basalt	116	7,1	23.0	–	11.0
6.	Mugearite	127	9.5	21.3	–	15.9
8.	Mugearite	129	6.2	22.6	–	13.0
10.	Trachy-Andesite-Basalt	133	14.5	25.0	–	5.9
7.	Mugearite	135	9.4	27.0	–	14.7
9.	Trachy-Andesite-Basalt	137	12.9	27.5	–	11.6
11.	Trachy-Andesite-Basalt	138	13.2	28.0	–	14.5
12.	Trachy-Andesite-Basalt	164	13.9	21.9	–	17.0
13.	Trachy-Andesite	167	16.7	27.8	–	6.2
14.	Trachy-Andesite	168	16.7	30.3	–	6.3
17.	Trachy-Andesite (Latite)	170	15.7	35.2	0.1	6.6
16.	Trachy-Andesite	171	13.5	27.3	–	8.9
15.	Trachy-Andesite	175	11.1	29.5	–	9.6
18.	Tephritic Trachyte	190	18.7	26.9	–	11.4
21.	Sodalite-Plag.-Trachyte	200	20.7	33.2	0.1	3.8
19.	Biotite-Trachyte	213	20.3	32.2	–	3.5
23.	Sodalite-Trachyte	225	21.3	39.2	0.5	2.5
20.	Biotite-Trachyte	243	17.5	33.7	–	5.1
22.	Sodalite-Trachyte	252	22.2	34.6	0.5	1.6

the trachyandesites. Compared to other basalts, these rocks have a much smaller quantity of Wo in proportion to other melanocratic molecules.

In the remaining trachyandesite-basalts (Nos. 9–12), the Or content is greater than in the alkali-basalts and mugearites, about the same as in the saturated trachyandesites, but less than in the other trachyandesites. The An content (except No. 10) is much higher than occurs in the trachyandesites, and is more comparable with that found in the alkali-basalts and mugearites. (As remarked earlier, DUNNE believed that the analyses given by CAMPBELL SMITH for Nos. 11 and 12 were unreliable, with too high an Ne value for No. 11, and that the SiO_2 and Al_2O_3 content was over-estimated in No. 12.) The above trachyandesite-basalts lie between trachybasalts and trachyandesites, having the more basic plagioclase of the former and the total dark component percentage of the latter, or vice-versa. The norms, however, justify the classification as trachyandesite-basalts.

As distinct from the above rocks, three of the trachyandesites Nos. 13, 14 and 17, are under-saturated, with a normative amount of nepheline and feldspathoid varying between 4.1 and 13.0. Compared to the trachyandesite-basalt, they have more Or and Ab but

	M				Q		k	mg	π	λ
Cs	Fs	Fa	Fo	Cp	Ru	Q				
18.8	11.9	11.8	21.1	–	3.7	10.3	0,06	0.47	0.59	0.30
12.0	14.3	4.3	16.2	–	3.5	16.4	0.12	0.46	0.69	0.25
12.1	5.2	10.3	20.1	0.8	3.1	16.9	0.25	0.56	0.35	0.25
6.3	8.2	2.8	12.7	4.0	2.5	22.4	0.27	0.54	0.56	0.19
7.0	2.2	9.0	20.9	0.5	2.3	25.1	0.18	0.65	0.42	0.18
7.8	3.5	9.7	9.7	–	2.6	23.1	0.34	0.42	0.31	0.25
7.4	3.2	8.9	10.3	2.2	3.2	22.9	0.30	0.46	0.24	0.23
8.1	3.2	8.4	9.7	–	2.8	24.3	0.34	0.45	0.30	0.28
7.2	3.6	7.8	12.7	1.0	2.2	24.4	0.24	0.53	0.27	0.22
1.7	4.6	6.8	8.3	1.7	1.4	28.8	0.31	0.42	0.34	0.08
3.6	3.9	7.3	11.3	1.3	2.3	28.5	0.21	0.50	0.31	0.13
7.0	3.7	6.4	7.6	1.7	2.1	25.1	0.37	0.41	0.13	0.26
–	8.7	3.8	6.3	–	1.9	28.2	0.26	0.33	0.28	–
1.9	2.3	6.2	6.5	1.9	2.3	26.9	0.32	0.43	0.22	0.10
2.9	0.1	6.9	5.2	–	1.7	27.5	0.32	0.42	0.26	0.19
0.1	4.2	2.7	4.3	–	1.4	34.5	0.39	0.38	0.32	0.01
3.0	3.2	4.4	5.2	1.4	1.7	30.4	0.38	0.41	0.12	0.17
4.8	3.3	3.2	3.2	0.8	1.5	29.9	0.35	0.33	0.12	0.31
1.9	1.6	3.9	2.7	1.6	1.8	28.9	0.31	0.33	0.12	0.33
1.5	3.2	4.0	6.0	1.2	1.5	32.9	0.33	0.45	0.18	0.10
1.2	3.2	4.9	4.3	1.1	1.4	33.7	0.27	0.35	0.19	0.08
0.5	1.8	2.5	2.5	–	0.8	34.9	0.41	0.37	0.20	0.07
3.0	2.3	1.6	1.7	0.4	0.9	32.3	0.38	0.30	0.06	0.34
1.9	3.0	1.5	2.0	0.4	0.8	34.4	0.39	0.31	0.06	0.23
0.1	2.2	0.7	0.0	0.1	0.3	33.1	0.34	0.00	0.04	0.03
0.1	1.8	1.1	1.6	–	0.5	38.6	0.34	0.36	0.09	0.02
0.8	1.8	0.8	0.1	0.1	0.5	36.9	0.38	0.09	0.03	0.22

much less An and melanocratic molecules. The Or, Ab and An values in these three specimens are about equal, and only the Ne of the leucocratic molecules undergoes change.

Compared to the trachyandesites, the trachytes (Nos. 19–23) show more Or and Ab but less An and melanocratic normative molecules. These rocks are more leucocratic, having less plagioclase but more alkali-feldspar. They consist chiefly of alkali-feldspar (70–85%) of which the normative potash feldspar content varies between 40% and 50%. Nos. 19, 20 and 22 are almost saturated, whereas Nos. 21 and 23 are distinctly under-saturated. Melanocratic molecules vary between 2.5% and 8.5%, but it must be added that quite a proportion of the leucocratic normative minerals are present in the dark components, e. g. the Or in biotite, the Na_2O of Ab in aegerine, etc. The Wo content must of course be treated with due caution in such rocks which contain large percentages of alkali and lime alumino-silicates. Specimen No. 20 contains 1.2 normative quartz, shown microscopically as small quantities of cristobalite. No. 22 yields 1.4 quartz, if the total possible sodalite is built, but it is evident that if the Cl_2 was incorrectly determined, then most of the 5 % sodalite should be converted to Ab. This Q excess may be due either to an error in analysis or then the sodalite, and perhaps the feldspar molecules too, are more silicified than the theoretical ones.

The essexitic-gabbro (No. 24) is much more saturated than any of the basalts, with no normative Ne, and as much as 7.5 En. As regards saturation, it is like the trachyandesite-basalts of Inaccessible Island. Its *mg* content of 0.65 (Table 64) is higher than any other rock discussed here from the Tristan group. Its *k* content of 0.18 is less than any others, with the exception of the coarse-grained inclusions.

Of the three analyses of xenoliths (Nos. 25–27), all show a small percent of normative nepheline, all of which occurs in the dark components, especially hornblende, and the Or, Ab and An also is attributed to melanocratic material. Hornblenditization is evidenced in the state of oxidation, for the Fe of the hornblende is almost entirely in the ferric state. Nos. 26 and 27 therefore only show Mt and Hm and no Fa in the norms.

Table 68   Norms (BAKER, GASS et al, 1964)

Sample Number	Q	Or	Ab	An	Ne	Di	Ol	Wo	Mt
28	–	8.68	6.73	17.95	7.17	32.18	10.76	–	8.09
29	–	12.06	6.95	20.16	8.93	28.01	5.48	–	8.47
30	–	17.73	6.03	22.23	15.99	20.80	6.84	–	3.63
31	–	18.32	11.58	18.64	11.88	20.12	5.11	–	5.41
32	–	18.15	16.16	19.39	8.66	17.25	6.12	–	5.97
33	–	18.38	13.27	19.15	11.33	17.24	7.21	–	5.49
34	–	16.08	21.63	21.13	5.43	13.73	4.03	–	11.03
35	–	18.68	13.63	19.22	10.99	15.71	9.09	–	3.76
36	–	18.68	11.88	20.18	13.13	19.16	2.53	–	6.57
37	–	18.62	17.92	17.06	8.49	16.51	5.51	–	5.31
38	–	19.51	8.70	18.55	16.83	19.95	4.47	–	4.49
39	–	20.10	15.45	18.42	10.33	17.15	4.85	–	4.77
40	–	19.98	22.80	17.86	9.37	13.15	2.73	–	5.48
41	–	22.93	22.48	14.72	10.46	12.68	4.17	–	3.70
42	–	26.78	33.37	15.85	5.02	8.09	0.45	–	4.89
43	–	26.78	32.25	16.33	6.64	7.71	1.30	–	3.77
44	–	29.27	29.99	13.57	9.66	9.60	0.94	–	2.34
45	–	28.55	31.61	12.68	9.26	8.03	–	0.88	–
46	–	28.88	31.67	11.74	9.23	8.96	–	0.77	4.56
47	–	33.10	22.20	10.66	15.94	10.64	0.96	–	2.90
48	–	29.72	31.14	14.10	9.03	6.22	–	1.29	4.45
49	–	31.33	40.12	8.38	8.06	5.36	0.55	–	2.46
50[1]	–	39.01	44.39	6.12	2.08	–	0.70	–	–
51[2]	–	39.60	43.57	6.75	4.82	–	0.35	–	0.49
52	–	19.33	24.79	20.32	5.50	11.31	4.74	–	3.83
53[3]	–	30.55	45.52	7.00	–	3.13	2.56	–	–
54	–	35.58	39.17	4.44	6.28	6.18	–	0.19	4.21
55[4]	–	41.85	46.34	5.70	2.08	–	0.67	–	0.44
56	–	15.37	13.19	23.12	10.73	17.54	9.07	–	4.78

Table 68 Norms (BAKER, GASS et al, 1964)

Sample Number	Il	Hm	Ap	Fr	Hl	H$_2$O+	H$_2$O−	Pf	X
28	7.08	–	1.39	–	–	0.13	0.11	–	12.3
29	7.81	–	1.37	–	–	0.32	0.44	–	16.3
30	6.27	–	0.71	–	–	0.09	0.20	–	35.0
31	6.93	–	1.98	–	–	0.05	0.12	–	22.2
32	5.89	–	2.12	–	–	0.53	0.07	–	25.5
33	5.37	–	2.50	–	–	0.05	0.02	–	28.8
34	4.16	–	2.78	–	–	0.04	0.08	–	1.8
35	5.85	–	2.88	–	–	0.07	0.06	–	35.2
36	6.53	–	1.23	–	–	0.15	0.05	–	14.8
37	6.72	–	3.35	–	–	0.23	0.19	–	22.0
38	6.65	–	1.18	–	–	0.18	–	–	22.4
39	6.53	–	1.77	–	–	0.34	0.27	–	22.6
40	5.65	–	2.78	–	–	0.09	0.03	–	14.2
41	6.04	–	2.51	–	–	0.24	0.15	–	23.1
42	3.36	–	1.75	–	–	0.16	0.28	–	4.0
43	3.44	–	1.77	–	–	0.08	0.19	–	20.1
44	3.00	–	1.01	0.09	0.15	0.02	–	–	34.7
45	2.91	4.85	0.91	0.17	0.18	0.02	0.02	0.12	0
46	3.08	–	0.84	0.14	0.21	0.03	0.03	–	8.8
47	3.23	–	0.71	–	–	0.19	0.30	–	25.1
48	3.03	–	0.67	0.16	0.37	0.00	0.00	–	4.5
49	2.28	–	0.47	–	–	0.19	0.10	–	20.4
50[1]	0.64	2.40	0.12	–	–	2.30	1.30	–	0
51[2]	0.95	1.96	0.07	–	–	1.00	0.40	–	0
52	4.96	–	2.05	–	–	1.82	1.30	–	21.8
53[3]	0.85	6.47	1.23	–	–	0.08	1.28	1.11	0
54	2.37	0.01	1.01	–	–	0.26	0.38	–	0
55[4]	0.72	–	0.24	–	–	0.64	0.69	–	92.7
56	4.94	–	0.47	–	–	0.79	0.90	–	25.7

[1] Includes 0.84 C., 0.16 Ru.
[2] Includes 0.58 C.
[3] Includes 1.21 Tu.
[4] Includes 0.89 C.

Column "X" = $\dfrac{\text{FeO}}{\text{MgO} + \text{FeO}}$ % in Di, Ol, Hy.

Table 68 shows the norms of the rocks studied by BAKER, GASS et al.

Table 69 (p. 266–269) gives spectrographic analyses of trace element content in parts per million, after BAKER, GASS et al. In the plotting of diagrams given by the above authors of the element concentrations, the implication was made that their samples formed part of a differentiation sequence – see later.

Additional Comments

BAKER, GASS et al considered the presence of leucite and the significance of the plutonic xenoliths worthy of further remarks, and the following are their generalizations.

Table 69   Spectrographic Analyses of Trace Element content in parts per million
(BAKER, GASS et al, 1964)

	28	29	30	31	32	33	34	35
Si	201	198	213	214	217	215	215	215
Ti	22	25	20	22	19	17	13	17
Al	64	75	97	88	88	90	89	90
Fe^{3+}	39	41	17	26	29	26	53	18
Fe^{2+}	64	66	65	57	57	58	42	65
Mn	1	–	–	–	–	–	–	–
Mg	62	40	28	29	28	29	29	28
Ca	90	85	71	71	67	68	67	67
Na	18	21	31	29	28	30	28	30
K	12	17	25	26	25	26	23	26
Fe (total)	103	107	82	83	86	84	95	83
Differentiation Index	22.6	27.9	39.7	41.8	43.0	43.0	43.1	43.3
Nb	20	35	100	110	120	130	95	130
Mo	<3	3	4	5	6	6	5	7
Zr	100	200	300	300	350	350	300	350
Ga	25	27	27	28	28	28	27	35
Cr	250	65	–	–	30	45	18	30
V	400	400	200	400	190	180	200	200
Y	10	15	50	40	60	55	40	60
La	<100	110	190	200	200	200	180	250
Be	–	–	–	–	–	–	–	–
Ni	150	50	–	–	–	10	–	10
Co	50	40	18	25	20	20	20	25
Mn	550	1100	1600	1600	1800	1800	1600	1700
Sr	700	1000	1200	1600	800	1100	1100	900
Pb	<10	10	18	11	12	18	18	21
Ba	700	750	1000	1200	800	1000	950	850
Li	<4	4	10	7	4	4	<4	6
Rb	90	110	300	170	180	170	110	170

a) Significance of leucite

No previous investigators of the Tristan rocks had observed leucite, although SCHWARZ (1905) no doubt was near the mark when he described greyish rocks with lighter rounded spots, but thought this was clear, residual glass.

Leucite usually forms early as euhedral phenocrysts or then a euhedral matrix, but in Tristan it is present interstitially as a late-stage product most frequently. Almost never is leucite present interstitially in rocks throughout the world.

Leucite occurs in rocks from the Cape Verde Islands and also from Kerguelan Island. However in the former there is a structural relation with Africa, whereas in the latter, geological evidence suggests an affinitiy with the Antarctic continent.

The Tristan leucite-bearing rocks show no other outstanding differences from suites in other oceanic islands. Continental leucite-bearing rocks show much more K_2O than Na_2O, but the converse is the case in Tristan. Nepheline is almost absent in these rocks,

Table 69. Spectrographic Analyses of Trace Element content in parts per million (BAKER, GASS et al, 1964)

	36	37	38	39	40	41	42	43
Si	215	217	216	220	227	231	252	253
Ti	21	22	21	21	18	19	11	11
Al	94	85	96	92	95	94	101	103
Fe^{3+}	32	26	22	21	26	18	24	18
Fe^{2+}	48	54	52	52	40	44	24	26
Mn	–	–	–	–	–	–	–	–
Mg	25	28	28	26	20	21	10	10
Ca	69	68	67	64	61	54	45	44
Na	32	29	35	30	35	37	37	39
K	26	26	27	28	28	32	38	38
Fe (total)	80	80	74	73	66	62	48	44
Differentiation Index	43.7	45.0	45.0	45.9	52.2	55.9	65.2	65.7
Nb	85	110	100	80	160	120	160	160
Mo	5	6	4	4	9	5	5	4
Zr	300	300	300	300	400	350	350	400
Ga	27	27	28	27	29	28	27	28
Cr	–	–	–	–	–	17	–	12
V	200	300	170	280	250	200	130	130
Y	30	40	45	25	50	40	45	50
La	170	200	170	160	250	180	250	250
Be	–	–	–	–	–	–	–	–
Ni	–	–	10	–	–	–	–	–
Co	14	20	18	15	14	10	–	–
Mn	1500	1500	1500	1300	1800	1400	1700	1700
Sr	1100	1400	1100	1100	1100	1500	1200	1300
Pb	15	8	35	10	16	10	17	15
Ba	950	1000	950	1000	950	1200	1100	1100
Li	10	7	6	6	<4	10	12	11
Rb	110	160	180	110	220	190	200	200

an interesting point, for in the majority of under-saturated oceanic lavas, either nepheline or analcime are the first under-saturated minerals to be produced.

The above authors speculate on two possible causes for the presence of leucite in these Tristan rocks: the desilicating effect of the resorption of the under-saturated basaltic amphibole, and the high volatile content of the lavas which might have aided to suppress leucite during early stages of crystallization.

b) Plutonic Xenoliths

These may be of accidental origin, they might be parts of the source rock from which the containing magma was derived or then they may have crystallized under plutonic conditions from the same magma as the extrusives in which they occur.

It is thought that amphibole, pyroxene, plagioclase and ore formed the stable mineral association at depth. Such xenoliths in lavas display amphibole converted to pyroxene,

Table 69 Spectrographic Analyses of Trace Element content in parts per million
(BAKER, GASS et al, 1964)

	44	45	46	47	48	49	50	51
Si	257	255	256	253	256	271	279	284
Ti	9	10	10	10	10	7	3	3
Al	104	102	101	106	105	103	104	108
Fe^{3+}	11	34	22	14	21	12	17	16
Fe^{2+}	26	9	23	26	21	17	16	3
Mn	–	–	–	–	–	–	–	–
Mg	9	9	9	11	7	6	2	1
Ca	41	41	40	37	40	24	9	10
Na	44	45	45	45	45	48	42	46
K	42	41	41	46	43	44	55	56
Fe (total)	37	43	45	40	42	29	33	19
Differentiation Index	70.1	70.4	70.7	71.2	71.5	79.5	85.5	88.0
Nb	140	170	150	130	140	130	160	230
Mo	6	6	6	7	6	7	<3	4
Zr	350	350	350	300	350	350	500	500
Ga	29	27	28	27	26	27	29	26
Cr	–	–	–	–	–	–	–	–
V	95	100	120	75	110	50	16	20
Y	40	55	45	45	45	35	20	40
La	200	250	200	200	200	190	120	250
Be	–	–	–	–	–	–	8	13
Ni	–	–	–	–	–	–	–	–
Co	–	–	–	–	–	–	–	–
Mn	1500	1800	1500	1300	1600	1200	1500	1900
Sr	1300	1400	1500	1000	1400	650	40	55
Pb	14	16	17	22	14	28	24	25
Ba	1200	1300	1400	1100	1300	1000	20	25
Li	10	12	13	12	13	15	15	20
Rb	230	220	210	270	260	350	400	280

whilst those from pyroclastic centres, sometimes the converse can be seen. It is possible for amphibole and pyroxene to co-exist at depth provided there is lack of water to provide all the hydroxyl, as also to provide necessary pressure conditions for the stability of amphibole. The change-over from pyroxene to amphibole as a magma slowly rises with constant water pressure can be favoured by any of three factors: 1. where there is equal water pressure at all levels, the content of water will be greatest at the summit, and so if the equilibrium or equipotential factors are maintained, water content will increase in the rising column, decrease in the reservoir beneath, and such an increase in water content would allow of further conversion of pyroxene to amphibole. 2. where the water pressure remains constant but there is a decrease of total pressure, then the magma could migrate from an area of pyroxene stability to amphibole stability. 3. heat loss by conduction could cause the magma to migrate into an amphibole stable area.

Table 69   Spectrographic Analyses of Trace Element content in parts per million
(BAKER, GASS et al, 1964)

	52	53	54	55	56
Si	227	268	271	291	217
Ti	16	10	7	2	16
Al	94	90	95	108	94
Fe^{3+}	18	45	20	2	23
Fe^{2+}	40	2	18	7	56
Mn	–	–	–	–	–
Mg	21	12	7	0	35
Ca	58	27	23	9	66
Na	31	40	45	44	29
K	27	43	50	59	22
Fe (total)	58	47	38	9	79
Differentiation Index	49.6	76.1	81.0	90.3	39.3
Nb	130	140	140	130	110
Mo	5	6	10	3	5
Zr	350	500	550	550	300
Ga	29	35	35	30	24
Cr	–	17	16		20
V	200	75	75	10	200
Y	40	35	15	10	50
La	200	180	<100	100	180
Be	–	–	–	–	–
Ni	–	45	–	–	80
Co	12	10	–	–	–
Mn	1500	1200	1000	650	1600
Sr	1000	650	750	180	800
Pb	16	14	17	18	32
Ba	950	700	700	160	850
Li	8	8	15	9	8
Rb	150	220	270	250	170

Where the magma is less hydrous but displaying effusive features, the amphibole would be less inclined to convert into pyroxene. If such a fluid lava erupts it might be that the amphiboles in the xenoliths are partically converted into pyroxene – a stable phase under surface conditions where the water pressure is negligible.

Magma Differentiation and Genesis

DUNNE believed that the Tristan rocks were good examples of magmatic differentiation by means of fractional crystallization, which was controlled by the settling of crystals under gravitational influence.

Studies of the majority of the volcanics of Atlantic association show only basaltic rocks – ca. 99 % of the total bulk – which, though showing great variation in themselves, are yet quite different chemically from lesser volcanic masses composed of trachytic and phonolitic rocks. The usual absence of true intermediate rock types or, if such are present, their specific grouping, does not find a ready explanation. DUNNE suggested that the answer lay in the speed of crystallization, which conceivably could vary at different stages of the process, and may indeed even be rhythmetical in nature.

The rocks of the Tristan group were all named alkaline by DUNNE, even those basalts of distinctly femic character. These volcanics showed no evidence of a primary 'normal olivine-basalt' such as interpreted by DALY, which latter may be taken as representative of the solid phase of the 'primary magma'. It was concluded therefore that the trachytic and intermediate lavas were differentiates of an alkaline basalt which itself of course could very well be a product of differentiation. DUNNE was of the opinion that many of the so-called plagioclase trachytes, trachyandesites, etc. were not 'true and direct' differentiation products, but that they represented rocks of auto-assimilation or auto-hybridization during the magmatic evolution. Thus they are are to be considered rather as trachytic melts into which have been introduced phenocrysts which crystallized directly from a basaltic magma. This trachytic magma must have been of high viscosity, witnessed by the tendency to form domes and flows of only small length, and was extruded during the last magmatic phase.

DUNNE summarizes the process and mechanism of differentiation with regard to the Tristan rocks as follows: 1. Differentiation occurred principally upon the withdrawal of augite, plagioclase and, to a lesser degree, of hornblende and olivine. The role of olivine in the process is of small significance, picritic lavas or peridotitic inclusions being absent. 2. In the development of the trachytic melts, diffusion on a large scale was important, being due to the dissolved volatiles. 3. The principal differentiation products were basalts and trachytes, but rocks intermediate between these groups also are present. 4. Many so-called intermediate rocks are really assimilation products of the basaltic phenocrysts by the trachytic melt. 5. The general sequence of eruptions was as follows:

Olivine-alkali-basalt	Trachyandesite-basalt
Hornblende-alkali-basalt	Trachyandesite
Trachybasalt	Biotite-trachyte
Mugearite	Sodalite-trachyte

6. This order is the same as that of the trend of differentiation.

Table 70, from BAKER, GASS et al gives the chemical composition of four average rock types and their norms, reference only to rocks from Tristan da Cunha. It is seen therein that SiO_2, Al_2O_3, Na_2O and K_2O increase through the series, whereas the other oxides decrease, a variation reflected in the norm minerals. Here there is an increase in Or, Ab and Ab + Ne from the basic to the acid end, whilst An shows decreases overall. Di and Ol decrease rapidly, the latter being absent in the average trachyandesite and trachyte. Il and Mt decrease steadily. All such changes are also found in many other suites of volcanics.

If it is assumed that xenoliths are obtained directly from crystallization in a differentiating magma, such low silica material crystallizing and separating at depth would inhibit further under-saturation in silica within the residual liquid and might be the reason why final members are trachytic instead of phonolitic. Had the magma been less hydrous one would look for greater under-saturation in the end members. It could also

Table 70   Composition of Average Rock Types (BAKER, GASS et al, 1964)

Weight %	1.	2.	3.	4.
SiO_2	43.1	46.7	54.9	60.0
TiO_2	4.1	3.6	1.8	0.9
Al_2O_3	13.1	17.3	19.6	20.2
Fe_2O_3	5.5	3.8	2.8	2.1
FeO	8.5	7.1	2.9	1.1
MgO	9.0	4.7	1.5	0.5
CaO	12.4	9.7	5.7	2.3
Na_2O	2.7	4.1	5.9	6.8
K_2O	1.6	3.0	4.9	6.1
Norm				
Or	9.5	17.7	29.0	36.0
Ab	4.6	11.2	29.8	41.1
Ne	9.9	12.7	10.9	8.9
An	18.9	20.0	12.5	6.6
Di	33.8	22.6	8.1	2.7
Ol	7.6	3.5	–	–
Wo	–	–	2.2	0.6
Il	7.8	6.8	3.4	1.7
Mt	8.0	5.5	4.1	0.9
Hm	–	–	–	1.4
$\dfrac{FeO \times 100}{FeO + MgO}$ (In femics)	12.7	20.4	0.7	0
Differentiation Index	24.0	41.6	59.7	76.0

1. Average Alkali Basalt (3 analyses)
2. Average Trachybasalt (10 analyses)
3. Average Trachyandesite (9 analyses)
4. Average Trachyte (4 analyses)

be argued that if the xenoliths came from pre-existing material, then their assimilation into a 'normal magma' would give rise to an under-saturated magma such as that of Tristan.

Trachytes and trachyandesites are repeated at various times in the eruptive history of Tristan. If the former are differentiation end-products, then this process has been repeated with new mother-material being added periodically. BAKER, GASS et al believed that such repetition in the volcanic sequence implied new material being introduced into the magmatic cycle on each occasion.

The rocks of Tristan da Cunha show higher normative nepheline than those from the other islands. The average content is as shown:

	% Normative Nepheline		
	Tristan	Nightingale	Inaccessible
Trachybasalt	10.0	6.6	5.7
Trachyandesite	10.4	2.1	0
Trachyte	6.7	3.4	0

The Tristan rocks display distinct differences in degree of saturation from those of the other two islands. As Tristan is younger than the other two, the composition difference amongst the islands is due, no doubt, to the time factor. But whether such differences occur in the original parent magmas or then due to differentiation processes, cannot be stated. We would note, however, that some Nightingale and Inaccessible basalts are quite as under-saturated as are Tristan basalts, and this fact, along with the identity in trace element compositions of the three islands presume it likely that the parental material has been equally under-saturated and very similar. The difference lies in the contention that the process of differentiation giving rise to the Nightingale and Inaccessible rocks was different from the later differentiation processes in Tristan in that more saturated derivatives were produced.

Gass (1965) remarked that although the rocks of the Tristan group belonged to the alkali-basalt type of magma, there was a difference from most other oceanic islands in the relatively large content of potassium, rubidium, strontium, barium and niobium. He believed that high-level magma chambers were present, now or earlier, beneath the three principal islands of the group, that fractionation of a parental trachybasaltic magma in these chambers gave rise to ankaramites, olivine-basalts, trachyandesites and trachytes. This trachybasaltic magma was thought to have had its source in the upper mantle. The means whereby the highly potassic rock assemblages were produced included: 1. partial fusion of abnormal mantle material, 2. primary differentiation of the magma in the mantle in an intermediate pressure environment, 3. enrichment of the magma within the mantle of those elements typical of residual liquids by means of a mechanism similar to zone-refining, and 4. very slight melting of mantle material. The absence of a known sialic crust prohibits assimilation of crustal rocks as providing the potassium and trace elements. Baker, Gass et al conclude by saying: "No matter how the parental magma has acquired its unusual characteristics, it is considered to have been tapped-off at intervals and admitted to the region of normal differentiation, in which trachyandesites and trachyte have formed". The rocks of the Tristan group therefore would appear to indicate a two-stage process: firstly, the basalt becomes highly concentrated in residual elements typical of the island group, and secondly, trachyandesites and trachytes form as the result of normal differentiation processes.

Provincial Relationship

Physiography

Gough and Ascension Islands are single volcanic peaks, St. Helena has the volcanic pile of the island itself and an accompanying submarine peak some 100 km to the W thereof, and Tristan has three volcanic peaks. Other comparative matters are shown in Table 71.

Vulcanology

There is no evidence of volcanic activity in historical times in Gough, St. Helena or Ascension, although the first-mentioned erupted some 2300 years ago (U. Hafsten, 1960). It should also be noted that Daly (1925, vd. Ascension references) remarked upon the young appearance of volcanic features on that island.

Table 71   Physiographic Relationships (Based on BAKER, GASS et al, 1964)

	Gough	Tristan	St. Helena	Ascension
No. of islands in group	1	3	1 (+1)	1
Approx. area (km^2)	65	108	118	98
Max. height. (m)	907	2062	823	859
Height above sea-floor (m)	3960	6095	5180	2740
Distance from supposed Mid-Atlantic Rift (km)	560 E.	480 E.	768 E.	112 W.
Distance from axis of Mid-Atlantic Ridge (km)	336 E.	368 E.	800 E.	144 W.

However, characteristic of the four islands referred to here, is the periodic eruption of trachytic lavas within basaltic eruptions. Gough and Tristan have abundant dykes, with a marked radial pattern; Ascension has very few dykes. Parasitic cinder cones are common features.

Table 72   Comparative Vulcanology (BAKER, GASS et al, 1964)

	Gough	Tristan*	St. Helena	Ascension
Recorded historical volcanic activity	None	Recent	None	None
Eruptive sequence	(T), B, B, B, T, B	No apparent rhythmic sequence	B, T	B, T, B
Volcanic form	Complex mass	Single cone with parasitic centres	Complex mass possibly a doublet: fissure eruptions?	Irregular shallow cone with parasitic cemtres
Erosional state	Deeply dissected	Very little erosion except for sea cliffs	Deeply dissected	Very little erosion
Dyke characteristics	Radial dyke patteren; swarms in lower horizons	Marked radial dyke pattern	Multiple linear dyke swarms; especially in pyroclastic compleyes	Very few

* Not including Nightingale and Inaccessible. B = Basalt phases. T = Trachytic phase.

Petrography

Table 73 shows petrographic relations between Tristan, Gough, St. Helena and Ascension. The rocks of these island belong to the Atlantic or soda-alkaline suite.

All types of rocks, from basalt to trachyte, are well represented in these islands, and from those of the latter composition, late differentiates tend towards either phonolites or then rhyolites. Basaltic rocks are more abundant than trachytic, with the exceptions of Gough and Nightingale Islands.

All except St. Helena have xenolithic material, the commonest types being gabbroic and peridotitic; dioritic, syenitic and granitic types are restricted to Ascension. In both

Table 73   Comparative Petrography (BAKER, GASS et al, 1964)

	Gough	Tristan	St. Helena	Ascension
Volcanic Rocks				
Ankaramites, etc.	X ↑	X ↑	X ↑	–
Alkali Basalts	X	X O	X	X ↑
Trachybasalts	X O	X ↓ ↑	X O	X O
Trachyandesites	X	X A	x	X ↓
Trachytes	X	X ↓	X ↑ A	X ↑
Phonolites	x ↓	X	X ↓ ↓	– A
Rhyolites	–	–	–	X
Obsidians	–	–	–	X ↓
Approx. % of basaltic rocks	45	95 Tr 5 Ni 90 In	98	88
Xenoliths				
Peridotitic	x	–	–	X
Gabbroic	X	X	–	X
Dioritic	–	–	–	X
Syenitic	–	–	–	X
Granitic	–	–	–	X
Charakteristic Minerals of Gabbroic Types				
Olivine	X	x	–	X
Hypersthene	X	–	–	X
Amphibole	–	X	–	X

X = Occurrence of rock type or mineral. x = Rare occurrence of rock type or mineral.
O = Crystallization range of Olivine. A = Crystallization range of Amphibole.
Tr = Tristan. Ni = Nightingale. In = Inaccessible.

Gough and Ascension, gabbroic types feature hypersthene, a mineral which does not crystallize from alkali magmas normally, but can do so if at depth and under pressure.

The Tristan rocks show close similarity to most other soda-alkaline provinces, and such minor differences as occur are due to such factors as likely slight differences in the primary magmas, and the method which differentiation pursued. It results that some minerals and rock types may be more or less prolific, perhaps even absent, in the various provinces. But, as DUNNE remarked: " ... the really remarkable similarities met with in the soda-alkaline volcanic rock associations are probably the best proof that they must all have been derived from a similar primary source by one and the same uniform process, which can only be that of magmatic differentiation controlled mainly by fractional crystallization".

Chemistry

Chemical comparisons between the above four islands are shown in Table 74 (p. 276) and Fig. 93, both of which are based upon averages of respective analyses.

A study of the CIPW norms brings out the chemical differences. Regarding the two Tristan averages, for a given SiO_2 content, the high alkalies are reflected in the large amounts of nepheline in the normatives, these two analyses lying well within the alkali-

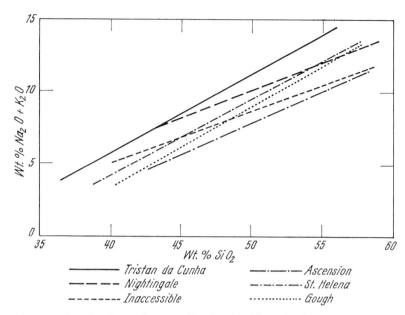

Fig. 93. Comparative chemistry of some Mid-Atlantic Ridge Islands. (BAKER, GASS, et al, 1964)

basalt field. Gough and St. Helena have small quantities of nepheline in their norms, being close to the critical plane of silica under-saturation. But Ascension has normative hypersthene, and thus would lie in the olivine-tholeite field. BAKER, GASS et al simplified the norm into various summations and compositional ratios for comparative purposes. Thus, comparing the Tristan trachy-basalt with the basalts of St. Helena and Ascension, a close similarity developes between the three ratios (FPO), the Fe percentage and the differentiation index. But the similarity cannot be pushed further, for the OrAbAn ratios for the Tristan trachybasalt (24/49/27) are quite different from those of St. Helena and Ascension (approx. 11/51/38), which but again reflects the high K_2O values for Tristan. It should also be noted that the FPO, Fe% and differentiation index ratios for Tristan and Gough show no similarity.

BAKER, GASS et al summarize their comparative studies as follows.

The Tristan rocks are higher in alkalies and relatively under-saturated with respect to silica compared to other islands on the Mid-Atlantic Ridge. There is some indication that island location with respect to the Ridge has some bearing on the compositional variation. Ascension, with its strongly rhyolitic late differentiates, lies close to the centre of the Ridge; Gough and Tristan, further removed from the Ridge, show trachytic late differentiates with slight phonolitic tendencies: St. Helena, furthest removed, has definite phonolitic late differentiates. One may surmise, therefore, that increased distance from the Ridge results in increased silica under-saturation. Carrying this concept well beyond the confines of the Ridge, the same would appear to hold true, for in Fernando do Noronha, Trindad and Martin Vaz, closer to the South American continent (as also the Canary Islands and the Cape Verde Islands, closer to the African continent), the rocks are all strongly phonolitic and alkaline in character.

Table 74   Chemical Relationships (BAKER, GASS et al, 1964)

	Gough 'basalt'	Tristan basalt	Tristan trachy-basalt	St. Helena 'basalt'	Ascension 'basalt'
SiO_2	47.7	43.1	46.7	47.9	50.1
TiO_2	3.2	4.1	3.6	3.6	2.8
Al_2O_3	15.2	13.1	17.3	16.4	16.3
Fe_2O_3	2.3	5.5	3.8	3.9	4.1
FeO	8.7	8.5	7.1	8.2	7.5
MgO	9.7	9.0	4.7	6.4	5.4
CaO	8.9	12.4	9.7	8.7	8.8
Na_2O	2.7	2.7	4.1	3.8	3.7
K_2O	1.6	1.6	3.0	1.1	1.3
Q	–	–	–	–	–
Or	9.46	9.46	17.73	6.50	7.68
Ab (F)	22.13	4.55	11.17	29.65	31.31
Ne	0.39	9.91	12.74	1.36	–
An	24.63	18.90	19.95	24.45	24.03
Di	15.67	33.79	22.59	15.03	15.82
Hy (P)	–	–	–	–	5.50
Ol	18.30	7.63	3.48	10.52	4.39
Il (O)	6.08	7.79	6.84	6.84	5.32
Mt	3.34	7.97	5.51	5.65	5.95
Total F	57	43	62	62	63
Total P	34	41	26	26	26
Total O	9	16	12	12	11
Mol. % Fe in P	22	13	20	22	25
Or	16	18	24	10	12
Ab (+Ne)	42	46	49	52	51
An	42	36	27	38	37
Diff. Index (Q + Or + Ab + Ne)	32	24	42	38	39

Economic Geology

Guano deposits occur in the Tristan group but are considered of no economic value. DOUGLAS (1930) gave chemical analyses of those on the N side of Nightingale Island and in a cave in Middle Island:

	Nightingale	Middle
Moisture	72.12	17.00
Organic matter, ammonia salts	24.70	15.15
Phosphoric acid	nil	3.85
Lime	nil	5.10
Magnesia, alkalis, etc.	1.60	10.20
Siliceous matter	1.58	48.70
Total	100.00	100.00

Geochronology of the Tristan Group

Other than outwash alluvial material, no sediments, no fossils occur in the islands.

DUNNE, writing before radiometric dating techniques were available, attempted utilizing indirect means for assessing the age of the group.

Obviously the islands are older than the eustatic changes in sea level. Even the late secondary vents on Tristan da Cunha must ante-date this event, for sea caves here and there are carved out of lava flows originating from such vents.

DUNNE, adopting DALY's approach for St. Helena (q. v.), was of the opinion that in Tristan da Cunha and Inaccessible Islands, the age of such was "several millions of years". (Vd. the radiometric datings for St. Helena indicating an age of more than 14 m. y.)

The great heights of the cliffs in Tristan da Cunha and Inaccessible Island led DUNNE to postulate an age of one or two million years for the islands, whereas the less dissected landscape of Nightingale-Stoltenhoff-Middle Islands suggested a considerably younger age.

MILLER (1964) carried out radiometric datings on samples collected by the Royal Society Expedition, herewith presented in Table 75 (p. 278). Basalt samples were studied, using the potassium-argon method, as also samples 17, 18, 19 and 20 which were collected by the Challennger Expedition in 1885 and housed in the British Museum. Sample 20 gave a much older dating than the other Tristan da Cunha values, and this, along with the fact that the precise locality from where it was taken is not known, caused MILLER to disregard the dating. On geomorphological grounds, the degree of dissection of the landscapes increase from Tristan da Cunha to Inaccessible to Nightingale, the last-mentioned being all that remains of a deeply eroded volcanic mass. Radiometric datings would confirm these age relationships, where the averages of the samples tested give values of 1 m. y., 6 m. y. and 15 m. y. respectively for the above three islands. (There seems little doubt that DUNNE was in error in believing that Nightingale had been subjected to less erosional wasting than the other two.)

GASS (1967) has an interesting publication dealing with the question of the ages of the Tristan group, and gives some added radiometric information.

Regarding Tristan da Cunha, he mentions a private communication from BAKER, DODSON & REX informing him that a specimen taken from the N cliff of the island, at an elevation of some 180 m above S. L., gave a probable age value between 30 000 and 70 000 years.

Two Carbon-14 determinations were made of a carbonaceous, brown, organic silt containing diatoms occurring in a freshwater pool on the sides of the main cone, formed after eruptivity had ceased here, which yielded ages of 10,770 ± 156 years B. P. and 11,310 ± 168 years B. P. These silts were later overlain by scoria and cinders erupting from the Big Green Hill parasitic cone. As this Hill represents but one of some 30 odd secondary centres formed on the constructional surface of the primary cone, some estimate can be made as to the age of the other centres by comparing their erosional state with that of Big Green Hill, realizing of course that such an estimate has no quantitative value. It is thus surmized that eruption via the central vent ended some 15 000 years ago and that during the past 10 000 years, vulcanism has occurred in some 11–15 parasitic centres. It is further deduced that subaerial eruptions forming the island took place over

Table 75 Radiometric Datings of Tristan Samples (MILLER, 1964)

$\lambda_\varrho = 0.584 \times 10^{-10}$ yr^{-1}
$\lambda\beta = 4.72 \times 10^{-10}$ yr^{-1}

Sample		K_2O (%)	Atmosph. contam. (%)	Age + est. error. (m. y.)	Island
D 1	Flow or large sill halfway between Big Pt. and Rookery Pt. 10 ft. above S. L.	2.54	–	Recent	
D 2		2.30	–	Recent	
D 3		2.73	12.2	3 ± 3	
*E 1	Flow halfway between Big Pt. and Rookery Pt. immediately beneath D 1, D 2, D 3	2.78	–	Recent	
E 2		2.79	–	Recent	
E 3		2.81	94.8	0.5 ± 1	
G 1	Sea-edge of flow 100 yds. extreme W of flow 10 ft. above S. L.	4.66	79.8	0.6 ± 1	
G 2		4.66	31.3	0.8 ± 1	
I 1	Flow forming 3rd. waterfall above largest waterfall, Hottentot Gulch. 1500 ft. above S. L.	2.65	–	Recent	Tristan da Cunha
I 2		2.55	85.9	1 ± 1	
I 3		2.51	–	Recent	
H 3	Flow on seashore between Settlement and E of Herald Pt. 0 ft. above S. L.	3.14	–	Recent	
H 2		3.21	–	Recent	
H 1		3.18	–	Recent	
J 1	Flow approx. 100 yds. W of pinnacle, Hottentot Gulch. 900 ft. above S. L.	2.38	–	Recent	
J 2		2.51	–	Recent	
906	Basalt	1.64	–	Recent	
BM	Settlement Plain 1927 1252 (22)	2.79	54.1	3 ± 1.5	
BM	Basalt from edge of lake. 1927 1252 (2)	3.3	37.5	3 ± 1	
BM 64747	Basalt	2.9	77.5	9 ± 2	
BM 64775	Basalt from dyke intruding rear of Blomby's Cove.	4.8	95.5	2 ± 1	Nightingale Is.
BM 64760	Basalt	1.9	43.2	6 ± 1.5	Inaccessible Is.
BM 1927	Basalt. 1252 41.	3.73	97.8	12 ± 4	Middle Island
BM 1927	Basalt. 1252 41.	3.73	96.6	18 ± 4	

* Using an omegatron type mass spectrometer, this sample gave ages of 0.80 ± 0.10 and 1.10 ± 0.15 m.y, in agreement with the above determination. A figure of 0.95 m.y. is got from the mean of the two values.

a period of some 150 000 ± 37 000 years. Assuming a constant rate of eruption, then the total volcanic mass of which Tristan da Cunha represents merely the uppermost 5%, was formed throughout a period of some 3 m. y.

Specimen No. 22 in Table 75, Inaccessible Island, likewise was collected by the Challenger Expedition, its locality not being precisely given. Another specimen collected by the Royal Society Expedition from one of the lowest lavas seen in the cliffs behind Salt

Bay gave a age value of 2.9 ± 0.3 m. y., and MILLER has suggested that this value, rather than that of the Challenger specimen (6.0 ± 1.5 m. y.) should have preference. As the Royal Society specimen came from the basal volcanic sequence, one might therefore postulate that during the past three million years, the island was constructed to a maximum diameter of some 16 km – the maximal extent of the shallow-water platform surrounding the island – and to a maximum height of some 2290 m. If these postulates are accepted, then between 90 and 95% by volume of this original island has been destroyed by marine erosion during the past 3 m. y.

The three determinations given in Table 75 for the Nightingale group are Challenger specimens. Basalts from dykes in Middle Island gave radiometric datings of 18 ± 4 m. y. and 12 ± 4 m. y. From Nightingale Island, a dyke of similar composition yielded an age of 2 ± 1 m. y. As previously mentioned, organic material in the raised beaches on Nightingale Island gave ages of 39 160 (+ 6090–3410) years and more than 36 900 years B. P. in near-by localities – say 40 000 years B. P. The above basaltic dykes cut the Older Pyroclastics and thus the latter are older than 18 ± m. y. (The 2 ± m. y. age for the Nightingale dyke is thought to be in error.) It could be postulated therefore that the first subaerial volcanic activity of Nightingale is far older than that on either Tristan da Cunha or Inaccessible. The 40 000 year age for the raised beach deposit would suggest that volcanic activity recurred on Nightingale after a long period of both quiescence and erosion, a period perhaps of the order of 10 m. y.

GASS (op. cit.) thought that the Tristan group of islands showed a two-phase cycle: 1. an eruptive phase, when they were volcanically formed and during which agents of erosion were of minor importance; 2. a phase of lessened or ceased volcanic activity – but perhaps spasmodic outbursts – with erosion now dominant. He believed that Tristan da Cunha was likely near the end of the first phase, for eruptions over the past 15 000 years have been restricted to parasitic cones, during which time marine erosion has made few changes. This constructional phase was of very short duration, perhaps some 150 000 years. Inaccessible and Nightingale Islands represent second phase islands, the former having experienced erosional attack for some 3 m. y., the latter for some 18 m. y. By analogy with Tristan da Cunha, it can be presumed that both had a relatively short constructional phase, lasting probably not more than one million years and later erosion operative for some 20–25 m. y. which also likely represents the life-span of these two islands.

Lastly it should be noted that in general oceanic islands show high positive gravity anomalies and are thus isostatically unstable, hence a tendency to sink. Islands constructed of lighter material, e. g. phonolitic, trachytic, may quite possibly have a somewhat longer life than the 20–25 m. y. years postulated for basaltic oceanic islands.

Geological Evolution

The geological history of Tristan da Cunha began when a volcanic complex was formed, chiefly through a central locus or loci of central type, from which olivine-basaltic flows in the main were extruded. Intercalated with these flows were a few eruptions of pyroclastics forming tuffaceous beds. Smaller vents of somewhat later date erupted ash, glass and thinner olivine-basaltic flows. These less powerful and more transitory eruptions were subsequently smothered by extrusions still continuing from the main central vent area. Such later eruptions from the central locus involved thinner, more viscous hornblende-alkali-basalt flows, building-up the Peak region. During this phase, explosive activity was more frequent, resulting in larger and thicker beds of pyroclastics. The next phase was the emission of a vast quantity of black and red ash which today forms the uppermost 200 m of the Peak. Pene-contemporaneous with this event, another vent slightly S of the main one went into action, so constructing the second peak, and thus in present-day Tristan da Cunha the Peak area is actually formed of eruptive products from these two focal localities. Within the cinders forming the Peak crater occur outcrops of hornblende-alkali-basalts similar to those mentioned above, as well as sodalite-trachytes and trachyandesites, present high up in Cave Gulch. We can thus presume that trachytic eruptions succeeded the basaltic outpourings but were formed prior to the formation of the black and red ash forming the central summit.

A relatively long period of quiescence followed the construction of the main cone and hence the main portion of the island. Erosion had probably not wrought much more change than is seen today, but then a relatively sudden onset of vulcanicity occurred which was characterized by the formation of small cinder cones and explosion craters aligned along definite trends. The cinder cones comprise tuffaceous ash and a few thin flows of hornblende-alkali-basalts; the explosion craters are built of ash, except in one instance when trachytes were emitted after the crater formation.

Inaccessible Island, structurally the most complex, began by the emission of olivine-basaltic flows which were rudely deformed by the intrusion of a great number of olivine-basalt, trachybasalt, trachyandesite and trachyte dykes. Following this phase, essexitic gabbros were formed but consolidated under a considerably thick mantle of rock now no longer evident. It is presumed that the flows issued from several vents but quite likely a main locus of activity. As the rate of eruptivity was not the same at these centres, various of them became inundated by either later or then more voluminous outpourings from other vents.

The eastern part of the island is formed of thick flows of trachybasalts and olivine-trachyandesites lying unconformably upon thinner olivine-basalts. Many dykes cut the latter but do not penetrate the former, indicating an age for the dykes between that of the thick and the thin flows. One important centre of eruption of the later flows is the breached crater in the SE part of the island, from which mugearites were outpoured, and in the vicinity are one or two smaller trachyte cones. DUNNE believed that the trachytic eruptions on Inaccessible were more or less contemporaneous with the main phase of vulcanicity in Tristan da Cunha, whereas GASS (1967) believed all essential construction had ended long ago on Inaccessible by the time Tristan da Cunha was building up.

The Nightingale group form a unit which can be treated together. Basal rocks exposed on Nightingale comprise volcanic agglomerates, consisting of fragments of various types of lava cemented by brown glass, ash and lapilli. It is significant that within this agglomerate occur pieces of all the lava types, e. g. sodalite-trachyte, alkali-trachyte, trachyandesite, trachybasalt, olivine-basalt, which were formed on the island subsequent to the formation of said agglomerate. Basic dykes which do not cut the later rocks are numerous in the agglomerate. Most of Nightingale is composed of biotite-soda-trachytes and hornblende-trachyandesites. The former occurs as lapilli in the basal agglomerate, as boulders on the abandoned shoreline on the S coast, and within the N-S ridge in the E. The trachyandesites are considered to be xenolithic admixtures of basaltic phenocrysts in a trachytic melt. It was the opinion of DUNNE that the trachytes and trachyandesites represented one large, composite, monolithic intrusion, with associated dome construction and lava flow formation. Preceding the extrusion of these trachytic magmas, much pyroclastic material was expelled, seen in the basal agglomerates, and much brecciation took place. Lying in unconformable relation upon the agglomerates is a 2 m thick sandy tuff. It is older than the trachytic flows, but it cannot be said with certainty that it is older than all the trachytic and trachyandesitic eruptions. To postulate here an age sequence basal agglomerates(oldest)-sandy tuff-trachytic masses would imply that the bulk of the island is of relatively late origin.

Such a scheme as outlined above for the evolution of the Tristan group is not in entire agreement with the radiometric datings. DUNNE believed that the main building of Inaccessible Island was more or less contemporaneous with that of Tristan da Cunha, whereas GASS, adopting radiometric datings, would claim that Inaccessible underwent construction a long time previously, that during the period of construction of Tristan da Cunha, Inaccessible was undergoing drastic erosion. DUNNE would claim that the Nightingale group are of relatively recent origin, whereas GASS considered the group to be the oldest, perhaps some 20 m. y. old.

Table 75 lists 24 radiometric datings. Of these, three are stated by MILLER-GASS to be of questionable value – the 9 m. y. basalt sample from Tristan da Cunha, the 2 m. y. sample from Nightingale Island and the 6 m. y. sample from Inaccessible Island, in other words, $12-1/2\%$ of the age determinations are not trust-worthy, by admission of those intimately concerned with said datings. We must further note the large role which supposition plays in the geochronological scheme of things as presented by GASS, quite as outstanding as the indirect deductions made by DUNNE. One cannot help but note that the 2 m. y. age for the Nightingale basalt dyke specimen, though thought to be erroneous by MILLER-GASS, would accord well with DUNNE's contention that the island is of relatively late origin. The older radiometric dates for the Nightingale group come not from Nightingale itself but from Middle Island. Agreed, the basalt dykes from which the specimens were studied in Nightingale and Middle Islands are said to show similar composition, but in spite of this, in spite of acknowledged propinquity, this does not argue, *ipso facto*, for similarity in age for the Nightingale group as a whole.

Considerable work dealing with the assessing of radiometric age determinations has led the writer to view with most cautious reserve all such findings. At this time, we must frankly confess that techniques involved are in need of much improvement.

The 1961 Eruption

The primary objective of the Royal Society Expedition was to make a geological survey of the Tristan group, study the new eruption and the effects on the vegetation and fauna.

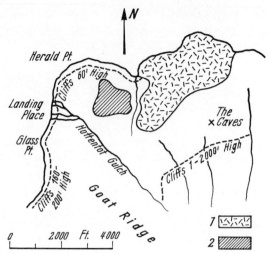

Fig. 94. Sketch Map of area of volcanic activity, Dec. 16–17, 1961, Tristan da Cunha. 1. Extent of lava flows, 2. Edinburgh. (HARRIS & LEMAITRE, 1962)

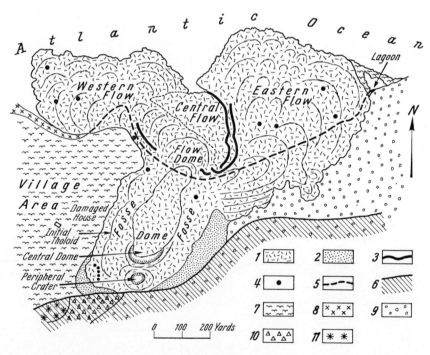

Fig. 95. The 1961 Eruptive Centre. 1. Lavas (1961), 2. Fragmentals, 3. Lava channels, 4. Fumaroles, 5. Former coastline, 6. 600 m (approx.) cliffs, 7. Alluvium, 8. Lavas from surface cinder cones, 9. Lavas, main volcanic sequence, 10. Pyroclastic centre, main volcanic sequence, 11. Intrusives. (BAKER, GASS et al, 1964)

Tremor Phase

Though a volcanic island, Tristan da Cunha had shown no indications of vulcanism since the Stony Hill parasitic cone was in action probably some 200–300 years ago. Earthquakes had never been known. But beginning in the first week of August, 1961, slight shocks were felt near the Settlement and at Sandy Point. During the next three weeks, these increased in intensity, and then there was a lull. Tremors began again in early September, now more frequent and more violent, thought to have grades, as per the Modified Mercalli Scale of: $A = 3$, $B = 4$, C 5, D 6. It was believed that the hypocentre lay at a depth of about 1.6 km, with a magnitude of 3 and energy release of perhaps 10^{16} ergs. Along with the heaviest tremors, many rock-falls occurred behind the Settlement and to the NE of the village.

These Tristan tremors were comparable to seismic disturbances commonly heralding vulcanicity. The shallow-focus and hence rapid horizontal decrease in intensity of such tremors was well exemplified here, for at distances of 6–7 km away from Settlement area, shocks were mild, and at a distance of 11 km, not noticed at all. Increasing intensity of the shocks as October approached was likely due to approach of the ascending magma towards the surface. The tremor phase reached a climax on October 9th., when great cracks, 200 m in length and 2 m broad, were formed parallel to the coastline. On the seaward side, land was upraised 9 m and from the cliffs behind, blocks were detached and rolled downward, such landslides being most evident just behind the Settlement.

Eruption Phase

Tremors ceased as soon as the eruptive phase began, which latter began by a reddish glow emanating from a mound in the early hours of the morning of October 10th. In but a few hours this had developed into a dome or tholoid measuring ca. 20 m in height and 50 m in diameter. During the day this continued to grow, and also a further disturbance took place some 200 m W of the tholoid. After four days the tholoid had grown to a height of some 75 m, covering an area of 12 500 m². Soon afterwards lava began pouring out and flowing down northwards, and by the 27th., the canning factory, NE of the Settlement had been overwhelmed. Data are lacking as to events in November, but on December 6 the volcano was erupting violently, the red glow of the lava being visible more than 30 km away at sea. On the 16th two geologists of the Expedition, HARRIS & LEMAITRE arrived by ship, and although unable to land, could observe events. Now the original plug had been breached on the seaward side and a stream of block lava coursed seawards. They estimated that the volcano had now reached a height of some 145 m above S. L., the lava field pushing beyond the original coastline for some 400 m and about 1000 m wide at the seaward margin. By January 5, 1962, the first crater was still active, emitting brownish lava, steam and smoke, and the lava front was now some 1300 m wide. The Expedition arrived on January 27th., by which time activity was very much less. The embryonic dome was now an elevated ridge which attained a maximum elevation of 147 m above S. L., thus showing a very slow rate of growth, compared say to Mont Pelé. (This maximum elevation was attained on March 19th.) Fumarole activity lessened during February and March. The Expedition left Tristan on March 20th., by which time relative quiet reigned.

Morphology

The land surface upon which volcanic products were formed had an inclination seawards of some 10°, and this inclination no doubt determined the form of the volcano and the lava field were directly influenced by this surface. The lava field achieved a maximum width of 1200 m with a maximum longitudinal extent of 1000 m approximately. The volcanic products cover a total area of some 585 280 m², two-thirds of which lies beyond the former coastline.

Fig. 95 shows the morphological features of the eruptive area. The source region plus the lava field constitute the two chief units of the eruptive centre. The source region is rimmed by a U-shaped ridge open seawards to the lava field. The ridge is the external vestage of the initial tholoid. In the central part of this marked subsidence occurred when the first lava poured out. In this area of subsidence rises the central cone to a height of some 137 m above S. L. The dome breaches the northern wall of the cone and extends downwards and seawards. Separating the central cone and the dome from the outermost ridge is an arcuate fosse. A smaller peripheral crater occurs at the southern extremity of this ridge where the latter meets the base of the main cliffs.

Petrography

All specimens analysed by the Expedition have a trachyandesite composition. The table below shows modal analyses of samples from various units of the parasitic volcano, which demonstrates their restricted mineral variation:

	Initial tholoid	Lava flows	Dome
Plagioclase	29.0	28.2	28.8
Pyroxene	4.7	7.6	6.4
Amphibole	2.3	2.1	1.5
Ore	2.6	4.0	4.1
Groundmass	61.4	58.1	59.2
No. points counted	3319	4597	5271

Hand specimens show the rocks to be finely vesicular, with now and then short prismatic crystals of plagioclase and elongate amphibole in an aphanitic base. Large plagioclase phenocrysts have a basic labradorite core, surrounded by a narrow zoned rim ranging in composition to sodic andesine. These plagioclases have strongly corroded cores which contain brownish glass and ore and pyroxene inclusions. Large clinopyroxene phenocrysts are much rarer, occurring poikilitically enclosing plagioclase laths and ore granules. The core of these crystals is aegerine-augite, and the wide rim likely a titanaugite. Strongly coloured amphibole phenocrysts show some degree of resorption. The amphibole is thought to be basaltic hornblende, but it may approach kaersutite in composition. The ore seems to be chiefly ilmenite. A most interesting constituent of some specimens, occurring as scattered, discrete grains, is haüyne. Leucite is sparingly present, and, unlike all other occurrences in the Tristan rocks, here it does not appear interstitially.

Occasional small plutonic xenoliths are present in the dome and lava field. Four were examined, from which the following modal analyses were made:

Sample No.	No. 645	No. 510	No. 519	No. 509	Average
Plagioclase	62	60	46	45	56
Olivine	1	0	0	0	--
Pyroxene	25	16	6	3	14
Hornblende	7	0	1	4	5
Biotite	0	16	33	33	15
Ore	4	6	9	10	7
Apatite	1	2	5	5	3

As plagioclase decreases, so also does pyroxene, whilst ore and apatite increase. Olivine only occurs in the rock with highest pyroxene content. Under the naked eye, these plutonic xenoliths have a gabbroic appearance and granular texture. In thin-section, some unzoned plagioclase occurs with a composition of ca. An_{75} whilst the strongly zoned ones have cores of about An_{65}. The clinopyroxenes are strongly coloured, ore inclusions are common. Subhedral, platy crystals of amphibole show no resorption, and appear to be chiefly basaltic hornblende. Some biotites are strongly oxidized. The granules of ore are thought to be mostly ilmenite. Apatite is abundant, present poikilitically in plagioclase, pyroxene and amphibole crystals.

On a mineralogical basis, these coarse-grained xenoliths show close resemblance to those found elsewhere in Tristan da Cunha.

Fumarolic minerals are plentiful in and around the small peripheral crater at the southern margin. Colours of these are grey, white, yellow and orange. They may occur as clear crystalline masses or then amorphous masses. Such minerals are present on blocks of lava which are bleached white. The following minerals were identified:

Sulphates: Gypsum, anhydrite, bassanite, alum, mirabilite(?), louderbackite, metavoltine, jarosite.

Fluorides: Fluorite(?), cryolite(?), ralstonite, thomsenolite(?), hieratite, cryptohalite.

Chlorides: Halite, erythrosiderite.

Carbonates: Natron.

Borates: Tincalconite(?).

Elements: Sulphur.

CHAPTER 14

Gough Island

General

Gough Island, or to use its lesser-known name, Diego Alvarez Island, lies in lat. 40° 20′ S, long. 9° 56′ W. The nearest land is Tristan da Cunha, 350 km to the NNW, and the mainlands of South Africa and South America are about 2400 km and 2800 km distant respectively. Gough is 13 km in length, 6.5 km broad, rising to a maximum height of 907 m in Edinburgh Peak.

Below 300 m there is luxuriant forest vegetation and thick undergrowth which makes penetration on foot up the narrow valleys and steep slopes extraordinarily difficult. Doubtless this is responsible for the fact that although visits, expeditionary and otherwise, have been made to Gough for more than 150 years (the island was annexed by the British Crown in 1806), the first adequate map was only compiled in 1956. (HOLDGATE et al, 1956, HEANEY 1957a, 1957b, HEANEY & HOLDGATE 1957.) Above an elevation of some 600 m mosses and peat bogs are plentiful.

Gough, lacking all human habitation, is the breeding ground for the giant wandering albatross and is the most northerly point at which the Antarctic giant petrel is found. There are vast numbers of penguins, also many seals. Guano deposits abound.

Physical Features

The rugged and deeply dissected nature of the island recalls that of St. Helena and Ascension rather than the nearest neighbour, Tristant da Cunha. The SW coastline is remarkably straight, those of the N, NE and SE slightly embayed, but there are no safe harbours. Small beaches occur here and there but in general imposing cliffs, 15 m to 300 m in height, drop precipitously down to the sea. Streams leap over these cliffs in magnificent waterfalls, the spray perpetually drenching the narrow, rocky strand, forever washed by turbulent seas.

The V-shaped valleys are actively working backwards towards the island centre. Only two valleys, the Glen and Sophora Glen, reach sea level at approximately grade; all other streams are hanging valleys, on occasion debouching 180 m above the level of the sea. Such impressive hanging valleys testify either to recent uplift or then immensely powerful marine erosion.

In the interior are one or two plateaux of which the largest is Tarn Moss, nearly 1.5 km broad and lying 600 m above sea level.

Edinburgh Peak and Expedition Peak, 905 m, are about 1.5 km distant in the north-central area, and Mount Rowett, 839 m, is almost in the island centre.

Fluviatile, marine and aeolian erosion are all active in attacking this small island. Marine erosion is seen in the cliffing, arches, stacks and caves. Tristania Rock, 165 m high, lying off the NW coast, is the largest stack. BARTH (1942) thought that the sea stacks might be monolithic pipe-intrusions of alkaline trachyte but LeMAITRE (1960) preferred to regard them rather as detached blocks from trachytic flows.

Aeolian erosion is seen where the soft tuffs have been blown into fantastic shapes.

The highly dissected nature of Gough, with its entrancing deep, narrow valleys, bear witness to the efficacy of river erosion.

Climate

During the scientific survey of Gough (HOLDGATE, 1956, HEANEY, 1957a), meteorological abservations were taken. The data refer only to a period of five months in 1955–56, as follows:

	Dec.	Jan.	Feb.	Mar.	Apr.
Temp. (°C)	15.2	14.6	15.4	13.9	13.7
Rain (mm)	197.7	304.5	152.2	101.0	261.9

During this period the highest temperature recorded was 23.7° C in February; the lowest, 5.5° C in January. Though well removed from equatorial climes a rainfall of some 1000 mm in five months, and this not in the 'wet season', is more typical of tropical islands such as those of the Gulf of Guinea. The riot of vegetation is indicative of high rainfalls.

The climate is oceanic, dominated by a series of warm and cold front systems. Orographic clouds are very frequent.

Geology

DOUGLAS (1923) has briefly given his topographic and geologic impressions of Gough as follows: "The island forms a monoclinal block, with dip slopes to the W and escarpments to the E. The highest point, on the long ridge which runs down the longer axis of the island, is ca. 2915′ above sea level. (This elevation is incorrect, as per the surveys of HEANEY, HOLDGATE, see references, RCM-T.) The W side of the ridge goes down in a long slope to cliffs bordering the sea. The escarpments of the E side are cut by three or four glens. The largest one, about half-way down the coast, gives access to the interior (The Glen – RCM-T). The most striking feature looking up the glen is a great stock of an acid intrusive which rises to 2270′. (Incorrect elevation for Hag's Tooth – RCM-T). The island is the result of a series of fissure lava flows of basaltic and trachytic nature. These flows have been intruded by the stock mentioned above, and many fissures were opened

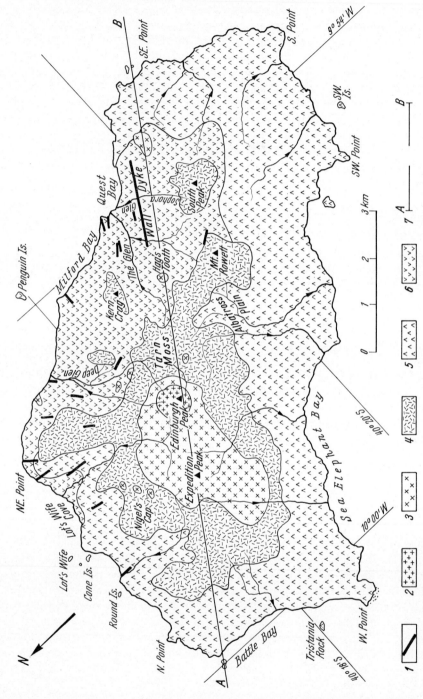

Fig. 96. Geological Map of Gough Island. 1. Basaltic and trachytic dykes, 2. Upper 'Basalts', 3. Upper Trachytes, 4. Middle 'Basalts', 5. Lower Trachytes (lavas and tuffs), 6. Lower 'Basalts', 7. Geologic Section. (LeMaitre, 1960)

by it. The rock forming the dykes is very hard with the result that they are now prominent features, protruding in some cases ca. 50′ above surrounding country. This is due of course to differential weathering".

The dykes of Gough are indeed impressive features, often as thick as 3.5 m. The best example of such is the Wall Dyke (Devil's Wall) which runs straight for some 1.8 km. These dykes are the cause for the majority of the many waterfalls, some of which are 6 m and more high.

Petrography

Only igneous rocks occur, forming what LeMaitre (1962) terms a continuous series in the alkali basalt-trachyte association. Studies of the Gough rocks have been made by Pirsson (1893), Campbell (1914), Campbell Smith (1930), Barth (1942) and LeMaitre (1960, 1962, 1965).

Pirsson divided the basalts into two varieties, 1. dark-grey, dotted with black phenocrysts of augite, yellow olivine, white feldspars, with a dense groundmass, 2. dark greyishbrown, porous in texture, thickly dotted with white, broad, lath-like phenocrysts of labradorite, often grouped in star-fashion, irregular-shaped olivine and rare augite. The latter variety contains somewhat more silica but has a lower specific gravity than the former.

Trachytes were represented by trachytic tuffs and obsidian. A chemical analysis of the obsidian is given below:

SiO_2	61.22
TiO_2	0.42
Al_2O_3	18.01
Fe_2O_3	1.32
FeO	4.51
MnO	Tr.
MgO	0.44
CaO	1.88
Na_2O	6.49
K_2O	5.93
H_2O	0.46
Total	100.68
S. G.	2.210

Table 76 Chemical Analysis of Obsidian (Pirsson, 1893)

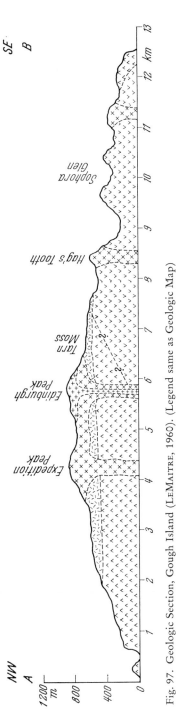

Fig. 97. Geologic Section, Gough Island (LeMaitre, 1960). (Legend same as Geologic Map)

CAMPBELL, who studied the collection made by the 'Scotia' Expedition of 1904, classed the lavas as of two types, trachytes and trachy-dolerites. The former showed the abundant presence of anorthoclase, and were of three categories: 1. biotite-trachyte, 2. sodalite-trachyte, 3. aegerine-augite-trachyte. The trachy-dolerites were more basic in character, olivine always present, the lime-soda feldspars being accompanied by varying amounts of albite and albite-oligoclase. A vesicular lava which he studied, containing phenocrysts of olivine, titanaugite and basic plagioclase in a groundmass of brown glass with microlites of plagioclase and augite, was said to closely resemble the vitrophyric basalts of Tristan da Cunha. The intrusives were represented by trachylite, basalt and essexite. The essexite resembled the pyroxene-rich varieties of Norwegian essexites described by BROGGER. The relative proportions of augite and olivine made a close analogy with the porphyritic essexites of Lennoxtown, Scotland, but nepheline was absent in the Gough rocks. Pyroclastics were represented by trachytic tuffs.

CAMPBELL summarized by saying that the rocks studied by him showed them all to have been derived from a soda-rich alkali magma, these rocks having a common origin with those of other islands of the Mid-Atlantic Rise.

CAMPBELL SMITH, who studied the 'Quest' rocks collected by DOUGLAS (vd. infra), made a more detailed investigation of the petrography. In a traverse up the Glen to the summit of Mt. Rowett by DOUGLAS, CAMPBELL SMITH identified the following (heights by aneroid barometer, after DOUGLAS): trachybasalt, mugearite, 16 m; trachybasalt, essexite of porphyritic type, 39 m; basaltic dyke, 60–90 m; basalt, 120 m; diabase dyke, Devil's Wall, 158 m; compact, jointed basalt, 174 m; vesicular basalt in mugearite countryrock, 182 m; basaltic dyke, 197 m; basalt, 709 m; trachybasalt (essexite porphyrite) 900 m. (DOUGLAS recorded the elevation of the summit of Mt. Rowett as 900 m, but the HEANEY survey gives an elevation of 839 m.) In most cases these specimens were collected from the S side of the Glen.

CAMPBELL SMITH described the main rock types as follows:

Alkali-trachytes. The aegerine-augite trachytes of CAMPBELL (op. cit.) were well represented, comprising one of the most outstanding topographic landmarks in the island, the Apostle, Gough Monument or, as it was later named, Hag's Tooth. On the N side of the island, most of the trachytes collected by DOUGLAS were of the same type. Chemical analyses of the trachytes are given below.

Trachybasalts. These more basic rocks are the effusive equivalents of the essexites mentioned by CAMPBELL, ranging from trachytoid trachybasalts to more basic basaltoid types. Comparisons can be made here with some trachydolerites (trachybasalts) of Madeira and East Africa, as classified by L. FINCKH (1913) in his publication referring to the essexite family and the effusive representatives. More specifically, CAMPBELL SMITH recognized trachytoid trachybasalts and trachybasalts – the porphyritic essexites of FINCKH.

Chemical analyses are shown in Table 77, Nos. 1, 2 and 3.

Olivine-poor Basalts. The more basic dykes of the island were classed as porphyritic basalts with abundant small phenocrysts of plagioclase in a darkish-grey, aphanitic, olivaceous groundmass.

From the chemical analysis (Table 78, No. 4) the rock compares favourably with olivine-poor basalt from St. Helena, as per DALY (1927). Likewise it shows a composition similar to the mugearite from Inaccessible Island, Tristan da Cunha (q. v.).

	1.	2.	3.
SiO_2	63.84	62.40	60.74
Al_2O_3	18.46	18.37	17.72
Fe_2O_3	1.54	1.83	3.19
FeO	1.79	2.64	2.41
MgO	0.78	0.65	0.51
CaO	2.45	2.30	1.67
K_2O	3.08	4.05	6.03
Na_2O	7.25	7.00	6.02
TiO_2	–	tr.	0.73
MnO	–	0.17	0.17
P_2O_5	0.24	0.07	–
H_2O+ 105 °C	0.39	0.37	0.52
H_2O- 105 °C	0.10	0.12	0.58
Total	99.92	99.98	100.55*
S. G.	–	–	2.50

Table 77 Chemical Analyses of Alkali Trachytes (CAMPBELL SMITH, 1930)

1. Biotite-bearing alkali trachyte, N side of island.
2. Aegerine-augite trachyte, cave at foot of Hag's Tooth.
3. Aegerine-augite trachyte, N side of island at sea level.

* Includes ZrO_2 0.26, S 0.002.

	1.	2.	3.	4.
SiO_2	55.80	54.85	48.85	49.49
Al_2O_3	17.53	16.92	14.39	16.47
Fe_2O_3	1.24	3.14	1.04	4.12
FeO	5.87	3.92	8.54	6.00
MgO	2.52	2.70	7.49	3.53
CaO	5.77	6.96	8.88	8.88
Na_2O	7.28	5.25	3.56	3.31
K_2O	2.98	2.55	1.77	2.43
TiO_2	tr.	1.03	3.37	3.76
MnO	tr.	tr.	0.24	0.17
P_2O_5	0.79	0.78	0.63	tr.
H_2O+ 105 °C	0.12	1.08	0.79	0.35
H_2O- 105 °C	0.20	0.80	0.40	1.24
Total	99.90	100.06*	99.95	100.08**

Table 78 Chemical Analyses of Trachybasalts and Basalts (CAMPBELL SMITH, 1930)

1. Trachytoid trachybasalt, dyke at sea level, near Glen beach.
2. Trachybasalt (essexite, porphyritic type), dyke known as Devil's Wall.
3. Trachybasalt (essexite, porphyritic type), summit, Mt. Rowett.
4. Olivine-poor basalt, dyke at sea level, near Glen beach.

* Includes S 0.08; ** Includes ZrO_2 0.01, Cl 0.28, S 0.04.

Mugearites. Difficulties of rock collecting presumed CAMPBELL SMITH to believe that dyke rocks were better represented than lavas, but three unmistakable flows were sampled in the Glen. The rocks were taken to be likely close associates of mugearite, more basic than the trachytoid trachybasalts.

Tuffs. Only one specimen was studied, taken from an occurrence overlying basalt on the shore N of the hut on Glen beach. It was classified as a crystal-vitric tuff with lapilli.

BARTH made a study of rocks collected from the island by LARS CHRISTIAN (1935) when one of his tankers stopped here to allow him to 'have a look around', as he cryptically remarked. BARTH classed the rocks as being dominantly basaltic, trachytic subordinate, with dykes of transitional character between these two.

Table 79 Chemical Analyses and Molecular Norms, Basaltic-type rocks (BARTH, 1942)

	Chemical Analyses			Molecular Norms			
	1.	2.		1.	2.	No. 3 Table 78	No. 4 Table 78
SiO_2	47.16	49.10	Q	–	–	–	0.7
TiO_2	3.62	3.59	Or	12.0	16.5	10.5	14.5
Al_2O_3	11.82	16.21	Ab	21.0	30.8	32.0	31.0
Fe_2O_3	5.96	2.87	An	16.8	20.5	17.5	23.5
FeO	7.16	6.84	Ne	–	0.7	–	–
MnO	0.09	0.05	C	–	–	–	–
MgO	9.83	5.04	Di^a	14.4	16.4	17.6	17.6
CaO	7.86	8.90	Hy	27.9	–	8.6	8.2
Na_2O	2.28	3.53	Ol	–	10.8	11.4	–
K_2O	1.96	2.76	Mt	6.3	3.2	1.1	4.5
P_2O_5	0.67	0.54	Ap	1.6	1.1	1.3	–
Cl	tr.	tr.					
SO_3	tr.	nil.	a) Including Ilmenite				
H_2O+	0.60	0.39					
H_2O-	0.24	0.08					
Total	99.35	99.90					

1. Olivine-basalt, shore inside Penguin Island.
3. Olivine-basalt (essexite-porphyrite type), summit of Mt. Rowett.
4. Olivine-poor basalt, dyke at sea level near Glen beach.

Basaltic-type rocks. Table 79, from BARTH, gives chemical analyses and molecular norms for two specimens of such rocks and molecular norms for two of CAMPBELL SMITH's specimens – see Table 78, Nos. 3 and 4. No. 1 represents the most typical kind of basalt in the collection. Megascopically it is a vesicular, grey, dense rock with a few olivine and plagioclase insets and some diopside-augite crystals. The matrix shows flow structure – parallel arrangement of small laths of plagioclase. Quartz occurs as a filler medium. Small olivine and tiny pigeonitic pyroxene and dusty ore minerals complete the groundmass. No. 2 is megascopically similar. The small labradoritic phenocrysts are zoned, some being more albitic in character. The groundmass comprises small grains of diopsidic augite, ore minerals and a network of plagioclase. As distinct from the first specimen, some alkali-feldspar certainly occurs interstitially.

BARTH established the modal mineral composition for the two specimens thus:

Table 80 Modal Mineral Composition of two Basalts (BARTH, 1942)

	No. 1	No. 2		No. 1	No. 2
Quartz	2%	–%	Olivine	6	11
Feldspar	50	68	Ore	6	3
Pyroxene	34	17	Apatite	2	1

The mode and norm of No. 2 is thus seen to be almost identical. The rock is undersaturated in silica, and the observed quantity of olivine corresponds to that of the norm.

No. 1 shows no normative olivine, as it is saturated with silica, but it does contain much olivine which is in balance with an equivalent quantity of interstitial quartz.

Trachytic-type rocks. Both CAMPBELL & CAMPBELL SMITH found representatives of trachytes. The three types mentioned by the former doubtless merge into each other, and according to BARTH should be considered as accidental varieties of essentially the same magma. Table 81 shows a chemical analysis of an olivine-trachyte studied by BARTH and molecular norms for the trachyte-obsidian of PIRSSON (vd. Table 76) and molecular

Table 81 Chemical Analyses and Molecular Norms of some Trachytic-type rocks (BARTH, 1942)

Chemical Analysis			Molecular Norms				
	1A		1A	Table 78			
				No. 1	No. 2	No. 3	No. 4
SiO_2	60.66	Q	–	3.1	–	–	–
TiO_2	0.40	Or	36.0	17.8	23.5	35.5	34.5
Al_2O_3	18.54	Ab	53.3	63.8	61.7	53.5	47.0
Fe_2O_3	2.25	An	5.0	8.4	6.7	3.3	2.2
FeO	2.26	Ne	0.1	–	–	–	6.0
MnO	tr	C	0.2	–	–	–	–
MgO	0.51	Di[b]	–	1.8	3.2	3.4[c]	5.6
CaO	1.34	Hy	–	3.2	3.0	–	–
Na_2O	5.99	Ol	2.5	–	–	–	3.3
K_2O	6.08	Mt	2.4	1.6	2.0	4.2[d]	1.3
P_2O_5	0.28	Ap	0.5	0.5	0.1	–	–
H_2O+	0.97						
H_2O-	0.55	a) Includes Cl, 0.08, SO_3, nil.			c) Includes Wo, 0.6			
Total	99.91[a]	b) Includes Ilmenite			d) Includes Hm, 0.6			

1A Olivine-trachyte, shore inside Penguin Island.
1. Felsophyric biotite-bearing trachyte, N side of island.
2. Aegerine-augite trachyte, cave at foot of Hag's Tooth.
3. Aegerine-augite trachyte, N side of island, at sea level.
4. Trachyte-obsidian of phonolitoid type, pebble from the shore.

norms for the four CAMPBELL SMITH specimens (vd. Table 77). The BARTH specimen is a light grey rock, aphyric, fine-grained, comprising essentially small laths of alkali-feldspar, a mix-crystal of orthoclase-albite and some anorthite, with imperfect trachytoidal arrangement. Aegerine-augite, olivine always decomposed along the edges and iron-rich biotite are also present. Specimen No. 1 of CAMPBELL SMITH (Table 77) was thought by BARTH to show a most unusual chemical analysis. In spite of there being present phenocrysts of both potassic feldspar and biotite, the rock contains less potash than the other specimens listed. As per CAMPBELL SMITH, the specimen was said to contain "a fair percentage of sodalite" which BARTH found most surprising in view of the chemical composition of the rock. Indeed BARTH seriously questioned the correctness of both the chemical analysis and the mineral composition of this CAMPBELL SMITH specimen. The modal mineral composition of BARTH's olivine-trachyte is given below:

Table 82   Modal Mineral Composition of Olivine-Trachyte (BARTH, 1942)

Feldspar	92%	Biotite	1.5
Aegerine-augite	2.5	Red Fe oxides and ore	2.0
Olivine	1.5	Apatite	0.5

Transitional rocks. Intermediate between the olivine-basalts and the olivine-trachytes is an olivine-bearing trachyandesite – vd. Table 78, Nos. 1 and 2. BARTH would name No. 1 an olivine-phonolite rather than a trachytoid trachybasalt, as per CAMPBELL SMITH. BARTH established the molecular norms of these two specimens (Table 83). Table 84 was compiled by BARTH to illustrate how the Gough rocks could be arranged in a series whereby certain characteristic properties are seen to change in a regular manner from member to member. The data presented in the table indicate that the rocks of Gough are co-magmatic, the more acid members having been derived from basaltic members by a process of fractional crystallization. If the contention is then that the trachytic rocks are derivatives through such crystallization of a basaltic magma, the basalts of Gough should, more or less, closely correspond to a parent magma, and indeed the chemical analyses presented in Tables 78 (Nos. 3, 4) and Table 79 are remarkably similar.

From his study of the publications of PIRSSON, CAMPBELL & CAMPBELL SMITH, coupled with his own investigations, BARTH concluded that the basaltic magma of Gough is parental, i.e. other island rocks are derivatives, and that it is a primary magma, i.e. directly derived from the basaltic substratum of the earth.

Table 83   Molecular Norms of Transitional Rocks (BARTH, 1942)

	Table 78			Table 78	
	No. 1	No. 2		No. 1	No. 2
Or	17.5	15.0	Di	13.2	11.6
Ab	40.0	47.5	Hy	–	3.8
An	6.2	15.5	Ol	6.3	1.5
Ne	14.1	–	Mt	1.3	3.3
C	–	–	Ap	1.6	1.6

1. Olivine-phonolite, dyke at sea level near Glen beach.
2. Olivine-trachyandesite, Devil's Wall dyke, Glen stream.

Table 84   Series Arrangement of Gough Island Rocks (BARTH, 1942)

	Basalt				Transitional		Trachyte
	No. 1 Table 79	No. 3 Table 78	No. 2 Table 79	No. 4 Table 78	No. 2 Table 78	No. 1 Table 78	No. 1A Table 81
SiO$_2$	47.2	48.9	49.1	49.1	54.9	55.8	60.7
$\frac{FeO}{MgO + FeO}$ (mol. %)	29	39	43	50	45	56	72
K$_2$O + Na$_2$O	4.2	5.3	6.3	5.7	7.8	10.3	12.0
CaO	7.9	8.9	8.9	8.9	7.0	5.6	1.3
An in norm. feld	34	30	30	34	20	10	5
Colour Index	48	40	32	30	22	22	8

LeMaitre (1962) has given a much more detailed account of the petrology of Gough Island. He claimed that classifications based on modal analyses were not practical with fine-grained rocks, and therefore named the rocks on the basis of their alkali contents, setting arbitrarily defined limits as follows for the Gough rocks:

	Total alkalis (%)	Soda (%)
Picrite basalt	< 3.5	
Olivine-basalt	3.5– 5.0	
Trachybasalt	5.0– 8.5	
Trachyandesite	8.5–10.5	
Trachyte	> 10.5	5.0–5.5
Aeg.-augite Trachyte	> 10.5	> 5.5

Picrite-basalts are relatively rare, found in only three places (Table 85, No. 1). Megascopically highly porphyritic, the rock consists of phenocrysts of black diopsidic augite and yellowish-green olivine, some grains of the latter having undergone iddingsitization. The intergranular groundmass comprises pyroxene, feldspar and iron ore in order of abundance. Magnetite and ilmenite are also present. These minerals all occur in an alkali-feldspar matrix with abundant tiny acicular needles of apatite.

Olivine-basalts, although of simple mineral content, have a wide textural range, from porphyritic to aphyric, vesicular to non-vesicular, aphanitic to crystalline. Porphyritic varieties are more common as flows, dykes and veins representing the more massive varieties. The porphyritic olivine-basalt (Nos. 2–5) under megascopic examination shows black pyroxene, yellow-brown olivine and a few colourless plagioclase phenocrysts. Microscopically the pyroxene indicates it is a slightly titaniferous augite. Olivine, partially altered to iddingsite, and iron ore occur as inclusions. The groundmass comprises feldspar, pyroxene, iron ore, olivine, magnetite and ilmenite laths and minute acicular apatite needles. Olivine-poor basalts (Nos. 6, 7) are megascopically finely vesicular rocks with 'phenocrysts' of Fe-Mg minerals in a dark grey matrix. Microscopically the 'phenocrysts' are seen to be glomerocrysts. The groundmass consists of plagioclase and pyroxene within a somewhat glassy iron ore-bearing medium.

Trachybasalts are the presumed most common mafic rocks on the island, and like the olivine-basalts, vary considerably texturally (Nos. 8–15). No. 8 in hand specimen is porphyritic, vesicular, with phenocrysts of black pyroxene, yellow-green olivine and some white plagioclase laths. Vesicles are large but not abundant. Microscopically the pyroxene has many inclusions of olivine and magnetite. The olivine shows only slight indications of iddingsitization. The intergranular groundmass comprises pyroxene, olivine and iron ore in interlocking plagioclase laths and interstitial alkali-feldspar. Equidimensional magnetite, lath-like ilmenite and subhedral pyroxene grains are present also. Some flakes of a purple-brown mineral (pseudo-brookite?) can be detected in the groundmass.

Trachyandesites (Nos. 16–18) also have various textures. No. 17 megascopically has abundant small phenocrysts of white plagioclase, black pyroxene and yellow-brown olivine in a light grey matrix. In places the rock shows vesicles. Microscopically the plagioclase phenocrysts often have narrow rims of alkali-feldspar. The pyroxene phenocrysts have inclusions of apatite, olivine and iron ore. The olivine phenocrysts are slightly

Table 85 Chemical Analyses and Norms (LeMaitre, 1962)

	1.	2.	3.	4.	5.	6.	7.	8.	9.	10.
SiO_2	46.57	46.04	49.73	47.85	46.72	51.16	51.05	48.89	45.90	48.82
TiO_2	1.85	2.97	3.30	3.40	3.29	2.91	2.55	3.07	3.25	3.09
Al_2O_3	8.20	14.16	15.53	15.05	16.36	15.90	16.79	16.15	17.25	17.25
Fe_2O_3	1.20	2.53	2.02	3.44	3.43	2.76	3.85	1.53	7.62	3.68
FeO	9.75	8.23	8.95	7.23	7.63	5.84	5.26	8.15	4.32	7.06
MnO	0.14	0.12	0.14	0.10	0.13	0.12	0.12	0.13	0.16	0.08
MgO	19.65	10.71	8.37	8.51	7.00	4.89	4.30	7.25	5.06	6.26
CaO	9.43	8.88	8.71	8.00	7.92	9.39	8.97	7.56	7.63	6.75
Na_2O	1.56	2.46	2.89	2.90	2.73	2.91	3.26	3.35	2.80	3.68
K_2O	1.18	1.45	1.70	1.97	1.29	2.25	2.37	2.61	2.50	2.52
P_2O_5	0.26	0.04	0.29	0.29	0.08	0.36	0.34	0.15	0.39	0.31
H_2O+	0.11	1.16	0.18	0.59	2.00	1.19	0.84	0.42	1.87	0.19
H_2O-	0.12	0.94	0.06	0.47	1.15	0.27	0.66	0.44	1.58	0.25
F	0.04	0.08	n.d.	n.d.	0.01	n.d.	n.d.	n.d.	0.07	0.15
Cl	n.d.	0.06	n.d.	n.d.	0.09	n.d.	n.d.	n.d.	n.d.	n.d.
	100.06	99.83	99.87	99.80	99.83	99.95	100.36	99.72	100.40	100.09
Less	0.02	0.04	–	–	0.02	–	–	–	0.03	0.07
Total	100.04	99.79	99.87	99.80	99.81	99.95	100.36	99.72	100.37	100.02
Q	–	–	–	–	–	1.68	0.96	–	–	–
Or	7.23	8.90	10.01	11.68	7.78	13.34	13.90	15.57	15.01	15.01
Ab	11.00	19.65	24.10	24.63	23.06	24.63	27.77	22.93	23.58	29.34
An	11.68	23.07	24.46	22.24	28.63	23.63	24.19	21.13	27.24	23.07
Ne	1.14	0.69	–	–	–	–	–	2.91	–	0.85
C	–	–	–	–	–	–	–	–	–	–
Di	26.84	16.22	13.69	12.36	7.78	17.12	14.78	12.58	6.26	7.10
Ol	36.11	19.53	16.57	11.20	4.70	–	–	15.03	2.66	12.32
Hy	–	–	0.86	4.49	13.23	7.97	6.18	–	6.00	–
Il	3.50	5.62	6.23	6.54	6.23	5.47	4.86	5.78	6.23	5.93
Mt	1.86	3.71	3.02	5.10	4.87	3.97	5.57	2.32	4.87	5.34
Ap	0.67	0.34	0.67	0.67	0.34	0.67	0.67	0.34	1.01	0.67
Hm	–	–	–	–	–	–	–	–	4.32	–
Wo	–	–	–	–	–	–	–	–	–	–
Remainder	0.25	2.20	0.24	1.06	3.23	1.46	1.50	0.86	3.49	0.52
Total	100.28	99.93	99.85	99.97	99.85	99.94	100.38	99.45	100.67	100.15

iddingsitized. Ilmenite and magnetite as equidimensional phenocrysts are often closely associated with colourless apatite phenocrysts. The intergranular groundmass comprises plagioclase, alkali-feldspar, pyroxene, olivine, iron ore and apatite. Probably pseudo-brookite also is present. Lining most of the vesicles is trachytic material comprising blocky crystals of alkali-feldspar and some acicular needles of apatite. Some flakes of biotite, small iron ore grains and perhaps amphibole occurs also in the matrix.

Trachytes can be divided into 1. trachytes, and 2. biotite-trachytes (Nos. 19–22). The former occur as highly vesicular flows, whereas the latter commonly occur as dykes.

Table 85 Chemical Analyses and Norms (LeMaitre, 1962)

	11.	12.	13.	14.	15.	16.	17.	18.	19.	20.
SiO_2	48.79	49.21	51.46	51.68	52.01	54.41	55.80	55.89	58.17	59.17
TiO_2	3.18	2.62	2.69	2.68	2.11	1.67	1.87	1.60	1.00	0.85
Al_2O_3	17.39	18.26	17.12	17.50	18.76	17.37	18.41	16.90	19.31	18.83
Fe_2O_3	2.48	6.82	2.96	3.41	2.46	4.02	3.07	1.41	1.98	1.46
FeO	7.39	2.32	6.05	6.30	5.99	3.29	3.78	6.71	3.45	3.65
MnO	0.10	0.13	0.15	0.12	0.05	0.12	0.09	0.13	tr.	0.14
MgO	4.00	4.24	4.03	2.90	3.11	2.27	2.13	2.26	1.05	0.93
CaO	8.97	5.37	5.94	6.92	6.60	4.36	5.07	4.48	1.84	2.08
Na_2O	3.28	3.75	4.06	3.91	4.34	4.94	4.57	4.44	5.05	5.13
K_2O	2.28	3.88	3.69	3.70	3.38	4.69	4.30	4.58	6.55	6.60
P_2O_5	0.26	0.54	0.26	0.19	0.44	0.46	0.23	0.34	0.36	0.15
H_2O+	0.98	1.66	1.09	0.54	0.34	0.86	0.34	0.99	0.84	0.66
H_2O-	0.76	1.00	0.26	0.43	0.14	1.50	0.26	0.26	0.50	0.28
F	n.d.	n.d.	n.d.	n.d.	0.16	0.13	n.d.	0.04	0.07	n.d.
Cl	n.d.	n.d.	n.d.	n.d.	n.d.	n.d.	n.d.	0.04	n.d.	n.d.
	99.86	99.80	99.76	100.28	99.89	100.09	99.92	100.07	100.17	99.93
Less	–	–	–	–	0.07	0.05		0.03	0.03	–
Total	99.86	99.80	99.76	100.28	99.82	100.04	99.92	100.04	100.14	99.93
Q	–	–	–	–	–	–	–	–	–	–
Or	13.34	22.80	21.68	21.68	20.02	27.80	25.58	27.24	38.92	38.92
Ab	27.77	31.05	31.05	30.39	33.54	39.30	38.77	37.73	41.65	39.82
An	26.13	21.41	17.51	19.46	21.68	11.40	16.68	12.51	6.39	8.62
Ne	–	0.50	1.91	1.42	1.70	1.42	–	–	0.43	1.99
C	–	–	–	–	–	–	–	–	1.53	–
Di	13.57	1.30	8.51	11.31	6.74	5.19	5.96	6.77	–	0.94
Ol	6.99	7.00	7.40	4.57	6.06	2.40	–	6.26	4.06	4.76
Hy	–	–	–	–	–	–	3.79	2.58	–	–
Il	6.08	6.88	5.17	5.17	3.95	3.19	3.65	3.04	1.98	1.52
Mt	3.71	–	4.41	4.87	3.48	5.80	4.41	2.09	3.02	2.09
Ap	0.67	1.34	0.67	0.34	1.01	1.34	0.34	0.67	1.01	0.34
Hm	–	5.02	–	–	–	–	–	–	–	–
Wo	–	–	–	–	–	–	–	–	–	–
Remainder	1.74	2.66	1.35	0.97	0.57	2.44	0.60	1.30	1.38	0.94
Total	100.00	99.96	99.66	100.18	98.75	100.28	99.78	100.19	100.37	99.94

Vesicular trachyte (No. 19) in hand specimen is abundantly vesicular, of elongated, flattened and arcuate shape. The sporadic white feldspar 'phenocrysts' seen megascopically are, when studied under the microscope, seen to be glomerocrysts consisting of interlocking crystals of alkali-feldspar and plagioclase. Associated with these are a few highly resorbed crystals of biotite, completely replaced by granular iron ore. Small phenocrysts of olivine are common. The orthophyric-trachytic groundmass comprises alkali-feldspar, magnetite, olivine and ferro-magnesian material. Biotite-trachyte (No. 22) megascopically is porphyritic with many phenocrysts of white feldspar, biotite, some

Table 85 — Chemical Analyses and Norms (LeMaitre, 1962)

	21.	22.	23.	24.	25.	27.	26.	28.
SiO_2	57.54	59.88	60.52	60.90	61.14	62.45	60.17	60.48
TiO_2	1.22	0.90	0.45	0.37	0.52	0.38	0.15	0.30
Al_2O_3	18.76	18.21	18.72	18.76	18.27	17.92	18.45	18.02
Fe_2O_3	4.54	1.68	2.67	1.30	2.48	2.06	3.32	2.46
FeO	1.56	3.54	2.85	3.67	2.53	3.11	2.53	2.76
MnO	0.25	0.12	0.19	0.21	0.15	0.16	0.23	0.26
MgO	1.08	0.97	0.52	0.23	0.57	0.09	0.04	0.26
CaO	2.42	2.50	1.87	1.66	1.72	1.17	1.44	1.86
Na_2O	5.01	5.35	5.83	6.30	5.78	6.47	7.52	6.71
K_2O	5.66	6.19	5.77	6.03	6.00	6.14	5.43	5.70
P_2O_5	0.15	0.29	0.19	0.10	0.17	0.09	0.04	0.07
H_2O+	0.89	0.21	0.32	0.25	0.24	0.16	0.31	0.90
H_2O-	0.54	0.10	0.07	0.07	0.18	0.27	0.30	0.26
F	0.21	n.d.	n.d.	n.d.	n.d.	n.d.	n.d.	0.05
Cl	0.07	0.02	n.d.	n.d.	n.d.	0.03	0.38	0.17
	99.90	99.96	99.97	99.85	99.75	100.50	100.31	100.26
Less	0.10	0.01	–	–	–	0.01	0.08	0.06
Total	99.80	99.95	99.97	99.85	99.75	100.49	100.23	100.20
Q	–	–	–	–	–	–	–	–
Or	33.92	36.70	34.47	35.58	35.58	36.14	32.25	33.92
Ab	42.44	43.74	48.86	46.37	49.25	52.00	48.73	47.50
An	11.12	7.23	7.51	5.00	5.84	1.39	0.56	1.95
Ne	–	0.71	0.21	3.83	–	1.63	7.95	4.90
C	0.20	–	–	–	–	–	–	–
Di	–	2.80	0.70	2.20	1.40	3.24	3.94	5.56
Ol	–	3.49	2.86	3.69	0.96	1.57	–	0.59
Hy	2.70	–	–	–	1.38	–	–	–
Il	2.28	1.67	0.91	0.76	0.91	0.76	0.30	0.61
Mt	2.55	2.55	3.71	1.86	3.71	3.02	4.87	3.71
Ap	0.34	0.67	0.34	0.34	0.34	0.34	0.34	0.34
Hm	2.72	–	–	–	–	–	–	–
Wo	–	–	–	–	–	–	0.58	–
Remainder	1.61	0.32	0.39	0.32	0.42	0.45	0.91	1.32
Total	99.88	99.88	99.96	99.95	99.79	100.54	100.43	100.40

pyroxene in a green-grey, aphanitic groundmass. In thin section the phenocrysts comprise alkali-feldspar, plagioclase (often these two are interlocked), biotite, olivine, iron ore and pyroxene. The trachytic-textured groundmass consists chiefly of abundant laths of alkali-feldspar, between which are granules of magnetite, olivine and alkali-feldspar. A few crystals of amphibole(?) are also present.

Aegerine-augite Trachytes differ from trachytes in that they contain aegerine-augite, sometimes along with sodalite. They are the sole rocks of the trachytic plugs and domes, but also occur as flows (Nos. 23–28). In hand specimen, No. 26 is a hard, compact rock

Table 86 Names, Types and Localitites of Analyzed Rocks

Sample No.	Name	Type	Locality
1.	Picrite Basalt	Dyke	Mouth of Deep Glen
2.	Porphy. Olivine-Basalt	Vein	W side, Archway Rock
3.	Porphy. Olivine-Basalt	Dyke	S side, Archway Rock
4.	Olivine-Basalt	Flow	1/2 mile up The Glen stream
5.	Porphyritic Basalt	?	Boulder, base of cliffs W of Buttress Rock.
6.	Olivine-poor Basalt	Dyke	The Glen stream, 70 ft. above S. L.
7.	Olivine-poor Basalt	Dyke	The Glen stream, 240 ft. above S. L.
8.	Porphy. Trachybasalt	Flow	Waterfall Point, under trachyte flows.
9.	Porphy. Trachybasalt	Flow	N shoulder of Mt. Rowett.
10.	Porphy. Trachybasalt	?	Boulder on beach S of Buttress Rock
11.	Trachybasalt	Dyke	Cliffs N of The Glen beach
12.	Vesicular Trachybasalt	Flow	Summit of Edinburgh Peak
13.	Trachybasalt	Flow	Edge of plateau, head of Deep Glen
14.	Trachybasalt	Dyke	Cliffs, mouth of The Glen
15.	Porphy. Trachybasalt	Dyke	N slopes of The Glen, 150 ft. from S. L.
16.	Trachyandesite	?	Block from tuff, Penguin Is.
17.	Porphy. Trachyandesite	Dyke	Wall Dyke, The Glen
18.	Trachyandesite Glass	Dyke	Selvage to dyke opposite to No. 7
19.	Vesicular Trachyte	Flow	1/4 mile S of South Peak
20.	Vesicular Trachyte	Flow	1-1/4 miles W of South Peak
21.	Biotite-Trachyte	Dyke	The Glen stream, 240 ft. above S. L.
22.	Biotite-Trachyte	Dyke	Mouth of The Glen
23.	Aeg.-Augite-Trachyte	Flow	1/4 mile E of Waterfall Point
24.	Aeg.-Augite-Trachyte	Dome	1/4 mile E of Edinburgh Peak
25.	Aeg.-Augite-Trachyte	Dome	1/2 mile NNW of Edinburgh Peak
26.	Aeg.-Augite-Trachyte	Plug	Hag's Tooth
27.	Sodalite-Aegerine-Augite-Trachyte	Plug	1/4 mile E of Nigel's Cap
28.	Sodalite-Aegerine-Augite-Trachyte	Plug	Head of Deep Glen.

involving black crystals in a mottled light grey groundmass. Under the microscope it has a trachytic-bostonitic texture, with abundant alkali-feldspar, some aegerine-augite, olivine and magnetite. Some of the larger alkali-feldspar crystals show a slight degree of unmixing. The aegerine-augite occurs in almost 'skeletal' form, filling the interstices between the alkali-feldspar; it may also occur however as discrete 'nonskeletal' crystals. The olivine, almost pure fayalite, is not so common as the aegerine-augite. Equidimensional magnetite grains are abundant. Acicular needles of apatite and very minor amounts of zircon are also present. Sodalite can locally be abundant in the groundmass in some of these trachytes (Nos. 27, 28). The colourless sodalite mostly occurs as discrete crystals in small vesicles, often bordered with aegerine.

From the chemical analyses and norms presented in Table 85, it can be stated that in general the Gough rocks are undersaturated as regards silica. No evidence from the specimens studied by LeMaitre gave any indication of tholeiitic tendencies.

Table 87 — Average Composition of Typical Rock Types (LeMaitre, 1962)

	Average Composition (Wt. %)								
	SiO_2	TiO_2	Al_2O_3	Fe_2O_3	FeO	MgO	CaO	Na_2O	K_2O
1.	46.8	1.9	8.2	1.2	9.8	19.8	9.5	1.6	1.2
2.	47.7	3.2	15.2	2.3	8.7	9.7	8.9	2.7	1.6
3.	51.1	2.8	17.6	2.8	6.8	4.8	6.9	4.0	3.2
4.	56.3	1.8	17.8	2.9	4.7	2.3	4.7	4.8	4.7
5.	59.5	0.9	19.4	1.7	3.6	1.0	2.0	5.2	6.7
6.	61.5	0.3	18.3	2.6	2.8	0.2	1.5	7.0	5.8

	Norms								
	Or	Ab	An	Ne	Di	Ol	Il	Mt	Salic (%)
1.	7.2	9.8	11.4	2.0	28.6	35.5	3.6	1.9	30.4
2.	9.5	21.9	24.5	0.6	15.8	18.4	6.1	3.2	56.5
3.	18.9	28.3	20.5	3.1	11.0	8.7	5.3	4.2	70.8
4.	27.8	38.4	13.3	1.1	8.1	3.6	3.5	4.2	80.6
5.	39.5	39.1	9.7	2.6	0.3	4.5	1.7	2.6	90.9
6.	34.5	48.9	1.1	5.6	5.5	0.1	0.6	3.7	90.1

	Molecular Composition Feldspar			Mol % Fa in Olivine	Mol % Ilmenite in Ore
	Or	Ab	An		
1.	22	44	34	17	75
2.	16	42	42	22	74
3.	25	48	27	25	66
4.	33	51	16	29	56
5.	41	29	10	53	50
6.	35	64	1	79	20

1. Picrite-basalt. (No. 1.)
2. Olivine-basalt. (Nos. 2, 3)
3. Trachybasalt. (Nos. 8, 10, 13, 14, 15)
4. Trachyandesite. (Nos. 16, 17, 18)
5. Trachyte. (Nos. 19, 20)
6. Aegerine-augite-Trachyte. (Nos. 26, 27, 28)

(Numbers in brackets refer to specimens in Table 85)

This author (1962) divided the rocks studied from Gough into six groups and gave the average composition of the typical rock types (Table 87). From this table it is seen that, with the exception of the picrite-basalt, orthoclase and albite increase gradually through the series, but normative anorthite, diopside, olivine and ilmenite decrease steadily. The percentage of salic minerals increases rapidly at first, then more gradually. The Fe-content of the olivine at first increases slowly, then more quickly. The ilmenite content of the total ore minerals also decreases.

LeMaitre (op. cit.) provided three lines of evidence to support his contention that the Gough rocks were formed from an alkali-olivine-basalt magma by means of crystal differentiation: 1. Many rock types have residual trachytic patches lining vesicles, e. g. in the picrite-basalt (Table 85, No. 1). This is actually the last liquid fraction to crystallize, the high volatile content in the vicinity of the vesicles lowering the melting point of the liquid. The trachytic nature of such patches is proven both texturally and mineralogically. Less obvious evidence also comes from the interstitial alkali-feldspar usually present in the rocks. 2. A table presented shows that when some materials are subtracted from an average olivine-basalt in appropriate quantities, there results average trachybasalt, trachyandesite, trachyte and aegerine-augite trachyte, said material compositions being determined graphically, using the addition-subtraction diagram of N. L. BOWEN (1928). LeMaitre admits that this method does not lead to a unique solution of the problem, and requires a certain degree of manipulation. He claimed however that another line of chemical evidence substantiates his views, viz. the 'petrogeny's residual system' of BOWEN (1937). According to this method, if the normative salic constituents of the entire trachytic rocks are re-calculated in terms of $NaAlSiO_4$, $KAlSiO_8$ and SiO_2 and accordingly plotted, then if they form a sequence of residual liquids derived from a common source, they should not only lie in the low-temperature part of the system but should also show a distinct trend, with the earlier liquids at a higher temperature than the later ones. A study of specimens 19–28 in Table 85 shows conformity with this principle. 3. When the normative compositions of the feldspars from the six average rock types given in Table 87 are plotted on an Or-Ab-An triangle, the numbers correspond to those given in this table. The normative feldspars, i. e. from olivine-basalt to trachyte, lie on a smooth curve, convex side towards the Ab apex, and the normative feldspar from the average aegerin-augite-trachyte lies off this curve towards the Ab apex. If the rocks belong to a series where crystal differentiation has been active, then the points plotted of these normative feldspars will represent approximately the composition of the feldspathic component of the successive liquids in the crystallizing melt. Such a study bears this out.

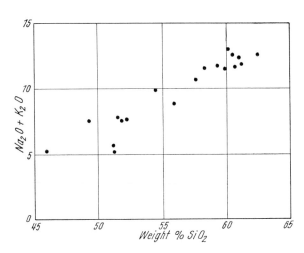

Fig. 98. Total Alkali-Silica Diagram. (LeMaitre, 1962)

LeMaitre (1965) also commented on the prevalence of gabbroic xenoliths in many basalt and trachybasalt flows, dykes, in some tuff horizons but absent in trachytic late differentiates. Peridotitic xenoliths are rare. The gabbroic ones, of various shape and

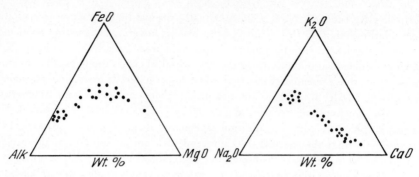

Fig. 99. MgO–FeO–Alk triangular variation Diagram. (LeMaitre, 1962)

Fig. 100. CaO–Na$_2$O–K$_2$O triangular variation Diagram. (LeMaitre, 1962)

measuring up to 20 cms across, have, on occasion, been entirely re-crystallized into hornfels. A chemical analysis of two gabbroic xenoliths, after LeMaitre, is given in Table 88. He suggests that the xenoliths find their origin in the mantle, from depths of 40–50 km below the surface, i. e. just above the transformation zone, and that the bulk composition of the mantle in this vicinity is essentially that of an olivine-tholeiite.

	1.	2.
SiO$_2$	47.89	52.42
TiO$_2$	1.19	0.68
Al$_2$O$_3$	7.64	18.57
Fe$_2$O$_3$	5.20	1.45
FeO	8.27	3.63
MnO	0.08	0.13
MgO	19.40	8.09
CaO	8.27	11.33
Na$_2$O	1.42	3.30
K$_2$O	0.36	0.45
P$_2$O$_5$	0.11	0.05
H$_2$O+	0.39	0.06
H$_2$O–	0.19	0.07
Total	100.41	100.23

Table 88 Chemical Analyses of Gabbroic Xenoliths (Le Maitre, 1965)

Isotope Geochemistry

Two papers, one by Gast, Tilton & Hedge (1964), the other by Tilton et al (1964) give some isotope lead and strontium analyses of Gough Island volcanics, as also some from Ascension Island. The Gough samples were from the LeMaitre collection (1962), the chemical analyses of which are given in Table 85, the types and localities in Table 86.

In Table 89 are listed the lead and strontium compositions of six LeMaitre samples by Gast et al, and in Table 90, five of the same samples are taken from Tilton et al (Numbers refer to those in Tables 85 and 86).

Table 89 Isotopic Composition of Lead and Strontium from some Volcanics (GAST et al, 1964)

	Isotopic Lead and Strontium Compositon				
	$\dfrac{Pb\ 206}{Pb\ 204}$	$\dfrac{Pb\ 206}{Pb\ 207}$	$\dfrac{Pb\ 206}{Pb\ 208}$	$\dfrac{Sr\ 87}{Sr\ 86}$	$\dfrac{Sr}{Rb}$
Ol.-poor basalt (No. 6)	18.36	1.171	0.4711	0.7045 / 0.7038	–
Trachybasalt (No. 15)	18.43	1.171	0.4695	0.7043	17
Trachyandesite (No. 17)	18.37	1.170	0.4694	0.7050	13
Trachyte (No. 24)				0.7069	0.05
Trachyte (No. 25)	18.71 / 18.73	1.193 / 1.191	0.4744 / 0.4738	0.7050	0.5
Trachyte (No. 27)	18.63 / 18.63	1.178 / 1.179	0.4688 / 0.46.99	0.7094	0.0155

Table 90 Isotopic Composition of Lead from some Volcanics (TILTON et al, 1964)

	Isotopic Lead Composition				
	$\dfrac{Pb\ 206}{Pb\ 204}$	$\dfrac{Pb\ 206}{Pb\ 207}$	$\dfrac{Pb\ 206}{Pb\ 208}$	$\dfrac{Pb\ 207}{Pb\ 204}$	$\dfrac{Pb\ 208}{Pb\ 204}$
Ol.-poor basalt (No. 6)	18.37	1.1713	0.4711	15.68	38.99
Trachybasalt (No. 15)	18.43	1.1712	0.4695	15.74	39.25
Trachyandesite (No. 17)	18.38	1.1708	0.4694	15.70	39.16
Trachyte (No. 25)	18.72	1.1921	0.4741	15.70	39.49
Trachyte (No. 27)	18.64	1.1791	0.4693	15.81	39.72
Basalt (Average)	18.40	1.171	0.470	15.71	39.15
Trachyte (Average)	18.68	1.185	0.472	15.76	39.60

The Pb 206/Pb 204 ratios are higher in the trachytes than in the basalts and trachyandesite. The uranium-lead ratio (1.6 ppm uranium, 8.4 ppm lead) of 14 for No. 25 indicates that it must have had a higher Pb 206/Pb 204 ratio than the basalts at the time of emplacement of the trachyte. For such a uranium-lead ration, 140 million years (GAST et al) or 125 million years (TILTON et al) would be needed to generate the difference of 0.3 noted in the Pb 206/Pb 204 ratios. A preliminary K-Ar age determination for the other trachyte (No. 27) suggests that the age of the rock is less than 2 million years. It is concluded that the Pb 206/Pb 204 difference was present when these rocks were emplaced, and hence the various rock types could not all originate from an isotropically homogeneous magma. It should also be noted that the higher Pb 206/Pb 204 ratios for the trachytes could be accounted for by contamination of basaltic lead with pelagic lead. It is seen too from Table 90 that the trachytes have likewise a higher Pb 208/Pb 204 ratio than the other samples.

As regards the strontium isotope ratios, GAST et al remark that the initial ratios for Nos. 24 and 27 are rather doubtful because of the high rubidium/strontium ratios and the uncertainty as to the age of these samples. If No. 27 is indeed considerably less than 2 million years, then the initial ratio in the rock is higher than that of the more basic rocks.

Geological Evolution

LeMaitre (1960) presumed that Gough is of Tertiary age, as per the rest of the Mid-Atlantic Ridge. He divided the chronologic sequence of events thus:

5. Upper 'basalts'
4. Upper 'trachytes'
3. Middle 'basalts'
2. Lower 'trachytes'
1. Lower 'basalts'

The sequence represents the result of a series of fissure lava flows. As the xenoliths (vd. supra) occur in the lowermost flow, the assumption is made that these form part of the cone below sea level. This cone, of which Gough is the visible expression, rises some 3000 m above the top of the Mid-Atlantic Ridge before reaching sea level.

The first event was the outpouring of basaltic-type flows, with local small occurrences of tuff. The lower trachytic phase was ushered-in by great explosive action which disintegrated most of the then island, leaving only a stump of basalts on the E coast. After the explosive phase, trachytic flows poured over and around the basaltic stump. The middle basalts, confined to the higher parts of the island, overlie both the older series. Petrographically, the lower and middle basalts are indistinguishable, but that there is a distinction is evidenced in the occurrence of basalts overlying trachytic flows, e. g. at Nigel's Cap, and whilst many dykes cut basalt flows in the Glen, none cut the flows at Mt. Rowett, Kern Crag or South Peak. The upper trachytes are well represented by plugs, e. g. Hag's Tooth at the head of the Glen, on the E side of Nigel's Cap. These plugs were forcibly intruded, on occasion tearing off xenoliths and brecciating the country rock. Domes are also present, likely supplied by separate feeders. No dykes are seen to cut this phase. Originally it was believed that the upper basalts were identical with the middle basalts, the upper trachytes being intrusive rather than extrusive. But because of similarities between the domes of Gough and Ascension, it appears more likely that the basaltic capping of Edinburgh Peak represents yet another extrusive phase of the same feeder which supplied the underlying domes, such as occurs in Ascension.

It thus appears that there was a cyclical sequence of basaltic-trachytic occurrences. The much greater exposure of trachytic rocks rather than basaltic is due to the first-formed basaltic sequence having undergone drastic disintegration at the time of the explosions prior to outpouring of the first trachytic phase. No faulting or folding can be detected on the island, but the relatively linear trend of the E coast, coupled with the westerly dip slopes, suggest that the E coast outlines a fault plane (Douglas, unpublished report). The NE part of Gough has possibly undergone comparatively recent uplift. Sea-worn caves, some 3 m above present high mean tide, and at ca. 5 m above the same datum, the occurrence of a pebble beach – raised sea beach or gravel bar formed by a stream from the interior? – these are suggestive of such movement.

A study of the basaltic-type rocks of Gough shows similarities with those of St. Helena and Tristan da Cunha. The late differentiates of the Azores, Ascension, St. Helena, Tristan da Cunha and Bouvet Island can all be correlated with similar rocks in Gough. Noting that these islands are the exposed summits of the Mid-Atlantic Ridge, the mode of origin of this immense world feature is obviously closely related petrologically to the intrinsic igneous events of these minute outposts of land set midst the vast ocean expanses.

Age Determinations

The publication by BAKER, GASS et al (1964) on the Tristan Group gives three age determinations for Gough. Only one of the samples (LEMAITRE number G-159) is recorded in the chemical analyses (Table 85) under No. 28.

Table 91   Age determinations on samples from Gough Island (BAKER, GASS et al, 1964)

	Sample	K_2O (%)	Atmospheric contamination (%)	Age and estimated error. (m. yr.)
82013	Olivine-basalt The Glen	2.33	21.5	6 ± 2
G–159	Trachyte plug, Head of Deep Glen	5.5	78.2	3 ± 1
G–80	Trachyte, near S. L., Waterfall Point	5.8	44.7	2 ± 1

$\lambda\varrho = 0.584 \times 10^{-10}$ yr^{-1} $\lambda\beta = 4.72 \times 10^{-10}$ yr^{-1}

CHAPTER 15

Falkland Islands

General

The Falkland Islands archipelago, numbering several hundred islands, large and small, lies between lats. 51° S and 52° 56′ S and longs. 57° 42′ W and 61° 19′ W. In latitude 52° S, the nearest point of the islands lie some 535 km distant from the Patagonian coast of Argentina.

The two principal islands are West Falkland or Gran Malvina and East Falkland or Isla Soledad, separated by the Falkland Sound or San Carlos Strait, the latter measuring up to 40 km in breadth. Both main islands are highly indented, with about 200 smaller islands lying close to the main ones, and several more distant, Ile Beauchêne being the most distant, some 60 km S of East Falkland.

The area of the Falklands is usually given as 11 960 km², but the Instituto Geografico Militar of Argentina lists this to be 11 718 km². West Falkland and adjacent islands have an area of 5278 km², East Falkland and adjacent islands, 6682 km². West Falkland is about 130 km long, 70 km broad, rising to a maximum elevation of 698 m in Mount Adam; East Falkland measures 166 km in length, 80 km in width, and has a maximum height of 685 m in Mount Osborne.

The Falkland Islands Colony includes the dependencies of South Georgia, South Sandwich Islands, South Orkneys, South Shetlands and other islands lying to the S in the Scotia Bay, and also part of Antarctica.

For long Argentina has laid claim to the Falklands, and even today does not recognize sovereignty of Great Britain over the archipelago. Because of this dual claim, there is a duality in geographic names, the islands being known as Islas Malvinas in Argentina.

Although the archipelago is treeless, there is actually quite a dense vegetation, tussock grass being profuse. Most of the soils are peaty, and many peat bogs are present. Native mammals are now extinct. The few land birds are migrants from South America. No cereal crops are grown except limited amounts of oaten hay. Some vegetables and berries are grown, but actually all foodstuffs are imported. Fishing is important locally, the mullet and smelt being used chiefly for local consumption. Nearly all the major islands are divided into sheep farms, each farm also supporting dairy cattle.

The principal items of export are wool, tallow, hides, whale and seal oil.

There are no aboriginal peoples. The present population is about 2100, three-quarters of whom live in East Falkland. The capital and only town and real harbour is Port Stanley, population about 1200, located in the extreme eastern part of East Falkland. Darwin is a small outpost also on East Falkland. Most of the peoples are of British stock, Scottish and Welsh largely.

There are no railways or inland navigation systems.

Physical Features

The topography of the islands is controlled largely by the geology, structure and tectonics. West Falkland is more hilly, with two chief ranges, the Hornby Mountains, trending parallel to the E coast, and the Byron Heights, forming the peninsula between Byron Sound and King George Bay. In East Falkland, the E-W trending Wickham Heights divide the island into a gently undulating northern region and a low-lying plain to the S, nowhere over 50 m above sea-level. Apart from these ranges, there are isolated mountains, and in both islands, highest elevations lie in the northern sectors.

The general low relief of the Falklands is geomorphologically very similar to the Chubut hill country of Central Patagonia.

Both West and East Falkland, in spite of the heavy rainfall, are lacking in well-developed drainage networks. The short streams are often decidedly sinuous and not well entrenched, extending into open rias, bays, inlets of various dimensions. Valleys are filled with sub-angular to angular blocks, known locally as 'stone rivers' or 'stone runs'. These, as well as other features, are interpreted as indicative of a peri-glacial environment during the Pleistocene. Large areas of both islands are poorly drained, and here extensive peat bogs occur.

The lengthy coastlines of submergence are highly indented, innumerable bays, sounds, rias penetrating far inland. The coastal aspect of the archipelago is very similar to that in Tierra del Fuego. Both the cliffed and subdued coastal topography is forever under strong marine erosional attack. Higher cliff scenery is found at Cape Meredith, and cliffed scenery finds its counterpart in many coastal stretches of Patagonia. Choiseul Sound and the embayment where Port Stanley lies have somewhat lower cliffs, and in general cliffs are low throughout the archipelago. Falkland Sound, trending NE-SW, separating the two major islands, is bordered on both sides by rather abrupt coastlines, especially the more rectilinear West Falkland coast. On the other hand, the SW side of the Sound is highly irregular, where softer Triassic sediments outcrop.

Climate

The Falkland Islands, windswept, drenched by cold, drizzly rains, bleak and treeless, presents a rather forbidding and inhospitable appearance. Mean average temperatures are 5.5° C, varying from 0.5° C in August to 10.5° C in January. However because of the maritime influences, temperatures below freezing are rare, even in winter. The average annual rainfall is 700 mm, rather evenly distributed throughout the year. Rain or snow are said to fall on an average of 260 days annually. Seas are often stormy, gales occur about once every five days. Strong winds are persistent all year round. At Port Stanley the average maximum daily temperature is 9° C, the average minimum daily temperature, 2° C; average annual rainfall, 680 mm.

Geology

Constituting an emergent part of the Patagonian Shelf, the Falkland Islands, though subject to geological enquiry for some 130 years, are still but imperfectly known. Most surveys have concentrated on coastal sections and most of the interior of the two main islands has been given only cursory examination. Added to this must be mentioned the fact that large tracts have a thick tussock grass and/or peat covering, concealing the rocks.

Opinion is unanimous that the oldest rocks, which have a small outcrop in the extreme S of West Falkland, are Archaean or Precambrian. The oldest sediments are classed as Lower Devonian, though palaeontological proof is lacking. These beds rest with marked discordance on the basement complex. The decipherable stratigraphical succession includes rocks ranging in age from Lower Devonian to Upper-Middle Triassic. Again there is a very great hiatus, for no further Mesozoic or Tertiary rocks have been discovered, and the Recent has extremely few occurrences.

Metamorphics are present in the most ancient rocks but are very sparse in younger sequences. The pre-Gondwana and Gondwana sediments total some 6500 m at least in thickness. Granitic and pegmatitic intrusions, also lamprophyric-type dykes occur in the Precambrian, and diabasic-doleritic dykes intrude the sediments.

Folding is common throughout, and several lengthy, tight folds occur. Small-scale faulting can be seen in the Devonian rocks, usually parallel to the fold axes, but of greater significance are the major fractures which, in certain places, outline coastal sections.

Stratigraphical correlations can be made between the archipelago and South Africa and South America. The folded ranges of the Falklands can be correlated with the Gondwanides of Argentina and the 'Cape Foldings' of South Africa.

A reasonably voluminous literature is available treating of various geological aspects, dating from the years 1833–34 when DARWIN, in the famous voyage of the "Beagle" landed on the archipelago. His publications, those of ANDERSSON, HALLE, BAKER, ADIE & BORRELLO, plus palaeontological writings of MORRIS & SHARPE, ETHERIDGE, NEWTON, NATHORST, CLARKE, SEWARD & WALTON form essential background material.

Sedimentary Rocks

The sedimentary succession comprises a somewhat monotonous series of siltstones, mudstones, fine and coarse standstone, quartzites, conglomerates, boulder beds, sandy shales and shales. Siltstones and mudstones are pretty well restricted to Permo-Triassic strata. Quartzites, quartzitic sandstones and quartz-conglomerates are more usually associated with Devonian and Lower Carboniferous(?) rocks. The tillites or glacial boulder beds comprise the Permo-Carboniferous. Arenaceous shales are commonest in the Middle Devonian. Most of the Fox Bay and Port Philomel sandstones are micaceous; feldspathic sandstones typify the Lafonian Sandstone, whereas quartzitic sandstones are commonest

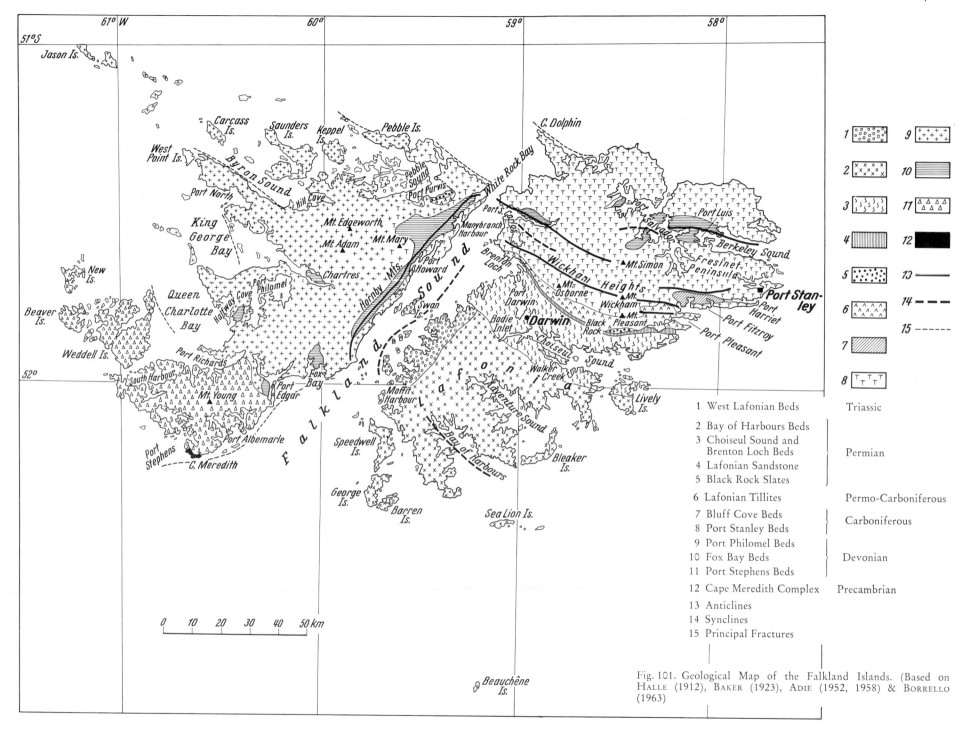

Fig. 101. Geological Map of the Falkland Islands. (Based on HALLE (1912), BAKER (1923), ADIE (1952, 1958) & BORRELLO (1963)

Beiträge zur Regionalen Geologie der Erde Band 10: MITCHELL-THOMÉ, S. Atlantic Islands
Gebrüder Borntraeger, Berlin · Stuttgart

in the Port Philomel and Port Stanley beds. A common feature in the Port Stephens and Lafonian Sandstone beds is the occurrence of current bedding of the sandstones.

The Lower Devonian is a glacio-marine facies, whereas the Middle and Upper Devonian or Lower Carboniferous are of terrestrial origin. The Upper Carboniferous may be either of marine or then fluvio-glacial facies, succeeded by Permo-Carboniferous terrestrial glacial deposits. The Lower Permian represents a terrestrial fluvio(?) and limno-glacial facies; Middle-Upper Permian sediments represent glacial deposition in a more typical freshwater environment. The Triassic represents continental sedimentation.

Long, sinuous accumulations of loose boulders along with clays, forming the 'stone rivers', shelly limestones with abundant angular detrital material, scattered pebbles and tree trunks belong to the Quaternary, the bouldery material dating from the Pleistocene, the other deposits being of Recent age.

Igneous Rocks

Unlike other islands of the South Atlantic treated here, the Falklands are not of volcanic constitution. Igneous rocks play a minor role in the development of the archipelago.

The Precambrian basement includes intrusions of microcline-granites invaded by microline-pegmatites, both occurring as laccoliths, dykes and veins within the crystallines. Lamprophyric dykes also are to be noted, which, as they do not occur in the Lower Devonian sediments, can only be classed as of pre-Middle Palaeozoic age.

The sedimentary sequence is invaded by many diabasic or then doleritic dykes. On a petrographic basis, the doleritic types are of three kinds: 1. Porphyritic olivine-dolerites, representative of the Fox Bay area. The rocks are of dark colour, medium-grained and porphyritic, with many idiomorphic olivine small phenocrysts in a groundmass of tabular plagioclase and pigeonite. The zoned plagioclase shows Pericline, Albite and Carlsbad twinning. Accessories include titaniferous magnetite, apatite, iron pyrites and some biotite. 2. Porphyritic olivine-free dolerites occur in the Brenton Loch-Port Sussex area. It is presumed that the porphyritic aspect is due to large phenocrysts of plagioclase, enstatite-augite and pseudomorphs after rhombic pyroxene. Accessories include biotite and much iron ore. 3. Ophitic and amygdaloidal dolerites can be observed at Fox Bay and Halfway Cove. The texture is ophitic, due to the intimate intergrowth of labradorite-bytownite and clino-pyroxene. The few phenocrysts showing most frequently are serpentinized olivines and pseudomorphs after rhombic pyroxene. The dolerites at Halfway Cove show a strong amygdaloidal character, the cavities being composed of chloritic and carbonate matter. At Fox Bay, the groundmass of the rocks comprises ophitic intergrowths of andesine-labradorite and pigeonitic augite.

In the Fox Bay area occurs an olivine-diabase dyke with many olivine phenocrysts which are slightly altered. The groundmass comprises augite and basic plagioclase. In this locality also occurs an olivine-free diabase-porphyrite dyke with plagioclase phenocrysts in an ophitic matrix of plagioclase and augite.

DARWIN mentioned "numerous basaltic dykes" in the Falklands but omitted to report specifically where these occur. RENARD (1889) described basaltic-type dyke rocks from the "Challenger" collection, but HALLE (1912) believed that these were boulders only. DARWIN himself did not observe his "numerous basaltic dykes", as these were reported to him by a crew member of the "Beagle".

Metamorphic Rocks

The Cape Meredith Complex is principally composed of crystalline schists and gneisses of various kinds. BAKER (1922b) described the schists as being entirely of green hornblende and quartz. Many gneisses are garnetiferous. Granitic and pegmatitic bodies are more usually intruded into the gneisses rather than the schists.

The Black Rock formation comprises dark grey and black slates of a distinct cherty character. These slates lie between a lower layer of quartzite pebbles and an upper layer of dark shales, the slates totalling some 10 m in thickness. It may be that these slates should rather be thought of as silicified shales rather than true metamorphics. The flaggy and slaty sediments in the Fox Bay formation also appear to be scarce true metamorphics. In both instances it is probable that sediments more prone to react under dynamic forces have become slightly altered without going so far as to create real metamorphic transformations.

Stratigraphy

Rocks as old as Precambrian and as young as Recent are recognized in the archipelago, but the bulk of these are of Middle and Upper Palaeozoic age.

Most of West Falkland comprises Middle Palaeozoic strata, as also in northern East Falkland. Upper Palaeozoic, ranging probably up into the Triassic, outcrop in the region known as Lafonia, the southern flatter part of East Falkland. Permo-Triassic beds occur in the NW region of Lafonia fronting Falkland Sound. Precambrian is restricted to a small southern area in West Falkland, but no Lower Palaeozoic is known in either West or East Falkland.

In West Falkland there is a larger stratigraphic representation, although to date post-Triassic beds are almost unknown here. However it is likely that here, as indeed in East Falkland too, the grassy vegetation and heavy peat formation conceal many rocks whose age is not known.

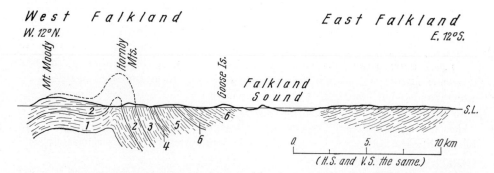

Fig. 102. Geologic Section across the Falkland Islands. 1. Fox Bay Group, 2. Monte Maria Group, 3. Lafonian Tillite Group, 4. Black Rock Group, 5. Lafonian Group, 6. Falkland Sound Group. (Based on BAKER, 1923)

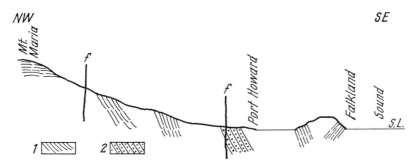

Fig. 103. Schematic Sketch of the E coast of West Falkland. Vertical scale much exaggerated. 1. Devonian slate and sandstone with invertebrates, 2. Devonian quartzitic sandstones. (HALLE, 1912)

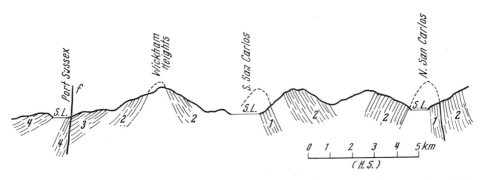

Fig. 104. Schematic Section from Port Sussex to North San Carlos. 1. Devonian quartzite, 2. Devonian sandstone and slate, 3. Lafonian tillites, 4. Lafonian sandstone and slate. (HALLE, 1912)

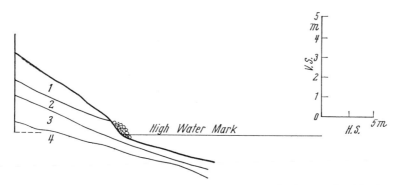

Fig. 105. Schematic Sketch of the forest bed at West Point Island. 1. Yellowish clay, 0–90 m, 2. Angular blocks and pebbles, cemented with clay, 0–60 m, 3. Clay, lenses of black soil, tree trunks, 0–90 m, 4. Black vegetal soils and densely packed tree trunks, 1 m +. (HALLE, 1912)

In both islands, angular blocks in drainage channels are phenomena associated with the Quaternary, and some very localized Recent sediments are present in West and East Falkland.

Precambrian

Rocks of this age are known to outcrop only in the Cape Meredith region, in the extreme S of West Falkland, and thus are known as the Cape Meredith Series or Complex, or then the Archaean Basement. The crystalline schists and gneisses, intruded by granites, pegmatites, acid and basic dykes, were first noted by ANDERSSON (1907), and because of the degree of crystallinity and their unconformable position below the Lower Devonian sediments, he assigned them to the Archaean. BAKER (1922) gave further details and descriptions of these same rocks, which extend on either side of the Cape and have a NNE-SSW trend. Outcrops in places are difficult to decipher, and this, along with the high cliff sections, make detailed study of the Precambrian somewhat hazardous, and even today our knowledge of these rocks in this area is far from precise. ANDERSSON & BAKER alike assigned the metamorphics and igneous rocks here as being of Archaean age, but BORRELLO (1963) thought that perhaps the granites and pegmatites should rather be referred to the Proterozoic. Without offering any explanations, FRAKES & CROWELL (1967) merely list the Cape Meredith Complex as "Precambrian(?) or L. Palaeozoic(?)".

Schists dipping NW and forming the W coast of the Cape were described by BAKER as composed entirely of green hornblende and quartz. The younger(?) gneisses of the eastern part of the Cape are intruded by medium-coarse grained, pinkish and yellowish microcline-granite, with large hornblende-schist xenoliths, and these in turn invaded by microcline-pegmatites with xenotime. The bluish-black schists disintegrate into sandy material, and hence the name 'Archaean Basement of the Sandstone Formation' used by ANDERSSON. Intrusions have caused the rocks to be severely folded, with here and there great distortion evident. In appearance, these ancient rocks are very similar to those outcropping in the western part of Chubut Province, Argentina, where a thick succession of the Upper Palaeozoic lies above the metamorphics. In this part of Argentina, it is not certain whether these crystallines are of Palaeozoic or Precambrian age. Because of terrain difficulties in the study of the Cape Meredith rocks, because the metamorphics have not been studied in detail, because of the uncomformable relations with the overlying Lower Devonian and because rocks similar to these old crystallines in Chubut have not definitely been placed in the stratigraphical column, at this time we cannot say with certainty whether these Falkland rocks are indeed Precambrian; they may be Infracambrian or even Lower Palaeozoic.

Granites and pegmatites intrude the Cape Meredith metamorphics, being associated more with the gneisses than the schists. The yellowish-coloured granites are holocrystalline and of coarse grain, comprising yellowish quartz and feldspar (chiefly microline) along with biotite altered to chlorite. Granites reddish in colour but otherwise quite similar intrude Palaeozoics of various age in the Sierra Australes of Buenos Aires Province.

Pegmatites, of rose colour, coarse-grained, with rose-tinted feldspars and quartz dominant, occur as laccoliths, dykes and veins. They show no signs of major structural deformation.

A lamprophyre dyke cutting the crystallines at an angle of 40° can be seen at Cape Meredith. This altered rock is composed of a reddish feldspathic and hornblende groundmass, along with olivine and augite accessories. Augite-hornblende lamprophyric dykes, possibly the youngest of the intrusions, are also present.

The relations of the above intrusives to the metamorphics have not been clearly determined, but it is the custom to consider them as also Precambrian in age.

Palaeozoic

Strata of Palaeozoic age, which outcrop over almost the entire archipelago, have been recognized since the time of DARWIN. Devonian to Permian beds are present, the succession extending up into the Triassic without any clear lines of demarcation.

Authors dealing with the stratigraphy of the Falklands have grouped the rocks on a purely stratigraphical-palaeontological basis, but BORRELLO placed the Middle Palaeozoic under the term "Sistema de Gran Malvina" and the Upper Palaeozoic as "Sistema de la Isla Soledad", thus rendering a geographical connotation to the terms, a method open to criticism.

The Palaeozoic of the archipelago totals some 6500 m in thickness at least.

a) Lower Devonian

The oldest recognized sediments are the Port Stephens Beds, occurring only in West Falkland. (The stratigraphic nomenclature adopted here is that of ADIE [1958], and nearly all stratigraphic references to this writer are concerned with this publication. References of whatever type to BORRELLO are almost exclusively to his 1963 publication.) Geological maps of the Islands show an indefinite boundary between these strata and the Middle Devonian Port Philomel Beds (q. v.) as running in a NW-SE direction between Port Richards and Port Edgar. At Cape Meredith the Port Stephens formation shows a thin basal red micaceous shale bed at the base in direct unconformable relation to the old crystallines.

The rocks are chiefly composed of coarse-grained, cream-coloured sandstones, quartz-conglomerates and quarzites. At Port Stephens the beds are well exposed in the imposing cliff faces and here total some 1500 m thick. The rocks are monotonously similar lithologically. Quite often the abundance of well-rounded vein-quartz pebbles in the sandstones gives rise to quartz-conglomerates. The sandstones themselves are frequently current-bedded and jointing is well developed. In those localities where secondary quartz deposition has resulted in quartzitic sandstones, weathering is almost entirely controlled by jointing and the extent of cementation. More friable quartzitic sandstones weather-out into some odd-shaped outcrops. At South Harbour, near Port Stephens, BAKER claimed to have found tillites and shales. Seemingly neither ADIE nor BORRELLO could confirm the presence of the former, but both are prepared to accept such, indicative of a Lower Devonian glaciation such as is recognized in the Table Mountain Series of Cape Province, South Africa, and in the San Juan Pre-Cordillera of Argentina. NIDDRIE (1953) indeed remarked that all the Port Stephens Beds very closely resembled the sandstones of the Table Mountain Series, especially the western and southern exposures of the latter. FRAKES & CROWELL (1967) state they searched for this tillite occurrence N of Port Stephens but did not find it.

The strata are usually horizontal, but dip gently to the NE between Cape Meredith and Port Richards. The unit has the appearance of a slightly disturbed block, tilted somewhat towards the NE.

To date no fossils have been recovered from these rocks which undoubtedly are of marine origin.

The Port Stephens formation has hithertofore been assigned to the Devonian, but FRAKES & CROWELL mention that A. J. BOUCOT had discovered marine fossils in the overlying Fox Bay Beds, "at a level several hundred metres above the unconformity overlying the complex" of about Early Emsian age, and "the intervening strata may be somewhat older than Devonian", and thus suggested that the Port Stephens Beds might be Silurian.

The Fox Bay Beds outcrop in many parts of the main islands and in many of the smaller ones, forming discontinuous exposures. Lithologically the rocks pass from abundantly fossiliferous shales through greyish-white micaceous quartzitic sandstones with numerable instances of false bedding, into micaceous flagstones such as are een at the type locality of Fox Bay, West Falkland. The formation is taken to be some 760–800 m thick.

BAKER claimed that the Port Stephens Beds passed conformably into the Fox Bay formation, and FRAKES & CROWELL speak of a gradation between the two, but ADIE claimed that the passage is apparently obscured, although he recognized the presence of passage beds near Port Richards and Port Edgar. The Fox Bay formation passes gradually upwards into the more arenaceous Port Philomel Beds.

Throughout much of the northern parts of both West and East Falkland the Fox Bay rocks outcrop on lower ground below ridges of the Port Stanley quartzites. In the eroded anticlinial axes of the highly disturbed rocks forming the Hornby Mountains, the Fox Bay Beds are dipping steeply, and the dip is also considerable in gently pluning folds forming the cores of the anticlines, especially in the eastern part of West Falkland.

Several workers noted the strong analogies, both lithological and palaeontological, between these rocks and the Lower Bokkeveld Series of sandstones and shales of South Africa, and Fox Bay outcrops at Chartres, West Falkland, cut by doleritic dykes, bear close resemblance to similar suites amongst the Karroo strata of South Africa.

Like the Port Stephens formation, the Fox Bay Beds are of marine origin.

FRAKES & CROWELL, quoting a personal communication from A. J. BOUCOT who was examining some marine fossil collections of the Fox Bay Beds, state that BOUCOT was of the opinion that two species present, *Australocoelia touteloti* and *Australospirifer antarcticus* were typical of the Malvino-kaffric province of the Southern Hemisphere, and were of about Early Emsian (late Early Devonian) age.

b) Middle Devonian

At the entrance to Port Philomel and near Port North, West Falkland, the Fox Bay Beds are seen to grade upwards into the Port Philomel Beds, of Middle Devonian age. The rocks comprise friable, yellowish sandstones, thin-bedded, micaceous sandstones and greyish-brown sandy shales. Towards the top of the formation the beds are coarser, take on a reddish hue, are more quartzitic and generally more shaley. The formation is estimated to be 150 m thick.

These shales and sandstones contain abundant distinct and indistinct traces and impressions of plants. NIDDRIE reported however that he found no plant remains in the upper beds of the formation. No other types of fossils have been found in the strata.

NIDDRIE linked the Port Philomel rocks, along with the Fox Bay formation, as being closely similar to the South African Lower Bokkeveld Series, and BORRELLO drew analogies between the Port Philomel formation and similar lithologies and plant remains in the San Juan Pre-Cordillera of Argentina.

c) Lower Carboniferous(?)

No distinct boundary has been detected between the Port Philomel Beds and the overlying Port Stanley Beds, and the transition is taken to be a gradual one.

The Port Stanley rocks are prominent in the northern areas of the Falklands, forming the major higher terrain and mountains, and are well represented in the Wickham Heights of East Falkland. They comprise massive, white, cross-laminated quartzitic sandstones quartzites and rare shales, and are believed to be well in excess of 600 m in thickness. The fine, crystalline quartzites are cut by a maze of large quartz veins. Outcrops facing the windward sides are deeply pitted by honeycomb weathering. The rocks provide boulders for the 'stone rivers'.

The beds are folded along axes trending WNW-ESE, showing both simple and isoclinal folding, also frequent faulting. It is this formation which is responsible for the structural complexity of much of East Falkland.

No fossils have been found and such carbonaceous traces as have been observed cannot positively be identified as plant material.

NIDDRIE claimed that the rocks closely resembled the Witteberg quartzites of SW Cape Province, South Africa, and ADIE likewise correlated tham with this South African series.

BORRELLO placed both the Port Philomel and Port Stanley Beds in his Monte Maria Group. (His use of the term "grupo" is purely descriptive, not the usual stratigraphical connotation.) This terrestrial group of sediments rests conformably on the Lower Devonian marine strata. He believed the age of the group to be not younger than Upper Devonian. He pointed out that SEWARD & WALTON, who examined the plant collection made by BAKER and compared them with those determined by HALLE, were of the opinion that the plant remains indicated a Devonian age, probably Middle rather than Lower. BORRELLO did not believe that the formation was of Eo-carboniferous age. The Middle Devonian terrestrial period of sedimentation is now represented in the Falklands by some 3000 m of rock of which the Port Stanley formation comprises about one-fifth. It was his belief therefore that his Monte Maria Group should, by preference, be assigned to the Middle Devonian. The opinion of SEWARD & WALTON, plus the absence to date of known Lower or Middle Carboniferous in the Falkland Islands, caused BORRELLO to adopt this view. However the discovery of older Upper Palaeozoic fossils in Chubut, Argentina lends distinct interest to the problem and indicates the necessity of more detailed stratigraphical work in the archipelago.

FRAKES & CROWELL also bracket the Port Philomel and Port Stanley formations together and are of the view that "Continuity in the sequence suggests to the present authors that a Devonian age is more likely" than the Early Carboniferous assignment of ADIE.

ADIE offered no objections to HALLE's two-fold division of the Devonian of the Falkland Islands into a lower marine and an upper terrestrial facies, of which the Port

Stephens and Fox Bay Beds represent the former, the Port Philomel and Port Stanley Beds the latter, yet in his description of the Port Stanley formation he lists it as Lower Carboniferous.

In view of what has been said above, therefore, we prefer to give a questionable Lower Carboniferous age for the Port Stanley formation.

d) Upper Carboniferous(?)

A 10 m thick succession of coarse-bedded, greenish-brown sandstones and shales, showing pronounced lamination towards the top, have been named the Bluff Cove Beds. They lie unconformably on the Port Stanley formation and pass upwards into the Lafonian Tillite. These rocks, of marine or fluvio-glacial origin, have only local occurrence in the Bluff Cove and Port Fitzroy areas of East Falklands. Elongated boulders are often seen, with their long axes perpendicular to the bedding planes of the laminated shales, which would suggest that here, at the beginning of the glacial transgression, detrital deposition occurred under lacustrine or marine conditions.

No fossils have so far been found, and the character of the rocks shows a direct passage into the tillites above.

ADIE correlated these beds with the Lower Dwyka Shales of Cape Province, South Africa.

FRAKES & CROWELL describe the formation as consisting of mudstones, fine-grained sandstones and poorly-sorted, fine-grained sandstones with cross-stratification in sets less than 5 m thick. They give a thickness of 100 m for the formation. To these writers the Bluff Cove Beds are likely pre-Early Carboniferous in age, probably not older than Middle Devonian.

Because of lack of fossil evidence for the age of these beds, a Late Carboniferous age assignment given by BAKER & ADIE cannot be certified, and the possibility that the Port Stanley and Bluff Cove formations are Devonian (Middle?) cannot be ruled out at this time.

e) Permo-Carboniferous

The Lafonian Tillites or Lafonian Diamictite of FRAKES and CROWELL has an appreciable distribution in the central part of East Falkland, the most important outcrops lying along the southern flanks of the Wickham Heights and eastwards to the coast at Port Pleasant, Port Fitzroy and Port Harriet. In West Falkland they are present in a coastal strip facing Falkland Sound, from Port Howard to near Fox Bay, and more isolated occurrences are found at Port Purvis, Hill Cove and Mount Edgeworth.

This glacial formation, measuring some 650 m thick at least (350–850 m according to FRAKES & CROWELL), is composed of a compact, somewhat massive, deeply-weathered clayey matrix in which are water-worn and subangular to angular pebbles and boulders ranging up to a metre and more in diameter – up to 7 m in the Hill Cove area, according to FRAKES & CROWELL. The boulders appear to be erratics, many of which show signs of striation and soling. According to the locality, there is variation in the proportion of rock types involved, and include quartzite, biotite-granite, granite-gneiss, garnetiferous gneiss, diabase, quartz-porphyry, slate, sandstone, conglomerates of quartzitic pebbles in a ferruginous matrix. The dominant rock type is a well-lithified, dark-coloured diamictite (vd. R. F. Flint, J. E. Sanders & J. Rodgers, 1960) comprising a sandy-clayey matrix and a scattered coarser part. According to Frakes & Crowell; "Pebbly graywacke is the most common rock, but cobble and boulder graywacke occur locally. There is a complete gradation from boulders to fine clay-sized particles...". According to Borrello not all the rock types of the erratics can be recognized as originating in the Falkland Islands.

The formation is pronouncedly massive when fresh (Frakes & Crowell) and shows a pseudo-stratigication as crude layering, this layering being regionally aligned with stratified intercalations. It is believed by the above authors that the layering is due to compaction during lithification and variation in porosity and pemeability rather than any primary cause. Thin bands of claystone or sandstone are rarely present, either singly or in sets, usually alternating with thin, well-sorted sandstone layers. Sometimes stratification is shown by the alignment of cobbles, boulders, closely packed along a single horizon.

In the central part of East Falkland, the Lafonian Tillite or Diamictite is folded, at times intensely so, and near Port Stanley, N-S striking faults occur. In West Falkland also faulting can be seen at Port Purvis. Further, in East Falkland the tillites show vertical cleavages, parallel to the fractures.

Lithologically the formation is similar to proven glacial marine rocks. Frakes & Crowell recognized three well-defined facies in the archipelago, distinctions being chiefly based on the kind of stratified intercalations, thus: 1. linear and sheet sandbodies are restricted to western West Falkland (Hill Cove); 2. tabular interbeds in East Falkland (Port Fitzroy); 3. fragmental and contorted sand bodies in eastern West Falkland (Port Purvis). The respective thicknesses of the Lafonian Diamictite in these three areas are 850 m, 400 m and 350 m. "The western facies is interpreted as marginal subglacial, and the eastern facies is interpreted as offshore marine(?) in origin. The contorted and disrupted slabs of the intervening third facies apparently originated from downslope subaqueous mass movement. It can be speculated that a fourth facies, representing distal marine deposits, is represented by the black shale of the Black Rock Member of the Port Sussex Formation" (Frakes & Crowell).

In some places the Bluff Cove Beds underlie the tillites, and Frakes & Crowell thought that this formation "may represent a local non-glacial facies developed at the base of the Lafonian Diamictite". Pebbles are lacking in the Bluff Cove formation, unlike most interbedded material of the Diamictite of East Falklands, and for the Bluff Cove Beds, unconformably resting on older strata and conformably overlain by the Diamictute, a shallow marine environment was postulated by these authors.

At Port Purvis and W of Hill Cove Baker had observed the tillites resting unconformably on smooth, striated Devonian quartzite pavements. As these 'striae' were orientated N-S, it was inferred that the principal movement of the ica-sheets was from S to N. Frakes & Crowell are extremely doubtful of this interpretation of Baker's in the first place because the features are slickensides due to faulting and are not striae. They claim that this, plus a combination of other factors such as trends of linear sand bodies, sedimentary structures in intercalated strata, variation in mean and maximum clast-size, diamictite clast-fabric, relationships among the facies of the intercalated strata, all suggest that glacial ice flowed from W to E or perhaps SW to NE, out into deep water E of the archipelago. In reconstructing the palaeogeography at the time of formation of the diamictites, they envisage a land mass W of the Falklands accounting for the sizable body of ice and great thickness of glacial deposits, this land area being situated on the South American continental shelf. Rocks exposed here on the continent and interveing continental shelf were likely similar to the pre-Carboniferous surface rocks of the Falklands but probably also included some not occurring in the Falklands, which could explain BORRELLO's contention that not all the Falkland erratics are to be found in the archipelago. The western diamictite facies at Hill Cove with its linear sand bodies, here suggested as eskers, were near the margin of a considerable ice body, with sub-glacial streams laying down well-sorted sand in channels and in small fans at their mouths, near the buoyancy line of the glacier. At Hill Cove and for some distance to the W, ice was grounded on moraine below sea-level, but to the E it likely was afloat as shelf ice. It may be that the major part of the unstratified diamictite originated by deposition as ground moraine but some also was probably re-deposited by downslope gravity processes. The submarine slope offshore likely received thick accumulations of material transported out from under the moving ice, forming the necessary conditions for subaqueous mass movement whereby submarine landsliding too place. This could account for the deformed and contorted deposits in the Port Purvis region. The graded bedding in poorly sorted graywacke in the eastern facies suggest marine conditions, from which it can be inferred that much of the diamictite here was deposited beneath the sea, as at Port Fitzroy, for example.

No fossils have as yet been obtained from these tillites, perhaps due, as FRAKES & CROWELL suggest, to the nearby ice margin lack of typical well-populated littoral environment, greatly reduced penetration of light and oxygen-producing plants because of the floating ice body above.

The exact stratigraphic position within the Upper Palaeozoic of these Lafonian beds is uncertain. BORRELLO suggested that the tillites may largely be of Upper Carboniferous age. With the exception of the Sierra de la Ventana region in Buenos Aires Province, where the Upper Palaeozoic is represented solely by the Permian, in the Pre-Cordillera of San Juan and Mendoza and also in Central Patagonia the tillites range from Lower to Upper Carboniferous, which might suggest a similar age for those of the Falklands. ADIE correlated the Lafonian Tillite with the Dwyka Tillite of South Africa and the Itararé Beds of South America and gave a Permo-Carboniferous age. According to FRAKES & CROWELL: "All evidence considered, the maximum possible age for the Lafonian Diamictite is Middle Devonian, but the great thickness of unfossiliferous strata lying unconformably below the diamictite suggests that it is probably not older than Early Carboniferous". BORRELLO placed the Tillite near the base of his East Falkland system, a sequence of beds which, in neighbouring Gondwana areas of South America and some parts of Antarctica, show a constancy of glacial facies prior to the pronounced development of typical continental Permian. It is to be noted that HARRINGTON (1956) who draws upon standard references for both Argentina and the Falkland Islands, makes no allusion to Carboniferous strata in the archipelago.

f) Permian

Lying apparently concordantly on the Lafonian Tillites are the Black Rock Slates, 15 m thick according to ADIE, about 125 m thick according to FRAKES & CROWELL. They are best developed at the type area between Mount Pleasant and Black Rock, on the southern side of the Wickham Heights, East Falkland. Poor outcrops prevent a definite recognition of the beds in the Port Purvis area of West Falkland.

The formation comprises black, carbonaceous shale and dark grey cherty shales with abundant smooth, water-warn quartzitic pebbles at the base. As per Frakes & Crowell, the lower metre or so contains abundant diamictite layers interbedded with dark shale, cross-laminated sandstone layers and laminations of siltstone and common sulphide staining also being present.

Adie mentioned that there was a gradual transition from the bouldery tillites up through dark shales to very fine-grained cherty shales. The transition appears gradual to Frakes & Crowell also but Borrello spoke of a sudden break between the coarse-bedded tillites and the fine stratification of the Black Rock Slates. According to Adie, a thin series of dark shales occur at the top, before the prominent Lafonian Sandstones are encountered, but Borrello, referring to Halle & Baker, gave the impression that there is a gradation from the argillaceous lower beds to the predominantly arenaceous upper formation. Frakes & Crowell claim that the Black Rock beds are the lower Member of their Port Sussex formation, the upper strata being named the Shepherds Brook Member, the latter consisting of grey shale and mudstone, unconsolidated, blue-grey, fine- to medium-grained sandstone, with porcellaneous claystone and plant-bearing claystone near the top. According to these authors, the type section of the Port Sussex formation, 275 m thick, is exposed along the N shore of Port Sussex from Hells Kitchen to the head of the bay, where the strata dip 50–80° to the S and are only slightly affected by minor folding.

Adie placed the Black Rock Slates in the Lowest Permian and correlated the beds with the 'White Band' (Upper Dwyka Shales) of South Africa and the Irati Shales of Brazil and Uruguay. Frakes & Crowell quote J. M. Schopf (personal communication) as reporting that fragmental material in the Shepherds Brook Member is likely of Permian age and that fossils collected from Bodie Creek Bridge are suggestive of a *Glossopteris* flora of Permian age.

The Lafonian Sandstone comprises fine bands of brownish feldspathic sandstones with rare shale interbeds, totalling 90–100 m thick. At the base small fragments and pebbles of granite and other igneous rocks occur locally, thus giving the formation the appearance of a conglomerate. (Frakes & Crowell state that quartz is dominant in the formation, but feldspar grains constitute ca. 25%, rock fragments, 20% and argillaceous material ca. 20%.) Only occasionally is graded or current bedding observed. Towards the top is a recurring alternation of fine bands of sandstone and laminated clays which were thought by Halle to be similar to varves in fluvio-glacial sediments. Adie defined the upper limit by the gradual transition from sandstones into variegated and striped siltstones of the succeeding formation.

The Lafonian Sandstone in outcrop closely parallels that of the Lafonian Tillites, i. e. along the southern flank of the Wickham Heights, with dips towards the SW, increasing in angle towards the western side of East Falkland.

Thick peat and vegetal covering render it difficult to delimit the outcrops of the formation.

Adie correlated the sediments with the lower part of the Middle Ecca Series of Natal, South Africa.

BORRELLO combined the Black Rock Slates and Lafonian Sandstones in his Black Rock Group of the East Falkland System. Following BAKER, he contended that his group constitutes a unit intervening between the glacial Lafonian Tillites and the plant-bearing beds of later Permian age, and took as more acceptable HALLE's grouping for the unfossiliferous argillo-arenaceous sediments the term "Fluvio(?)- and Limno-Glacial Beds", of Lowest to Middle Permian age. The earliest sediments of the group relate to a marine environment, whereas the later arenites are a continental facies, representing the partial accumulation of the de-glaciation stage which has some relation to the succeeding plant-bearing rocks. BORRELLO was in agreement that his group should be placed in the Permian rather than the Upper Carboniferous. He pointed out the correlation in lithology and cyclical development between his Black Rock Group and the Piedra Azul Series of the Sierra de la Ventana, Buenos Aires Province, which latter is assigned to the upper part of the Upper Palaeozoic, namely, the Permian.

FRAKES & CROWELL named the Lafonian Sandstones the Terra Motas Sandstone, of Permian age.

Interbedded siltstones and soft sandy shales, totalling at least 300 m thick, lying conformably on the Lafonian Sandstone are named the Choiseul Sound and Brenton Loch Beds. They outcrop along the southern side of the Wickham Heights and across Choiseul Sound into Lafonia and Lively Island. Both this and the succeeding formation are not known to date in West Falkland.

The soft, fine-bedded sediments, lacking in compaction, weather easily and crumble into greyish silt or then coarse, arenaceous clay. Doubtless it is due to this ease of weathering that has allowed Choiseul Sound and Brenton Loch to carve a passage far inland, indeed almost severing Lafonia from the northern part of East Falkland.

Phyllotheca and *Glossopteris* flora and fossil woods have been obtained from Bodie and Walker Creeks in Lafonia.

Minor thrust faulting and folding can be seen here and there. The beds dip gently towards the SW.

The Choiseul Sound formation passes upwards into the Bay of Harbours Beds, from which they can be distinguished solely by the flora.

ADIE believed that the Choiseul Sound Beds did not comprise varved sediments such as HALLE & BAKER described, but were to be considered rather as correlatives of the upper part of the Middle Ecca-Lower Beaufort Beds of South Africa, and assigned a Middle to Upper Permian age to the formation. On the other hand, NIDDRIE claimed it would be a difficult matter to correlate any of the sediments on the basis of appearance with any South African member. Unfortunately fossils collected by him from the Falkland rocks were lost in transit to South Africa, and hence he can say little about matters palaeobotanical. Evidently from his appraisal of the publication by SEWARD & WALTON, based upon the palaeobotany he would correlate the Choiseul Sound formation with the upper beds of the Middle Ecca of South Africa.

Palaeozoic-Mesozoic

There appears to be no clear distinction between the Palaeozoic and Mesozoic in the archipelago, Palaeozoic strata conformably passing upwards, with scarce any lithological distinction, into the Mesozoic.

Permo-Triassic

In conformable relation to the underlying Choiseul Sound formation are the 300 m thick Bay of Harbours Beds. They comprise coarsely bedded, soft sandstones and compact mudstones, dipping gently southward. There is an abundant *Glossopteris* flora and also fossil woods. ADIE mentioned that as the flora has a relatively long time-range, dating of the beds cannot be too reliable, and he ranged them from Upper Permian to Lower Triassic. It is most difficult to distinguish these beds lithologically from the succeeding West Lafonian Beds of Middle to Upper Triassic age.

BORRELLO placed the Choiseul Sound and Bay of Harbours Beds in his Lafonia Group, together occupying almost the entire Lafonia region. This Group is distinguished from the Black Rock Group on the basis of its abundant flora, whereas in the latter, plant remains are not recognizable. He recognized a close connection between the Permian of the Falklands and the mainland of Patagonia, based chiefly upon the floral characteristics of the two regions.

On the basis of hurried examinations only, NIDDRIE could merely state that the rocks of the southern part of Lafonia are predominantly of Upper Karroo age.

Mesozoic

Sediments of this age are restricted to western Lafonia and the islands in Falkland Sound. However BORRELLO merely mentioned but did not elaborate upon the presence of the Mesozoic along the western side of the Sound, in West Falkland.

Triassic

Strata lithologically similar to and only distinguishable on a palaeobotanic basis from the Bay of Harbours Beds, are restricted to the western side of Lafonia, from Dos Lomas S to Moffitt Harbour and the islands in the Sound. These strata, totalling about 2100 m thick, are known as the West Lafonian Beds, or then the Falkland Sound Group of BORRELLO.

The formation comprises hard, siliceous sandstones, greenish-grey mudstones and a prominent horizon of conglomerate interbedded with soft sandstones and shales. The rocks dip at angles of 12° to 15° towards the NW, whereas in the islands of the Falkland Sound on the western side, they are seen to dip to the SE at much greater angles. It is on the islands of the Sound that the highest members of the formation outcrop.

Structural complexity in this general region confuses matters, but BORRELLO was of the opinion that these Mesozoics were originally considerably thicker than the present estimate, and further, ADIE left open the possibility that an unconformity lies between this formation and the lower Bay of Harbours rocks.

The Sound marks the site of a submerged syncline, with gentle flanks on the SE side and much steeper dips on the NW side of the structure.

The strata have yielded *Glossopteris*, *Gangamopteris* and *Phyllotheca* species. However the one fossil of significance is *Neocalamites carrerei*, occurring elsewhere only in the Upper Beaufort Series and Molteno Beds of South Africa. On this basis, ADIE inferred that the West Lafonian Beds were at least Middle Triassic, and believed they could be placed as Middle to Upper Triassic.

Borrello remarked that this species of plant fossil had been discovered more recently in the Triassic of the Upper Rio Bermejo basin, in the southern part of La Rioja Province, Argentina (J. Frenguelli, 1948), and is also known in the N of Mendoza Province. At these localities the strata are grouped in the Triassic-Rhaetic of the San Juan-Mendoza Pre-Cordillera.

As remarked earlier, many dykes are present in the Falkland Islands which have strikes generally parallel to the fold axes, NNE-SSW being the prominent direction. The doleritic dykes, which do not invade the Gondwana succession, show petrographic features similar to those of the Middle Jurassic Karroo Dolerites of South Africa. Diabasic dykes are also present in West Falkland cutting Devonian sediments.

Dykes are thus seen to invade rocks dating from the Devonian to Permo-Triassic. As they cut some of the highest Gondwana series, they are obviously of post-Middle Triassic age. Borrello suggested they might be post-Triassic – comparable in age then to the Karroo Dolerites? – and Adie assigned to them a Rhaetic(?) age.

To complete comments on the older rocks of the archipelago, we might note that Frakes & Crowell have, in their geologic sketch map of the Islands, combined all the post-Lafonian Diamictite formations into what they term the Upper Lafonian Series.

Quaternary

Recent sediments are represented by shelly limestones, clay lenses, tree trunks, fossil soils with angular detrital material, and boulders and blocks of the 'stone rivers'. The relief of the Falklands shows a distinct evolution, indicative of a mature stage. Quaternary deposits testify to rapid morphological evolution conditions associated with epeirogenic movements continuing into the Recent.

Adie recognized two Recent formations. The Shell Point Limestones comprise a lower unconsolidated and an upper compact shelly limestone, one metre in thickness, occurring at Shell Point, Port Fitzroy, East Falkland. The beds rest on a 6 m wave-cut platform of Late Monastirian age. Sub-fossils *Gastropoda* and unidentifiable *Pelecypoda* have been encountered in the upper limestones.

At West Point Island, West Falkland, a 2 m thick bed comprising tree trunks, clay lenses and black fossil soils, with yellowish clays and much angular detrital matter and scattered pebbles was named the West Point Island Forest Bed. Small twigs and branches up to 2 m in length, along with attached roots, pseudomorphed by marcasite, occur here, but no trunks have been observed in their original vertical position. Pollen grain spores have also been noted, believed to be indigenous. According to Halle, these deposits are terrestrial and formed *in situ*, but Baker considered them to be allochthonous deposits, the tree trunks having been carried thither by marine currents.

Two species of conifer now typical of temperate areas in South America have been recognized.

Adie believed that the rocks probably represent an early Recent age.

Quite large areas of the Falklands are covered with peat deposits, occurring both in valleys and on high mountain land. According to Andersson, they attain a thickness of 2.70 m over argillo-arenaceous rock material. The peats are thought to have originated in Holocene time, after the dissipation of the Pleistocene vegetal covering.

Mention has been made of the 'stone rivers' or 'stone runs', those peculiar, winding accumulations of boulders filling valleys and giving the impression of "streams of stones" as Darwin remarked. Most of the rocks comprising these bodies consist of quartzitic sandstones of Devonian age. In size they very from about 0.50 m to 2–3 m and even larger, may extend for as much as 3000 m in length along lower, sinuous courses. They are angular to sub-angular in shape, indicating some water-wear, and show no evidences of true sedimentation. Occasionally there are depressions filled with these boulders and here there may be vegetal coverings. The loose blocks when viewed from a distance show a uniform aspect – Renard (1885) stated that "les coulées de roches font l'effect d'un gigantesque glacier" – but from near at hand display amazing confusion. At the base of the boulders filling the valleys water circulates, flowing into the main longitudinal depressions from the sides.

Darwin related the development of these 'stone rivers' to positive epeirogenic movements of the Islands, rapid deposition resulting from more violent disturbances and vibratory movements transporting the blocks further inland up the valleys. Wyville Thomson (1877), on the other hand, thought the blocks were products of disintegration of the hard quartzitic Devonian sediments when these latter became exposed through erosion of the softer, intermediate rock layers. These harder layers, now lacking support, would split along joint surfaces and tumble down into the valleys, being partially buried in the vegetal-soil covering. The partly exposed parts of the blocks would be subjected to expansion and contraction as a result of atmospheric agencies, thus detaching fragments, grains, etc. and forming a matrix. Meanwhile lateral rivulets would assist in moving the disintegration products, thus again rendering the position of the blocks as insecure, which would then gradually move downwards towards the main valley.

Renard (op. cit.) offered no explanation of his own, but seemed content with the hypothesis of Wyville Thomson, drawing attention to the fact that neither Darwin nor Wyville Thomson invoked glacial action as a means of transportation.

Andersson (1906) put forward an idea which found more general acceptance, namely, solifluxion. These "streams of stones" were essentially currents of blocks originating under sub-glacial climatic conditions during the Quaternary. The accumulation of snow during colder periods, and its rapid disappearance during less rigorous times, would result in blocks being fused, as it were, with the snow-ice, yielding semi-fluid masses of rock, soil, mud and water, and in this condition there was available a medium of transport for the blocks.

Baker in general accepted Andersson's hypothesis. However, he believed the climatic conditions to have been more intense than Andersson postulated, resulting in a more restricted genetic environment. He presumed a lowering of sea level, with only the folded, higher areas of the islands protruding as 'nunataks' midst expanses covered in deep snow. As the boulders did not show indications of major transport and are derived from localities near at hand, as per Baker, glacier formation was reduced. In the bottoms of the valleys, the action of running water resulted from thawing, thus taking away some of the sharp edges and corners of the blocks. The boulders occur in those places where the slow descent of the pasty mass was least hindered.

Joyce (1950) devoted a paper to these unusual boulder accumulations. It was his contention that solifluxion played only a minor role in the development of the stone runs.

He emphasized rather the close relationship of the sites of these accumulations to the structural trends of the exposed quartzites and concluded therefrom that many of these boulder accumulations were nought else than the litter or débris which remained after the weathering of the quartzite outcrops.

MALING (1951) was of the view that whilst the above opinion of JOYCE may account for some of the 'stone runs' on hillsides and near summits, the hypothesis was inadequate in explaining extensive valley accumulations. He showed that the maturing of the landscape and the close adjustment of the drainage pattern to the structure were both suggestive that the degradation and the retreat of the interfluves had been operative long before the 'stone runs' existed.

According to BORRELLO, the maritime conditions and the distance separating the archipelago of the Falklands from the South American mainland probably militated against the expansion of the Quaternary glaciation, whose morainic fronts extend through southern and western Patagonia, and these factors no doubt explain the differences in the glacial facies of the Quaternary in southern Argentina and the Falkland Islands.

It remains to be added that RENARD (1885), who studied samples of the blocks collected by WYVILLE THOMSON, was greatly impressed by an amphibolized diabase, "dont elle offre un type des plus remarquables".

The origin of these unusual 'rivers of stone' has not so far been satisfactorily explained. Though known for some 135 years, though reported upon by various travellers and scientists, to date no systematic study has yet been taken. It must also be noted that up till the present, most study has been given to those 'stone runs' occurring between Stanley Harbour and Port Salvador, but inland in less accessible localities, extremely little is known on this topic.

Palaeontology

During DARWIN's visit to the archipelago fossils were collected and reported upon by MORRIS & SHARPE, the first palaeontological publication to appear concerning these islands. ETHERIDGE wrote on some fossil collections made during the "Challenger" expedition. ANDERSSON gave palaeontological information relating to the Swedish Antarctish National Antarctic Expedition, and CLARKE & HALLE have also reported on collections made by the same. SEWARD & WALTON, ADIE & BORRELLO have contributed to our fossil knowledge of these islands also. The recent publication of FRAKES & CROWELL mentions that palaeontological studies are going on at present, presumably on collections made by these authors by J. M. SCHOPF, A. J. BOUCOT and perhaps others, but at this time their findings have not been published. Unfortunately collections made by NIDDRIE were lost in transit to South Africa.

To date, the Port Stephens, Port Stanley, Bluff Cove, Lafonian Tillite, Black Rock Member and Lafonian Sandstone Beds have not yielded fossil remains, nor yet have any vestiges of life been noted in the Cape Meredith Complex.

The Late Lower Devonian Fox Bay Beds are richly fossiliferous, and the following have been recognized:

Table 92

Stratigraphic Correlation Chart (ADIE, 1952a, 1952b)

		Falklands	South Africa	Brazil - Uruguay - Argentinia
Gondwana Succession	Upper Triassic	West Lafonian Beds	Molteno Beds to U. Beaufort Series	Rio do Rastro Beds, U. Estrada Nova Beds
		Bay of Harbours Beds	M. Beaufort to L. Beaufort Series	~~~~ d ~~~~?
	Upper Permian	Choiseul Sound-Brenton Loch Beds	L. Beaufort to Upper M. Ecca Series	
	Lower Permian	Lafonian Sandstones	Lower M. Ecca Series	
	Lowest Permian	Black Rock Slates	Lower Ecca to Upper Dwyka Shales	L. Estrada Nova, Irati Shales, Rio Bonita Beds
	Upper Carboniferous	Lafonian Tillite	Dwyka Tillite	Itararé Beds
		Bluff Cove Beds ~~~~ d ~~~~	Lower Dwyka Shales	
Pre-Gondwana Succession	L. Carboniferous	Port Stanley Beds	Witteberg Series	Barreiro Sandstone (B)
	Middle Devonian	Port Philomel Beds	Upper Bokkeveld Series	
				Ponta Grossa (B), Rincon de Alonso (U)
	Late L. Devonian	Fox Bay Beds	Lower Bokkeveld Series	
	Lower Devonian	Port Stephens Beds ~~~~ d ~~~~	Table Mountain Series ~~~~ d ~~~~	Furnas Sandstone (B), Carmen Sandstone (U), Sierra de la Ventana Quartzites (A) ~~~~ d ~~~~
	Archaean	Cape Meredith Complex	Basement Complex	Basement Complex

A = Argentinia, B = Brazil, U = Uruguay.

Beiträge zur Regionalen Geologie der Erde Band 10: MITCHELL-THOMÉ, S. Atlantic Islands
Gebrüder Borntraeger, Berlin · Stuttgart

ad p. 324

Table 93 Stratigraphic Correlation Chart between the Falkland Islands and Various Regions in Argentina (BORRELLO, 1963)

System	Falklands	Chubut	Sierra de la Vertana (Buenos Aires)	Pampa Range	PreCordillera (San Juan, Mendoza)
Triassic	Dykes / Falkland Sound Group (Neocalamites carrerei, Glossopteris flora)			Gualo Group / Ischigualasto Group (Neocalamites carrerei) / Los Rastros Group (Neocalamites carrerei) / Ischichuca Group / Esquina Colorada Group / Patquiense, and/or Catunense (Trias-Rhaetic)	Rio Blanco Group / Cacheuta Group / Potrerillos Group (Thinnfeldia flora.) / Las Cabas Group / Zuberia cf. calamites Carrerei flora.) / Las Pircas Conglomerate (Potrerillos System)
Permian	Lafonia Group (Glossopteris flora) / Black Rock Group / Lafonian Tillite Group	Salitralense (Glossopteris flora)	Las Tunas Ser., Bonete Ser. (Glossopteris flora, Eurydesma fauna) / Piedra Azul Series (Murchisonia sp.) / Sauce Grande Series (Astarte pusilla) (Pillahuinco System)	Barakaria dichotoma flora / Tupense (Rhacopteris ovata flora) / Penoniano (Septosyringothyris Keideli fauna.)	Santaclarense (Glossopteris flora) / Montanense / Mollarense / Cielense / Barrealense (Linoproductus fauna) / Leoncitense (Septosyringothyris Keideli fauna.) (St. Clara System / Barreal System)
Carboniferous	? (West Falkland System)	Shotlense (Gangamopteris-Rhacopteris flora) / Loma Chata Group (Septosyringothyris Keideli fauna) (Te System)	Hiatus	Jagueliano	Qa. de la Chavela Group (Hadrorhacus sp., Hostimella, etc. flora.) / Lomas de los Piojos Group (Australospirifer antarcticus fauna.)
Devonian	Monte Maria Group / Fox Bay Bay Group (Australospirifer antarcticus fauna) / Port Stephens Group	Hiatus	Lolen Group (Australospirifer antarcticus) / La Providencia Group / Naposta Group / Bravard Group (Ventana Series)	Hiatus	
Precambrian	Discordant Veins / Granites / Pegmatites / Metamorphics / Dykes	Arroyo Pescado Beds / Esquel (?) Beds	Quartz-Porphyry Granite	Lower to/ and Middle Palaeozoic	Granite Crystallines / ?

Table 94 Stratigraphical Correlation table showing Terms used by Adie (1958), Borrello (1963) and Frakes & Crowell (1967)

ADIE (1958)			BORRELLO (1963)		FRAKES & CROWELL (1967)	
Quatern.	Recent	Shell Point Limestones and West Point Islands Forest Bed	Quatern.	Holocene	'Stone rivers', peats	
Jurassic	Rhaetic (?)	Diabase Dykes and Karroo Dolerites	Triassic to Upper Permian	East Falkland	Dykes	
Triassic to Permian	Upper-Middle	West Lafonian Beds				West Lafonian Beds
	L. Trias-U. Perm.	Bay of Harbours Beds			Falkland Sound Group	Bay of Harbours Beds
	Upper-Middle	Choiseul Sound-Brenton Loch Beds	Permian to Upper Carboniferous	West Falkland Group	Lafonia Group	Choiseul Sound-Brenton Loch Beds
	Lower	Lafonian Sandstones				Terra Motas Sandstone
		Black Rock Slates			Black Rock Group	Port Sussex Fm. / Shepherds Brook Mem. / Black Rock Member
Permo-Carboniferous		Lafonian Tillite			Lafonian Tillite Group	Lafonian Diamictite
Carbon-iferous	Upper	Bluff Cove Beds				Bluff Cove Beds
	Lower	Port Stanley Beds			Monte Maria Group	Port Stanley Beds and Port Philomel Beds
	Middle	Port Philomel Beds				
Devonian		Fox Bay Beds	Devonian		Fox Bay Group	Fox Bay Beds
	Lower	Port Stephens Beds			Port Stephens Group	Port Stephens Beds
Precambrian		Cape Meredith Complex	Precambrian		Granite and Pegmatite Dykes, Metamorphics	Cape Meredith Complex

Frakes & Crowell age column: West Lafonian Beds — Lower Mesozoic to Upper Palaeozoic (Upper Lafonian Series); Port Sussex Fm. — Early Permian to Early Carboniferous?; Bluff Cove Beds / Port Stanley Beds — Middle?; Fox Bay Beds — Late Early Devonian; Port Stephens Beds — Silurian; Cape Meredith Complex — Lower Palaeozoic? or Precambrian?

Beiträge zur Regionalen Geologie der Erde Band 10: Mitchell-Thomé, S. Atlantic Islands
Gebrüder Borntraeger, Berlin · Stuttgart

Trilobites
 Dalmanites falklandicus Clarke
 D. accola Clarke (S)
 D. (Mesembria) acacia Schwarz (A)
 D. africanus (Salter) Lake (A)
 Cryphaeus australis Clarke (A, S)
 C. allardyceae Clarke
 Acaste (Calmonia) ocellus (Lake) (A, S)
 Calmonia signifer Clarke (S)
 C. sp.
 Homalonotus (Burmeisteria) herscheli Murchison (A)
 Proetus sp.

Brachiopodes
 Prothyris (Paraprothyris) knodi Clarke (S)
 Australospirifer antarcticus (Morris & Sharpe) (A. S.)
 Spirifer hawkinsi Morris & Sharpe (A)
 Leptocoelia flabellites (Conrad) (A, S)
 Derbyina sp.
 Coelospira? sp.
 Schuchertella sulivani (Morris & Sharpe) (A, S)
 S. agassizi Hartt & Rathbun (S)
 Leptostrophia concinna (Morris & Sharpe) (S)
 L? mesembria Clarke (S)
 Chonetes falklandicus Morris & Sharpe (A, S)
 C. skottsbergi Clarke
 C. hallei Clarke
 Cryptonella? baini (Sharpe) (A, S)

Rensellaeria falklandicus Clarke
 R. sp.
 Orbiculoidea baini (Sharpe) (A, S)
 O. cf. bodenbenderi Clarke (S)
 O. sp. (Large)

Lamellibranchs
 Nuculites sharpei Reed (A, S)
 N. reedi Clarke (S)
 N. cf. branneri Clarke (S)
 Leptodomus cf. ulrichi Clarke (S)
 Janeia sp.
 Palaeoneilo (Large sp.)
 Toechomya?
 Cardiomorpha? colossa Clarke (S)

Gastropods
 Diaphorostoma baini Sharpe
 D. allardycei Clarke (A)
 Bellerophon (Pelctonotus) quadrilobata (Salter) (A)
 Ptomatis moreirai Clarke (S)
 Trophidocyclus antarcticus Clarke
 Loxonema? sp.

Cephalopods
 Orthoceras cf. gamkaensis Reed (A, S)

Annelids
 Tentaculites crotalinus Salter (A, S)
 Conularia africana Sharpe (A, S)

Sponges
 Clionolithus priscus (McCoy) (A)

Fish plates and crinoidal stems have also been recognized. (The letters (A) and (S) refer to South African and South American forms respectively, as per Borrello.)

The principal fossil localities listed are: Port Louis, Port Salvador, Port San Carlos in East Falkland: Fox Bay, Port Howard, Manybranch Harbour and the Chartres River area, West Falkland.

Baker remarked that the marine invertebrates tended to be congregated in certain localities or then levels, the trilobites, for example, being distinctly more abundant at certain horizons. On the other hand, Australospirifer antarcticus (Morris & Sharpe) is present at all horizons within these Lower Devonian strata. Of these Devonian fossils, 24 forms are common in South America and 19 in South Africa, with 12 types common in the three regions, and only 6 belonging exclusively to the Fox Bay formation.

As already mentioned earlier, Boucot recognized Australocoelia touteloti and Australospirifer antarcticus from Port San Carlos in the Fox Bay formation overlying the Cape Meredith Complex.

From the Port Philomel Beds, Halle obtained fragmentary plant remains from Halfway Cove and Port Purvis districts. His Lepidodendroid stem fragments were compared by SEWARD & WALTON with *Hornea Liguieri* KIDSTON & LANG, and an unknown plant fragment of HALLE's was compared to branched axes from the Lower Old Red Sandstone of Caithness, Scotland. SEWARD & WALTON recognized *Lepidodendroid* stems from the Port Purvis region. It was their opinion that although these plant remains were not sufficiently well-preserved to be assigned to previously recorded species, they indicated a Devonian age, Middle rather than Lower.

The Choiseul Sound and Brenton Loch Beds have yielded the following plants:

Glossopteris indica SCHIMPER
G. indica SCHIMPER cf. var. WILSONI
G. indica SCHIMPER var. *G. decipiens* FEISTMANTEL
G. browniana BROGNIART

Dadoxylon lafoniense
D. bakeri SEWARD
D. cf. *angustum* FELIX
Phyllotheca sp.

Also a perfectly preserved insect wing of the genus Palaeodictyoptera was discovered by HALLE at the head of Port Darwin.

The succeeding Bay of Harbours formation includes the following:

Glossopteris indica SCHIMPER var *G. decipiens* FEISTMANTEL
G. indica SCHIMPER
G. browniana BROGNIART
G. angustifolia

Phyllotheca australis BROGNIART
P. deliquescens (GOEPPERT)
Desmiophyllum sp.
A species close to *Voltzia heterophylla*

Principal fossil plant localities for the Choiseul Sound fossils include the head of Bodie Inlet, Arrow Harbour, Tranquilidad, Walker Creek, Fanny Cove, Darwin, Goose Bay and Low Bay. The chief sites for the Bay of Harbours plant remains are at North Arm, Dos Lomas, Speedwell and George Islands. BORRELLO however was doubtful of HALLE's Dos Lomas locality, HALLE remarking upon faults here which were not observed by BAKER, and further, the outcrops more likely are of later age, belonging to BORRELLO's Falkland Sound Group = West Lafonian Beds, a view seemingly shared by ADIE.

The following fossils have been determined from the West Lafonian Beds:

Neocalamites carrerei (ZELLER)
Glossopteris browniana BROGNIART
G. indica SCHIMPER
G. damudica

Gangamopteris cyclopteroides var. *major* FEISTMANTEL
Phyllotheca cf. *deliquescens*

The significant plant remain here is the first-mentioned which also occurs in the Upper Beaufort Series and Molteno Beds of South Africa. In Argentina it has been discovered in the sediments of the Upper Rio Bermejo basin, La Rioja Province, and also in Mendoza Province where a Triassic-Rhaetic age is postulated. ADIE placed these West Lafonian Beds as being at least Middle Triassic.

SEWARD & WALTON investigated plant fossils from the Falklands and also made critical comments on the findings of HALLE. They listed some fossil localities in areas where the Choiseul Sound, Bay of Harbours and West Lafonian strata are present, but placed all these fossil plants as being Permo-Carboniferous in age. As already remarked, these beds are now assigned rather a Middle Permian to Upper Triassic age.

From the Shell Point Limestones, of Recent age, the following sub-fossil gastropods include:

Trophon philippianus DUNKER
T. geversianus (PALLAS)
Xymenopsis liratus (GOULD)
Crepipatella dilatata (LAMARCK)
Pachysiphonaria lessoni (BLAINVILLE)
Kerguelenella lateralis (GOULD)
Patinigera aenea (MARTYN)

Margarella expansa (G. B. SOWERBY)
Adelomelon ancilla (SOLANDER)
Pareuthria fuscata (BRUGIÈRE)
Rissoa sp.
Mytilus ovalis LAMARCK
M. ovalis L MARCK var. *M. edulis* LINNÉ
Samarangia exalbida (CHEMNITZ)

Some unrecognizable pelecypoda have also been observed in the upper limestones.

The West Point Island Forest Bed has yielded two conifer species closely similar to *Podocarpus chilina* and *Libocedrus chilensis,* both characteristic of temperate areas in the South American continent. *Pteridophyta* spores and *Podocarpus* sp. pollen grains have also been recorded. This Forest Bed has only been identified at Clifton Station, West Point Island.

Structure

Folding is characteristic of the sediments, and marked synclines and anticlines are evident. Small faults generally trend parallel to the anticlinal-synclinal structures. Larger fractures appear to control the coastal morphology in several areas. Dykes of doleritic and diabasic nature transect the sediments, and lamprophyric-type dykes occur in the Precambrian crystallines.

The folded Palaeozoic-Mesozoic sediments represent an extremity of the Patagonian Fold Belt of southern Argentina.

In both main islands, asymmetric foldings are common. In West Falkland strong folds strike parallel to the eastern coast; in East Falkland folds trend more nearly W-E in the central part, and decrease both in number and intensity towards the N. Folding is still significant in northern Lafonia, but towards the S these decrease in magnitude and degree of disturbance. In the western areas of West Falkland the Devonian beds are only moderately deformed, though somewhat more so than in the southern part of Lafonia where only gentle flexuring occurs. In both West and East Falkland, Permo-Triassic strata are folded into a major NE-SW striking syncline which is now submerged beneath the waters of Falkland Sound.

The E-W striking folds of greatest intensity in central East Falkland coincide with the highest terrain. All these folds in plan are arranged convexly towards the S. The trend of Choiseul Sound follows very closely this convex pattern. Between Port Stanley and Port Fitzroy, Devonian strata occupy the deepest parts of the syncline: the anticlinal crests in this vicinity are composed of Carboniferous and Permo-Carboniferous sediments. In the Berkeley Sound area the anticlinal fold strikes E-W, with Lower Devonian Fox Bay Beds forming the crest and Lower Carboniferous(?) Port Stanley strata lower down the flanks. It is assumed this anticline plunges eastwards under the Sound. In the Port San Carlos district of northwest East Falkland, a NW-SE striking antiline occurs, with Fox Bay Beds outcropping at the culmination of the fold. Possibly the Bay of Harbours marks the site of a sunken syncline, such as occurs in Falkland Sound.

Tight folding occurs in eastern West Falkland, striking ENE-WSW. The principal structure here is an asymmetric anticline whose major topographic expression is the Hornby Mountains. The eastern flank has beds dipping almost vertically; the western flank is slightly flexured, to flatten-out into a sub-horizontal position some 5 km from the axis. Southwards this anticline fades out beneath the highest Devonian; to the N, it veers towards the E, meeting the sea at White Rock Bay, only a short distance from the western extremity of the NW-SE striking anticline in the Port San Carlos area of East Falkland. BORRELLO was of the opinion that these two anticlines were part of the same major structure, the West Falkland anticline undergoing torque and twisting almost through a right-angle to link up across Falkland Sound with the Port San Carlos anticline. He also surmised that the syncline running NW-SE through Bay of Harbours may similarly link up with the syncline running through Falkland Sound, these anticline-syncline structures describing a sub-parallel arcuate plan. No faulting occurs where the presumed sharp bend of the West Falkland anticline enters the sea in the vicinity of White Rock Bay. Both the Fox Bay Beds and the Port Philomel Beds forming the exposed parts of the anticline in this locality comprise soft, incompetent sandstones and shales which, within a distance of some 5–10 km could very well have their strikes altered by 90°. The Port San Carlos anticline, on the other hand, has outcropping Fox Bay Beds but principally Port Stanley Beds, the latter comprising strong, hard, massive quartzites and quartzitic sandstones which, one would assume, would resist bending in plan to the extent demanded here. In other words, that the two anticlines are one and the same structure is indeed likely, but it is the West Falkland one, at its northern extremity, which has submitted to the right-angled bending rather than the East Falkland anticline.

One may postulate that during the processes of folding, major compressive stresses acted towards the S and SSE in East Falkland, and towards the SE in West Falkland. These directions point towards Lafonia, but the exposed sediments here give no evidence of unusual strength and are only slightly deformed. It may be, however, that the crystalline basement occurs at not too great a depth here, which basement lithologically presented strong rocks more able tectonically to withstand powerful stresses directed towards it. As the stronger folds of the Palaeozoics occur approximately where the Wickham Heights trend NW-SE across East Falkland, it may be that here a greater thickness of these Palaeozoics are present, the basement lying much deeper in northern East Falkland, these sediments being thrust and folded against the higher-lying crystalline complex lying below the surface in Lafonia.

Significant NW-SE striking fractures coincide with rectilinear coastal stretches, as is seen on the S shore of Byron Sound, N coasts of Pebble Island and continuing to Falkland Sound, and SE from Cape Dolphin. More northward-veering strikes are postulated for major fractures in King George and Queen Charlotte Bays. In the southern region of West Falkland, a major fracture runs from S of Cape Meredith to Fox Bay. It may be that this fracture passes into the almost-vertical beds forming the common flank of the Mount Hornby anticline-Falkland Sound syncline. If this is so, perhaps the Precambrian basement, of which only a fragment is exposed at Cape Meredith, underwent more profound downthrow in this longitude, whereas NE of here, along Falkland Sound, the extent of downward movement lessened as the fracture passed into a tight fold.

Whether Falkland Sound is a fold or a fault feature is not definite: HALLE favoured a fault hypothesis, but BAKER & ADIE considered the feature rather as a deep synclinal fold.

Post-Triassic strata being absent in the archipelago, the dating of disturbances giving rise to folding and fracturing cannot be clearly stated. BORRELLO suggested that the chief period of tectonism corresponded to the Early Kimmeridgian phase of STILLE (M. Jurassic) and contemporaneous with the Palisade orogeny of the U. S. A. Eo-Kimmeridgian movements are recognized in the Patagonides of Argentina, where Liassic lies unconformably on Permian, and in the San Carlos Gulf region of Patagonia, a discordance lies between the Permian and Triassic-Lower Jurassic sediments.

Economic Geology

The Falkland Islands are poor in mineral resources, and peat deposits represent the sole item of economic interest.

Small quantities of siderite and limonite, in lenticular masses, have no economic value, and the same holds for the kaolin and graphite occurrences.

Some claystones have been used locally for construction purpose, taken from a quarry in the Choiseul Sound formation near Darwin.

Frequently along the coasts there are occurrences of white beach sands, almost exclusively of pure silica, which may be of some commercial value in glass manufacture.

Peat is abundant throughout the archipelago, and doubtless totals several million tons. At the present time this is exploited by the islanders only for their own use.

At Port Sussex, at the base of the Black Rock Slates, are some fine-grained sands and shales impregnated with carbonaceous matter, but of no economic significance. Carbonaceous and graphitic material is also present at the contacts of the doleritic dykes with sediments in the western part of East Falkland. Apparently there are no favourable indications pointing to any oil occurrences.

Geological Evolution

The oldest rocks exposed in the Falklands, schists, gneisses, granites and pegmatites, probably date from the Precambrian, but whether Lower, Middle or Upper cannot be stated. The archipelago represents an emergent part of the Patagonian Shelf, and in Argentina no unquestionable early Precambrian rocks are known. Perhaps the Cape Meredith rocks are not older than Middle Precambrian.

During a very long time-interval of some 300+ million years, these ancient rocks appear to have undergone intensive erosion, with the development of a pronounced relief, the discordance separating them from the younger sediments being a marked one. During this time-interval it is presumed that the basement was emergent, or at least if any submergence took place, all traces of such have long since been destroyed.

By Lower Devonian times, subsidence had occurred, the 2000+ m Port Stephens and Fox Bay strata being a marine facies. The presence of glacio-marine sediments in the Lower Devonian of the Falklands, also present in the San Juan Precordillera of Argentina and in the Table Mountain Series of South Africa, testify to a widespread period of glaciation at this time. This earlier Devonian marine transgression likewise is characteristic of austral regions of the Southern Hemisphere. The Lower Devonian sediments of the Falklands have been considered by some writers to be a regressive facies, but if this is so there is neither general agreement on this here nor in Argentina and South Africa.

Succeeding the transgressive phase, tectonic movements associated with the Acadian-Bretonic phase caused emergence and folding, such as also took place in the Sierras Australes of Argentina. This interruption in marine sedimentation gave rise to the clastic terrestrial sequence of the Port Philomel and Port Stanley formations, which probably bridge the time-gap of Middle and Upper Devonian, perhaps even into the Lower Carboniferous. The Upper Carboniferous Bluff Cove Beds are only some 10 m thick, these argillo-arenaceous sediments lying discordantly on the preceding beds. They are of marine and fluvio-glacial origin, and it may thus be presumed that some time around the boundary of the Devonian-Carboniferous there was slight subsidence, giving rise to very shallow seas and the development of ill-drained lakes and rivers which resulted from the gentle flooding of the lands.

Contemporaneous with this submergence there was a change to glacial conditions which ushered-in the development of the Lafonian Tillites. The 800 m thick sequence of partially Carboniferous and Permian strata represented by these Tillites, the Black Rock Slates and the lower part of the Choiseul Sound Beds, constitute what HALLE termed the Fluvio-(?) and Limno-Glacial Beds, terrestrial accumulations during a period of continental glaciation, which can be closely correlated with events in Argentina.

By late Middle Permian times the climate ameliorated, and slow subsidence began, giving rise to what BORRELLO termed paralitic geosynclinal (Exo-paralitic geosyncline of Marshall Kay) sedimentary deposition, represented by the Middle Permian-Triassic beds. Whether or not sedimentation continued uninterruptedly throughout this interval is not definite. ADIE suggested the possibility of an unconformity separating the Lower Triassic Bay of Harbours rocks from the Upper-Middle Triassic West Lafonian Beds, and remarked upon the structural complexity of the area where the latter outcrops.

In Argentina it is presumed that this period of sedimentation ceased sometime between the Lower and Middle Triassic, and it may well be that more detailed field investigations will substantiate a discordant relationship at this period in the Falklands. In Argentina tectonism sometime between Lower and Middle Triassic is suggested, and as previously mentioned, in that country, Liassic beds lie unconformably upon Permian and also Triassic-Lower Jurassic sediments rest discordantly on the Permian. Structural complexity in the Middle to Upper Triassic rocks of the Falklands indicates tectonic disturbances of a younger date which, as previously stated BORRELLO suggested may have occurred during the Middle Jurassic.

After this last phase of folding which affected these youngest Mesozoics in the Falklands, the area was upraised to some extent, but at what period exactly cannot be stated.

More definite are vertical upward and downward epeirogenic oscillations of Quaternary times. ANDERSSON postulated that before the onset of the Quaternaty the archipelago stood higher above sea level by some 46–73 m than at present, thus uniting into one block the majority, probably all, of the off-shore smaller islands. Pre-Pleistocene rivers which once occupied larger depressions on the land, e. g. Falkland Sound, Choiseul Sound, Berkeley Sound, etc. were subsequently transformed into submerged valleys. At the beginning of the Quaternary, the relative height of the archipelago was similar to that of today, but during the Holocene or post-glacial period, the archipelago block underwent a submersion varying from 70 m to 117 m, as per ANDERSSON.

Both HALLE & BAKER accepted such oscillatory movements for the island region, pointing to marine terraces occurring inland in both major islands, where bones of vertebrate marine animals are found at considerable altitudes above present sea level.

Such epeirogenic oscillatory movements postulated for the Falklands can also be recognized in Patagonia, beginning in the Upper Tertiary and continuing until Recent times, testimonies of basculatory movements which can be recognized well into the interior not only of Patagonia but also in Tierra del Fuego. Unfortunately, to date no studies have been made in trying to co-ordinate epeirogenic movements in the archipelago and in Patagonia-Tierra del Fuego.

ADIE mentioned a late Monasterian wave-cut platform 6 m above present sea level, occurring to N and S of Port Stanley. The sub-fossil fauna here were held to indicate climatic changes taking place in the late Pleistocene.

Finally we would mention the strong cliff development seen in many places along the lengthy coastlines, indicative of powerful marine erosion which is no less evident today in this windy, exposed and gale-ridden archipelago.

Palaeogeographic Connexions with South America - chiefly Argentina

The discovery in more recent years of Upper Palaeozoic sediments in SW Chubut and to the S of San Jorge Gulf in Patagonia (Argentina) make it possible to establish palaeogeographic relationships between Patagonia and the Falkland Islands. The findings of these Gondwana beds make it appear probable that the archipelago is but part of a tectonic arc extending from about lat. 43° S in Patagonia, trending S then SE towards the Falklands (Fig. 106). This old mountain chain corresponds to the Patagonides dorsal, formed and folded as a result of deformations in late Jurassic times. In lat. 46° S on the mainland, this chain lies concealed beneath Jurassic volcanics which extend southwards and eastwards to the Atlantic, the strike of these beds curving in the same sense, thus suggesting that the old chain curves in a similar fashion towards the Falklands. Further, we would note that the broad valley of the Rio Chico runs parallel to the arc on its western side.

Major fractures within the Patagonide arc begin at about lat. 46° on the mainland, and are also present in the submarine platform between here and the archipelago. These fractures are of radial pattern, dissecting the arc into blocks, of which the Falklands represent one. These blocks have been subjected to up and down movements resulting from tectonism in Late Tertiary-Quaternary times. The positive blocks, comprising a Precambrian-Lower Palaeozoic basement on which lie younger Palaeozoics and Mesozoics, were named 'nesocratons' by BORRELLO, two of which occur on the mainland and the third is represented by the Falklands. The nesocratons lie on the inner or concave side of the Patagonide arc.

A feature of note of the Patagonide front is the reduced section of Precambrian occurring at the base of the folded edge and on the exterior side thereof. The small

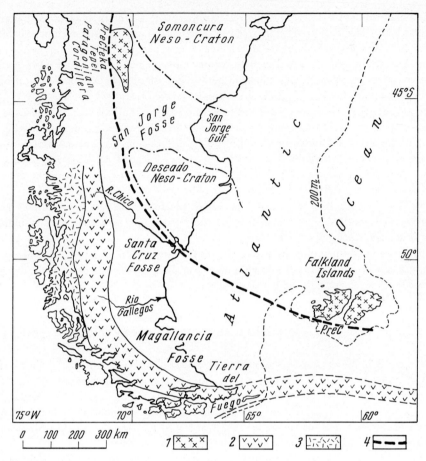

Fig. 106. Structural features of Southern Patagonia – Falkland Islands region. 1. Neo-Palaeozoics, 2. Mesozoic-Kainozoic fold belt, 3. Batholith, 4. Patagonide arc. (Modified after BORRELLO, 1963)

occurrences of ancient crystallines in the Tecka zone of Patagonia finds its counterpart in the Cape Meredith exposures in West Falkland.

As previously remarked, it is postulated that the orogenesis responsible for the Patagonides dates from early Kimmeridgian disturbances. In the Tepel-Nueva Lubecka region of Patagonia, early Mesozoic strata are not known, but are present in the Falklands, from which it may be presumed that tectonism was manifest somewhat earlier in Patagonia than in the archipelago.

In the vicinity of lat. 46° in Patagonia, the Patagonide arc was downthrown, so allowing Liassic transgressing seas to breach the mountain range and extend for a short distance to the W thereof. In compensatory movement, the Deseado nesocraton to the S of this embayment was upraised, and succeeding epeirogenic movements maintained its higher position until the Tertiary. The Falkland block likewise was upraised in early Jurassic and attained an apparently static state, perhaps limited eastwards by the South

Atlantic fosse – indeed HALLE considered the Falklands as a typical horst structure. Disturbances during the Cretaceous affected certain parts of the Patagonides and in some regions subsidence occurred (The author's predilection for the term 'disturbance' (perturbation) has been explained in: MITCHELL, R. C. (1958)). Palaeogene movements may possibly explain the extension of the arc barrier to the N of lat. 45° on the mainland, thus limiting the pan-Patagonian transgression in western Chubut.

The Patagonide arc constitutes a type of foreland related to which a mountain chain of Upper Jurassic-Tertiary age was integrated to form the Patagonian-Tierra del Fuego Andes. HALLE thus considered the Patagonide structure as a prolongation of the Precordilleras which, as regards tectonic relationship, form an old chain parallel to the modern Andes of western Argentina. The Southern Andes thus form a second arc, a product of Tertiary orogenesis, which lie sub-parallel and exterior to the southern Patagonides. This younger chain extends southwards through Tierra del Fuego, thus curving eastwards in keeping with the Patagonide curve eastwards through Falklands, the former linking-up with chains in Antarctica. The western border of these Southern Andes forms a batholith of Upper Cretaceous-Tertiary granites and granodiorites. Fragments of this batholith are to be found in the Precambrian and/or Lower Palaeozoics of the main cordillera.

Lying between the Patagonian Andes and the Patagonide arc are fosses or basins – negative tectonic units in the sense of movement direction. Stratigraphically the two basins, those of Santa Cruz and Magallanica, are similar, but structurally are separated by extrusives which welled-up to the surface along fractures, sub-parallel to the Andean foldings in the Rio Gallegos region. Within these basins strata ranging in age from Precambrian to Upper Palaeozoic form the basement, superincumbent rocks including Jurassic volcanics, Upper Jurassic-Cretaceous sediments, and in the Tertiary, paralitic sediments of great thickness.

In about lat. 46° is the San Jorge basin, lying between the Somoncura and Deseado nesocratons. The fosse is limited westwards by fractures responsible for the lowering of the Patagonides in this latitude, and to the E it is presumed the basin extends out into the Patagonian Shelf. In this basin, Cretaceous-Tertiary sediments show a different tectonic milieu from the two basins to the S. In the latter, after the Jurassic igneous episode there was continuous marine sedimentation, but in the San Jorge fosse, Upper Jurassic and Neocomian are unknown – or at least marine facies thereof. This would suggest that epeirogenesis has not been of like magnitude on either side of the intervening nesocraton, nor were the movements synchronous.

As regards the character of the block lying between the Patagonian coast and the Falklands, between the two arcs and now part of the Patagonian Shelf, only merest inferences can be made on the basis of what is exposed in the southern nesocration and fosse. Similarities in the stratigraphical relationships between here and West Falkland can be noted, but analogies cannot be pressed too far.

K. E. CASTER (1952) claimed that the Falklands were an emergent part of Patagonian Shelf, covered by a cratonic expression of the Furnas Sandstone (Lower Devonian) and succeeding fossiliferous shales, there being no indication of a geosyncline on either side. The archipelago agrees, both lithologically and faunally to the hypothetical features of the missing S. Africa craton. CASTER further remarks that ecologically, in spite of the

Falkland position so far S and across the axis of the pre-Andean trough, the Devonian of the archipelago matches the Parana basin.

A. L. Du Toit (1927), in his scheme of continental translations, would 'float' the archipelago northwards so that it was placed between the Parana basin and the Cape system of South Africa. Thus the Patagonide arc would be considered as an old infrastructure associated with such 'drift', the edge of the 'floating' block, as it were. On the other hand, Caster, were he to visualise 'floating' continents, would prefer to leave the Falklands where they are with respect to Argentina, for they would neatly fit S and W of present remnants of the Cape system, providing a fraction of the shield over which the cratonic element (Parana) of the South African Bokkeveld fauna transgressed in order to reach the Cape. However, neither Caster nor Borrello favoured the continental 'drift' hypothesis, and in whatever manner linkage between South America and South Africa was established, Caster categorically stated that "there was no 'South Atlantic basin' in existence in any dimension or condition which would justify applying the modern name to it" in Lower Devonian times.

Palaeogeographic Connexions with South Africa

The above descriptions of relationships with Argentina are largely taken from Borrello who throughout stresses the similarities with his native country but has little to say regarding South Africa.

Adie is the foremost protagonist is stressing former relationships between the Falklands and South Africa. Indeed, he goes so far as to say: "The Falklands bear no stratigraphical or structural relation to Patagonia, as Du Toit (1927) suggested" (1952b). Although in his papers of 1952 Adie shows correlations between the Falklands, South Africa and South America, yet in his stratigraphical publication of 1958 he makes only two references to strata in South America, and throughout correlates his stratigraphic sequences of beds with those in South Africa.

According to Adie, in the former reconstructions of Gondwana (the term 'Gondwanaland' is redundant), the Falklands should be placed some 250 km to the E of Eastern Province, South Africa, i.e. in the present Indian Ocean. It was his contention that, taking recognition of the descriptive aspect of Gondwana, as well as the movements involved, it appeared more reasonable to postulate that the archipelago had 'travelled' westwards from this position rather than the NNW 'drift' as proposed by Du Toit towards the South American coast, although both schemes would involve a relative displacement of about the same magnitude. (As seen above, such a view would also be in disagreement of Caster's contention.)

Caster (op. cit.) remarked that the Devonian fauna of the Falklands contained the largest percentage of African species of any of the Western Hemisphere developments of the 'austral' fauna.

Opposing Views of Palaeogeographic Relationship

Undoubtedly as regards the Falkland Islands, Adie speaks with the voice of authority; Borrello knows his Argentina and Du Toit was well versed in South African geology.

Table 95 — Principal Data regarding the South Atlantic Islands (ad p. 335)

Island	Lat.	Long.	Area (km²)	Max. Elev. (m)	Distance to nearest mainland (km)	Mean Annual Temp. (°C)	Mean Annual Rain. (mm)	Pop. (approx.)	Distance to crest of Mid-Atlantic Ridge (km)	Principal Rock Types	Principal Physiographic Characteristics	Age: Stratigraphic (and estimates) (V = Volcanics; S = Sedimentaries)	Age: Radiometric Datings (m.y.)	
St. Paul Rocks	0°56'N	29°22'W	?	23	950 (S. Amer.)	--	--	None	80 S	Dunites. No sediments	Steep, rocky islets. Deep waters near shores.		4500	
Fernando de Noronha	3°50'S	32°15'W	184	321	345 (S. Amer.)	25.4	1318	1,030	750 SW	Pyroclastics, Phonolites Trachytes. Few sediments	Peripheral hills, central plain, extending westwards into coastal plain. Very slight stream dissection. Coasts steep and rocky	Upper Cretaceous-Neogene (V)	Quixaba Formation Sao José Formation Remedios Formation	1.7 - 3.2 9.49 8.91 - 11.79
Rocas	3°52'S	33°49'W	?	13	215 (S. Amer.)	--	--	None	820 SW	Coral Reef	Circular-shaped coral islet with central lagoon.	Recent (S)		
Trindade	20°30'S	29°19'W	8	600?	1140 (S. Amer.)	23.2	806	None	1760 W	Phonolites, Limburgites, Analcitites, Nephelinites, Tambuschites, Pyroclastics, Alluvial cones, talus, etc.	Narrow, central highland area. Slopes abrupt. Deep, narrow valleys. Rocky coasts. Prominent monoliths.	Upper Cretaceous/Tertiary to Post-Glacial (V)	Morro Vermelho Fm. Desejado Formation Complex	< 27,000 yrs. 2.27 1.1 - 3.63
Martin Vaz	20°30'S	28°51'W	2.1	175	1188 (S. Amer.)	--	--	None	1712 W	Haüynites, Ankaratrites	No information except the inaccessible nature of the coasts.	Upper Cretaceous-Neogene (?) (V)		
Sao Tome	0°12'N	6°36'E	854	2023	260 (Africa)	24.1	965	60,000	2100 NE	Alkali-Basalts, Phonolites, Andesites. Few Sandstones	Mountainous, steep slopes, strong relief except in the NE. Valleys deep and narrow. Coasts strongly cliffed.	Aquitanian (S) Pre-Miocene (V)		
Principe	1°37'N	7°24'E	110	948	230 (Africa)	24.2	950	4,500	2300 NE	Basalts, Phonolites, Tephrites, Trachytes, Nephelinites. Some Shelly Limestones	Low, rolling landscape in the N. Elsewhere, irregular, mountainous aspect. Coastline extensively embayed.	Aquitanian-Vindobonian (S) Pre-Miocene (V)		
Fernando Poo	3°30'N	8°42'E	2009	2972	51 (Africa)	27.0	1798	62,000	2450 NE	Alkali-Basalts, Basalts. No sediments	Central, high mountain range. Deep, narrow valleys. Rocky, cliffed coastlines.	Cretaceous (?) (V)		
Annobon	1°26'S	5°37'E	17	700	360 (Africa)	--	--	1,400	2000 NE	Oceanites, Picritic Basalts. No sediments	Strong relief. Most valleys usually dry. S. and W coastlines indented.	Cretaceous (?) (V)		
Ascension	7°57'S	14°22'W	98	859	1450 (Africa)	16.7	635	1,400	144 W	Basalts, Trachyandesites, Trachytes, Rhyolites, Trachydolerites, Pyroclastics. No sediments	High land in E, steeper slopes to E and S. Ravines deep, confined. E Coast strongly cliffed.	Pleistocene (V)		
St. Helena	15°57'S	5°42'W	118	823	1900 (Africa)	20.1	995	4,600	800 E	Basalts, Phonolites, Trachytes, Pyroclastics. Calcareous Sands	NW-SE trend of highest land. Centrifugal drainage pattern. Valleys steep and narrow. Strong cliffing in some areas. Inclined rock benches.	Eocene-Miocene (V)	Intrusives Late Lava Flows Lower Main Shield Flows SW and NE Volcanoes.	6.8 - 8.4 8.1 - 9.6 10.1 - 10.9 11.3 - 14.0
Tristan Group	37°05'S	12°17'W	108	2062	2800 (Africa)	16.0	1650	300	370 E	Alkali Basalts, Trachy-Basalts, Trachyandesites, Trachytes, Pyroclastics. Outwash Alluvium	T. da C.: Steep central area, developing into lower gentler slopes, ending in great cliffs. Inaccess: Strong coastal cliffing. Night.: N-S highland ridge in E; elsewhere rolling topography	Night. Group: Considerably younger than other isles. (V) T. da C.) "Several million years". (V) Inaccess.)	Tristan da Cunha Inaccessible Nightingale Group	0.5 - 3.0 6 12 - 18
Gough	40°20'S	9°56'W	55	907	2400 (Africa)	--	--	None	336 E	Alkali Basalts, Alkali-Trachytes, Trachyandesites. No sediments	Rugged, deeply dissected. Deep, narrow, hanging valleys, Imposing coastal cliffs.	Tertiary (V)	2 - 6	
Falkland Islands	51°00'- 52°56'S	57°42'- 61°19'W	11,718	698	535 (S. Amer.)	5.5	700	2,100	3200 W	Siltstones, Sandstones Quartzites, Conglomerates, Shales. Igneous-Metamorphic Rocks of minor importance	Rolling landscape, mild relief, few hill ranges. Lacking in well-developed drainage network. Coastlines highly indented.	Precambrian-Upper Triassic. (S)		

Beiträge zur Regionalen Geologie der Erde Band 10: MITCHELL-THOMÉ, S. Atlantic Islands
Gebrüder Borntraeger, Berlin · Stuttgart

The geographical nearness of the archipelago to Argentina very naturally invites speculations as to possible various correlations, but because of political reasons, Argentine geologists are not welcome in the Falklands, and hence the views of BORRELLO as regards these islands are based solely on his appreciations of available literature.

ADIE has also visited South America and South Africa, as did NIDDRIE. In neither case were travels extensive or prolonged, and as regards NIDDRIE, his fossil collection from the Falklands was lost in transit, thus prohibiting him from making a more complete palaeogeographical study and appraisal. However, he was impressed by the marked similarity in rock types and sequences encountered in the Falklands and South Africa, which latter country he had best acquaintance.

Lastly, the writer must confess he has visited neither the Falklands nor South Africa but has worked quite extensively in parts of Argentina, Uruguay and Brazil.

In general, Argentine geologist, DU TOIT and CASTER see close geological connexions between the Falklands and Patagonia or then southern South America. ADIE and NIDDRIE point rather to the relationship between the Falklands and South Africa, ADIE adopting an extreme view in this matter.

The difficulty of reaching a satisfactory appraisal of the true ligations, rests on the fact that seemingly no one can claim an intimate knowledge of southern South America, Falklands and South Africa. Perusal of the literature, discussions at conferences, armchair speculations, authority in one region – none of these can compensate for this lack of overall appreciation of the geological factors and conditions, based upon field investigations, pertaining to these three areas of the world. Rather than take a stand, rather than show bias to one or the other contention, the author prefers to present such facts as are known, such opinions as are held, and leave it at that.

Abstract

The South Atlantic islands here discussed have a total area of some 16 000 km², less than 0.05 % of the area of this southern part of the Ocean. Areal insignificance, however, is totally outweighed by the geological importance of these tiny specks of land, for here factual evidence can be obtained to corroborate or otherwise, hypotheses based upon oceanographical studies.

With the exception of the Falkland Islands, all the islands are volcanic, or, in the case of Rocas Atoll, has a volcanic foundation. All, again with the exception of the Falklands, appear to have maximum ages not much greater than the Tertiary. Fourteen islets, islands and islands-groups are discussed, being treated on a national basis-Brazilian, Portuguese, Spanish and British.

The volcanic islands show a somewhat monotonous rock assemblage, with basalts, phonolites and trachytes as characteristic extrusives. Pyroclastics are prominent only in some of the islands. Plutonic xenoliths are significant in Ascension and the Tristan da Cunha group. In Trindade and Tristan da Cunha we can observe vulcanism of Recent date, and in the latter eruptivity occurred a matter of some ten years ago.

Sedimentary occurrences on the volcanic islands, other than alluvial outwashes and littoral deposits, are negligible. However in Sao Tomé and Principe, small outcrops of sandstones and limestones have yielded significant fossil finds which enable us to give relative datings of the sediments and igneous events.

The Falkland Islands are a group apart, not merely as regards far greater area, lower altitudes, milder relief, but also in their fundamental constitution and structural development. Rudaceous, arenaceous and argillaceous rocks outcrop, ranging in age from Lower Devonian to Upper Triassic, and totalling at least 6700 m in thickness. Here and there the sediments are intruded by diabasic and Karoo doleritic dykes, of presumed Rhaetic (?) age. At Cape Meredith are outcrops of gneisses and schists, invaded by granites, pegmatites, other acidic and basic dykes. This complex is presumed to be of Precambrian age.

Characteristic of the volcanic islands, with the exception of Rocas, Martin Vaz and Fernando de Noronha, is the pronounced relief, steep slopes and prominent sea cliffing. Indeed some sea cliffs are as high and imposing as to be found anywhere in the world, rising sheer from the water's edge to almost 700 m. No less remarkable are some topographic forms, such as, for example, the inselberg-type prominences in Trindade, rising in bare, vertical walls for over 400 m.

As would be expected in these South Atlantic islands, exposed as they are to the long fetch of winds and waves, marine erosion is taking drastic toll.

Some islands, e. g. Fernando de Noronha, Trindade, Ascension, represent but the exposed parts of immense submarine composite cones built up from oceanic plains. Some of these volcanic piles rise 5500 m above the floor of the ocean. The volume and area of these structures is truly impressive; e. g. the area of the base of the St. Helena cone is some ten times greater than the area of the base of Mount Etna.

Rocas Atoll, rising a mere few metres above sea level, represents the exposed coralline superstructure of a large seamount which rises from a depth of 3800 m. Seamounts characterize the stretch of waters lying between the Brazilian islands and the South American mainland.

Radiometric datings are available at present for St. Paul Rocks, Fernando de Noronha, Trindade, St. Helena, the Tristan da Cunha group and Gough. A basalt analysis from a dyke in Nightingale Island in the Tristan group, gave a value of 18 ∓ 4 m. y., which, as of the present, is the oldest radiometric dating for the volcanic islands, St. Paul Rocks excepted.

St. Paul Rocks, a group of microscopic, rocky, uninhabited islets, whose total area is to be measured in mere square metres, have long interested the petrologist, being essentially of dunite formation. But it is only within very recent times that these specks of land have acquired an added interest. Radiometric datings on some samples give values of ca. 3500 m. y., and one sample has yielded an age determination of 4500 m. y., the oldest radiometrically dated rock specimen in the world, as of this time. These peridotitic rocks are considered to represent unaltered mantle material, rocks of great depth origin, now exposed as mere specks of land in a waste of seas.

As vulcanism typifies the islands in general, emphasis is placed upon the petrology, petrography, petrochemistry of the rock associations. In the post-war years, considerable attention has been given to most of such islands, and the petrological, petrochemical and vulcanological evolution of these islands is reasonably well understood. This enables us to integrate and correlate the igneous series, sequences and events with those of other islands of the North Atlantic.

In the case of the Falklands, which are not true oceanic islands, correlations are made with the Gondwana neighbours of South America and South Africa.

Zusammenfassung

Die hier besprochenen Südatlantischen Inseln haben insgesamt eine Oberfläche von etwa 16 000 km², weniger als 0,05 % der Oberfläche dieses Südteils des Atlantik. So unbedeutend ihre Fläche ist, so bedeutend sind diese winzigen Landflächen vom geologischen Stadtpunkt aus, denn hier können Beobachtungen gemacht werden, die imstande sind Hypothesen, die auf ozeanographische Studien beruhen, zu bestätigen oder zu widerlegen.

Mit Ausnahme der Falkland Inseln sind alle Inseln vulkanisch oder haben, wie das Rocas Atoll, ein vulkanisches Fundament. Alle, wieder mit Ausnahme der Falkland Inseln, scheinen höchstens tertiären Alters zu sein.

Vierzehn Inselchen, Inseln und Archipele werden besprochen, gruppiert nach ihrer nationalen Zugehörigkeit – Brasilien, England, Portugal, Spanien.

Zusammenfassung

Die vulkanischen Inseln zeigen eine einigermaßen monotone Gesteins-Vergesellschaftung, mit Basalten, Phonolithen und Trachyten als den charakteristischen Extrusiva. Pyroklastische Gesteine sind nur auf einigen Inseln von Bedeutung. Plutonische Xenolithe sind wichtig auf Ascension und dem Tristan da Cunha Archipel. Rezenter Vulkanismus kann auf Trindade und Tristan da Cunha beobachtet werden; auf letzterer Insel fand der letzte Ausbruch vor etwa 10 Jahren statt.

Auf den vulkanischen Inseln sind Sedimente, außer Alluvionen und Strandablagerungen, kaum von Bedeutung. In Sao Tomé und Principe haben jedoch kleine Aufschlüsse von Sandsteinen und Kalken fossile Faunen geliefert, die es gestatten, das relative Alter der Sedimente und der vulkanischen Ereignisse anzugeben.

Die Falkland Inseln sind eine besondere Gruppe, nicht nur wegen ihrer größeren Fläche, geringeren Höhe und sanfterem Relief, sondern auch wegen ihrer grundsätzlichen Zusammensetzung und ihrer Struktur-Geschichte. Psephitische, psammitische und pelitische Gesteine beißen in Mächtigkeiten von wenigstens 6700 m aus, und im Alter von Unter-Devon bis Ober-Trias. Hier und da finden sich Intrusionen von Diabasen und Karru Doleriten, für die ein Rhätisches (?) Alter angenommen wird. Am Kap Meredith sind Gneise und Schiefer aufgeschlossen, durchdrungen von Graniten, Pegmatiten und anderen sauren und basischen Gängen; für diesen Komplex wird ein präcambrisches Alter angenommen.

Kennzeichnend für die Vulkan-Inseln, mit Ausnahme von Rocas, Martin Vaz und Fernando de Noronha, sind das ausgeprägte Relief, steile Hänge und hohe Steilküsten. Einige dieser Steilküsten gehören zu den größten und imposantesten der Welt, mit Abstürzen von fast 700 m. Nicht weniger auffallend sind einige topographische Formen, wie zum Beispiel die Inselbergartigen Erhebungen auf Trindade, deren nackte vertikalen Wände mehr als 400 m hoch werden.

Wie man bei dem langen Einzugsgebiet für Wind und Wellen erwarten kann, sind diese Südatlantischen Inseln einer besonders starken marinen Erosion ausgesetzt.

Einige der Inseln, z. B. Fernando de Noronha, Trindade, Ascension, stellen lediglich die sichtbaren Teile riesiger, zusammengesetzter submariner Kegel dar, die den Ebenen der Tiefsee aufgesetzt sind. Manche dieser vulkanischen Massen erheben sich 5500 m über den Meeresboden. Volumen und Fläche dieser Strukturen sind äußerst eindrucksvoll; so z. B. ist die Grundfläche des Kegels von St. Helens zehnmal größer als die Grundfläche des Ätna.

Das Rocas Atoll, das sich nur wenige Meter über den Meeresspiegel erhebt, stellt den Oberbau eines Korallen-Riffs auf einem großen seamount dar, der aus einer Tiefe von 3800 m aufsteigt. Seamounts kennzeichnen die Zone zwischen den Brasilianischen Inseln und dem Südamerikanischen Festland.

Radiometrische Altersbestimmungen gibt es augenblicklich für den St. Pauls Felsen, Fernando de Noronha, Trindade, St. Helena, den Tristan da Cunha Archipel und Gough. Die Analyse eines Basalts von einem Gang auf der Nightingale Insel des Tristan Archipels gab einen Wert von 18 ± 4 Mio. Jahren; dies ist augenblicklich das höchste bekannte radiometrische Alter für die Vulkan-Inseln, den St. Pauls Felsen ausgenommen.

Die St. Pauls Felsen, eine Gruppe winziger, felsiger unbewohnter Inselchen, deren Gesamtfläche nach Quadratmetern zählt, sind seit langem von besonderem petrographischen Interesse, da sie im wesentlichen aus Dunit bestehen. Erst in neuester Zeit haben

Zusammenfassung

diese sehr kleinen Inselchen erneutes Interesse gefunden: Radiometrische Messungen einiger Proben zeigten ein Alter von etwa 3500 Mio. Jahren, und bei einer Probe ergab die Altersbestimmung 4500 Mio. Jahre; dies ist das höchste bisher auf der Erde festgestellte Alter. Man betrachtet diese peridotitischen Gesteine als unverändertes Mantel-Material, als Gesteine, die in großer Tiefe entstanden, und nun als winzige Fleckchen Land im Meer auftauchen.

Da die Inseln im allgemeinen vulkanische Bildungen sind, wurde besonderer Wert auf die Petrographie und Chemie der Gesteins-Gesellschaften gelegt. In den Jahren nach dem Kriege wurden die meisten dieser Inseln stark beachtet und ihre petrographische, chemische und vulkanologische Entwicklung ist verhältnismäßig gut bekannt. Das gestattet uns, die vulkanischen Gesteine, ihre Aufeinanderfolge und die zugehörigen vulkanischen Ereignisse mit denen anderer Inseln im Nord-Atlantik zu vergleichen.

Im Falle der Falkland Inseln, die keine eigentlichen ozeanischen Inseln sind, wurden Vergleiche zu ihren Gondwana-Nachbarn in Süd-Amerika und Süd-Afrika gezogen.

General Bibliography

ABDEL-MONEM, A.; WATKINS, N. D. & GAST, P. W. (1967): Volcanic History of the Canary Islands. – Trans. Amer. Geophys. Union. Abstr. **48**, No. 1, Washington.
ANDERSSON, J. G. (1906): Solifluction, a component of subaerial denudation. – J. Geol. **14**: 94–96, Chicago.
BARRAT, C. (1895): Sur la géologie du Congo français. – Ann. Mines. (9), **7**: 379–510, Paris.
BEBIANO, J. B. (1932): A Geologia do Arquipelago de Cabo. Verde. – Serv. Geol. de Portugal. Comun. **18**: 275 pp., Lisboa.
BERGEAT, A. (1907): Staukuppen. – N. Jb. Miner., Geol. Palaeont. Festband: 310–329, Stuttgart.
BORN, A. (1923): Isostasie und Schweremessung. – Berlin.
BOWEN, N. L. (1928): The Evolution of the Igneous Rocks. – Princeton Univ. Press. 332 pp., Princeton, N. J.
– (1937): Recent high temperature research on silicates and its significance in igneous geology. – Amer. J. Sci. **33**: 1–21, New Haven.
BOWIE, W. (1917): Investigation of Gravity and Isostasy. – U. S. Coast and Geod. Surv. Spec. Publ. No. **40**, Washington.
BROGGER, W. C. (1906): Eine Sammlung der wichtigsten Typen der Eruptivgesteine des Kristiangebietes. – Nyet. Mag. for Naturfid. **44**, Oslo.
– (1920): Die Eruptivgesteine des Kristiangebietes. IV. Das Fen-Gebiet in Telemarken, Norwegen. – Vid. Skr. I. Mat.-nat. Kl. No. 9, Oslo.
BUCHANAN, J. W. (1874): On Geological Work done on Board of H. M. S. "Challenger". – Proc. Roy. Soc. **108**, London.
BURRI, C. & NIGGLI, P. (1945): Die jungen Eruptivgesteine des mediterranen Orogens. – Publ. Vulkaninst. I. Friedländer. No. 3, Zürich.
CASTER, K. E. (1952): Stratigraphic and Palaeontologic Data relevant to the Problem of Afro-American Ligation during the Palaeozoic and Mesozoic. In: The Problem of Land Connections across the South Atlantic, with special reference to the Mesozoic. – Bull. Amer. Mus. Nat. Hist. **99**, Art. 3: 105–152, New York.
CHRISTIAN, L. (1935): Such is the Antarctic. – London.
CLOOS, H. & CLOOS, E. (1927): Die Quellkuppen des Drachenfels am Rhein. Ihre Tektonik und Bildungsweise. – Z. Vulk. **9**, Pt. 1: 36–40, Stuttgart.
CROSS, W.; IDDINGS, J. P.; PIRSSON, L. V. & WASHINGTON, H. S. (1902): The Quantitative Classification of Igneous Rochs. – J. Geol. **10**: 555–690, Chicago.
DAVIS, W. M. (1933): Glacial epochs of the Santa Monica Mountains, California. – Bull. Geol. Soc. Amer. **44**: 1041–1133, New York.
DU TOIT, A. L. (1927): A Geological Comparison of South America with South Africa. – Carnegie Inst. Publ. No. **381**: 158 pp., Washington.
ESENWEIN, P. (1929): Zur Petrographie der Azoren. – Z. Vulk. **12**: 108–227, Stuttgart.
FINCKH, L. (1913): Die Gesteine der Inseln Madeira und Porto Santo. – Z. d. geol. Ges. **65**: 453–517, Berlin.
FLINT, R. F.; SANDERS, J. E. & RODGERS, J. (1960): Diamictite, a substitute term for symmictite. – Bull. Geol. Soc. Amer. **71**: 1809–1810, New York.
FRENGUELLI, J. (1948): Estratigrafia y edad del llamado "Retico" en la Argentina. – G. A. E. A. **8**: 159–309, Buenos Aires.
FURON, R. (1956): Lexique stratigraphique international. Vol. IV. Afrique. Fasc. **6**: 18–19. C. N. R. S. – Paris.
– (1960): Géologie de l'Afrique. – Payot et Cie. 380 pp. Paris.
HARRIS, P. G. (1957): Zone refining and the origin of potassic besalts. – Geochem. et Cosmochen., Acta. **12**: 195–208, London.

HIBSCH, J. E. (1898): Erläuterung zur geologischen Karte des böhemischen Mittelgebirges. – Tscherm. Miner. Petr. Mitt. **17**: 1–96, Wien.
– (1904): Geologische Karte des böhmischen Mittelgebirges. – Tscherm. Miner. Petr. Mitt. **23**: 305–383, Wien.
JOHANSSEN, A. (1938): A Descriptive Petrography of the Igneous Rocks. Vol. 4. – University Chicago Press. 522 pp., Chicago.
– (1939): A Descriptive Petrography of the Igneous Rocks. Vol. 1 (2nd. edit.) – University Chicago Press. 318 pp., Chicago.
JUNG, J. (1955): Une nouveau type de diagramme pour la répresentation des caractères chimiques des associations régionales de laves. – C. R. Acad. Sci. **240**: 799–800, Paris.
KAY, MARSHALL (1951): North American Geosynclines. – Geol. Soc. Amer., Mem. **48**: 143 pp., New York.
KNORR, H. (1932): Differentiations und Eruptionsfolge in Böhmischen Mittelgebirge. – Tscherm. Miner. Petr. Mitt. **42**: 318–370, Wien.
KÖPPEN, W. (1923): Die Klimate der Erde: Grundriß der Klimakunde. – W. de Gruyter. 369 pp. Berlin.
KUNO, H. (1959): Origin of Cenozoic Petrographic Provinces of Japan and surrounding areas. – Bull. Volcan. Ser. 2, **20**: 37–76, Naples.
KUSHIRO, I. & KUNO, H.(1963): Origin of primary basalt magma and classification of basaltic rocks. – J. Petr. **4**: 74–89, Oxford.
LACROIX, A. (1904): La Montagne Pelée et ses éruptions. – Masson et Cie. 662 pp., Paris.
– (1916): La constitution des roches volcaniques de l'extrême nord de Madagascar et de Nosy bé; les ankaratrites de Madagascar en général. – C. R. Acad. Sci. **163**: 253–256, Paris.
– (1917): Les laves à häuyne d'Auvergne et leurs enclaves homoeogènes: importance théorique de ces dernières. – C. R. Acad. Sci. **164**: 588, Paris.
– (1923): Minéralogie de Madagascar. – Soc. d'Edit. Géogr. Mar. et Colon. **3**: 450 pp., Paris.
– (1928): La constitution lithologique des îles volcaniques de la Polynésie australienne. – Acad. Sci. Mém. **59**: 1–80, Paris.
MITCHELL-THOMÉ, R. C. (1958): Terminologies relatives aux perturbations tectoniques. – Cahiers Géol. Nr. **47**: 461–465, Seyssel (Ain).
– (1964): The Sedimentary Rocks of the Cape Verde Archipelago. – Serv. Géol. Luxembourg. **14**: 229–253, Luxembourg.
NIGGLI, P. (1931): Die quantitative mineralogische Klassifikation der Eruptivgesteine. – Schweiz. miner. petr. Mitt. **11**: 296–364, Zürich.
– (1935): Die quantitative mineralogische Klassifikation der Eruptivgesteine. – Schweiz. miner. petr. Mitt. **15**: 295–318, Zürich.
– (1936): Die Magmentypen. – Schweiz. miner. petr. Mitt. **16**: 335–399, Zürich.
PEACOCK, M. A. (1931): Classification of Igneous Rock Series. – J. Geol. **39**: 54–57, Chicago.
REYER, E. (1888): Theoretische Geologie. – E. Schweizerbart'sche Verlag. 867 pp., Stuttgart.
RINNE, F. (1921): Gesteinkunde. – Jänecke Verlag. 428 pp., Leipzig.
RITTMANN, A. (1952): Nomenclature of Volcanic Rocks. – Bull. Volcan. Ser. 2, **12**: 75–102, Naples.
ROSENBUSCH, H. (1877): Mikroskopische Physiographie der Mineralien und Gesteine. Bd. II. Massige Gesteine. – E. Schweizerbart'sche Verlag. 596 pp., Stuttgart.
RUDOLPH, E. (1887): Gerland's Beitr. Geophy. Bd. **1**.
STILLE, H. (1924): Grundfragen der vergleichenden Tektonik. – Gebr.Borntraeger. 443pp., Berlin.
STOCK, J. (1888): Die Basaltgesteine des Löbauer Berges. – Tscherm. Miner. Petr. Mitt. **9**: 429–469, Wien.
TRÖGER, W. E. (1935): Spezielle Petrographie der Eruptivgesteine. Ein Nomenklatur-Kompendium. – D. Miner. Ges. 350 pp., Berlin.
– (1938): Eruptivgesteinsnamen. – Fortschr. Miner. Krist. Petr. B. **XXIII**: 41–90, Jena.
WACE, N. M. & HOLDGATE, M. W. (1958): The Vegetation of Tristan da Cunha. – J. Ecol. **56**: 593–620, London.
WALLACE, A. R. (1895): Island Life. 2nd. edit. – London.
WASHINGTON, H. S. (1899): The Petrographical Province of Essex County, Massachusetts. – J. Geol. **7**: 53–56, Chicago.
– (1906): The Roman Camagmatic Region. – Carnegie Inst. Publ. Nr. 57: 199 pp., Washington.
– (1913): J. Geol. **21**. Chicago.
WINCHELL, H. (1947): Honolulu Series, Oahu, Hawaii. – Bull. Geol. Soc. Amer. **58**: 1–48, New York.
WOLFF, F. VON (1931): Der Vulkanismus. Spezieller Teil **2**. Die Alte Welt. – F. Enke Verlag.: 829–1111, Stuttgart.

Bibliography relating to the South Atlantic Islands

ABREU, J. DE L. (1890): Apontamentos sobre a ilha de Fernando de Noronha en 1857. – Arq. Inst. Hist. e Geogr. de Pernambuco, Rev., No. **38**: 3–17, Recife.
ADIE, R. J. (1952a): Representatives of the Gondwana System in the Falkland Islands. – Symposium sur les séries de Gondwana. **19e** Congr. intern. géol.: 385–392, Alger.
– (1952b): The Position of the Falkland Islands in a Reconstruction of Gondwanaland. – Geol. Mag. **89**: 401–410, Hertford, Herts.
– (1953): New evidence of sea level changes in the Falkland Islands. – F. I. Dependencies Surv. Scient. Reports. No. **9**: 8 pp., Port Stanley.
– (1958): Falkland Islands. In: Lexique stratigraphique international. Vol. 5. Amérique latine. Fasc. 9c: 35–59. – C. N. R. S., Paris.
ALMEIDA, F. F. M. (1958): Geologia e Petrologia do Arquipelago de Fernando de Noronha. – Div. de Geol. e Min., Depto. Nac. Prod. Min., Minist. da Agric., Monog. **13**: Revised edition. 181 pp., Rio de Janeiro.
– (1960): Quelques aspects sousmarins au large de la côte brésilienne. – 21st. Inter. Geol. Congr., Pt. **10**: 23–28, Copenhagen.
– (1961): Geologia e Petrologia da Ilha da Trindade. – Div. de Geol. e Min., Depto. Nac. Prod. Min., Minist. das Minas e Energia. Monog. **18**: 197 pp., Rio de Janeiro.
ANDERSSON, J. G. (1902): Antarctic expeditionen arbeten pa Falklandsoäarna och Eldslandet 1902. – Ymer. No. **22**: 515–528, Stockholm.
– (1903): The scientific work of the Swedish Antarctic Expedition at the Falkland Islands and in Tierra del Fuego. – Geogr. J. **21**: 159–162, London.
– (1907): Contributions to the Geology of the Falkland Islands. – Wiss. Ergebn. schwed. Südpolarexped., 1901–03. Bd. 2, Liefg. 2: 1–38, Stockholm.
Anonymous (1805): Description of the Island of St. Helena, containing observations on its Singular Structure and Formation. – London.
ASSUNÇAO, C. F. T. (1956): Lavas feldspathoidicas de Sao Tomé. – Conf. internat. Afric. Ocid. **2**: 11–17, Lisboa.
– (1957): Algunas aspectos de petrografia da Ilha de Sao Tomé. – Garcia de Orta. Vol. 5, No. **3**: 497–515. Lisboa.
ATKINS, F. B.; BAKER, P. E.; BELL, J. D. & SMITH, D. G. W. (1964): Oxford Expedition to Ascension Island, 1964. – Nature. 204, No. 4960: 722–724, London.
BAKER, H. A. (1922a): Geological Investigations in the Falkland Islands. – Proc. Geol. Soc. No. **1093**: 12–13, London.
– (1922b): Final Report on Geological Investigations in the Falkland Islands 1920–22. – Gov't Press. 38 pp., Port Stanley.
BAKER, I. (1968): Compositional variation of minor intrusions and the form of a volcano magma chamber. – Quart. J. Geol. Soc., **124**: 67–79, London.
BAKER, I.; GALE, N. H. & SIMONS, J. (1967): Geochronology of the St. Helena Volcanoes. – Nature. **215**: 1451–1456, London.
BAKER, P. E. & HARRIS, P. G. (1963): Lava Channels on Tristan da Cunha. – Geol. Mag., **100**: 345–351, Hertford, Herts.
BAKER, P. E.; GASS, I. G.; HARRIS, P. G. & LEMAITRE, R. W. (1964): Vulcanological Report on the Royal Society Expedition to Tristan da Cunha, 1962. – Phil. Trans. Roy. Soc., (A) **256**: 439–578, London.
BARROS, J. A. L. (1950): Relatorio prévio sobre a Expediçao Joao Alberto a Ilha da Trindade. – 275 pp., Rio de Janeiro.
BARROS, L. A. (1960): A Ilha do Principe e a "Linha dos Camaroes". – Mem. Junta Invest. Ultramar. No. **17**: 127 pp., Lisboa.

BARTH, T. W. F. (1942): Lavas of Gough Islands. Result of the Norwegian Antarctic Expedition, 1927–28. Vol. 2, No. 20: 19 pp. – Oslo.
BELL, J. D. (1965): Eruption-mechanism on Ascension Island. Proc. Geol. Soc. No. **1626**: 145–146, London.
– (1967): The Occurence of Obsidian and other Natural Glasses in Ascension Island. – Proc. Geol. Soc., No. **1641**: 179–181, London.
BEMMELEN, R. W. VAN (1966): Le mécanisme de la dérive continentale. – Scientia. Sér. **7**, Mars–Avril: 1–10, Como.
BERG, G. (1903): Gesteine von Angola, Sao Tomé und St. Helena. – Tscherm. Miner. Petr. Mitt. **22**, Wien.
BERTHOIS, L. (1955): Contribution à l'étude lithologique des roches sédimentaires des îles de Sao Tomé et Principe. – Bol. Mus. Lab. Miner. Geol., Fac. Cien. Sér. **7**. No. **23**: 45–69, Lisboa.
BESNARD, W. (1951): Resultados cientificos do cruzeiro do "Baependi" e do „Vega" a ilha da Trindade. – Inst. Paulista de Oceanografia. Bol., T. **2**, Fasc. **2**: 37–48, Sao Paulo.
BOESE, W. (1912): Petrographische Untersuchungen an jungvulkanischen Ergußsteinen von Sao Tomé und Fernando Poo. – N. Jhb. Miner. Geol. Paläont. **34**: 253–320, Stuttgart.
BORRELLO, A. V. (1956): Combustibles Solidos Minerales. – Rec. Min. de la Rep. Argentina. Pt. **3**, Inst. Nac. Invest. Cien. Natur. y Mus. Argen. Cien. Nat. "Bernadino Rivadavia" Cien. Geol., Buenos Aires.
– (1963): Sobre la Geologia de las Islas Malvinas. – Ed. Cultur. Argentinas, Min. Educ. y Justicia. 70 pp., Buenos Aires.
BRANNER, J. C. (1888): Notes on the islands of Fernando de Noronha. – Amer. Naturalist. **22**: 861–871, Philadelphia.
– (1889): The Geology of Fernando de Noronha. Pt. 1. – Amer. J. Sci. **37**: 145–161, New Haven.
– (1890a): The aeolian sandstone of Fernando de Noronha. – Amer. J. Sci. **39**: 247–257, New Haven.
– (1890b): Geologia de Fernando de Noronha. – Rev. Hist. Geogr. de Pernambuco, 20–22, Recife.
– (1893): Os gres eolicos de Fernando de Noronha. – Rev. Inst. Hist. Geogr. de Pernambuco. No. **44**: 161–171, Recife.
– (1903): Is the Peak of Fernando de Noronha a volcanic plug like that of Mont Pelée? – Amer. J. Sci. **16**: 442–444, New Haven.
– (1919): Outlines of the Geology of Brazil, to accompany the Geologic Map of Brazil. – Bull. Geol. Soc. Amer., **30**: 189–337, New York.
BRITO, P. T. X. (1877): Memoria historica e geografica da ilha da Trindade. – Rev. Inst. Hist., Geogr. Etnog., Brasileira. **40**, Pt. 2: 249–275, Rio de Janeiro.
BUCHANAN, J. Y. (1876): Preliminary Report to Prof. Wyville Thompson .... on work done on board H. M. S. "Challenger". – Proc. Roy. Soc., **24**: 614–615, London.
– (1885): In: Narrative of the cruise of H. M. S. "Challenger" .... 210–214, London.
CAMPBELL, R. (1914): Rocks from Gough Island, S. Atlantic. – Trans. Roy. Soc. Edinburgh. **50**, Pt. 2: 397–404, Edinburgh.
CARDOSO, J. C., & GARCIA, J. S. (1962): Carta dos Solos de Sao Tomé e Principe. – Mem. Junta Invest. Ultramar. No. **39**: 306 pp., Lisboa.
Carnegi Institute (1964): Annual Report of the Director of the Geophysical Laboratory, 1963–64. pp. 43. – Washington.
CARVALHO, A. F. DE (1921a): As rochas da Ilha de Sao Tomé. – Mem. e Not., Mus. Lab. Miner. Geol. No. **1**: 9–24, Univ. Coimbra.
– (1921b): Estudo microscopico de algunas rochas da Ilha de Sao Tomé. – Publ. Mus. Miner. Geol. No. **1**: 24 pp., Univ. Coimbra.
– (1929): Noticia sobre o estudo geologico das Ilhas de Sao Tomé e Principe. – Bol. Agencia-Geral Colon. No. **43**: 126–142, Lisboa.
CAVACO, J. F. (1921): Analise quimica da rocha do Cao Grande da Ilha de Sao Tomé. – Mem. e Not., Mus. Lab. Miner. Geol. No. **1**: 24–25, Univ. Coimbra.
CHEVALIER, A. (1906): L'île de San Thomé. – La Géographie. **13**, Paris.
CLARKE, J. M. (1913): Fosseis Devoninos do Parana. – Serv. Geol. Miner. Brasil. Monogr., **1**: 55–79, 326–332, Rio de Janeiro.
– (1919): Falklandia. – Proc. Nat. Acad. Sci. **5**: 102–103, Washington.
Conselho Nacional de Geografia (1950): Atol das Rocas. Map. Scale 1 : 50 000. – Rio de Janeiro.
CORDANI, U. G.: Potassium-Argon ages of rocks from the Brazilian South Atlantic Islands. – Symposium on Continental Drift. UNESCO, Montevideo, 1967. Montevideo (InPress).
COSTA, D. (1938): A Ilha da Trindade. – Rev. Maritima Brasileira. Nos. **5** e **6**, Rio de Janeiro.

Costa, F. A. P. (1887): A ilha de Fernando de Noronha. Noticia historica, geografico e economico. – 117 pp., Pernambuco.
Daly, R. A. (1922): The Geology of Ascension and St. Helena Islands. – Geol. Mag., **59**: 146–156, Hertford, Herts.
– (1925): The Geology of Ascension Island. – Proc. Amer. Acad. Arts and Sci., **60**: 3–124, Philadelphia.
– (1927): The Geology of St. Helena Islands. – Proc. Amer. Acad. Arts and Sci., **62**, No. 2: 31–92, Washington.
Darwin, C. (1839): Journal of researches into the geology and natural history of the countries visited by H. M. S. "Beagle" . . . H. Colburn. 615 pp., London.
– (1846a): Geological observations on South America, being the third part of the Geology of the Voyage of the "Beagle" during 1832 to 1836. – London.
– (1846b): On the Geology of the Falkland Islands. – Quart. J. Geol. Soc. **2**: 267–274, London.
– (1876): Geological Observations on the Volcanic Islands. – Smith, Elder and Co. 2nd. edit. 674 pp., London.
– (1901): Journal of Researches. New edit., London.
Davidson, C. (1889): On the Origin of Stone Rivers of the Falkland Islands. – Geol. Mag. **3**. – London.
Davies, T. (1890): The Natural History of the Island of Fernando de Noronha. – J. Linn. Soc. **26**: 86–94, London.
Derby, O. A. & Barros, L. M. (1881): Jazidas de fosfate de cal na ilha Rata, do Arquipelago de Fernando de Noronha. – Min. Agric., Rel., Rio de Janeiro.
Diniz, M. de (1959): Contribuçao para o conhecimento des Foraminiferos do Arquipelago de Sao Tomé e Principe. – Garcia de Orta. **7**, No. 3: 471–482, Lisboa.
Douglas, G. V. (1923): Appendix in: Shackleton's Last Voyage: The Story of the "Quest". – F. Wild. Cassel and Co.: 314–328, London.
– (1930): Topography and Geology of the Tristan da Cunha Group. In: Report on the Geological Collections made during the Voyage of the "Quest", Shackleton-Rowett Expedition in 1921–22. – Brit. Mus. (Nat. Hist.): 67–71, London.
Dunne, J. C. (1941): Volcanology of the Tristan da Cunha Group. Results of the Norwegian Scientific Expedition to Tristan de Cunha, 1937–38. – Norske. Vid.-Akad. No. 2: 145 pp., Oslo.
Etheridge, R. (1885): Report on the Scientific Results of the voyage of H. M. S. "Challenger", Narrative. Vol. **1**, Pt. 2: 892–893. (On Devonian Fossils). London.
Feruglio, E. (1949–50): Descripcion geologica de la Patagonia. – Dir. Yac. Petrol. Fisc. **1–3**. – Buenos Aires.
Frakes, L. A. & Crowell, J. C. (1967): Facies and Palaeogeography of Late Palaeozoic Diamictite, Falkland Islands. – Bull. Geol. Soc. Amer. Vol. **78**: 37–58. New York 1967.
Fuster, J. M. C. (1950a): Aportaciones a la petrografia de la Isla de Fernando Poo, Guinea Espagnola. – Arch. Inst. Estud. Afric. No. **11**: 27–37, Madrid.
– (1950b): Las rocas ultrabasicas de Annobon y su relacion con los magmas basalticos de otras islas del Golfo de Guinea. – Arch. Inst. Estud. Afric., No. **13**: 37–54, Madrid.
– (1954): Estudio petrogenetico de los volcanes del Golfo de Guinea. – Inst. Estud. Afric. 155 pp., Madrid.
Gass, I, G. (1965): Magma-types of the Tristan da Cunha Group. – Proc. Geol. Soc. No. **1626**: 147–148, London.
– (1967): Geochronology of the Tristan da Cunha Group of Islands. – Geol. Mag. **104**: 160–170, Hertford, Herts.
Gass, I. G. et al (1962): Royal Society Expedition to Tristan da Cunha. – Nature. **194**: 1119–1122, London.
Gast, P. W.; Tilton, G. R. & Hedge, C. (1964): Isotopic Composition of Lead and Strontium from Ascension and Gough Islands. – Science, **145**, No. 3637: 1181–1185, Washington.
Geikie, A. (1882): A Search for Atlantis with the Microscope. – Nature. **27**: 25–26, London.
Gill, A. C. (1888): Petrographical notes on a rock collection from Fernando de Noronha. – Johns Hopkins Univ., Circular. **7**: 71-72, Baltimore.
Grant, B. (1883): A Few Notes on St. Helena and Descriptive Guide. 34–39. Jamestown, St. Helena.
Guimaraes, D. (1924): Contribuçao a petrografia do Brasil. – Serv. Geol. e Miner. do Brasil. Bol. **6**: 47 pp., Rio de Janeiro.
– (1932): Notas petrograficas. – Acad. Bras. de Cien. Anais. **4**, No. 1, Rio de Janeiro.

GUMBEL, C. W. (1880): Lithologisch-mineralogische Mitteilungen. I. Gesteine der Kerguelen-Inseln: II. Das weiße mineral der pflanzen-versteinerungen aus d. Tarentaise. – Tscherm. Miner. Petr. Mitt. **2**: 186–191, Wien.

HAFSTEN, U. (1960): Pleistocene development of vegetation and climate in Tristan da Cunha and Gough Islands. – Arbok. for. Univ. i. Bergen. Mat-Naturv. Serie No. **20**: 45 pp., Bergen.

HALLE, T. G. (1908): Note on the Geology of the Falkland Islands. – Geol. Mag., **5**: 264–265, Hertford, Herts.

– (1912): On the geological structure and history of the Falkland Islands. – Bull. Geol. Inst., Univ. Upsala. **11**: 115–229, Upsala.

HARRINGTON, H. J. (1956): Argentina. In: Handbook of South American Geology. Edit. by W. F. Jenks. – Geol. Soc. Amer., Mem. **65**: 131–165, New York.

HARRIS, P. G. & LEMAITRE, R. W. (1962): Volcanic Activity on Tristan da Cunha on December 16–17. – Nature. **193**: 719–721, London.

HARTT, C. F. (1870): Geology and Physical Geography of Brazil. – Fields, Osgood Co. 620 pp., Boston.

HEANEY, J. B. (1957a): Gough Island Scientific Survey 1955–56. – Polar Res. **8**, No. **55**: 338–345, Cambridge.

– (1957b): The Survey of Gough Island. – Empire Surv. Rev. **14**, No. **104**: 63–73, London.

HEANEY, J. B. & HOLDGATE, M. W. (1957): Gough Island Scientific Survey. – Geogr. J., **123**, Pt. 1: 20–31, London.

HIRST, T. (1951): Observations on the Geology and Mineral Resources of St. Helena. – Colon. Geol. Miner. Resources. **2**, No. **2**: 116–128, London.

HOLDGATE, M. W. (1957): Gough Island. – Geogr. Mag. **29**, No. **9**: 423–435, London.

– (1964): The Eruption of Tristan da Cunha, 1961–62. – Polar Res. **11**: 292–295, Cambridge.

HOLDGATE, M. W. et al (1956): The Gough Island Scientific Survey 1955–56. – Nature. **178**: 234–236, London.

IHERING, H. VON (1895): As ilhas oceanicas do Brasil. – Rev. Brasileira. **3**: 256–260, Rio de Janeiro.

IMBIRIBA, M. F. (1948): Relatorio para 1947. Territorio de Fernando de Noronha. – 120 pp. Rio de Janeiro.

Jérémine, E. (1943): Contribution à l'étude pétrographique du Cameroun Occidental. – Mém. Mus. Nat. d'Hist. Nat. **17**, Fasc. 1: 273–320, Paris.

JOYCE, J. R. F. (1950): The Stone Runs of the Falkland Islands. – Geol. Mag. **87**, No. **2**: 105–115, Hertford, Herts.

KITSON, A. (1931): Geological Notes on St. Helena, by various writers, with Remarks on the Economic Geology of that Island and Geological Map. – Colon. Off. No. **66**, London.

KNIGHT, E. F. (1887): The Cruise of the "Falcon". – Sampson Low, London.

– (1890): The Cruise of the "Alert". – Longmans Green Co. London.

LEA, T. S. (1888): The Island of Fernando de Noronha. – Proc. Roy. Geogr. Soc. **10**: 424–435, London.

LEMAITRE, R. W. (1960): The Geology of Gough Island, South Atlantic. – Overseas Geol. Miner. Resources. **7**. No. **4**: 371–380, London.

– (1962): Petrology of Volcanic Rocks, Gough Island, South Atlantic. – Bull. Geol. Soc. Amer. **72**: 1309–1340, New York.

– (1965): The Significance of the Gabbroic Xenoliths from Gough Island, South Atlantic. – Miner. Mag. **34**: 303–317, London.

LEMAITRE, R. W. & GASS, I. G. (1963): Occurrence of leucite in volcanic rocks from Tristan da Cunha. – Nature. **198**: 779–780, London.

LOBO, B. (1919): Ilha da Trindade. – Arq. do Mus. Nac. **22**: 105–158, Rio de Janeiro.

LOBOCH, M. Z. (1962): Noticia de Annobon. Geografia, Historia y Costumbres. – Publ. Diput. Prov. de Fernando Poo. 89 pp., Fernando Poo.

MACHADO, L. B. (1951): Resultados cientificos do cruzeiro do "Baependi" e do "Vega" a ilha da Trindade. – Inst. Paulista de Oceanografia. Bol. **2**: 96–103, Sao Paulo.

MACPHERSON, J.: Enumeracion y estudio de la colecciones recogidas en su viaje (a Fernando Poo y el Golfo de Guinea) por el Dr. Ossorio. – Anal. Soc. Esp. de Hist. Nat. **15**: 312–316, Madrid.

MALING, D. H. (1951): Falkland Islands. B. A. thesis. – Geogr. Dept., Univ. Durham. Unpubl., Durham.

MAYER, E. S. (1957): Trindade, ilha misteriosa do tropico. Liv. Tupa. – 158 pp., Rio de Janeiro.

MELLISS, J. C. (1875): St. Helena; a Physical, Historical and Topographical Description of the Island, including the Geology, Fauna, Flora and Meteorology. – Reeve and Co. London.

MENDELSOHN, A. C. (1942): Geological Report on the Investigation for Oil and Gas in the Islands of S. Tomé and Principe. – Junta Invest. do Ultramar. Report, Lisboa.
MILET-MURREAU, M. L. A. (1797): Voyage de La Pérouse autour du mond. – Imp. de la République. **2**: 26–29, Paris.
MILLER, J. A. (1964): Age determinations made on samples of basalt from the Tristan da Cunha group and other parts of the Mid-Atlantic Ridge. In: Vulcanological Report on the Royal Society Expedition to Tristan da Cunha, 1962. Appendix II. – Phil. Trans. Roy. Soc. (A), **256**: 565–569, London.
MORAES, L. J. (1928): Estudos geologicos no Estado de Pernambuco. – Serv. Geol. Miner. do Brasil. Bol. **32**: 100 pp., Rio de Janeiro.
MORRIS, J. & SHARPE, D. (1846): Description of eight species of Brachiopodus shells from the Palaeozoic rocks of the Falkland Islands. – Quart. J. Geol. Soc. **2**: 274–278, London.
MOSELY, H. N. (1892): Notes by a Naturalist. 487–489, London.
MURRY, J.; TIZARD, T. H.; MOSELY, H. N. & BUCHANAN, M. A. (1885): Narrative of the cruise of H. M. S. "Challenger", with a general account of the scientific results of the expedition. – London.
NATHHORST, A. G. (1906): Phyllotheca-Reste aus den Falkland Inseln. – Bull. Geol. Inst., Univ. Upsala. **7**: 72–76, Upsala.
NEGREIROS, A. (1901): L'île de San Thomé. – Paris.
NEIVA, J. M. C. (1946): Notas sobre o quimismo das formacoes eruptivas da ilha de Sao Tomé. – Bol. Soc. Geol. Portugal. **5**, Fasc. **3**: 151–158, Lisboa.
– (1954a): Quelques laves vacuolaires de l'île de St. Thomé et de l'îlot de Rolas. – Garcia de Orta. **2**, No. **1**: 53–59, Lisboa.
– (1954b): Chimisme des roches éruptives des îles de St. Thomé et Prince. – C. R. 19e Congr. intern. géol. **21**: 321–333, Alger.
– (1955a): Phonolites de l'île du Prince. Mem. e Not. No. **38**: 46–52. Univ. Coimbra.
– (1955b): Phonolites de l'île du Prince. – Garcia de Orta. **3**, No. **4**: 505–515, Lisboa.
– (1956a): Contribuçao para a petrografia das ilhas de Sao Tomé e Principe e dos ilheus das Rolas, das Cabras e Boné-de-Jockey. – Conf. intern. Afric. Ocid. **2**: 155, Lisboa.
– (1956b): Contribuçao para a Geologia e Geomorfologia da Ilha do Principe. – Conf. intern. Afric. Ocid. **2**: 157–162, Lisboa.
– (1958): Contribuçao para o estudo geologico e geomorfologico da Ilha de Sao Tomé e dos Ilheus das Rolas e das Cabras. – Conf. intern. Afric. Ocid. **2**: 147–153, Lisboa.
NEIVA, J. M. C. & PUREZA, F. G. (1956): Contribuçao para o conhecimento das areais das praias das ilhas de Sao Tomé e Principe. – Conf. intern. Afric. Ocid. **2**: 163–167, Lisboa.
NEIVA, J. M. C. & NEVES, J. M. C. (1957): Laterites da ilha do Principe. – Conf. intern. Afric. Ocid. **2**: 169–176, Lisboa.
NEIVA, J. M. C. & ALBUQUERQUE, C. R. DE (1962): Geologia e petrografia do ilheu das Rolas. – Mem. e Not., Mus. Lab. Miner. Geol. No. **53**: 3–19, Univ. Coimbra.
NEWTON, E. T. (1906): Notes on fossils from the Falkland Islands brought home by the Scottish National Antarctic Expedition in 1904. – Proc. Phys. Soc. **16**, No. **6**: 248–257, Edinburgh.
NIDDRIE, D. L. (1953): Falkland Islands. Appendix to: Necessity for Continental Drift, by L. C. KING. – Bull. Amer. Assoc. Petr. Geol., **37**: 2175–2177, Tulsa, Okla.
OLIVEIRA, A. I. & LEONARDOS, O. H. (1943): Geologia do Brasil. 2nd. edit. – Min. da Agric., Ser. Didatica. No. **2**: 813 pp., Rio de Janeiro.
OLIVEIRA, E. (1930): Rochas das ilhas oceanicas da bacia do Atlantico. – Cien. e Educ. Ano **1**. Nos. 11, 12. Rio de Janeiro.
OLIVER, J. R. (1869): Geology of St. Helena. – Jamestown, St. Helena.
PEREIRA, J. S. (1943): Subsidos geologicos e petrologicos para o conhecimento da Ilha de Sao Tomé. – Bol. Soc. Geol. Portugal. **3**: 125–144, Lisboa.
PIRIE, J. H. H. (1906): Geology of Gough Island. – Proc. Roy. Phys. Soc. **16**, No. **21**: 258–263, Edinburgh.
PIRSSON, L. V. (1893): Notes on some volcanic rocks from Gough Island, South Atlantic. – Amer. J. Sci., **45**: 380–384, New Haven.
PISSARRA, J. B.; CARDOSO, J. C. & GARCIA, J. S. (1965): Mineralogia dos Solos de Sao Tomé e Principe. – Est. End. e Docum., Junta Invest. do Ultramar. No. **118**: 141 pp., Lisboa.
POUCHAIN, E. B. (1948): Recursos economicos do Territorio de Fernando de Noronha. Calcarios Fosfatos. In: Relatorio anual do Director, 1947. – Div. Fom. Prod. Min., D. N. P. M. Bol. **83**: 66–71, Rio de Janeiro.
PRATJE, O. (1926): In: Berichte der deutschen Atlantischen Expedition. – Z. Ges. Erdkunde, Berlin.

PRIOR, G. T. (1897): Note on the occurence of rocks allied to Monchiquite in the island of Fernando de Noronha. – Miner. Mag. **11**: 171–175, London.
– (1900): Petrographical notes on the rock specimens collected from the little island of Trindade, South Atlantic, by the Antarctic Expedition of 1839–1843 under Sir JAMES CLARK ROSS. – Miner. Mag. **12**: 317–323, London.
– (1903): Contribution to the Petrology of British East Africa. – Miner. Mag. **13**: 228–263, London.
RAMOS, J. R. DE A. (1950): Expediçao a ilha da Trindade. – Rev. da Escola de Minas. Ano **15**, No. 6: 5–14, Ouro Preto.
RATTRAY. A. (1872a): On the Geology of Fernando de Noronha. – Quart. J. Geol. Soc. **28**: 31–34, London.
– (1872b): A visit to Fernando de Noronha. – J. Roy. Geogr. Soc. **42**: 431–437, London.
REED, F. R. C. (1949): The Geology of the British Empire. – Ed. Arnold. 2nd. edit. London.
REINISCH, R. (1912): "Geologie und Geographie". Deutsche Südpolar-Expedition 1901–1903. Teil 2, H. 7, Berlin.
RENARD, A. (1879): Peridotit von der St. Paul's-Insel im Atlantischen Ocean. – N. Jb. f. Miner. etc.: 389–394, Stuttgart.
– (1882a): Description lithologique des récifs de Saint-Paul. – Soc. belge de Microscopie. **6**: 13–65, Bruxelles.
– (1882b): "Challenger" Report. – Narrative. Vol. **2**. Appendix B: 29 pp. London.
– (1882c): Notice sur les roches de l'île de Fernando de Noronha. – Bull. Acad. roy. de Belgique. 3e Sér. **3**, No. 4: 352–361, Bruxelles.
– (1883): A propos des roches de St. Paul. – Soc. belge de Microscopie. **9**, No. 10: 165–178, Bruxelles.
– (1885a): Note sur la géologie du groupe d'îles de Tristan da Cunha. – Bull. Acad. roy. de Belgique. Sér. 3, **9**: 15 pp., Bruxelles.
– (1885b): Notice sur quelques roches des "fleuves de pierres" aux îles Falkland. – Bull. Acad. roy. Sci. Sér. 3, **10**, Nos. 9–10: 1–10, Bruxelles.
– (1885c): "Challenger" Report. **1**. – London.
– (1886): On some rock specimens from the islands of the Fernando de Noronha group. – Geol. Mag. **3**: pp. 33, London.
– (1887a): Notice sur les roches des îles Inaccessible et Nightingale (Groupe de Tristan da Cunha). – Bull. Acad. roy. de Belgique. Sér. 5, **13**: 28 pp, Bruxelles.
– (1887b): Notice sur les roches de l'île de l'Ascension. – Bull. Mus. roy. d'Hist. nat. de Belgique. **5**: 5–56, Bruxelles.
– (1889): Report on the rock specimens collected on Oceanic Islands during the voyage of H. M. S. "Challenger" during the years 1873–76. – Report on the Scientific Results. **2**. London.
RIBEIRO, P. A. (1951): Expediçao a ilha da Trindade. – Rev. Bras. de Geogr. Ano **13**, No. 2: 293–314, Rio de Janeiro.
RICHARDS, C. (1928): Newberyite and other Phosphates from Ascension Island. – Amer. Miner. **13**: 397–401, Washington.
RIDLEY, H. N. (1890): Notes on the Geology of Fernando de Noronha. In: T. Davies, The Natural History of the Island of Fernando de Noronha. – J. Linn. Soc. **26**: 86–94, London.
– (1891): The raised reefs of Fernando de Noronha. – Amer. J. Sci., **41**: 406–409, New Haven.
RIGGI, A. E. (1938): Las Islas Malvinas. Resena geografica y geologica. – Bol. Centr. Naval. **63**, No. 531: 26 pp., Buenos Aires.
– (1951): Geologia y Geografia de las Islas Malvinas. In: Soberania argentina en el Archipelago de las Malvinas y en la Antartida. – Univ. Nac. La Plata,
ROEDDER, E. & COOMBS, D. S. (1967): Immiscibility in Granitic Melts indicated by Fluid Inclusions in ejected Granitic Blocks from Ascension Island. – J. Petr. **8**, No. 3: 417–451, Oxford.
ROSS, J. C. (1847): A Voyage of Discovery and Research in the Southern and Antarctic Regions during the years 1839–1843. 2 Vols. – London.
ROTEIRO (1958): Brasil. Partes I and II. Marinha do Brasil, Hidrografia e Navigaço. Pte. **I**: 170 pp., Pte. II: 205 pp. – Rio de Janeiro.
SCHULTZE, A. (1913): Die Inseln Annobon. – Peterm. Mitt. **59**: 131–133, Gotha.
SCHUSTER, N. (1887): Petrographische Untersuchungen der von Oskar mitgebrachten Gesteine – Baumann aus Fernando Poo. – Peterm. Mitt. 268–269, Gotha.
SCHWARZ, E. H. L. (1905): The rocks of Tristan da Cunha brought back by H. M. S. "Odin", 1904, with their bearing on the question of the permanence of ocean basins. – Trans. S. Afr. Phil: Soc. **16**: 6–51, Johannesburg.

SCHWARZ, E. H. L. (1906): The Former Land Connections between Africa and South America. – J. Geol. **14**: 81–90, Chicago.
SCORZA, E. P. (1964): Duas Rochas Alcalinas das Ilhas Martin Vaz. Notas Preliminares e Estudos. – Div. de Geol. e Miner., Depto. Nac. da Prod. Min. Num. **121**: 1–7, Rio de Janeiro.
SERRALHEIRO, A. R. (1957): Novos elementos para o conhecimento da fauna fossil do Mioceno da Ilha do Principe. – Garcia de Orta. **5**, No. 2: 287–296, Lisboa.
Servicio Hidrografia da Marinha (1950): Atol das Rocas. Map. Scale 1:50000. – Rio de Janeiro.
Servicio Hidrografia da Marinha (1963): Ilhas Martin Vaz. Map. Scale 1:50 000. – Rio de Janeiro.
SEWARD, A. C. & WALTON, J. (1923): On a collection of fossil plants from the Falkland Islands. – Quart. J. Geol. Soc. **79**: 313–333, London.
SILVA, G. H. DA (1956a): La faune miocène de l'île du Prince. – Rev. Fac. Cien. **25**: 5–30, Univ. Coimbra.
– (1956b): O Miocenico da Ilha do Principe. – Conf. intern. Afric. Ocid. **2**: 231–256, Lisboa.
– (1958a): Nota sobre a microfauna do Miocenico marinho da Ilha do Principe. – Mem. e Not. No. **45**: 3–6, Univ. Coimbra.
– (1958b): Contribuçao para o conhecimento da microfauna do Miocenico maritimo da Ilha do Principe. – Garcia de Orta. **6**, No. 3: 507–510, Lisboa.
SMITH, A. CAMPBELL (1930): Report on the Geological Collections made during the Voyage of the "Quest", Shackleton-Rowett Expedition in 1921–22. – Brit. Mus. (Nat. Hist.) 22–38, 72–87, 105–107, 108–116, London.
SMITH, A. CAMPBELL & BURRI, C. (1933): The Igneous Rocks of Fernando de Noronha. – Schweiz. Miner. Petr. Mitt. **13**: 405–434, Zürich.
SOARES, L. DE C. (1944a): Territorio de Fernando de Noronha. – Cons. Nac. Geogr., Bol. Geogr. Ano 2, No. 19: 1019–1035, Rio de Janeiro.
– (1944b): Achegas para uma bibliografia sobre Fernando de Noronha. – Cons. Nac. Geogr., Bol. Geogr. Ano 2, No. 19: 1096–1102, Rio de Janeiro.
– (1958): Arquipelago de Fernando de Noronha. In: Enciclopedia dos Municipios Brasileiros. Vol. 4, Nordeste. – Inst. Bras. de Geogr. e Eststt. 451–477, Rio de Janeiro.
STUBBS, P. (1965): The Oldest Rocks in the World. – New Scientist. No. **426**: 82 pp., London.
SUERO, T. & CRIADO, R. P. (1955): Descubrimento del Palaeozoico Superior al Oeste de Bahia Laura, (Terr. Nac. de Santa Cruz) y su importancia palaeografica. – Notas, Mus. La Plata. **18**. Geol. No. 68: 157–168, La Plata.
TEIXEIRA, C. (1948–49): Notas sobre a geologia das ilhas de Sao Tomé e Principe. – Estud. Colon. **1**, No. 1: 37–46, Lisboa.
– (1949): Geologia das Ilhas de Sao Tomé e do Principe e do territorio de S. Joao Baptista de Ajuda. – Anais. Junta Invest. Colon. **2**, Fasc. 2: 5–20, Lisboa.
– (1955): Les roches sédimentaires des îles de Sao Tomé et Principe. – Bol. Mus. Lab. Miner. Geol., Fac. Cien. Sér. 7, No. 23: 41–42, Lisboa.
TENREIRO, F. (1961): A Ilha de Sao Tomé. Estudo Geografico. – Mem. Junta Invest. do Ultramar. No. 24. 279 pp., Lisboa.
THOMSON, W. (1877): The Atlantic. Part of the "Challenger" Report. – Narrative. **2**: 245–249, London.
THOMSON, C. WYVILLE (1878): The Atlantic. **2**: 221–229. – London.
– Narrative of the Cruise of the "Challenger". – 1, Pt. 2: 927–929, 944, London.
TILLEY, C. E. (1922): Density, refractivity and composition relations of some natural glasses. – Miner. Mag. **19**: 275–294, London.
– (1947):The dunite-mylonites of St. Paul's Rocks (Atlantic). – Amer. J. Sci. **245**: 483–491, New Haven.
– (1966): A Note on the Dunite (Peridotite) Mylonites of St. Paul's Rocks (Atlantic). – Geol. Mag. **103**: 120–123, Hertford, Herts.
TILLEY, C. E. & LONG, J. V. P. (1967): The Porphyroclast-Minerals of the Peridotite-Mylonites of St. Paul's Rocks (Atlantic). – Geol. Mag. **104**: 46–48, Hertford, Herts.
TILTON, G. R.; DAVIS, G. L.; HART, S. R.; ALDRICH, L. T.; STEIGER, R. H. & GAST, P. W. (1964): Geochronology and Isotope Geochemistry: Isotopic Composition of Lead in Volcanic Rocks from the Mid-Atlantic Ridge. In: Annual Report of the Director of the Geophysical Laboratory, Carnegie Institute, 1963–64. – 240–244. Washington.
TYRELL, G. W. (1934): Petrographical Notes on Rocks from the Gulf of Guinea. – Geol. Mag. **71**: 16–23, Hertford, Herts.
VELTHEIM, R V. (1950): Contribuçao a geologia da ilha da Trindade. – Anais da Acad. Bras. de Cien. **22**, No. 4: 463–469, Rio de Janeiro.

WADSWORTH, M. E. (1883): The Microscopic Evidence of a Lost Continent. – Science. **1**, No. 21: 590–592, Washington.
WASHINGTON, H. S. (1929): Dahllite from St. Paul's Rocks. – Amer. Miner. **14**: 369–372, Washington.
– (1930): The Petrology of St. Paul's Rocks (Atlantic). In: Report on the Geological Collections made during the Voyage of the "Quest", Shackleton-Rowett Expedition in 1921–22. – Brit. Mus. (Nat. Hist.) 126–144, London.
WEBSTER, W. H. B. (1834): Narrative of a Voyage to the Southern Atlantic Ocean in the years 1828–1830, performed in H. M. Sloop "Chanticleer". 2 Vols. – London 1834.
WHEELER, P. J. F. (1962): Death of an Island. – Tristan da Cunha. – Nat. Geogr. Mag. **121**, No. 5; 678–695, Washington.
WILLIAMS, G. H. (1889): Geology of Fernando de Noronha. Pt. II. Petrography. – Amer. J. Sci. **37**: 179–189, New Haven.
WINDHAUSEN, A. (1929–31): Geologia Argentina. – Ed. Peuser. 2 Vols. Buenos Aires.
WISEMAN, J. D. H. (1965): Petrography Mineralogy, Chemistry and Mode of Origin of St. Paul Rocks. – Proc. Geol. Soc. No. **1626**: 146–147, London.
WRIGHT, R. (1965): Ramification of Extrusive Age of St. Peter and St. Paul Rocks. – Bull. Amer. Assoc. Petr. Geol. **49**: 1709–1711, Tulsa, Okla.

Subject Index

Aa 166
Abrasion Platform 21, 32, 77, 78, 322
Acadian Orogeny 330
Agglomerate 17, 23, 73, 75, 76, 81, 131, 148, 168, 171, 172, 199, 218, 219, 222, 231, 238, 239, 240, 246, 257, 280, 281
Akerite 259
Algae 28, 29, 143, 202
Alluvial Cone 74, 77, 78, 79, 83, 225, 227
Alluvium 26, 28, 31, 78, 230, 232, 233, 277
Alochetite 20, 36, 47, 48
Ambohivorona 49
Analcitite 81, 82, 85, 86, 87, 100
Andesite 35, 109, 111, 112, 114, 115, 120, 126, 127, 132, 137, 138, 139, 173, 178, 179
Ankaramites 152, 153, 154, 234, 237, 251, 257, 272, 274
Ankaratrites 14, 20, 22, 23, 24, 25, 27, 28, 29, 31, 32, 33, 34, 36, 39, 46, 47, 48, 49, 50, 54, 55, 62, 63, 73, 74, 75, 76, 78, 81, 83, 89, 90, 92, 93, 103, 110, 112, 114, 134, 135, 149, 150, 249, 259, 260
Aquitanian 62, 124, 144
Archaean 308, 312
Arenite 26, 27, 28, 29, 32, 33, 69, 79
Ash 17, 70, 73, 75, 99, 100, 166, 167, 169, 170, 172, 232, 238, 239, 240, 279, 280
Asphalt 122, 141
Atoll 65
Augitite 20, 22, 35, 36, 39, 43, 46, 54, 86, 161
Avezacite 249

Barreiras Series 62
Barremian 62
Basalt 8, 9, 10, 16, 19, 20, 21, 22, 32, 35, 36, 39, 45, 47, 48, 50, 51, 52, 61, 62, 63, 64, 86, 87, 89, 97, 98, 99, 106, 107, 108, 109, 110, 111, 112, 113, 114, 115, 118, 119, 120, 122, 125, 126, 127, 128, 129, 130, 132, 135, 136, 137, 138, 139, 141, 143, 144, 145, 148, 149, 150, 151, 152, 153, 154, 155, 157, 158, 159, 160, 164, 165, 166, 167, 168, 169, 170, 171, 172, 173, 174, 175, 177, 178, 179, 186, 188, 189, 190, 191, 192, 193, 194, 198, 199, 200, 201, 202, 203, 204, 205, 209, 210, 211, 212, 215, 216, 219, 221, 222, 223, 229, 234, 235, 236, 237, 238, 241, 242, 243, 244, 246, 248, 249, 250, 251, 252, 257, 260, 262, 264, 270, 271, 272, 273, 274, 275, 276, 277, 278, 279, 280, 281, 288, 289, 290, 291, 292, 294, 295, 299, 300, 301, 303, 304, 305, 309
Basanite 9, 16, 25, 34, 35, 36, 39, 50, 52, 54, 55, 62, 63, 82, 85, 97, 98, 99, 111, 112, 113, 114, 115, 118, 119, 120, 129, 134, 135, 137, 138, 139, 161, 205, 249
Basanitoid 51, 111, 112, 113, 114, 120, 128, 137, 154, 192
Bay of Harbours Beds 320, 321, 326, 330
"Beagle" 3, 4, 6, 16, 308, 309
Bebedourite 91
Berondite 249, 259
Black Rock Formation 310
Black Rock Group 310, 320, 321
Black Rock Member 317, 324
Black Rock Slates 318, 319, 320, 329, 330
Bluff Cove Beds 316, 317, 324, 330
Bokkeveld Series 334
Bomb 17, 23, 35, 70, 72, 73, 74, 75, 76, 110, 166, 170, 172, 173, 177, 193, 231, 249, 251
Borolonite 117
Bostonite 55, 111, 119, 259
Bouguer Anomaly 217
Breccia 17, 18, 23, 53, 63, 70, 74, 75, 83, 90, 169, 170, 171, 172, 181, 200, 219, 246, 281
Bretonic Orogeny 330
"Bristol" 16
Burdigalian 142

Caldera 167, 169, 170, 171, 190
Caltonite 134
Camptonite 20, 22, 36, 39, 43, 51, 111
'Cameroun Line' 115
Cape Foldings 308, 334
Cape Meredith Complex 310, 311, 324, 325, 329, 332
Cape Meredith Series 312
Caracas arenite 26, 29, 33
Carboniferous 308, 309, 315, 316, 318, 320, 326, 329, 330
"Challenger" 6, 8, 16, 35, 172, 186, 203, 241, 277, 278, 279, 309, 324
Challengite 10
Choiseul Sound-Brenton Loch Beds 320, 321, 326, 327, 330
Cinder Cone 167, 169, 237, 249, 251, 252, 273, 280, 282
Clay 143, 309, 311, 317, 319, 320, 322
Claystone 218, 317, 319, 329

Subject Index

Comendite 184, 185
Composite Cone 165, 166, 190, 198, 229
Cone of Dejection 74, 75
Conglomerate 28, 143, 187, 308, 313, 316, 319, 321
Continental Drift 189, 334
'Coverhead' 77
Cretaceous 62, 63, 99, 103, 122, 145, 333

Delta 197
Descado Neso-Craton 332, 333
Desejado Sequence 71, 72, 73, 87, 88, 90, 92, 99, 100, 101
Devonian 308, 309, 311, 312, 313, 314, 315, 316, 317, 318, 322, 323, 324, 325, 326, 327, 328, 329, 330, 333, 334
Diabase 20, 173, 187, 188, 290, 308, 309, 316, 322, 324, 327
Dialaguite 152
Diamictite 316, 317, 318, 319, 322
Diatoms 277
Diorite 119, 177, 188, 205, 259, 273, 274
Dolerite 35, 49, 118, 119, 130, 134, 135, 178, 204, 243, 308, 309, 314, 322, 327, 329
Dome 18, 19, 20, 69, 70, 71, 72, 73, 76, 100, 166, 167, 168, 169, 170, 171, 172, 181, 185, 190, 199, 200, 206, 207, 213, 219, 235, 236, 246, 247, 255, 281, 282, 283, 284, 298, 299, 304
Dune 26, 27, 29, 32, 33, 34, 78, 202, 234
Dunite 6, 7, 8, 10, 62, 91
Dwyka Tillite: 318
Dykes 16, 18, 19, 20, 21, 22, 23, 24, 25, 39, 41, 44, 46, 47, 49, 52, 53, 61, 70, 71, 74, 75, 76, 77, 82, 84, 85, 86, 87, 89, 90, 98, 100, 101, 126, 145, 157, 172, 198, 199, 200, 201, 206, 207, 212, 215, 228, 230, 231, 233, 234, 235, 236, 237, 240, 242, 243, 244, 246, 249, 252, 257, 273, 278, 279, 280, 281, 287, 289, 290, 291, 294, 296, 299, 301, 304, 308, 309, 312, 314, 322, 327, 329

Earthquakes 217, 225, 283
East Falkland System 320
Ejectamenta 76, 172
Ejectiles 51, 52, 61, 83, 90, 92, 98, 99
Eocene 62, 219
Epeirogeny 107
Esker 318
Essexite 19, 20, 21, 22, 36, 39, 44, 45, 51, 52, 55, 61, 63, 64, 119, 126, 129, 134, 135, 152, 154, 177, 187, 205, 241, 244, 248, 259, 260, 262, 264, 280, 290, 291, 292
Etindite 50, 87, 114, 115

Falkland Sound Group 310, 321, 326
Fasinite 46, 249
Fault 71, 74, 172, 215, 226, 227, 228, 233, 304, 308, 315, 317, 320, 327, 328, 329
Fire-fountaining 74, 75, 76
Flagstone 314
Fluvio(?)-Limno-Glaçial Beds 320, 330

Fox Bay Beds 310, 314, 315, 316, 324, 325, 327, 328, 329
Fox Bay Group 310
Foyaite 119, 208, 259
Fumaroles 164, 282, 283, 285
Furchite 19, 20, 21, 22, 36, 39, 41, 43, 51, 54, 55, 86
Furnas Sandstone 333

Gabbro 20, 52, 61, 91, 118, 119, 134, 152, 154, 173, 177, 187, 188, 205, 241, 248, 259, 260, 262, 264, 273, 274, 280, 301, 302
Gauteite 39, 40, 41, 55
Gibelite 182
Glacial Period 76, 99, 100, 196, 202, 217, 223
Glass 10, 71, 92, 99, 112, 172, 174, 175, 178, 185, 186, 204, 238, 266, 279, 280, 299
Glenmuirite 20, 24, 36, 39, 45, 47, 48, 51, 52, 61, 119, 129, 134, 135
Glimerite 92
Gneiss 122, 203, 310, 312, 316, 329
Gondwana 308, 318, 322, 331, 334
Granite 16, 122, 170, 173, 182, 184, 187, 188, 189, 194, 203, 273, 274, 308, 309, 310, 312, 316, 329, 333
Granitite 173
Granodiorite 333
Gravity 216, 279, 318
Graywacke 187, 203, 317, 318
Grazinite 72, 76, 81, 83, 88, 92
'Grès sublittoraux' 122
Guano 4, 10, 11, 29, 30, 194, 276, 286

Harzburgite 9
Haüynite 103
Holocene 34, 322, 331
Honolulu Series 61
Hornblendite 91, 150, 152, 158, 205, 249, 260
Hornitos 166

Ijolite 55, 134
Infracambrian 312
Infusorial Earth 169
Inselberg 106
Insular Platform 15, 69, 79, 80, 105, 279
Irati Shale 319
Isostasy 216, 217, 279
Issite 249, 259, 260
Itararé Beds 318

Jacupiranguite 52, 91
Jurassic 322, 329, 330, 331, 333

Kainozoic 332
Kali-Gauteite 19, 20, 21, 22, 36, 39, 40, 41, 43, 52, 53, 55, 61, 81, 82, 85, 87, 93
Karoo Dolerite 322
Karoo System 314, 321
Kassaite 259, 260
Kaulaite 150, 152, 158, 205
Kiirunavaarite 91
Kimmeridgian 329, 332

Kivite 117
Knotty Ridge and Sandy Bay Complex 200

Laccolith 309, 312
Lafonian Diamictite 316, 317, 322
Lafonian Group 310, 321
Lafonian Sandstone 308, 309, 319, 320, 324
Lafonian Tillite 310, 311, 316, 317, 318, 319, 320, 324, 330
Lamprophyre 19, 21, 22, 36, 48, 52, 86, 308, 309, 312, 327
Lapilli 17, 23, 70, 72, 73, 75, 110, 169, 170, 172, 178, 238, 246, 280, 281, 291
Lardalite 55
Larvikite 259
Laterite 107, 127, 138, 139, 140, 141, 144, 145, 172, 202
Lava Channels 166, 234, 282
Lava Levés 234
Lava Moraines 234
Lava Tunnels 234
Lherzolite 6, 8
Liassic 329, 330
Lignite 143
Limburgite 20, 21, 22, 35, 36, 39, 46, 47, 81, 110, 111, 112, 113, 114, 115, 119, 120, 128, 129, 135, 137, 138, 139, 158, 161, 189
Limestone 26, 28, 33, 62, 127, 141, 142, 143, 144, 145, 202, 309, 322, 327
Limonitization 52
Linosaite 61
Lower Beaufort Series 320
Lower Bokkeveld Series 314, 315
Lower Complex 200, 201, 202
Lower Dwyka Shales 316

Magallancia Fosse 332, 333
Main Massif 198, 199, 200, 215, 222
Main Shield Volcano 200
Manganese Ore 217, 218
Mantle 8, 9, 10, 99, 193, 194, 272
Marine Platform 33, 67, 69, 75, 78, 197, 228, 233
Marine Terrace 15
Melanobasalt 128, 130
Melanobasanite 129, 136
Melteigite 48, 83, 90, 91, 93
Mesozoic 308, 320, 321, 327, 330, 331, 332
"Meteor" 4
Meteorites 11
Mid-Atlantic Ridge 7, 11, 93, 96, 221, 273, 275, 290, 304
Middle Ecca Series 319, 320
Mineral Springs 164, 218
Miocene 62, 127, 142, 145, 219, 221, 222
Molteno Beds 321, 326
Monasterian 322, 331
Momchiquite 19, 20, 21, 29, 36, 39, 41, 44, 52, 54, 81, 82, 86, 87
Monzonite 115, 117, 119, 188
Moraine 318, 324

Morro Vermelho Formation 73, 74, 75, 76, 89, 99, 100
Monte Maria Group 310, 315
Mudstone 308, 316, 319, 321
Mugearite 237, 241, 244, 246, 259, 260, 262, 270, 280, 290, 291
Murite 88
Mylonite 7

Neck 19, 63, 69, 70, 71, 72, 73, 76, 90, 92, 100, 102, 107, 110, 126, 198, 199, 206, 228, 231, 232, 236, 246
Neocomian 333
Neo-Cretaceous 34
Neogene 63, 103, 221
Neolithic 223
Nephelinite 20, 24, 34, 36, 39, 47, 49, 50, 54, 55, 72, 76, 81, 82, 83, 86, 87, 88, 92, 98, 100, 101, 127, 129, 132, 133, 138, 139
NE Massif 198, 200, 215, 222
Nordmarkite 182, 259, 260
Nosicombite 55, 259
Nuée ardente 70
Nullipores 26, 27

Obsidian 19, 173, 175, 184, 185, 187, 190, 249, 274, 289, 293
Oceanite 149, 152, 158, 159
"Odin" 241
Oil 109, 111, 122, 127, 143, 329
Old Red Sandstone 326
Olivinite 9, 39, 52, 55, 61, 91, 99
Ordanchite 115
"Owen" 4, 9
Owenite 10

Pa-hoe-hoe 166, 199, 234
'Pain du sucre' 106
Palaeogene 333
Palaeozoic 68, 309, 310, 312, 313, 315, 318, 320, 327, 328, 331, 332, 333
Palisade Orogeny 329
Pantellerite 174, 182, 185
Patagonides 329, 331, 332, 333, 334
Patagonian Fold Belt 327
Patagonian Shelf 308, 329, 333
Paulite 10
Peat 225, 286, 306, 307, 319, 322, 329
Pechstein 184
Pegmatite 91, 188, 308, 309, 310, 312, 329
'Pelée Tears' 73
Peridotite 5, 6, 7, 8, 9, 91, 92, 99, 188, 273, 274, 301
Perkinite 61, 90, 91
Permian 308, 309, 310, 313, 316, 318, 319, 320, 321, 322, 326, 327, 329, 330
Phenobasalt 128, 130
Phonolite 5, 14, 16, 17, 18, 19, 20, 21, 22, 23, 25, 28, 29, 30, 31, 32, 34, 36, 39, 40, 48, 51, 52, 53, 61, 62, 63, 64, 69, 70, 71, 72, 75, 76, 78, 81, 82, 83, 84, 85, 87, 88, 90, 91, 92, 93, 100, 102, 105, 106, 109, 110, 111, 115, 116,

117, 118, 119, 120, 122, 125, 126, 127, 129, 132, 133, 134, 136, 137, 138, 139, 144, 145, 154, 161, 189, 198, 200, 203, 206, 207, 208, 209, 210, 211, 212, 213, 219, 255, 270, 273, 274, 275, 279, 294
Phosphate 6, 10, 29, 30, 172, 194, 218
Phyllite 203
Picrite 149, 152, 153, 154, 158, 159, 204, 251, 295, 299, 300, 301
Piedra Azul Series 320
Pillow Lava 73, 75, 251, 253
Plant Remains 239, 240, 309, 311, 314, 315, 319, 320, 321, 322, 327
Pleistocene 34, 62, 63, 80, 99, 100, 189, 190, 202, 222, 307, 309, 322, 330, 331
Pliocene 62, 64, 101
Plug 18, 19, 21, 84, 107, 110, 117, 231, 283, 298, 299, 304, 305
Polzenite 55, 59
Port Philomel Beds 309, 313, 314, 315, 316, 325, 328, 330
Port Stanley Beds 309, 315, 316, 324, 328, 330
Port Stephens Beds 309, 313, 314, 316, 324, 329
Port Sussex Formation 317, 319
Pot-hole 4
Precambrian 308, 309, 310, 312, 313, 327, 328, 329, 331, 333
Proterozoic 312
Psammite 29
Psephite 26, 28, 29
Pulaskite 248, 259, 260
Pumice 17, 172, 175, 181, 185, 189, 190, 234
Pyroclastics 16, 17, 19, 23, 24, 25, 27, 28, 32, 52, 64, 68, 69, 70, 71, 72, 73, 74, 75, 76, 77, 85, 86, 90, 92, 107, 148, 149, 165, 166, 169, 171, 172, 173, 185, 199, 200, 201, 222, 229, 230, 231, 232, 233, 234, 235, 237, 238, 239, 240, 242, 253, 268, 279, 280, 281, 282
Pyroxenite 52, 61, 83, 91, 92, 249, 260, 262

Quartzite 121, 122, 308, 311, 313, 314, 315, 316, 317, 319, 323, 324, 328
Quartz-Porphyry 186, 316
Quaternary 309, 311, 322, 323, 324, 330, 331
Quellkuppe 18, 19, 71
"Quest" 4, 6, 8, 172, 182, 184, 203, 206, 241, 243, 290
Questite 10
Quixaba Formation 18, 22, 23, 24, 25, 29, 31, 34, 39, 50, 52, 63

Radiometric Datings 10, 11, 63, 99, 100, 219, 220, 221, 233, 239, 277, 278, 279, 281, 302, 303, 305
Recent 308, 309, 310, 311, 322, 327, 331
Reef 14, 26, 29, 34, 65, 67, 69, 78, 196
Regolith 15, 17, 70, 76
Remedios Formation 16, 18, 20, 21, 22, 23, 24, 28, 30, 31, 34, 39, 40, 47, 51, 52, 61, 63, 64
Rhaetic 322, 326

Rhyolite 172, 173, 174, 182, 183, 184, 185, 189, 190, 191, 273, 274, 275
Rouvillite 259

Sanaite 20, 36, 48
Sandstone 26, 27, 109, 121, 122, 186, 203, 218, 308, 309, 311, 313, 314, 315, 316, 317, 319, 321, 323, 328
San Jorge Fosse 332, 333
San José Formation 25, 34, 50, 62, 63, 64, 98
Santa Cruz Fosse 332, 333
Schist 187, 203, 310, 312, 329
Schlackenagglomerat 74
Scoria 165, 166, 167, 169, 172, 174, 227, 229, 231, 232, 233, 237, 252, 277
"Scotia" 290
Sea Caves 239, 240, 304
Seamount 65
Seismicity 7, 18, 68
Shale 308, 310, 313, 314, 315, 316, 319, 320, 321, 328
Shell Point Limestone 322, 327
Shepherds Brook Member 319
Shield Volcano 201
Shonkinite 48
Silt 277, 320
Siltstone 308, 319, 320
Silurian 314
Sistema de Gran Malvina 313
Sistema de la Isla Soledad 313
Slate 310, 311, 316
Solifluxion 323
Somoncura Neso-Craton 332, 333
South Atlantic Fosse 333, 334
Spatter Cone 166
Springs 170, 218, 227
Staukuppe 72
Stock 287
Storm Beach 28, 32
Strombolian-type Volcano 76
Syenite 51, 55, 83, 90, 119, 187, 188, 189, 259, 273, 274

Table Mountain Series 313, 329
Tahitite 40, 115, 116, 117, 120
Talus 29, 69, 70, 73, 74, 75, 76, 77, 81, 106, 239
Tannbuschite 20, 21, 22, 36, 39, 43, 46, 47, 54, 55, 70, 71, 74, 75, 77, 81, 87, 89, 90, 92, 93
Tautirite 116, 117, 120
Tephrite 109, 119, 125, 126, 127, 128, 129, 131, 132, 133, 135, 136, 137, 138, 139, 144, 145, 161, 241, 244, 246, 262.
Terra Motas Sandstone 320
Terrace 15, 26, 28, 29, 31, 69, 75, 79, 80, 107, 238, 240
Tertiary 33, 99, 100, 304, 308, 331, 332, 333
Teschenite 20, 22, 36, 39, 45, 47, 48, 54
Theralite 55, 118, 119, 134, 135, 205, 259, 260
Tholeite 99, 275, 299, 302
Tillite 308, 316, 317, 318, 319
Tinguaite 81, 82, 84, 100

Tombolo 14, 25, 29, 32, 34
Trachyandesite 111, 128, 129, 132, 136, 161, 167, 171, 175, 179, 190, 191, 192, 193, 211, 230, 234, 236, 237, 238, 239, 241, 242, 243, 244, 245, 246, 251, 254, 257, 260, 262, 263, 264, 270, 271, 272, 274, 280, 281, 284, 294, 295, 299, 300, 301, 303
Trachybasalt 111, 161, 174, 179, 211, 212, 215, 219, 229, 232, 234, 235, 237, 238, 240, 241, 242, 243, 244, 250, 251, 252, 253, 257, 260, 270, 271, 272, 274, 275, 276, 280, 290, 291, 294, 295, 299, 300, 301, 303
Trachydolerite 111, 112, 114, 118, 119, 129, 135, 149, 164, 166, 169, 175, 177, 178, 189, 190, 192, 198, 199, 203, 205, 210, 243, 244, 290
Trachylite 290
Trachyphonolite 129
Trachyte 16, 17, 19, 21, 22, 25, 29, 34, 35, 36, 39, 41, 51, 52, 54, 59, 61, 62, 64, 72, 84, 88, 109, 115, 116, 117, 118, 119, 120, 126, 127, 129, 131, 132, 135, 136, 137, 138, 139, 145, 148, 157, 158, 159, 160, 161, 164, 165, 166, 167, 168, 169, 170, 171, 172, 173, 174, 175, 178, 179, 180, 181, 182, 183, 184, 185, 186, 188, 189, 190, 191, 192, 193, 194, 198, 199, 200, 201, 206, 207, 211, 212, 215, 219, 231, 235, 236, 238, 239, 240, 241, 242, 245, 246, 247, 248, 251, 255, 256, 257, 260, 262, 263, 269, 270, 271, 272, 273, 274, 275, 279, 280, 281, 287, 288, 289, 290, 291, 293, 294, 295, 296, 297, 298, 299, 300, 301, 303, 304, 305
Triassic 307, 308, 309, 310, 313, 321, 322, 326, 327, 329, 330
Trindade Complex 70, 71, 72, 73, 81, 82, 85, 86, 87, 88, 89, 90, 92, 98, 100, 101
Tuff 14, 16, 17, 18, 19, 20, 21, 22, 23, 25, 28, 29, 30, 31, 32, 35, 39, 47, 49, 51, 52, 53, 61, 70, 72, 73, 75, 78, 83, 91, 92, 102, 107, 110, 122, 157, 167, 169, 170, 171, 172, 173, 175, 179, 181, 186, 188, 196, 198, 199, 200, 201, 204, 217, 218, 221, 222, 228, 234, 236, 239, 240, 245, 279, 280, 281, 287, 288, 289, 290, 291, 301, 304
Tuff-Breccia 17, 72, 73, 75
Tumuli 166

Ultrabasics 16, 34, 55, 71, 77, 85, 92, 118, 152, 157
Umprekite 208
Unconformity 74, 321
Upper Beaufort Series 321, 326
Upper Complex 198, 200, 201
Upper Dwyka Shales 319
Upper Lafonian Series 322

Valado Formation 74, 75, 82, 87, 89, 100
Vent 74, 75, 163, 164, 169, 188, 189, 198, 227, 229, 231, 233, 234, 277, 280
Vindobonian 142
Volcanic Breccia 16, 84, 111, 126

Wehrlite 6, 188
West Lafonian Beds 321, 326, 330
West Point Island Forest Bed 322, 327
Witteberg Quartzite 315
Worm Tubes 29
Würm Glaciation 80, 100

Xenolith 9, 18, 22, 25, 39, 51, 52, 61, 70, 73, 83, 84, 87, 90, 91, 92, 95, 98, 99, 172, 173, 174, 179, 185, 187, 188, 189, 192, 194, 199, 233, 241, 242, 249, 253, 256, 260, 262, 264, 265, 267, 269, 270, 271, 273, 274, 281, 284, 285, 301, 302, 304

Yamaskite 249
Yardang 78

Zwiebelstruktur 18

Index of Fossil Names

Acaste 325
Actaeon 143
Adeloaina 144
Adelomelon 326
Amphistegina 123, 124, 143, 144
Amphiora 33
Annelida 325
Arca 143, 144
Australocoelia 214, 235
Australospirifer 325

Bairdia 123
Bannara 68
Bellerophon 325
Bolivina 144
Brachiopoda 123
Bryozoa 28, 29

Calmonia 325
Cardiomorpha 325
Cardita 143, 144
Cardium 144
Cephalopoda 325
Cerithium 144
Chlamys 143
Chonetes 325
Cibicides 124, 144
Ciperaceae 68
Clionolithus 325
Clypeaster 143
Codokia 143
Coelenterata 143
Coelospira 325
Conularia 325
Conus 143
Coral 26, 29, 65, 67, 107, 196
Corbula 144
Crepipatella 327
Cribroelphidium 123, 124
Crinoidea 325
Crustacea 28, 29
Cryphaeus 325
Cryptonella 325
Cuspidaria 143
Cyathea 68
Cyclichna 143
Cyperus 68

Dadoxylon 326
Dalmanites 325

Derbyina 325
Desmiophyllum 326
Diaphorostoma 325
Dosinia 143, 144

Echinodermata 29, 123, 143

Felicineae 68
Fish 143, 144, 325
Foraminifera 26, 28, 29, 34, 123, 124, 141, 143, 144

Ganganopteris 321, 326
Gastropoda 34, 123, 141, 143, 144, 325, 327
Globigerina 124, 143, 144
Glossopteris 319, 320, 321, 326
Glycimeris 144

Homalonotus 325
Hornea 326

Janeia 325

Kerguelenella 327

Laevicardium 143
Lamellibranchiata 28, 123, 141, 143, 144, 325
Lepidodendron 325, 326
Leptocoelia 325
Leptodomus 325
Leptostrophia 325
Libocedras 327
Lithothamnium 14, 28, 29, 69, 143
Lithophyllum 34
Loxoconcha 123
Loxonema 325
Lucina 143, 144

Mactra 143
Margarella 327
Melobesias 29
Mollusca 29, 141
Murex 143, 144
Mytilus 327

Neoalveolina 144
Neocalamites 321, 326
Nuculites 325

Index of Fossil Names

Orbiculoides 325
Orthoceras 325
Ostracoda 123
Ostrea 143, 144

Pachysiphonaria 327
Palaeodictyoptera 326
Palaeoneilo 325
Pareuthria 327
Paspalum 68
Patinigera 327
Pelecypoda 327
Phacoides 144
Phyllotheca 320, 321, 326
Pisonia 68
Podocarpus 327
Polychaetae 29
Porifera 325
Proteus 325
Prothyris 325
Pteria 143
Pteridophyta 327
Ptomatis 325

Quinqueloculina 123, 143, 144

Rapanea 68
Rensellaeria 325

Rissona 327
Rotalia 123, 124, 143, 144

Samarangia 327
Scaphander 143
Schuchertella 325
Siphoninoides 123, 144
Sphyraena 143
Spirifer 325
Spiroloculina 123
Strombus 143
Stylophora 143

Tapes 143
Tellina 143, 144
Tentaculites 325
Terebra 143
Textularia 143, 144
Toechomya 325
Tritonidae 144
Trilobita 325
Triloculina 124, 143, 144
Triphon 327
Trophidocyclus 325
Turitella 143, 144
Tympanotonus 143

Virgula 123
Voltzia 326

Waltheria 68

Xymenopsis 327

Locality Index

Abade 126, 129, 131, 138, 141, 144
Abade R. 118, 120, 121
Abreu Bay 17, 18, 20, 22, 39, 45, 46, 61
Abrolhos Rocks 62
Actaeon Mt. 196
Aden 180
Africa 1, 4, 12, 50, 87, 117, 122, 146, 148, 162, 163, 195, 225, 266, 275, 290
Agua Castelo 106
Agua Grande R. (Principe) 125
Agua Grande R. (Sao Tomé) 106
Agua de Guadelupe 105, 122
Agua Izé 118
Agua Petroleo 143
Agua S. Joao 105, 124
Agua de S. Joao dos Angolares 106
Agua Tomé 121
Agulha de Joao Dias Pai 126
Agulha de Joao Dias Filho 126
Agulhas Bay 126
Agulhas R. 126, 131
Alfonso R. 124
Algarve 142
Almas 109
Amelia 110
Ana Chaves Bay 107
Anambo R. 107
Anchorstock Gulch 233
Andes 333
Andrade 73, 78
Angola 122
Annaberg 103
Annobon 135, 156, 157, 158, 161
Anselmo de Andrade 126
Antarctica 266, 286, 306
Antonio Dias 110
Apostle 290
Archway Rock 299
Argentina 306, 312, 313, 315, 318, 320, 322, 324, 326, 327, 329, 330, 331, 333, 334, 335
Arrow Harbour 326
Ascension 67, 93, 119, 135, 162, 163, 164, 165, 170, 172, 173, 174, 179, 180, 182, 188, 189, 190, 192, 193, 194, 195, 199, 202, 210, 225, 256, 272, 273, 274, 275, 276, 302, 304
As-Duas-Aguas R. 143
Asses Ears 213
Atalaia 13, 14, 17, 18, 20, 30, 43, 44, 51, 53

Atalaia Bay 17, 20, 23, 24, 26, 28, 29, 30, 31, 48
Atalaia Pt. 20, 22, 39, 41, 43, 44, 46, 47, 48, 51, 52, 64
Atlantic Ocean 7, 11, 55, 59, 63, 93, 95, 96, 97, 99, 106, 115, 117, 152, 189, 222, 243, 260, 269, 273, 282, 331, 332
Australia 225
Auvergne 41, 209, 243, 244, 249
Azores 76, 93, 99, 135, 304

Baleia Pt. 106
Bambouto Mts. 119
Bandeira 14, 31
Banks 202
Banks R. 200
Barriga Branca 126
Barro Bobo 110
Barro Vermelho Pt. 33
Basakata 153
Base 226, 227, 231, 233, 237, 250, 251
Basilé 153
Basuola 154
Batete 153
Bay of Harbours 327, 328
Bears Back 166, 167
Bella Vista 138
Belmonte 129, 131
Bencoolen 207
Berkeley Sound 327, 330
Bibi R. 126
Biboca 17, 21, 22, 23, 24, 39, 41
Big Pt. 278
Big Green Hill 233, 277
Big Gulch 232
Big Sandy Gulch 251
Binda 106, 109, 110
Black Rock 318
Blenden Hall 236, 244, 246, 247, 248
Blineye 251, 254
Blomby's Cove 278
Bluff Cove 316
Boa Vista 14, 17, 18
Boat Harbour Bay 251
Boatswain Bird Island 162, 163, 167, 171, 184
Bodie Creek 319, 320
Bodie Inlet 326
Bohemian Mittelgebirge 40, 45, 47, 59, 61, 62
Boldro 17, 19, 22, 49

Locality Index

Bombom Island 126, 128, 143
Booby Hill 166
Bouvet Island 304
Branco (F. de N.) 17, 19, 41, 51
Branco (Trindade) 70, 78
Brazil 4, 12, 15, 16, 26, 62, 67, 68, 158, 319, 335
Brenton Loch 309, 320
Britain 196, 225
Broad Gut 213
Buenos Aires Province 312, 318, 320
Burnt Hill 231
Burntwood 232
Burras 125, 131
Butt Crater 166
Buttress Rock 299
Byron Heights 307
Byron Sound 307, 328

Cabeluda Island 17, 32
Cabritos Bay 78, 82, 84, 85, 87, 90
Cachoeira Bay 70, 77, 90
Cachorro 17, 20, 23, 28, 39, 53
Cagungué 110
Caieira Bay 15, 17, 18, 19, 20, 21, 23, 29, 33, 39, 46, 47, 51, 52, 61
Cais General Fonseca 126
Caithness 326
Caixo 126
Caldas Monchiquite 44, 87
Calheira 73
Calheta 78, 83, 89
Cameroun Rep. 50, 87, 115, 117, 119, 146, 154
Canary Islands 59, 61, 62, 93, 99, 115, 119, 135, 158, 221, 275
Cangelolo R. 126
Cantador R. 107
Cao Grande: 107, 110, 116, 117, 118, 119, 120
Cao Pequeno 110, 115, 116, 120
Cape Dolphin 328
Cape of Good Hope 195
Cape Meredith 307, 312, 313, 328
Cape Province 313, 315, 316
Cape Verde Islands 55, 59, 60, 61, 62, 63, 93, 95, 99, 115, 119, 135, 158, 266, 275
Capim Açu Pt. 15, 23, 24, 29, 32, 33, 39, 49
Capitao Pt. 131
Captain Kidd Grotto 24
Caracas Pt. 15, 26, 33
Caraço Island 126, 129, 131, 132
Cariote 125
Carreiro de Pedra Bay 23, 25
Cascalheira 141
Castelo 129, 131
Castelo R. 105
Castle Rock 199, 207, 213
Cat Hill 166
Cave Gulch 246, 251, 280
Cemetery Hill 73
Cemeterio Island 65, 66
Chad Lake 115
Chapel 199, 207

Chapeu de Nordeste Island 25, 26, 27, 32
Chapeu de Sueste Island 26, 32
Chartres 341
Chartres R. 325
Charuto 131
Chile 248
Choiseul Sound 307, 320, 327, 330
Chubut 307, 312, 315, 331, 333
Cinco Farilhoes Island and Pt. 71, 78
Cliff Pt. 226
Clifton Station 327
Cocoanut Bay 167, 171
Cole's Rock 199
Cook Islands 88
Conceicao 24, 29, 39, 47, 51, 52
Conceicao Island 17, 32
Concepcion 147, 153
Crater Cliff 166
Cricket Valley 166, 170, 173, 175, 177, 178, 190
Crista de Galo 68, 71, 76, 82, 84, 85
Cristina 153
Cross Hill 167, 169, 172, 175, 178, 181, 190
Crown Pt. 215
Cuscuz Island 9, 14, 25, 32, 52
Czechoslovakia 40, 45, 47, 59, 61, 62

Dakar 14, 62
D. Amelia 109
Dark Slope Crater 166, 172, 175, 177, 188
Darwin 307, 326, 329
Deadwood Plain 217, 218
Deep Glen 299, 305
Desconhecido 71, 72, 74
Desejado 67, 69, 70, 71, 72, 83
Devil's Ashpit 182
Devil's Cauldron 167, 170, 171, 183, 185
Devil's Tower 71
Devil's Wall 289, 290, 291, 294
Diana Peak 195, 196, 204, 215
Diego Alvarez Island 286
Diogo Vaz 107, 109
Dois Abraços 14, 31
Dois Irmaos Island 32
Dois Irmaos 126
Donkey's Plain 181
Dos Lomas 321, 326
Dos Ovos Island 17
Drachenfels 19, 72
'Drip' 170 181
Dry Gut 218
Dry Water Course 173

East Crater 166, 167, 185
East End Gulch 251
East Falkland Island 306, 307, 310, 311, 314, 315, 316, 317, 319, 320, 322, 325, 327, 328, 329
East Molly Gulch 251
Eastern Province 334
Edinburgh 225, 226, 282
Edinburgh Peak 286, 287, 289, 299

Egg Island 199
Elba Island 162
English Bay 163
Erzgebirge 103
Esprainha 110
Etinde 119
Etna 196
Europe 225
Expedition Peak 287, 289

Falkland Islands 1, 306, 307, 308, 309, 313, 315, 317, 318, 320, 321, 322, 324, 326, 329, 330, 331, 332, 333, 334, 335
Falkland Sound 306, 307, 310, 311, 316, 321, 327, 328, 329, 330
Fanny Cove 326
Far East 195
Farol Island 65, 66
Fernao Dias 106, 107, 123
Fernando de Noronha 9, 12, 13, 14, 15, 16, 18, 19, 23, 29, 30, 31, 33, 34, 35, 39, 40, 44, 45, 46, 47, 49, 50, 51, 52, 55, 59, 60, 61, 62, 63, 65, 85, 93, 95, 98, 99, 103, 158, 275
Fernando Poo 1, 118, 119, 135, 146, 147, 148, 149, 151, 152, 154, 156, 157, 158, 205
Figo Pt. 106
First Lagoon Gulch 244
Fisher's Valley 218
Flagstaff Hill 200, 217, 218
Focinhoa de Cao 126, 131
Fogo (Annobon) 156, 157
Fogo Island (Cape Verde) 62, 115
Fora Island 9, 14, 25, 32, 52
Forca R. 126, 141, 142
Fortaleza 129
Forte 51
Forte de Remedios 18, 39, 41
Fox Bay 308, 309, 314, 316, 325, 328
Frade Island 14, 32
France 10, 11, 41, 209, 243, 244, 249
Frances 14, 31
Francis Plain 217
Frank's Hill 251
Fria R. 129, 131
Funda R. 106
Fundao 126
Furada Pt. 106

Gato 17
George Island 326
Geogetown 162, 166, 173, 182
Germany 47, 103
Giant's Causeway 109
Glen Beach 291, 292, 294
Goat Ridge 282
Goose Bay 326
Gough Island 93, 96, 182, 193, 194, 224, 225, 256, 272, 273, 274, 275, 276, 286, 287, 288, 289, 290, 294, 295, 300, 302, 304, 305
Gough Monument 290
Grazinos 72
Gran Cordillera 147

Gran Malvina 306
Great Basin 196, 198, 199, 215, 306
Great Hollow 196, 198, 215
Great Stone Top 199, 207
Green Mt. 162, 163, 164, 167, 168, 169, 170, 172, 173, 175, 178, 180, 181, 182, 184, 185, 187, 188, 190
Guadelupe Sela R. 106
Guanabara 69
Gulf of Guinea 1, 2, 99, 105, 112, 115, 148, 157, 161, 192, 210, 287
Gumwood 217

Hag's Tooth 287, 289, 290, 291, 293, 299, 304
Halfway Cove 317, 318
Hawaii 61, 62, 166, 243
Hayes Hill 166, 175, 177
Hell's Kitchen 319
Herald Pt. 243, 278
High Hill 199, 204, 207, 213, 216
High Knoll 199, 216
High Peak 196
High Ridge 251, 257
Hill Cove 316, 317, 318
Hillpiece 231, 253
Holdfast Tom 218
Hollands Crater 166, 167
Hooper Rock 196, 199, 206, 207, 213
Hornby Mts. 307, 310, 314, 328
Horse Pt. 217, 218
Horseshoe Crater 166
Horta Velha 129, 131
Hottentot Gulch 278, 282

Ile Beauchêne 306
Ilheu das Rolas 108, 110, 112, 113, 114, 120, 124
Inaccessible Island 224, 225, 228, 235, 236, 240, 243, 244, 245, 246, 251, 256, 257, 260, 264, 271, 272, 274, 275, 277, 278, 280, 281, 290
India 195
Indian Ocean 225, 243, 334
Inhame 111
Io 106
Io Grande 107
Iogologo Bay 106, 109
Ireland 109
Isla Soledad 306
Islas Malvinas 306
Italy 6
Ivory Coast 122
Izé R. 129, 131, 143

Jamestown 196, 197, 200, 217
James Valley R. 196, 197, 199
Juan Fernandez Island 248

Katmai 18
Kerguelen Island 243, 266
Kern Crag 304
Kilimanjaro Mt. 243
King George Bay 307, 328

Locality Index

Kivu Lake 117
Koupé Mt. 154
Knotty Ridge 204, 215, 222

Lady Hill 166
Ladder Hill 203, 204, 205
Lafonia 310, 320, 321, 327, 328
Laguna 159
Lama-Porco 114
Landing 239, 248, 282
Landing Pier 175, 177
Lapa 125, 126
La Rioja Province 322, 326
Leao 19, 29
Leao Island 17, 32
Lemba 110
Lemba-Lemba 126
Lennoxtown 290
Liberia 12
Lisbon 142
Little Stone Top 199, 207
Little White Hill 167, 168, 171, 183
Lively Island 320
Lobau 47
Long Ridge 234
Longwood 197, 217, 218
Lot 199, 206, 207
Lot's Wife 199, 206, 213
Lot's Wife Ponds 202, 213
Lourdres 75, 78
Low Bay 326
Lucena Island 32

Macaronesia 142
Madagascar 46, 47, 49, 243
Madeira Island 93, 96, 99, 119, 135, 222, 244, 290
Madeira 14, 17
Mae Marta 128
Maio Island (Cape Verde) 62
Malanza 106
Malaya 225
Mamelles (Dakar) 62
Manengouba Mts. 119
Manuel Jorge R. 105, 106, 113
Manybranch Harbour 325
Maria Fernandes 110, 118, 120
Maria Isabel 110
Maria Luiza R. 107, 120
Martinique Island 18, 283
Martin Mendes R. 105
Martin Vaz Island 67, 102, 275
Massachusetts 44
Matemba 126
Matens Sampaio 111
Meio Island 26, 27, 32
Meio 17, 20, 39, 43, 46
Melao 106
Mencorne 125, 129, 131
Mendoza 318
Mendoza Province 322, 326
Mexico 76

Middle Island 224, 225, 229, 240, 241, 244, 246, 247, 248, 276, 277, 278, 279, 281
Middleton Peak 172, 181, 185
Mizambu 110
Moffit Harbour 321
Moka 146, 147, 148, 153
Mont Doré 41, 243, 249
Mont Grand 154
Mont Pelée 18, 283
Monte Café 108, 114, 116, 118, 119, 120
Monte Maculo 112, 114, 120
Monte Muquinque 112
Monte da Praia Joana 114
Monte Sinai 110
Monumento 69, 70, 71, 82, 83, 85, 86
Monumento Pt. 71
Morgenberg 103
Mount Adam 306
Mount Edgeworth 316
Mount Maria 311
Mount Moody 310
Mount Osborne 306
Mount Pleasant 318
Mount Red Hill 166, 168, 175
Mount Rowett 287, 290, 291, 292, 299, 304
Mull 244
Munden Pt. 196
Musola 148

Natal 319
Netley Gut 217, 218
Nendorf 103
New Zealand 173
Nigel's Glen 299, 304
Nightingale Island 221, 224, 225, 229, 237, 238, 240, 241, 244, 245, 247, 248, 249, 251, 256, 271, 272, 274, 275, 276, 277, 278, 279, 280, 281
Noisy Beach 251
North Arm 326
North Island 102, 103
North Pond 251
North San Carlos 311
Northeast Pt. 46
Norway 10, 11, 90, 290
Nova Island 76
Novo Brasil 110
Nueva Lubecka 332

Oahu Island 61
Obelisco 68, 70, 76
Oque Boi 141
Oque Nazaré 126
Oque Tres 138
Os Dois Irmaos 129, 131
Ouro R. 106, 112, 113
Ovos Island 32

Pacific Ocean 194, 243, 244, 260
Palmeiras 121
Pantellaria Island 182
Pao de Açucar 69, 74, 75, 78

Papagaio 126, 129, 131
Papagaio R. 125, 141
Paracutin 76
Parana 334
Paredao Volcano 69, 73, 74, 75, 76, 79, 83, 89, 90, 92, 99
Patagonia 306, 307, 318, 321, 324, 329, 331, 332, 334, 335
Pebble Island 328
Pedra Alta Pt. 24, 49
Pedra Pt. 73, 79
Pedro Furada 124, 128
Peixo 106, 122
Peixo R. 105
Penguin Island 292, 293, 299
Pernambuco 12, 26
Pico (F. de Noronha) 13, 14, 18, 31, 39
Pico Principe 125, 126
Pico Sta. Isabel 146, 147, 148
Pico Tomé 105
Piedmont 6
Pigbite Gulch 251
Pillar Bay 167, 171
Pirimides 138
Pontinha Pt. 24, 33, 39, 43, 47, 49, 50
Pontuda 69, 72, 82, 90
Port Darwin 326
Port Edgar 313, 314
Port Fitzroy 316, 317, 318, 322, 327
Port Harriet 316
Port Howard 311, 316, 325
Port Louis 325
Port North 314
Port Philomel 308, 314
Port Pleasant 316
Port Purvis 316, 317, 318, 319, 325, 326
Port Richards 313, 314
Port Salvador 324, 325
Port San Carlos 325, 327, 328
Port Stephens 313
Port Stanley 307, 308, 317, 327, 331
Port Sussex 309, 311, 319, 329
Portao da Sapata 32
Porto 28
Portela Fundao 131
Porto Alegre 113, 114, 120
Portugal 44, 87, 105, 142
Portugueses Bay 70, 71, 72, 82, 83, 85
Potato Bay 202
Potato Patches 225, 231
Potsdam 216
Powell's Valley 213
Praia Abade 144
Praia Ana Chaves 123
Praia Angra Furada 124
Praia Baixa 109
Praia Burras 126, 129, 144
Praia Cabana do Pai Tomas 124
Praia Cabinda 126, 143, 144
Praia Caixo 126, 144
Praia Calio 123
Praia Campanha 131

Praia Cobo 107
Praia Diogo Nunes 124
Praia Grande (Principe) 126, 141, 143
Praia Grande (Sao Tomé) 106, 124
Praia Grande Norte 143, 144
Praia Grande Sul 144, 145
Praia Guegue 124
Praia Lagarto 107, 123
Praia Lanca 124
Praia Lapa 129, 131
Praia Lemba 107
Praia Norte 124
Praia Neves Ferreira 144
Praia Paciencia 126, 131
Praia Pantufo 107, 123
Praia Periquito 144
Praia Rainha 144
Praia Rei 124
Praia Rib. Fria 144
Praia S. Antonio 123
Praia Sta. Catarina 107
Praia Sta. Joana 123
Praia Sudoeste 124
Praia Sundy 129, 131, 144
Praia Uba 129, 131
Praia Ubabudo 113, 120
Praia Zale 124
Precipicao Norte 125
Preto 69, 71
Principe Bay 69, 78, 83, 84, 90
Principe Island 117, 119, 123, 125, 126, 127, 128, 129, 141, 143, 144, 145
Prosperous Bay 218
Puy de Dôme 18
Pyramid Rock 228

Queen Charlotte Bay 328
Queen Mary's Peak 225
Quinas 110
Quioveo 157
Quixaba 14, 15, 23, 30

Ragged Hill 167, 168, 175, 180
Rapta Island 26
Raratonga Island 88
Rasa Island 14, 25, 26, 27, 32
Rata Island 14, 23, 26, 29, 32, 48
Recife 14, 26
Red Hill 173, 177, 178, 186, 187
Remedios 14, 20, 39, 48, 51
Réunion Island 6
Rhein R. 19, 72
Ridge Beach 202
Riding School 167, 168, 169, 173, 175, 177, 180, 181, 182, 184, 185, 186
Riding Stones 207, 213
Rio Bermejo 322, 326
Rio Chico 331, 332
Rio Gallegas 332, 333
Rio de Janeiro 69
Roça Agua Izé 112, 113, 120
Roça D. Augusta 120

Roça Diogo Nunes 113, 120
Roça Diogo Vaz 120
Roça Esperança 129
Roça Granja 119
Roça Infante D. Henrique 129, 131
Roça Pontafigo 120
Roça Rio Ouro 120
Roça Sundy 128
Rocas Atoll 65, 66
Rofe Rock 207
Rookery Pt. 278
Roumpi Mts. 117, 119
Round Hill 166
Rupert Beach 202
Rupert R. 197, 200, 222

S. Antonio 14, 15, 17
S. Antonio Palea 157
S. Antonio Peninsula and Pt. 24, 25, 27, 29, 32, 49
S. Francisco 114
S. Jeronimo Pt. 107
S. Joaquim 126
S. Jorge 141
S. José 126
S. Miguel 106
S. Joao R. 106
Saddle Crater 166
Salt Beach 251, 278
Samoa Island 180, 189
San Carlos 146, 147, 153
San Carlos Gulf 329
San Carlos Strait 306
Sancho 23
Sandy Bay 196, 202, 205, 206, 207, 213, 217, 222
Sandy Bay Barn 207
Sandy Pt. 226, 232, 233, 240, 249, 251, 283
San Juan 318, 322, 329
San Juan Pre-Cordillera 313, 315
San Joaquim 147
San Jorge Gulf 331, 332
Santana 129, 131, 143
Santamina 156
Sao Bonifacio 67, 69, 71, 72
Sao José 9, 25, 26, 27, 32, 39, 50, 52
Sao Tomé 93, 105, 106, 107, 108, 110, 111, 112, 113, 114, 115, 116, 117, 118, 119, 120, 122, 123, 125, 126, 128, 137, 144, 145
Sapata Peninsula 15, 23, 33
Saxony 47, 103
Schanzburg 47
Scharfenstein 61
Scotland 244, 290, 326
Sea Hen Rocks 239, 240, 251
Seal Bay 235, 240
Sela Gineta Island 14, 17, 32, 39
Senegal 14, 62
Serranias 106
Settlement 225, 227, 229, 231, 232, 234, 251, 278, 283
Sheep Knoll 199, 206, 207, 213

Sheep's Pound Gut 217, 218
Shell Pt. 322
Sidepath 217
Sierra Australes 312, 330
Sierra Ventana 318, 320
Sister's Peak 166, 167, 170
Sister's Red Hill 166, 167, 170
Sol Pt. 138, 144
Sophora Glen 286, 289
South Africa 196, 286, 308, 313, 314, 315, 316, 318, 319, 320, 321, 322, 324, 325, 326, 329, 330, 333, 334, 335
South America 1, 4, 162, 163, 195, 225, 275, 286, 306, 308, 322, 324, 325, 327, 331, 334, 335
South Atlantic Ocean 1, 2, 14, 146, 197, 225, 309
South Gannet Hill 166, 167
South Georgia Island 306
South Harbour 313
South Hill 240
South Island 102, 103
South Orkney Island 306
South Peak 299, 304
South Pt. (Innaccessible) 244, 247
South Pt. (Trindade) 78
South Red Crater 166
South San Carlos 311
South Sandwich Island 306
South Shetland Island 306
Southeast Bay 14, 17, 18, 19, 21, 23, 26, 29, 32, 33, 39, 41, 43, 45, 46, 51, 55, 61
Southeast Crater 166, 168
Southeast Head 166, 167, 171, 175, 179, 180, 182, 196
Southwest Bay 166, 168, 172, 186
Southwest Pt. 236, 247
Southwest Red Hill 166
Speedwell Island 326
Speery Island 207
Spoon Crater 167, 168, 172
Springs Valley 217, 218
St. Helena Island 93, 96, 99, 119, 135, 161, 162, 163, 195, 196, 197, 198, 199, 200, 202, 203, 205, 206, 208, 209, 210, 215, 216, 217, 218, 219, 221, 222, 223, 225, 243, 244, 272, 273, 274, 275, 276, 277, 290, 304
St. Paul Rocks 1, 4, 5, 6, 7, 8, 9, 10, 11, 52, 172, 194
Sta. Barbara Island 62
Sta. Catarina 107, 109
Sta. Isabel 146, 148, 153
Sta. Jenny 107
Sta. Rita 126, 138, 143, 144
Sto. Antonio Bay 125
Sto. Antonio 125, 126, 138
Stanley Harbour 324
Stoltenhoff Island 224, 225, 229, 240, 246, 247, 248, 277
Stonetop Gut 218
Stony Beach 240, 249, 251
Stony Hill 226, 230, 232, 234, 235, 283

Street Crater 166
Sugarloaf Hill 202, 215
Sundy 138

Table Crater 166, 167
Tahiti 40, 115, 117
Tanandaré 14, 24, 26, 28, 33, 39
Tarn Moss 289
Tartarugas 69, 75, 78, 84
Tartarugas Bay 69, 75, 78, 79
Tautira 117
Tecka 332
Tepel 332
Terreiro Velho 126, 129
The Barn 196, 200, 204, 215, 216, 222
The Crags 167, 170, 181
The Glen 286, 287, 290, 291, 299, 304, 305
The Glen R. 299
The Peak (Ascension) 162, 166, 167, 169, 170, 181, 188
The Peak (Tristan) 225, 226, 227, 229, 230, 231, 234, 248, 250, 280
The Tarn 178
Thistle Hill 166, 167, 181
Thompson's Valley 213
Tierra del Fuego 307, 331, 332, 333
Tom Yalla 118, 153
Tranquilidad 326
Traveller's Hill 166, 167
Tres Paus 20
Trindade Island 13, 67, 68, 69, 70, 76, 77, 79, 80, 81, 84, 85, 86, 87, 88, 90, 91, 92, 93, 95, 96, 97, 98, 99, 101, 102, 103, 275
Trindade 67, 69, 72, 83
Tripe Bay 213
Tristan da Cunha Island 93, 96, 99, 119, 166, 195, 221, 224, 225, 226, 229, 230, 231, 237, 240, 241, 243, 244, 245, 246, 247, 249, 256, 270, 271, 272, 273, 274, 275, 276, 277, 278, 279, 280, 281, 282, 283, 285, 286, 290, 304, 305
Tristania Rock 287
Turks Gap 217
Tutuila Island 189

Two Boats 162, 163
Tyrol 48

Uba 141
Ubabudo 109, 121, 122
Upper Valley Crater 166
Ureka 148
Uruguay 319, 335

Valado Pt. 74, 79, 80
Vale Verde 70, 82, 86
Valley Tank 181
Velho 131
Vermelho 73, 76, 77
Vermelho Valley 73, 76, 83
Vesuvius 166
Vigia (Sao Tomé) 109
Vigia (Trindade) 71, 72, 75, 83, 90
Viuva Island 17, 32
Viracao 14, 23, 33

Walker Creek 320, 326
Wall Dyke 289
Waterfall 200
Waterfall Pt. 299, 305
Weather Post Hill 167, 170, 171, 175, 180, 182, 183, 185
West Falkland Island 306, 307, 308, 310, 311, 312, 313, 314, 316, 317, 319, 320, 321, 322, 325, 327, 328, 332
West Indies 163
West Pt. 246, 248
West Point Island 311, 322, 327
White Hill (Ascension) 167, 170, 171, 175, 182, 184, 185
White Hill (St. Helena) 196, 213
White Rocks 213
White Rock Bay 328
Wickham Heights 307, 311, 315, 316, 319, 320, 328
Wideawake Valley 185
Wiesenthal 103
Wig Hill 167, 168, 183
Wild Cattle Pound 213
Wyoming 71

Author Index

ABDEL-MONEM, A. 221
ADIE, R. J. 3, 308, 313, 316, 318, 319, 320, 321, 324, 326, 329, 330, 334, 335
ALBUQUERQUE, C. R. de 110, 113, 114, 121, 122
ALMEIDA, F. F. M. 3, 9, 16, 18, 19, 21, 24, 25, 27, 29, 30, 31, 33, 34, 35, 40, 41, 44, 46, 47, 48, 49, 51, 53, 55, 59, 60, 63, 64, 67, 68, 69, 70, 71, 75, 79, 81, 86, 91, 93, 96, 97, 98, 99, 100
ANDERSSON, J. G. 308, 312, 322, 323, 324, 330, 331
ASSUNÇAO, C. F. T. 3, 111, 112, 113, 114, 115, 116, 117, 118, 119, 120, 128
ATKINS, F. B. 164, 165, 166, 190

BAKER, H. A. 310, 312, 313, 314, 316, 317, 319, 320, 322, 323, 325, 326, 329, 331
BAKER, I. 3, 99, 197, 200, 201, 202, 210, 212, 213, 214, 215, 216, 219, 220, 221
BAKER, P. E. 164, 165, 166, 190, 225, 226, 227, 228, 230, 231, 232, 233, 234, 235, 236, 237, 238, 239, 240, 241, 242, 248, 250, 251, 256, 257, 264, 265, 266, 267, 268, 269, 270, 271, 272, 273, 274, 275, 276, 305
BARRAT, C. 122
BARROS, J. A. L. 69
BARROS, L. A. 3, 119, 120, 128, 129, 130, 131, 132, 133, 135, 136, 137, 138, 139, 140, 154, 158, 161, 183, 192, 197, 203, 210
BARROS, L. M. 16, 30
BARTH, T. W. F. 3, 287, 289, 291, 292, 293, 294
BEBIANO, J. B. 61, 62, 99
BELL, J. D. 3, 164, 165, 166, 185, 186, 190, 191
BEMMELEN, R. V. van 10
BERG, G. 115
BERGEAT, A. 72
BERTHOIS, L. 121, 141, 142
BESNARD, W. 79
BOESE, W. 112, 115, 118, 119, 120, 149, 150, 151, 153, 154
BORN, A. 217
BORRELLO, A. V. 3, 308, 312, 313, 315, 317, 318, 319, 320, 321, 322, 324, 325, 326, 328, 329, 330, 331, 332, 334, 335
BOWEN, N. L. 60, 61, 301
BOWIE, W. 216, 217
BRANNER, J. C. 14, 16, 18, 26, 27, 30, 35
BRÖGGER, W. C. 44, 90, 290
BUCHANAN, J. W. 10, 16, 35
BURRI, C. 16, 35, 40, 41, 44, 46, 47, 49, 50, 51, 53, 55, 59, 62, 150, 205

CAMPBELL, R. 289, 290, 293, 294
CARDOSO, J. C. 145
CARVALHO, A. F. de 111, 115, 120
CAVACO, J. F. 118
CASTER, K. E. 333, 334, 335
CHEVALIER, A. 115
CLARKE, J. M. 308, 324
CLOOS, H. and E. 18, 72
CORDANI, U. G. 3, 63, 100
CROWELL, J. C. 312, 313, 314, 315, 316, 317, 318, 319, 320, 322, 324

DALY, R. A. 162, 163, 164, 165, 166, 167, 168, 169, 170, 171, 172, 173, 174, 175, 176, 177, 178, 179, 180, 181, 182, 183, 184, 185, 186, 188, 189, 190, 192, 194, 195, 196, 197, 198, 200, 201, 202, 203, 204, 205, 206, 207, 208, 209, 210, 215, 216, 217, 219, 222, 223, 240, 272, 277, 290
DARWIN, C. 4, 6, 10, 16, 40, 62, 166, 169, 172, 173, 179, 184, 186, 187, 194, 196, 197, 200, 202, 203, 215, 216, 221, 222, 308, 309, 313, 323, 324
DAVIES, T. 16, 27
DAVIS, W. M. 77
DERBY, O. A. 16, 29
DINIZ, M. de 123, 141, 144
DOUGLAS, G. V. 4, 10, 169, 206, 234, 239, 241, 276, 287, 290, 304
DUNNE, J. C. 224, 226, 227, 231, 233, 234, 236, 237, 238, 239, 240, 241, 242, 243, 244, 245, 247, 248, 249, 256, 257, 258, 259, 260, 261, 262, 269, 270, 277, 280, 281
DU TOIT, A. L. 334, 335

ESENWEIN, P. 55, 61
ETHERIDGE, R. 308, 324

FINCKH, L. 290
FLINT, R. F. 317
FRAKES, L. A. 3, 312, 313, 314, 315, 316, 317, 318, 319, 320, 322, 324
FURON, R. 62, 122
FUSTER, J. M. C. 3, 112, 115, 118, 119, 122, 148, 149, 150, 151, 153, 154, 155, 157, 158, 159, 160, 161, 176, 177, 178, 182, 184, 189, 191, 192, 197, 203, 205, 206, 207, 208, 209, 210, 211

GALE, H. N. 99, 200, 201, 202, 210, 219, 220, 221

Author Index

GARCIA, J. S. 145
GASS, I. G. 3, 99, 221, 225, 226, 227, 228, 230, 231, 232, 233, 235, 236, 237, 238, 239, 240, 241, 242, 248, 250, 251, 256, 257, 264, 265, 266, 267, 268, 269, 270, 271, 272, 273, 274, 275, 276, 277, 279, 280, 281, 305
GAST, P. W. 3, 193, 194, 221, 302, 303
GEIKIE, A. 6
GILL, A. C. 16, 35
GUIMARAES, D. 9, 16, 52
GUMBEL, C. W. 16, 35

HAFSTEN, U. 3, 225, 229, 272
HALLE, T. G. 308, 309, 311, 315, 319, 320, 322, 324, 325, 326, 329, 330, 331, 333
HARRINGTON, H. J. 318
HARRIS, P. G. 98, 227, 228, 230, 231, 232, 233, 234, 235, 236, 237, 238, 239, 240, 241, 242, 248, 250, 251, 256, 264, 265, 266, 267, 268, 269, 270, 271, 272, 273, 274, 275, 276, 282, 283
HARTT, C. F. 62, 63
HEANEY, J. B. 286, 287, 290
HEDGE, C. 193, 194, 302, 303
HIBSCH, J. E. 40, 47, 59, 61
HIRST, T. 195, 197, 198, 200, 201, 202, 217, 218
HOLDGATE, M. W. 225, 286, 287

IMBIRIBA, M. F. 29

JÉRÉMINE, E. 112, 115, 117, 118, 119, 154
JOHANSSEN, A. 41, 44, 47, 48, 81
JOYCE, J. R. F. 323, 324
JUNG, J. 120, 137, 154, 161, 192, 210

KNIGHT, E. F. 81
KNORR, H. 59, 61
KUNO, H. 8, 97
KUSHIRO, I. 8

LACROIX, A. 18, 40, 46, 48, 49, 88, 128, 149, 249
LEA, T. S. 16, 26, 27
LEMAITRE, R. W. 3, 192, 211, 227, 228, 230, 231, 232, 233, 234, 235, 236, 237, 238, 239, 240, 241, 242, 248, 250, 251, 256, 264, 265, 266, 267, 268, 269, 270, 271, 272, 273, 274, 275, 276, 282, 283, 287, 288, 289, 295, 296, 297, 298, 300, 301, 302, 304, 305
LOBOCH, M. Z. 156, 157

MACPHERSON, J. 149
MALING, D. H. 324
MELLISS, J. C. 202, 216, 217, 218, 219
MENDELSOHN, A. C. 111, 115, 128, 141
MILLER, J. A. 277, 278, 279, 281
MITCHELL-THOMÉ, R. C. 62, 333
MORAES, L. J. 16
MORRIS, J. 308, 324

NATHHORST, A. G. 308
NEGREIROS, A. 115
NEIVA, J. M. C. 3, 107, 109, 110, 112, 113, 114, 115, 118, 119, 120, 121, 122, 127, 131, 132, 134, 135, 136, 137, 138, 139, 141
NEVES, J. M. C. 138, 141
NEWTON, E. T. 308
NIDDRIE, D. L. 313, 314, 315, 320, 321, 324, 335
NIGGLI, P. 54, 55, 59, 60, 119, 120, 121, 135, 136, 137, 149, 150, 158, 159, 178, 179, 182, 184, 192, 205, 208, 259, 260

OLIVER, J. R. 196, 197, 200, 203, 216, 219

PEACOCK, M. A. 96, 97
PEREIRA, J. S. 119
PIRSSON, L. V. 289, 293, 294
POUCHAIN, E. B. 16, 27, 30
PRATJE, O. 7, 10
PRIOR, G. T. 16, 35, 81, 172, 179, 180, 184, 197, 206

RATTRAY, A. 16
REINISCH, R. 103, 169, 172, 174, 175, 177, 178, 179, 180, 181, 182, 184, 185, 189, 192, 197, 203, 204, 205
RENARD, A. 4, 5, 6, 7, 8, 10, 11, 16, 35, 50, 172, 173, 177, 178, 179, 180, 182, 184, 185, 186, 187, 188, 194, 241, 246, 247, 309, 323, 324
REYER, E. 18
RIBEIRO, P. A. 67
RICHARDS, C. 194
RIDLEY, H. N. 16, 25, 26, 27, 46, 49, 51
RINNE, F. 91
RITTMANN, A. 121, 128, 129, 132, 135, 136, 137, 140, 154, 161, 192, 210
RODGERS, J. 317
ROSENBUSCH, H. 47, 62, 179
RUDOLPH, E. 7

SANDERS, J. E. 317
SCHULTZE, A. 157
SCHUSTER, N. 149
SCHWARTZ, E. H. L. 241, 266
SCORZA, E. P. 3, 103
SERRALHEIRO, A. R. 3, 141, 142, 144
SEWARD, A. C. 308, 315, 320, 324, 326
SHARPE, D. 308, 324
SILVA, G. H. da 141, 142, 143
SIMONS, J. 99, 200, 201, 202, 210, 219, 220, 221
SMITH, CAMPBELL A. 16, 35, 40, 41, 44, 46, 47, 49, 50, 51, 53, 55, 59, 62, 172, 174, 177, 179, 182, 184, 185, 187, 188, 197, 205, 206, 207, 241, 243, 244, 245, 246, 249, 256, 262, 289, 290, 291, 292, 293, 294
SMITH, D. G. 164, 165, 166, 190
SOARES, L. de C. 31
STOCK, J. 47
STUBBS, P. 10